Domestic Animal Behaviour and Welfare

5th Edition

Domestic Animal Behaviour and Welfare

5th Edition

Donald M. Broom

Department of Veterinary Medicine and St Catharine's College, University of Cambridge, UK

and

Andrew F. Fraser

formerly Memorial University of Newfoundland, Canada

CABI is a trading name of CAB International

CABI
Nosworthy Way
Wallingford
Oxfordshire OX10 8DE
UK

Tel: +44 (0)1491 832111
Fax: +44 (0)1491 833508
E-mail: info@cabi.org
Website: www.cabi.org

CABI
38 Chauncy Street
Suite 1002
Boston, MA 02111
USA

Tel: +1 800 552 3083 (toll free)
E-mail: cabi-nao@cabi.org

A catalogue record for this book is available from the British Library, London, UK.

Library of Congress Cataloging-in-Publication Data

Broom, Donald M., author.
Domestic animal behaviour and welfare / Donald M. Broom and Andrew F. Fraser. -- 5th edition.
p.; cm.
Includes bibliographical references and indexes.
ISBN 978-1-78064-539-1 (hardback : alk. paper) -- ISBN 978-1-78064-563-6 (pbk. : alk. paper) 1. Domestic animals--Behavior. 2. Pets--Behavior. 3. Animal welfare. I. Fraser, Andrew F., author. II. C.A.B. International, issuing body. III. Title.
[DNLM: 1. Animals, Domestic. 2. Behavior, Animal. 3. Animal Welfare. 4. Veterinary Medicine--methods. SF 756.7]

SF756.7.B76 2015
636--dc23

2014041431

ISBN-13: 978 1 78064 539 1 (hbk)
978 1 78064 563 6 (pbk)

Commissioning editor: Caroline Makepeace
Editorial assistant: Alexandra Lainsbury
Production editor: Tracy Head

Typeset by SPi, Pondicherry, India.
Printed and bound by Gutenberg Press, Tarxien, Malta.

Contents

Preface

All of those who have an interest in livestock production or companion animal management and breeding, including every farmer, pet owner and veterinary surgeon, need to know about domestic animal behaviour in order that they can carry out their jobs and care for their animals properly. All of these people and all consumers of farm animal products have to consider their moral stance in relation to domestic animal welfare and require precise information about that welfare in order to do so. This book is a comprehensive guide to the behaviour and welfare of domestic animals. It provides practical information for those involved with farming, pet animals and veterinary work and reviews scientific information about the assessment of animal welfare, and the evaluation of the effects on animals of genetic selection and of different management methods and housing conditions. Assessing welfare necessarily involves measurement of physiology, disease state and production as well as behaviour.

Ethology developed rapidly after 1950 and became fully established as a part of academic courses on biological subjects by 1980–1990. Ethology is the observation and detailed description of behaviour with the objective of finding out how biological mechanisms function. Some years after its expansion in zoology and psychology teaching, animal behaviour has become part of veterinary and animal science or agriculture courses. Indeed, it is an essential part of professional knowledge for all who use and care for animals and is a subject of widespread interest to the general public. There have been exciting developments in our knowledge of the behaviour of domestic animals and these are incorporated in this new edition.

Animal welfare science arose as a scientific discipline after 1980 and the major increase in its academic study is still occurring. The welfare of an individual is its state as regards its attempts to cope with its environment. This includes the state of mechanisms in its brain and other parts of the body including feelings and systems for dealing with disease. Coping with the environment involves a wide range of interacting biological systems and so the subject is a fundamental biological science. Its development has been rapid. There have long been researchers working on animal health, which is a key part of animal welfare. Other than these, there were only 20–30 animal welfare scientists before 1990 but there are now several thousand. In recent years, the EU Welfare Quality and Animal Welfare Indicators (AWIN) projects have provided much valuable information about welfare outcome assessment in practical situations. The use of animal-based welfare outcome indicators by welfare inspectors, veterinarians and farmers has developed, facilitated by the European Food Safety Authority reports and guidelines. This dynamic area of increasing knowledge has changed what we know about the many animal welfare topics discussed in this book.

The evaluation of behaviour and welfare is also important in human biology and medicine. The detailed scientific study of behaviour was developed for non-humans before it was applied to humans. It provides information that cannot be accurately obtained by asking people questions. The methodology developed in animal welfare science is also being applied to human psychiatry and medicine. The rapidly developing field of anthrozoology, which concerns interactions between humans and other species, has depended greatly on studies involving the behaviour and physiology of pets and their owners.

A further area of rapid development in science has been the study of brain function. Since the brain has a controlling effect on all behaviour, and all methods of coping with the environment of an individual are centred in the brain, neuroscience – including studies of cognitive functioning – is of central importance to the subject of this book. Studies of how clever animals are, and of the representations of events that they have in their brains, alter the way in which people think of the animal species. These studies are directly relevant to understanding animal needs and hence welfare. Some of the burgeoning literature on animal cognition feelings, emotion and sentience is discussed in this new edition.

This book extends the coverage of the fourth edition in that much new information is provided, for example on equids other than horses and on draught animals. Chapters have been added on welfare during stunning and slaughter, the welfare of sheep and goats, and the welfare of additional pet species. Hence all major farmed animals and companion animals are considered. Both authors have planned this updated edition and provided new illustrations, while the text revision has been carried out by Donald Broom. The concepts of behaviour and welfare are introduced, and then behaviour description, learning, cognition, motivation, evolution and welfare assessment are considered. Sections on the various aspects of individual, social and reproductive behaviour then follow. Welfare during transport and slaughter, welfare and disease, and the various kinds of abnormal behaviour are then described. In the final 11 chapters, the welfare of different species is discussed. The book is illustrated with many photographs and, since an understanding of the meaning of concepts is so important in this subject area, it includes an extensive glossary.

Donald M. Broom MA, PhD, ScD, Hon DSc, HonDr
Professor of Animal Welfare (Emeritus)
Centre for Animal Welfare and Anthrozoology
Department of Veterinary Medicine and St Catharine's College
University of Cambridge, UK

Andrew F. Fraser MRCVS, MVSc, FIBiol
Formerly Professor of Surgery (Veterinary)
Memorial University of Newfoundland, Canada

Acknowledgements

We thank first the many authors, mentioned in the reference list, whose published work forms the basis for this book. We thank especially the following, some now deceased, who, for one or more of the editions of the book, have given encouragement to the authors, provided illustrative material or helped in book preparation: A.M. Aitchison, Jack Albright, Graham Arnold, Bob Baldwin, D. Bieger, M. Bieger, Harry Bradshaw, Sally Broom, Alex Brownlee, D. Dooley, Ingvar Ekesbo, Michael Fox, David Fraser, C. George, Temple Grandin, Stephen Hall, Ruth Harrison, H. Hastie, Ken Johnson, Ron Kilgour, Erina Kirby, A. Littlejohn, Frank Loew, Mike Mendl, B. Payton, Anthony Podberscek, Irene Rochlitz, Hans-Hinrich Sambraus, James Serpell, Elizabeth Shillito-Walser, Barbara Sommerville Alex Stolba, Ray Stricklin, Sue Tennant, Tarjei Tennessen, C. Thorne, Klaus Vestergaard, Piet Wiepkema and David Wood-Gush. We are grateful for the efforts of the staff at CABI to improve this book.

Introduction to Ideas and Measurement

1 Introduction and Concepts

A significant part of our interest in the world around us is focused on non-human animals, not just as resources but also because of our empathy for them and our fascination with what they can do and how they function (Podberscek *et al.*, 2000; Serpell, 2004; D. Fraser, 2008; Broom, 2014b). The farming of animals has played an important part in the development of human civilization. Food, clothing and transport are obtained by man from a wide variety of species. It is thought that humans have had an even longer relationship with wolves, or dogs as we now call one form of the Middle-Eastern grey wolf (Clutton-Brock, 1999; von Holdt *et al.*, 2010). This relationship, and we may speak of wolves domesticating humans just as correctly as humans domesticating wolves (Broom, 2006a), seems likely to have been mutually beneficial to both species, as has other domestication (see Chapter 5). Dogs, cats, other companion animals and many farmed animals have long been treated as companions and viewed with affection by those whose job it was to care for them. Good stockmanship has always involved knowing how to respond to the behaviour of animals when handling them or identifying their problems.

By the start of the 20th century, farm animal use had increased with the expansion of the human population and consumption of animal products. Animals began to be kept in concentrated populations and, prior to 1970, intensive animal husbandry had arrived in the form of close confinement for cattle, pigs and poultry under new husbandry systems. The innovations in management are characterized principally by larger livestock numbers kept together in markedly reduced space. Such conditions have effects on disease transmission and they require considerable physiological and behavioural adaptation by the animals (Broom, 2006a). It was assumed that the animals could adapt to the environmental restrictions, but both adaptation and failure to adjust have come to be recognizable when welfare is assessed. As Mason (2010) explains, animals kept in captivity vary greatly in how well they can adapt and in the extent of poor welfare resulting from the captivity. Knowledge of the behaviour of livestock under intensive husbandry systems is needed to assess these systems of management, just as information about behaviour is needed to manage animals extensively. This knowledge can then be applied in the agriculture industry in order to improve production and welfare. Many animal husbandry problems are not soluble by investigation of nutrition, body physiology or disease control, but require investigations of the behaviour of the animals before progress can be made towards a solution.

The attitude of people to dogs (Serpell, 1995) ranges from viewing the dog as a cause of vicious and unprovoked attacks on children, of pollution of our streets and of serious disease risks, to considering the dog as a family member, an archetype of affectionate fidelity and a source of unconditional love. People who use animals as companions, or for some form of work or entertainment, are aware of the behaviour of the animals. In some cases, the behaviour is not what the people want and is viewed as a problem. In other cases the behaviour is the reason why the animal is useful, whether or not this use results in good welfare in the animal. Behaviour can be an indicator of good or poor welfare in any animal. The term ethology means the observation and detailed description of behaviour with the objective of finding out how biological mechanisms function.

The scientific study of animal behaviour has proceeded very rapidly during the last 50 years. Some of the changes in ideas that have occurred during this development are described by Jensen (2009). There have been substantial recent advances in the precision of behaviour description and the understanding of behaviour organization in relation to physiological and evolutionary processes. Modern techniques in ethology and in experimental psychology mean that we now have a much more extensive knowledge of sensory analysis, motor control, hormonal effects, motivation, body maintenance behaviour in good and difficult conditions, reproductive behaviour and social structure. This knowledge and many other methods of animal welfare assessment are used in animal welfare science and applied to domestic animals.

Sentience and Animal Protection

Animals vary in the extent to which they are aware of themselves (DeGrazia, 1996) and of their interactions with their environment, including their ability to experience pleasurable states such as happiness and aversive states such as pain, fear and grief. This capacity may be referred to as their degree of sentience. The term *sentience* has generally been used to mean that *the individual has the capacity to have feelings* (Kirkwood, 2006). This raises the question of what abilities are needed in order to have this capacity. Sentience implies a range of abilities, not just having feelings. A definition is: *a sentient being is one that has some ability to evaluate the actions of others in relation to itself and third parties; to remember some of its own actions and their consequences; to assess risks and benefits; to have some feelings; and to have some degree of awareness.* This definition, slightly modified after Broom (2006c), and various aspects of sentience are discussed further by Broom (2014b).

Human opinion as to which individuals of our own and other species are sentient has generally changed over time in well-educated societies to encompass first all humans instead of just a subset of humans, and then: certain mammals that were kept as companions; animals that seemed most similar to humans such as monkeys; the larger mammals; all mammals; all warm-blooded animals; then all vertebrates; and now some invertebrates.

The general public has been ready to accept some guidance about evidence for sentience from biologists who have collected information about the abilities and functioning of the animals. Animals that are shown to be complex in their organization, capable of sophisticated learning and aware are generally respected more than those that are not, and such animals are less likely to be treated badly. However, some people view animals solely on the basis of their effects on, or perceived (extrinsic) value to, humans and have little concern for them as individuals.

Ethics

Something is moral if it pertains to right rather than wrong and *ethics is the study of moral issues.* Humans and other animals, especially social animals, have many biological mechanisms that enable them to behave in a moral way. It is not possible to live successfully in a social group unless the individuals have the ability to avoid harming others (see Fig. 1.1) and perhaps to collaborate. As a consequence, morality has evolved and natural selection has favoured genes that promote abilities such as recognition of individuals and memory of moral and immoral actions (de Waal, 1996; Ridley, 1996; Broom, 2003, 2006). Two other biological mechanisms which may promote moral actions are empathy and compassion (Würbel, 2009). People are more likely to feel empathy with fellow humans and other animals perceived to have a capacity for feelings similar to those of humans. They are likely to show compassion to those whom they perceive to need compassion.

Two underlying approaches in thinking about how to behave towards people or other animals are known as deontological and consequentialist. The deontological approach to the organization of human conduct is one in which the structure is a set of duties pertinent to all individuals. Hence, the individual should assess what action duty dictates using rational thought and carry out that action.

Consequentialism in ethics implies that the extent to which an act is morally right is determined solely by the goodness of the act's consequences. This approach was extended into utilitarianism by J.S. Mill (1843), who argued that the right act or policy is that which will result in the maximum utility, or expected balance of satisfaction minus dissatisfaction, in all the sentient beings affected.

Although many aspects of utilitarianism are helpful when deciding what is morally right, as a general approach it may be viewed as incomplete (Broom, 2003). Acting in such a way that general happiness or general good is promoted will be entirely desirable in some circumstances, but following such a philosophy implies that decisions are taken only on the basis of the average or overall good of collections of individuals. This view does not take account of the fact that humans and other animals interact with and have concerns for individuals. The mechanisms underlying moral codes are based on effects on individuals as well as on collections of individuals. An example of the flaw in the extreme utilitarianism approach is that, following this approach, an individual could be caused extreme pain or other poor welfare, or could be killed if the overall

Fig. 1.1. These Longhorn cattle carry dangerous weapons and spend much time close together but they are very careful not to harm one another (photograph D.M. Broom).

effect on a collection of individuals was good. This individual might be a dangerous criminal or an entirely innocent person, but should they be tortured, caused prolonged misery or killed? Most people would not wish an innocent person to be killed, however great the resulting good, and those who hesitated on the issue might be swayed towards that view if the person were their neighbour, their mother or themselves.

Criticisms of the utilitarian position have been made by many, including Williams (1972) and Midgley (1978). Those who would consider themselves deontological ethicists would maintain that certain rights, rules, principles or obligations take precedence over utility. However, a wholly deontological approach also has flaws. Elements of both deontological and utilitarian approaches are necessary in order to act in a moral way.

It is my view (Broom, 2003, 2006b, 2010, 2014b) that all human behaviour and laws should be based on the obligations of each person to act in an acceptable way towards each other person and to each animal with which we interact. If we use a living animal in a way that gives us some benefit, we have some obligations to that animal. We have some obligations to any individual considered to have an intrinsic value and other obligations whenever we have concern for the individual's welfare. We should avoid causing poor welfare in the animal except where the action leads to a net benefit to that animal. In some cases, we might cause poor welfare because of a net benefit when we also take account of other animals, including humans, or of the environment.

It is better for strategies for living to be based on our obligations rather than to involve the concept of rights. This is because many so-called rights can result in harm to others. Arguments about the importance of freedom to control one's life led to the idea that such freedom is a 'right' which all should have. Strong proponents of a rights structure for determining what are proper actions regard the stated rights as absolute, so they cannot be mitigated by other circumstances. A key issue here is the establishment of what is a right. There are few so-called rights that would be accepted as valid in all circumstances. The oft-proclaimed right to free speech can cause great harm to certain individuals and hence can be morally wrong, in my view, as can the 'right' to drive a car as fast as you wish, or to carry a gun, or to determine the sex of your children. The concept of rights causes many problems. All behaviour and laws should be based on the obligations of each person to act in an acceptable way towards each other person or other sentient individual. Arguments based on obligations are better than any attempts to assert a 'right'. Laws and other such statements should provide guidelines for the behaviour of each person rather than stating what the individual who is the object of an action can demand.

The ethics aspects of animal behaviour research are discussed by M. Dawkins and Gosling (1992).

Sustainability and Animal Welfare

Members of the public now ask whether or not any system for exploiting resources is sustainable (Aland and Madec, 2009). The fact that something is profitable and there is a demand for the product is not now sufficient reason for the continuation of production (Broom, 2010a). *A system or procedure is sustainable if it is acceptable now and if its expected future effects are acceptable, in particular in relation to resource availability, consequences of functioning and morality of action* (Broom, 2001d, 2014b). An animal usage system might not be sustainable because: (i) it involves so much depletion of a resource that this will become unavailable to the system; (ii) a product of the system accumulates to a degree that prevents the functioning of the system; or (iii) members of the public find an action involved in it unacceptable. Where there is depletion of a resource or accumulation of a product, the level at which this is unacceptable, and hence the point at which the system is unsustainable, is usually considerably lower than that at which the production system itself fails.

A system could be unsustainable because of harms to: the perpetrator, other people, the environment or other animals. Agriculture is the main activity on most of the land in the world and has resulted in much loss of biodiversity, so it must be considered in most discussions of wildlife conservation (Balmford *et al.*, 2012) and the sustainability of many of its practices is questionable. Currently in the world, animal welfare is one of the major reasons why certain methods used in animal production and management are unsustainable. Aspects of production, such as the welfare of the animals, are now important when evaluating the quality of animal products. Consumers expect supermarket companies and restaurant chains to guarantee that the methods used to produce products are not negative in relation to animal welfare or any other aspect of sustainability and product quality. Efforts are now being made to design animal production systems that are sustainable in all ways.

Welfare Concepts

The scientific study of animal welfare hardly existed 40 years ago, with the exception of studies of animal disease, but has developed rapidly during the last 30 years (Broom, 2011). The concepts have been refined and a range of methods of assessment have been developed. Substantial challenges to animal functioning include those resulting from: (i) pathogens; (ii) tissue damage; (iii) attack or threat of attack by a conspecific or predator; (iv) other social competition; (v) complexity of information processing in a situation where an individual receives excessive stimulation; (vi) lack of key stimuli such as a teat for a young mammal or social contact cues; (vii) lack of overall stimulation; and (viii) inability to control interactions with the environment. Hence, potentially damaging challenges may come from the environment outside the body (e.g. many pathogens or causes of tissue damage), or from within it (e.g. anxiety, boredom or frustration that come from the environment of a control system). Systems that respond to or prepare for challenges are coping systems and *coping means having control of mental and bodily stability* (Broom and Johnson, 1993).

Coping attempts may be unsuccessful in that such control is not achieved but, as soon as there is control, the individual is coping. Systems for attempting to cope with challenge may respond to short-term or long-term problems, or sometimes to both. The responses to challenge may involve activity in parts of the brain and various endocrine, immunological or other physiological responses as well as behaviour. However, the more that we learn about these responses, the clearer it becomes that these various types of response are interdependent. For example, not only do brain changes regulate bodily coping responses, but adrenal changes have several consequences for brain function; lymphocytes have opioid receptors and a potential for altering brain activity; and heart rate changes can be used to regulate mental state and, hence, further responses.

Some coping systems include feelings as a part of their functioning, for example pain, fear and the various kinds of pleasure, all of which are adaptive (Broom, 1998; D. Fraser, 2008; Chapter 4). Bad feelings that continue for more than a short period are referred to as suffering. Other high- or low-level brain processes and other aspects of body functioning are also a part of attempts to cope with challenge. In order to understand coping systems in humans and other species, it is necessary to study a wide range of mechanisms including complex brain functioning, as well as simpler systems. Investigations of how easy or difficult it is for the individual to cope with the environment, and of how great is the impact of positive or negative aspects of the environment on the individual, are investigations of welfare. If, at some particular time, an individual has no problems to deal with, that individual is likely to be in a good state, including good feelings and indicated by body physiology, brain state and behaviour. Another individual may face problems in life that are such that it is unable to cope with them. Prolonged failure to cope results in failure to grow, failure to reproduce, or death. A third individual might face problems but, using its array of coping mechanisms, be able to cope but only with difficulty. The second and third individuals are likely to show some direct signs of their potential failure to cope or difficulty in coping and they are also likely to have had bad feelings associated with their situations.

According to Broom (1986c) *the welfare of an individual is its state as regards its attempts to cope with its environment*, and this includes feelings and health. Welfare is a characteristic of an individual at a certain time; the state of the individual can be assessed so welfare will vary on a range from very good to very poor. Welfare concerns how well the individual fares, or goes through life. Some other authors place sole emphasis on feelings when defining welfare (Duncan and Petherick, 1991). Health, like welfare, can be qualified as good or poor and varies over a range. It refers to body systems, including those in the brain, that combat pathogens, tissue damage or physiological disorder (see Chapter 23) so *health* can be defined as *the state of an individual as regards its attempts to cope with pathology* (Broom, 2006b). All of this is encompassed within the broader term welfare, so health is a part of welfare. Many years before the refinement of our definitions of such concepts, the World Health Organization (1946) said 'health is a state of complete physical, mental and social well-being and not merely the absence of disease or infirmity'. The word welfare was not being used in a scientific way at this time, but welfare is essentially the same concept as well-being so WHO was defining health as an aspect of welfare. Their definition is not easy to use scientifically, as both health and welfare can be poor as well as good and measurement of both depends on the concept being a scale from positive to negative. The WHO definition is also contrary to normal public usage in that most people limit health to conditions related to pathology, physical or mental, as explained above. Hence the WHO definition has been confusing for many people as a result and this is why it is not used here. Even those who see some of its limitations sometimes accept it and then conclude that health is the wider term, e.g. Nicks and Vandenheede (2014), but this does not follow scientific logic.

Where does naturalness fit with the concept of welfare? D. Fraser (1999) pointed out that when members of the public talk about animal welfare, their ideas often include the functioning of the animals, the feelings of the animals and the naturalness of the environment. The feelings, referred to by Fraser and others, fit comfortably into Broom's definition of welfare as they are important components of coping mechanisms and of biological functioning. Rollin (1989, 1995) advocated that 'animals should be able to lead reasonably natural lives' and both Rollin and D. Fraser (Fraser *et al.*, 1997; Fraser, 1999, 2008) refer to the importance of understanding animal needs. These authors did not say that naturalness contributes to a definition of welfare or should be part of welfare assessment. The state of an individual trying to cope with its environment will necessarily depend upon its biological functioning or, put another way, upon its nature. Natural conditions have affected the needs of the animal and the evolution of coping mechanisms in the species. The environment provided should fulfil the needs of the animal but does not have to be the same as the environment in the wild.

The assessment of welfare should be carried out in an objective way, taking no account of any ethical questions about the systems, practices or conditions for individuals that are being compared. Once the scientific evidence about welfare has been obtained, ethical decisions can be taken. Much of the evidence used in welfare assessment indicates the extent of poor welfare in individuals, but it is also important to recognize and assess good welfare (i.e. happiness, contentment, control of interactions with the environment and possibilities of exploiting abilities). Good welfare in general, and a positive status in each of the various coping systems, should have effects which are a part of a positive reinforcement system, just as poor welfare is associated with various negative reinforcers. We need to identify and quantify indicators of good welfare as well as those of poor welfare.

The term 'well-being' is often used interchangeably with 'welfare', but well-being is often used in a looser, less precise way. Welfare is the word used in English versions of modern European legislation (Broom, 2009). Some other languages have only one word that can be used to translate either welfare or well-being. The words that are equivalent to welfare in other languages, and that are used in identical legislation, have similar origins: for example, *welzijn* in Dutch, *bien-être* in French, *bem estar* in Portuguese, *bienestar* in Spanish, *velfaerd* in Danish and *dobrostan* in Polish. *Welzijn*, *bien-être*, *bem estar* and *bienestar* are very similar to well-being in origin, but are used by scientists and legislators in much the same way as English speakers use welfare. *Dobrostan* is close in use to welfare as defined in this chapter, and *velfaerd* has a wider meaning but is used specifically in legislation. In German, *Wohlbefinden* and *Wohlergehen* have similar meanings to welfare but *Tierschutz* means animal protection.

Most people who speak of stress refer to a situation in which an individual is subjected to a potentially or actually damaging effect of its environment. However, the usage of the term has sometimes been confusing, as it has been used to mean three different things: (i) an environmental change that affects an organism; (ii) the process of affecting the organism; or (iii) the consequences of effects on the organism. Some people have limited the use of the term stress to one kind of physiological response mechanism, hypothalamic–pituitary–adrenal cortex (HPA) activity or to mental rather than physiological responses.

However, it was demonstrated by Mason (1971) and in many other studies that several different responses to challenges could occur: HPA activity is temporarily increased during courtship, mating, active prey catching and active social interaction, none of which would be considered to be stressful by the majority of the general public or by scientists. To equate stress with HPA axis activity renders the word redundant and is considered unscientific and unnecessary by most scientists working in the area. Another meaning that has been ascribed to stress makes it largely synonymous with stimulation. If every impact of the environment on an organism is called stress, then the term has no value. Many stimuli that affect individuals in beneficial ways would never be called stressors by most people. *Stress is an environmental effect on an individual which overtaxes its control systems and results in adverse consequences and eventually reduced fitness* (Broom and Johnson, 1993, 2000). The ultimate measure of fitness is the number of offspring reaching future generations, and there are many different ways in which challenges overtax control systems and have such effects.

In her influential book *Animal Machines*, Ruth Harrison (1964) pointed out that those involved in the animal production industry were often treating animals like inanimate machines rather than living individuals. As a consequence of this book, in 1965 the British government set up the Brambell Committee, a committee chaired by Professor F. Rogers Brambell, to report on the matter. One of its members was W.H. Thorpe, an ethologist in Cambridge University. Thorpe emphasized that an understanding of the biology of the animals is important and explained that animals have needs with a biological basis, including some needs to show particular behaviours, and that animals would have problems if the needs were frustrated (Thorpe, 1965). The environment of an animal is appropriate if it allows that animal to satisfy its needs. Animals have a range of functional systems controlling body temperature, nutritional state, social interactions, etc. (Broom, 1981). Together, these functional systems allow the individual to control its interactions with its environment and, hence, to keep each aspect of its state within a tolerable range. The allocation of time and resources to different physiological or behavioural activities, either within a functional system or between systems, is controlled by motivational mechanisms (see Chapter 4). When an animal is actually or potentially homeostatically maladjusted, or when it must carry out an action because of some environmental situation, we say that it has a need. Unsatisfied needs are often, but not always, associated with bad feelings, while satisfied needs may be associated with good feelings. When needs are not satisfied, welfare will be poorer than when they are satisfied.

The term quality of life is principally used to refer to people, or companion animals, who are ill or recovering from illness. In judging quality of life, the impact on the functioning of the individual, including physiological and behavioural responses and especially indicators of pain or other suffering, should be evaluated. As explained in detail in Chapter 6, the measures of welfare include all of the scientific measures of quality of life. Both quality of life and welfare can be positive or negative, good or poor. There is some difference in the use of the terms as it would not be normal to talk about quality of life over a very short time scale such as a few hours or days. Welfare, on the other hand, can refer to short-term situations. *Quality of life means welfare during a period of more than a few days* (Broom, 2007, 2014b). Hence quality of life can be assessed used the wide range of indicators that are available for assessing welfare. If more subjective measures of quality of life are utilized, either for humans or for non-humans, their value is questionable unless rigorously verified (Green and Mellor, 2011) and this will often not be possible.

The concept of 'a life worth living' (e.g. Edgar *et al.*, 2013) is ethical rather than scientific, although the ethical judgements would be based on scientific information. Worth depends on a concept of value and raises the question of who decides when it is worth living or not worth living (Broom, 2014b). There is a problem with the term if the individual under consideration is non-human for it is a human evaluation rather than an

Fig. 1.2. This beaver was caught in a trap that clamps around the neck or thorax. The effect is to break bones and cause other injuries that often cause prolonged and extreme pain (photograph F. Wassenberg).

evaluation by the subject. Investigation of welfare refers to a measurable quality of the animal and this is assessed in an objective way, trying to take account of what that animal needs and its current state in relation to those needs. A decision about whether or not life is worth living might not be made in the same way by the person judging and by the animal itself. As explained by Broom (2014b) a person might judge that a pet animal is in pain and that its life is not worth living. However, animals in pain still strive to survive and carry out assessments of risks to themselves. Would the animal choose to live or die? As Green and Mellor (2011) point out, the idea of 'a life worth living' is not scientifically usable. It may be useful in order to criticize some housing systems or procedures on animals but objective measurement of welfare should be carried out on all occasions when evaluation is required, so that the worth of the life can be properly considered. Some ideas about how animal welfare science and practice might progress in the future are explained by a wide range of authors in Mellor and Bayvel (2014).

Animal Welfare: Scientific Assessment

Domestic animals have to contend with a complex environment and they have a variety of methods for attempting to cope with it. That environment includes physical conditions, social influences and predators, parasites or pathogens that may attack the individual. The coping methods include physiological changes in the brain, adrenal glands and immune system and, linked to some of these, behavioural changes. Some factors that affect an animal may result in it having great difficulty in coping. It may fail to cope in that its fitness is reduced and either it dies or it fails to grow, or its ability to reproduce is reduced in some direct way.

The assessment of welfare can be carried out in an objective way that is quite independent of any moral considerations. Mortality rate, reproductive success, extent of adrenal activity, amount of abnormal behaviour, severity of injury, degree of immunosuppression or level of disease incidence can all be measured (see Chapter 6). Our knowledge of each of these welfare indicators has improved rapidly in recent years as people with backgrounds in zoology, physiology, psychology, animal production and veterinary medicine have investigated the effects of difficult conditions and good conditions on animals.

In the early years of consideration of what constituted poor welfare, people tended to think especially of pain as a reason why welfare would be unacceptable. Figure 1.2 is an example of a situation where there would be pain. However, the welfare of this beaver might also be poor because of fear, adverse temperature conditions or starvation. Long-term problems, such as those resulting from having to live in housing where the needs of the individual are not met, can be more important than a short period of pain.

Animal welfare scientists can now compare welfare in different systems of management, designs of housing, methods of handling or transportation, and procedures in operations or in slaughter. This knowledge can also be applied to the care of companion animals and laboratory animals (Hubrecht and Kirkwood, 2010). In addition to measurements of poor welfare, it is possible to investigate the preferences of animals and the value that they place on various resources or other aspects of their environment. Such studies and a wide range of work on the basic biology of animals give information about the biological needs of animals. If these are not met there will often be indicators of poor welfare that can be measured but, in some circumstances, we have not yet acquired the expertise to evaluate adverse psychological effects on animals.

When scientific evaluation of welfare has been carried out, there remains the moral question of how good welfare should be in order that it is acceptable. In practice, it is often the threshold of unacceptably poor welfare that is considered. This is an issue where the farmer, the veterinary surgeon, the welfare research worker or the member of the general public are equally entitled to have an opinion. One person might say that a certain degree of poor welfare in an individual domestic animal is acceptable, given the human requirements involved, while another might consider that degree of poor welfare to be unacceptable. Moral positions in such matters have changed as people have come to know more about the complexity of animal functioning, the sophistication of animal behaviour and the degree of similarity between domestic animal species and man. Both recent research and media coverage of such research have contributed to this change of attitude. The feeling that humans have a moral obligation to ensure that the welfare of animals that are kept is never very poor has become widespread. The idea that, when decisions are taken about methods in animal husbandry, animals should be considered as individuals and their responses to their environment should be evaluated and understood, is now held by many in the agriculture industry, by those who care for companion animals and by those in the veterinary profession.

Animal welfare can be affected by a wide range of factors. The effects of disease, injury and starvation are negative. However, there can be beneficial stimulation and success in actions will have positive consequences. The effects of social interactions and housing conditions may be positive or negative. In a summary of information from animal welfare research, D. Fraser *et al.* (2013) list ten factors to consider in order to promote good welfare, focusing on animals in production systems. Implicit in this list is the necessity to consider scientific information about the needs of animals of the species and background under consideration. However, it is missing from their list so it has been put at (1) in Box 1.1. The words 'natural' (see above) and 'euthanasia' (see Chapter 22) have also been changed.

Behaviour and Animal Production

A major theme of this book is that farm animal behaviour research is relevant and necessary for animal production enterprises to be carried out effectively and economically. The stockman, the farm manager, the animal transporter, the abattoir worker and the designer of animal accommodation and equipment have to be aware of well-established facts and recent research on the ethology of farm animals.

An appreciation of how to handle animals necessitates a knowledge of behaviour that those managing animals in the past gradually acquired through personal experience. This information can be taught and is learned more easily if general principles of animal behaviour are known. Feeding behaviour is an example of an important topic for those who have to manage animals. The control of feeding, food selection at pasture or when composite feeds are offered, and learning about food and behaviour in competitive feeding situations, are all relevant to intakes and good feed conversion efficiency. For many companion animals and some farm animals it is useful to find out about the temperament of individuals. Using a simple test situation in which ease of handling young beef cattle was evaluated, MacKay *et al.* (2013) found that those steers with high flight speed in the test spent longer standing and less time lying, were more active in the home pen and were more frequently observed displacing others at the feeder than steers with low flight speed. Turner *et al.* (2011) categorized 16–18-month-old beef cattle temperament according to how they reacted to humans when put into and removed from a crush. They found that the calmer animals had higher daily weight gains during the

Box 1.1. Factors to consider in order to promote good welfare in domestic animals (after Broom, 2014b, modified after D. Fraser *et al.*, 2013).

1. Scientific information about the needs of animals of the species and background under consideration.
2. How genetic selection affects animal health, behaviour and temperament.
3. How the environment influences injuries and the transmission of diseases and parasites.
4. How the environment affects resting, movement and the performance of adaptive behaviour.
5. The management of groups to minimize conflict and allow positive social contact.
6. The effects of air quality, temperature and humidity on animal health and comfort.
7. Ensuring access to feed and water suited to the animals' needs and adaptations.
8. Prevention and control of disease and parasites, with humane killing if this is not feasible or recovery is unlikely.
9. Prevention and management of pain.
10. Creation of positive human–animal relationships.
11. Ensuring adequate skill and knowledge among animal handlers.

whole fattening period. An even more rapid test, found useful by Core *et al.* (2009), was the amount of eye white visible when the animal was in a squeeze chute. Applying knowledge of such individual differences in behaviour can aid those managing the animals and improve the animals' welfare.

Reproductive behaviour is of great importance to those managing a stock unit. Behaviour assessment is the major method of oestrus detection in dairy cows and pigs. Work on mating preferences and factors affecting libido is of critical importance in the management of sheep, goats, beef cattle and horses where a high proportion of successful matings is desired. Each animal whose offspring production fails, or is delayed, costs the farmer money. The frequency with which maternal behaviour fails in domestic animals and problems with the survival of the young, especially piglets, lambs or calves, can all be reduced by a knowledge of behaviour and consequential improvements in stockmanship.

Wherever animals have to be grouped or decisions have to be taken about the housing and density of animals, information about social behaviour is important. Farm animal management, which leads to fighting, injury or extreme fear, can result in reproductive failure, poor food conversion, reduced carcass value or increased mortality. Such losses can be substantially reduced by knowledge of social behaviour. As pointed out by Broom *et al.* (1995) and Jensen (2009), through use of ethological knowledge, technical equipment can be designed and management methods can be utilized to work better for the animals.

These very wide-ranging applications of behaviour study to farm animal production, together with the relevance of behaviour work to the understanding of animal welfare, emphasize the importance of including a series of lectures on animal behaviour in courses on animal science, animal production and applied biology.

Behaviour and Pet Management

In developed nations, about half of all households own a pet and pet-ownership increases with the age of children in families (Westgarth *et al.*, 2010). Many people decide to buy a pet for themselves or for a child and then discover that the complex being that they have brought into their home requires a wide variety of care and has responses and abilities that the owner may not be able to manage. Owners sometimes reach the stage of having to contend with difficult and embarrassing behaviour before resorting to advice or buying a book that may help them. Behavioural signs may tell you that your cat is sick or that its welfare is poor for some other reason, that your fish is about to die from a lack of oxygen or by being eaten by another fish, or that your dog or horse has interpreted the various signs that you have given to it as a clear indication that it can do what it wants rather than what you want.

The training of companion animals requires knowledge of learning and motivation. People do not have this knowledge unless they have studied animal behaviour or psychology, at least to the extent of reading a good-quality book. Some of those who teach animal training are very knowledgeable about the principles of learning and motivation, as well as having relevant practical experience. However, some of those who appear authoritative to the uninitiated may present incorrect and confusing ideas. It is important that all of those who offer professional advice on animal behaviour are properly trained. The identification of behavioural problems and the ability to propose methods of reducing or eliminating these is another subject where expert, helpful people may be difficult to distinguish from those who are unlikely to help in solving problems. As a consequence, accreditation schemes have been developed, for example by the Association for the Study of Animal Behaviour in the UK and elsewhere. Pet behaviour consultants, like other experts in complex subjects, should be accredited and should not be employed unless they are accredited.

Behaviour and Veterinary Medicine

The practising veterinary surgeon (referred to in many parts of the world as veterinarian) uses knowledge of behaviour frequently but, in the past, much of this knowledge has been acquired after formal training is completed, unless an animal behaviour course has been attended. Situations in which such expertise is important include: (i) handling animals; (ii) using behaviour as a sign in diagnosis; (iii) advising on animal husbandry methods; and (iv) dealing with behaviour problems and assessing welfare. Behaviour is also important as a part of the general biology of the animals so it is relevant when considering, for example, feeding responses to adverse temperature conditions, or disease transmission.

The importance to a veterinary surgeon intending to handle an animal, of recognizing the signals that indicate that the animal is about to attack or will kick if handled, are obvious. It is also very beneficial to the veterinary treatment if handling procedures can be modified according to observed behaviour in such a way that the animal is not adversely affected by the handling itself. Poor welfare during veterinary treatment is a major obstacle to success in disease treatment and in retaining client confidence. Studies of behaviour are also especially relevant where it is necessary to move animals from place to place along races, up ramps, into vehicles or into strange rooms (Grandin, 2014). Many domestic animals show behaviour that indicates fear of humans, and it is important to veterinary surgeons and others handling animals that such behaviour is recognized (Beaver, 1994).

The working veterinary surgeon is regularly presented with clinical cases having histories that are symptomatically behaviour-based. In fact, it is common for animal illness to be first manifested behaviourally, such as in loss of appetite, altered activity or diminished body care – for example, in colic and other painful conditions in horses (Mair *et al.*, 1998). Clinical veterinary work has a very real and special relationship with pathological behaviour in animals. Those who practise professionally the art and science of clinical veterinary medicine and surgery acquire competence at the interfaces between illnesses and their behavioural signs through years of training, experience and witness. This takes much time and training in behaviour is essential. Studies of the welfare of animals after veterinary treatment often incorporate neither our knowledge of behavioural nor scientific indicators of welfare (Christiansen and Forkman, 2007) wherever they should do so.

Behavioural signs of impairment and histories of behavioural symptoms give invaluable help to the veterinary clinician in the initiation of a clinical appraisal of the animal's condition. With such orientation, further points are sought out for special investigation and detailed examination. On systematic examination, the behavioural feature of a veterinary problem often presents as a screen over a generalized and mixed array of physical signs of illness. As substantive correlates of the behavioural manifestation are found in physiological and pathological factors, the behavioural picture fades into the background of the clinical problem. Treatment and restorative action then become focused physically on lesions and infections. The behavioural problem thus becomes resolved by a transformation through veterinary medical concepts into an identifiable clinical condition, which can then be given appropriate case management and therapy.

This is an important logical mistake, as the behavioural signs are often a key part of the information used in diagnosis. Sometimes the animal's irregular behaviour does not translate into a physical condition, but this does not mean that the associated pathology is not real. Some disorders are distinguishable only by their effects on behaviour, while others are manifested in a variety of ways.

A necessary prerequisite for the recognition of abnormal behaviour, whether as a sign of pathogen presence or as an indicator of poor welfare which is not due to a pathogen, is a knowledge of normal behaviour. The normal behaviour for a species can be described during a behaviour course, but requires practical experience in order that it can be understood well.

If a horse is seen to be pacing in its stable and kicking at its belly, this is identifiable as a sign of colic, but someone with no experience of horse behaviour would not recognize it as such. More subtle signs, such as the partially hunched posture adopted by a sheep with abdominal pain, require clear observation of both the normal and affected animals (see Chapter 22).

Expertise in behaviour and the assessment of animal welfare are areas where the veterinary surgeon is expected to be able to give advice. Behaviour problems that affect the practicalities of managing animals are also frequently presented to the veterinary surgeon for solution. One area of general behavioural knowledge that helps in the explanation of such problems, in both farm and companion animals, concerns the ways in which early experience affects behaviour development. Another area is learning. The veterinary surgeon should be able to advise on how to avoid behaviour problems and what training or other procedures to utilize in order to deal with them. The preparation that is required in order that every veterinary student has an adequate knowledge of the fundamental principles and veterinary application of animal behaviour is a course of lectures in the subject and reference to specific techniques and literature at various points in the clinical part of the course.

Questions about Behaviour

There are two kinds of question that can be asked when trying to understand a particular behaviour. The first of these is: 'How does it work?' The answers to this question refer to the mechanisms underlying the behaviour that cause it to occur at the time of observation and with the form in which it is seen. What changes are occurring within the body of that animal that result in the movements that are shown? Some of these changes are physiological processes which we know quite a lot about, such as those involved in sensory reception, impulse conduction along nerves or muscle contraction. Changes within the brain involving emotional variables, learning, decision making and control of actions have been extensively investigated. Although we still have much to discover about them, there have been significant developments in knowledge in recent years.

The second kind of question about behaviour is: 'Why does it happen?' The answers to this question refer to the way in which this behaviour has arisen in the species under observation. In order to try to appreciate how the pattern and use of a behaviour have evolved, it is necessary to consider what the selective advantage of the behaviour is. Put another way, in what way will the effects of a gene that affects behaviour promote the spread of that gene in the population? In practice, 'why?' and 'how?' questions are linked because questions about evolution depend upon a knowledge of the mechanism underlying a behaviour, and questions about causation are often helped by an understanding of the evolutionary origins of the system. For both kinds of questions we need to consider behaviour in relation to the general biology of the animal.

Behaviour, like physiology and anatomy, is part of the general functioning of an animal. As mentioned above, the various aspects of life can be classified into functional systems that include behaviour as a component (Broom, 1981). These are: (i) obtaining oxygen; (ii) osmoregulation; (iii) temperature regulation; (iv) cleaning the body surface; (v) feeding; (vi) avoiding chemical hazards; (vii) avoiding physical hazards; (viii) predator avoidance; and (ix) reproduction. Some of the ways in which behaviour has a role in homeostatic and other life control mechanisms, and its consequences for welfare, are shown in Box 1.1. It is clear from this box that there are various ways in which needs are met and welfare improved by carrying out particular behaviours. Behaviour often serves more than one function; for example, exploration and establishing social relationships may be relevant to almost all of them, so they become objectives in themselves. The role of behaviour in each of these functional systems is discussed in detail in this book, but the key to understanding the behaviour of an animal is an appreciation of how resources are apportioned and decisions taken about which activity to show and when.

The study area that deals with decisions about the timing and nature of changes in behaviour is that of motivation. This important subject is essential to an understanding of all aspects of behaviour and also to questions about animal welfare. Many welfare problems result from frustration or environmental unpredictability, and a knowledge of motivational state is needed in order that these can be recognized.

When trying to answer questions about how behaviour works, investigatory methods are used in which the experience of an animal is controlled and its effects are assessed. Some such studies are carried out during the development of the individual. Learning occurs when some experience modifies later behaviour (see Chapter 3). Learning has an effect on all aspects of life, so some reference must be made to it in discussions of every aspect of behaviour, and therefore it is relevant to the material in every chapter of this book. All systems develop in animals as a consequence of interactions between the genetic information in the animal and influences from the environment of that genotype. Since all behaviour depends on the genetic information in an animal and environmental factors will always affect the expression of genes, it is not useful to try to distinguish between instinctive or innate behaviour and that which is environmentally determined. The interesting questions concern how differences in genotype and in the environment result in differences in behaviour. Behaviour genetics and the processes occurring during behaviour development are both exciting topics that will be discussed in this book.

The various questions posed above are answered here, but with especial reference to the requirements of those who need to know about the management, housing, veterinary treatment and general biology of domestic animals.

The Sensory Worlds of Domestic Animals

Humans are particularly aware of information from vision and from the narrow range of sounds associated with speech. However, apart from taste, which humans can use with precision, olfaction is of less importance than in our domestic mammals. We are atypical mammals because, for most, olfaction is a major source of information about the surrounding world. Smells are often relatively long lasting and can continue after the source has departed. They exist in enormous variety and can be detected with great sensitivity, so they are used for recognition and assessment of foods, animal species, animal individuals and even animal mood changes by most animals, including the mammals and fish mentioned in this book. Birds are generally less olfactorily oriented, but use vision and hearing in sophisticated ways.

Another way in which humans are atypical in the animal world is that we are relatively large and slow-moving. We often have the idea that small, fast-moving animals are in some way less important and less capable because of their size. When Healy *et al.* (2013) compared the perception of temporal information in animals, small animals performed considerably better than large animals and humans were not particularly impressive. The critical flicker fusion frequency determines capability to evaluate rapid changes and small mammals and most birds are better than humans. Flies are more able still and live at a much more rapid pace than humans do, as do mice and hummingbirds.

In order to understand the behaviour of the members of any domestic animal species, we need to consider their sensory world and especially the role of odours. We should also be aware of their differences from humans in the range of detection and sensitivity using each sense. Rodents can hear sounds of above 60 KHz while humans hear little above 15 Khz (see Table 1.1). Dogs can also hear higher-pitched sounds than can humans, hence the use of dog whistles that are ultrasonic for the human ear. Animals can be severely disturbed by sounds that humans cannot hear. Some fish can hear and be disturbed by sounds that are very low frequency in comparison with the human range. Birds and some other animals can detect low-frequency vibrations when standing on land or another solid surface because of the sensory corpuscles in their leg joints. Reports of animals responding to an approaching tsunami provide evidence of the ability of these animals to detect and respond to the low-frequency sound that a tsunami would produce. Birds also have the ability to resolve complex sounds with components close together in time much better than can humans. Hence, elaborate bird songs or calls appreciated by other birds may sound like a buzz to humans.

Potentially detectable visual stimuli vary in the brightness, wavelength, plane of polarization and pattern of light. Birds and fish can use all of these aspects when evaluating their visual environment. Humans and some other mammals do not detect plane of polarization so cannot see the many polarized light patterns, for example those of a great variety of flowers and insects (Broom, 1981). Poultry and fish kept for farming or as pets, as well as seeing colours, may see complex polarized light or ultraviolet (UV) patterns in objects that appear white or plain to humans. Domestic mammals have limited or no ability to discriminate light of different wavelengths, so are largely or partly colour-blind (Piggins, 1992). The basic ability of vertebrate animals to discriminate visual patterns made up of blocks or lines is similar, but there are differences in ability to resolve the grain of patterns. For example, sheep can resolve a grating of black and white lines, from a grey stimulus, where the grating has 12–14 cycles per degree

Table 1.1. Hearing and sound frequency in humans and some domestic animals (from Heffner and Heffner, 1992).

Species	Lowest frequency detected (Hz)	Greatest sensitivity detected (KHz)	Highest frequency detected (KHz)
Human	31	8	17
Dog	68	8	40
Cat	50	8	70
Ferret	36	12	45
Horse	55	2	33
Pig	40	8	40
Sheep	125	10	40
Cattle	24	8	40
Goat	70	2	40
European rabbit	120	16	60
Brown rat	400	8	68
House mouse	3000	16	80
Domestic pigeon	8	3	6
Mallard duck	100	2	6
Turkey	200	2	6

and hence have lower visual sensitivity in this respect than humans who can resolve 40 cycles per degree (Sugnaseelan *et al.*, 2013). This visual ability of sheep allows them to use vision to have sophisticated grazing preferences, effective recognition of social group members, lambs by ewes and individual humans (see later in this chapter and Chapters 14 and 19). The light sources used by humans may vary in the extent and kind of brightness fluctuation as well as in the variable mentioned above, and these fluctuations are detected and lead to preferences, for example in chickens (Kristensen *et al.*, 2007). The visual spatial contrast sensitivity of chickens is substantially poorer than that of people (Jarvis *et al.*, 2009), but human visual acuity declines more steeply than that of chickens as light intensity drops to about 5 lux (Gover *et al.*, 2009). Below 5 lux, the chickens can see very little but humans can still see, after adaptation to the low light, so those who provide conditions for chickens may believe that there is enough light for them when there is not.

The visual and auditory world of domestic animals is a little different from that of humans, and we need to take account of that when we try to appreciate what it is that the animals of a particular species are responding to. Humans discriminate better than domestic animals in some visual and auditory modalities but worse in others. Some visual patterns cannot be seen by humans and some sounds cannot be heard. The situation in which this is particularly important is when a stimulus leads to poor welfare in animals but the people who are causing it are unaware of it. For example, high-frequency sounds that are inaudible to humans may be extremely unpleasant to some rodents, birds or dogs. We now know that some electronic switches and heating equipment cause noises that are very aversive to several non-human species (Sales and Pye, 1974). The frequency range for hearing in humans and some domestic animals is shown in Table 1.1.

One of the most dramatic differences in sensory appreciation that exists between humans and the animals that we keep is that between humans and fish. While fish respond to visual and auditory stimuli, their worlds are much more complex than just those two sensory modalities. The world of fish also includes that resulting from lateral line organs, electric receptors and olfaction, which is considered in the next section. The lateral line provides information about localized and general pressure changes. If an individual comes close to a fish, the fish detects it efficiently in complete darkness and without any sounds being involved. The newly arrived animal causes localized pressure changes that are transmitted through the water to the lateral line organs passing down the side of the fish. A moving but non-living object also initiates such changes and, when the fish swims up to a solid object like a rock, plant or the side of a tank, it readily detects the increase in pressure caused by the damming effect when water is compressed between the fish and the object. Fish can move around their environment without colliding with solid objects and without using vision, hearing or olfaction.

An approaching fish is likely to be detected by another fish long before it affects the lateral line organs because of its electrical output. All fish have some degree of electrical receptivity and some are outstandingly sensitive to electrical changes. For many years the function of the ampullae of Lorenzini under the skin of the head of a dogfish, a set of fluid-filled interconnected vesicles, was unknown. It then became clear that these were electric organs. When a muscle contraction occurs it produces an electrical charge in the immediate environment. The air dweller does not detect it because of the poor conductivity of air. However, most animals in the world live in water and the water dweller can be affected by the muscle contraction because water conducts electricity very much better than does air. Fish that live in water with poor visibility – perhaps because of suspended matter – are the most likely to have efficient electroreceptors that allow them to detect the approach of a live animal whose muscles are contracting. However, all fish have some ability in this respect. Sensitivity to electrical change is quite different from the ability to produce electrical fields, which is possessed by a very small number of fish with specialized muscles whose function is not to contract but to produce a large electrical output.

Olfactory communication is very important for most living animals, including all domestic mammals and fish. As explained by Manteca (2002), the olfactory mucosa of a dog is 75–150 cm^2 in area, as compared with that of a cat, which is 20 cm^2 and that of a human, at 2–10 cm^2. In terms of receptors, the dog has 200–300 million while the human has 5 million and the cat is intermediate. The region of the brain that analyses olfactory information, the olfactory bulb, is much bigger in dogs than in humans and, while in humans only a small proportion of inspired air is passed over the olfactory mucosa, in dogs the majority of inspired air is subjected to olfactory analysis. A further difference between dogs and cats is that, while dogs can detect odours without auxiliary movements, cats perform the flehmen movement (see below), in which the mouth and nose apertures are opened widely to maximize the olfactory epithelium contact with the odour under investigation. A useful method for evaluating olfactory discrimination in dogs is described by Salvin *et al.* (2012).

The human or other animal walking down a path is readily detectable by a following dog. Dogs can detect extremely small quantities of the complex mixture of odoriferous substances carried by every animal. A droplet of material from a moving animal will leave a pattern of deposit that allows the follower to identify the individual and to calculate the direction of movement. Dogs can track animals with an efficiency that is extraordinary to us humans. Dogs can also discriminate between different individuals of the same species and can pick out an individual human's armpit odour sample from an array of other such samples, even distinguishing between identical twins (Sommerville *et al.*, 1990, 1993; Settle *et al.*, 1994). A dog can identify a human from a urine sample and can indicate to a person that an epileptic seizure (Strong *et al.*, 1999) or a diabetic crisis (Wells, 2007) is imminent. They can also draw the attention of an owner to a melanoma that is later found to be malignant (Williams and Pembroke, 1989; Church and Williams, 2001). Several studies have reported that it is possible to train a dog to discriminate odours indicative of bladder cancer (Willis *et al.*, 2004), skin melanoma (Pickel *et al.*, 2004), lung cancer (McCulloch *et al.*, 2006), breast cancer (Gordon *et al.*, 2008) and ovarian cancer (Horváth *et al.*, 2008). However, these studies have some methodological shortcomings and, while dogs have the ability to discriminate relevant odours, they can also remember many individual odours that they have experienced and training them to generalize odours from human patients with specific disease conditions has been found to be difficult (Moser and McCulloch, 2010; Elliker *et al.*, 2014).

Social recognition by many mammals, and probably by many fish, depends on identifying an odour. There are therefore mechanisms in the brain facilitating olfactory recognition systems. For example, the hormone vasopressin acts on the mitral cells in the olfactory bulb of the rat brain in the course of social recognition (Tobin *et al.*, 2010). It is likely that there are similar mechanisms in most mammals. The sensory world of our domestic animals is sufficiently different from ours that we should try to consider carefully what the animal might be perceiving in any circumstance where we are trying to provide for its needs or to manipulate its behaviour.

Pheromones

A pheromone is a substance which is produced by one animal and which conveys information to other individuals by olfactory means. Humans respond to olfactory information, even if they are unaware that they do (Stoddart, 1990). Experimental studies have shown that, in humans, synchronization of oestrus, preferences for places to sit in a dentist's waiting room, the duration of telephone calls at a public telephone and attention to photographs of members of the opposite sex can all be influenced by exposure to pheromones. However, it is clear that responses to odours play a much more substantial part in the life of most animals. The world of a dog, pig, cow or tilapia is, in substantial part, an olfactory one. Odours remain in places where individuals of the same or different species have been. Frightened individuals leave a different odour from calm individuals, sexually active individuals have different odours from those who are sexually inactive and there may be differences between more aggressive or more friendly individuals and those who are less aggressive or friendly. Mice were more disturbed by the odour of a strange male mouse than by the odour of a female mouse and more disturbed by the odour on a shirt worn by a male human than by that on one worn by a female human (Sorge *et al.*, 2014). Most mammals can distinguish the odour of an individual from every other individual. For a review of the role of pheromones in behaviour see Wyatt (2003).

There is variation among species in the sources of pheromones. Some animals have scent glands within which the secretion of specific chemicals occurs. Often, these secretions are held for some time so the final composition and odour of glandular products depends upon bacterial action. Many different substances will be present in the odoriferous product. Other sources of pheromones are body products such as urine, faeces and saliva. For example, male mice have the involatile protein darcin in their urine which stimulates female attraction to known males (Roberts *et al.*, 2010). Female and male mice learned the location of sites marked with darcin in one trial and remembered for at least 14 days (Roberts *et al.*, 2012). The effects of pheromones may be the rapid initiation of action, as in alarm pheromones, or the production of sustained behavioural responses.

The important effect of pheromones is the stimulation of olfactory centres in the brain, so pheromone detection depends on macrosmatics, which is the possession of keenness of smell to an extremely high degree. Most animals' olfactory centres are more differentiated than are those of man, who is microsmatic, and the olfactory portion of the telencephalon of cattle and sheep is about 20 times larger than that of the human.

A special organ, apparently involved with pheromone reception through its mucous membrane, is the vomeronasal, or Jacobson's, organ (Meredith, 1999). This organ is an olfactory receiver in the form of a pair of blind-ended tubes located within the nasal cavity and linked to the roof of the mouth. It is connected to the centres of olfaction in the brain with its own mechanism of conduction and its own reactive behaviour. It is instrumental in an olfactory reflex act, which is known as flehmen. In flehmen, the head is elevated and extended, the upper lip is curled up with the mouth slightly open and the nostrils constricted. The flehmen reaction (Fig. 1.3) indicates that the male is testing the urine of the female. The concentration of pheromones can reflect levels of sex hormones in the individual and, in this case, the male is monitoring the oestrous state of the female. Oestrus synchronization in female mammals can result from common olfactory urine checking. Male urine can also provide olfactory information about hormone levels. Urine is used as a territory marker in some mammalian species, such as badgers, and some animals mix faeces, urine and scent gland products in acquiring a distinctive individual odour.

The two types of scent-producing glands are: (i) sebaceous glands, such as the ventral gland of the Mongolian gerbil, which is used for territory ownership marking; and (ii) apocrine glands. The two types of glands are often mixed in glandular areas. The axillary glands in man produce pheromones whose release is facilitated by both the presence of axillary hair, which provide a larger surface area, and by the action of arm raising. Pigs, ruminants and horses possess specialized gland complexes located in the skin. The skin, as the primary protector of the body surface, is covered with glands serving temperature control, excretion, lubrication and maintenance of the pH in fending off microorganisms. This, in its totality, produces a specific body odour. In these animals the convoluted type of skin glands, with their apocrine excretions, produce a volatile substance which is moved to the skin's surface and mixed in a product which is particularly suited to contain, or bond, odoriferous substances. The occurrence and location of skin

Fig. 1.3. Flehmen in a stallion. The reaction is shown in response to the odour of a mare in oestrus and results in the pheromone being brought in contact with the stallion's vomeronasal organ. The mare has elevated her tail (photograph A.F. Fraser).

regions that produce odour relate to certain forms of behaviour (Vandenbergh, 1983).

Widely studied glands include those of many insects and mammals. The chin gland of rabbits is used to mark territory, to warn others and to mark young for recognition. Dogs and their relatives have anal glands, and the secretion in the female fox contains about 12 volatile components including trimethylamine and several fatty acids. The chemical produced by bitches inducing sexual behaviour in male dogs is para-methyl hydroxy benzoate. Male wolves and dogs mark their territory using urine, and this indicates who the individual is and, according to elevation from the ground, how big that individual is. Cats have glands on the side of the face and at the base of the tail. These are often coloured differently in tabby and other variably coloured cats; they use these to mark objects in their environment, including their 'owners'. When cats rub their head and tail base against you they may well be merely establishing their ownership of you! The cat facial pheromone F3 can be made synthetically and used to prevent the occurrence of urine-spraying by male cats (Mills *et al.*, 2011). Deer have elaborate glands, that of the musk deer containing as much as 120 g of sebum. The tarsal tuft on the leg of the black-tailed deer has specially adapted hairs that trap volatile substances from the tarsal gland and from urine that is deposited on the tuft (Müller-Schwarze, 1999). The odours from the tarsal gland in this species and from the metatarsal gland in the roe deer (Broom and Johnson, 1980) attract much attention from other members of the species and probably allow individual recognition.

Many wild ruminants are endowed with inter-digital glands on all four feet, particularly below the dewclaws. The skin around these interdigital glands is richly endowed with convoluted glands, which produce a mixed secretion of pheromones. It appears that an alarm pheromone is produced and stored there as a colloid. The domestic ruminants have remnants of analogous glands that may still function as sources of specific odours.

Saliva can be involved in chemical communication, and behavioural evidence of this is notable in the pig in pre-mating and nursing situations and in social events in the horse. Experimentally, saliva has been found to play a part in organizing the suckling behaviour of the rat and the aggressive behaviour of mice. Rat pups reportedly engage in long periods of licking and nuzzling the mother's oral region, and this suggests that a chemical mechanism operates via saliva in regulating nursing interactions. Sows and their young piglets show essentially similar nosing of the maternal oral area. Young and older animals show a tendency to pursue olfactory examination of the mouths of associates (Figs 1.4 and 1.5).

Fig. 1.4. Young pigs investigating nose and mouth area (photograph D.M. Broom).

Fig. 1.5. Horse and dog investigating nose and mouth area. There may be transfer of salivary pheromones (photograph R. Harnum).

Experimentally, juvenile gerbils were found to respond preferentially to females carrying their littermates' saliva rather than to those carrying non-littermate saliva. Adult male gerbils spend more time contacting the facial regions of recipients carrying saliva from an oestrous female. They exhibited no difference in their contact times with the ventral scent glands or anogenital regions. Social preferences were evident in the behaviour following salivary exposures. These findings suggest that saliva-related cues may act as chemosignals in all stages of social behaviour in gerbils (Block *et al.*, 1981). Such salivary factors probably function in other animals, including farm livestock. Among farm animals many social interactions take the form of nose-to-mouth contacts, in which it is likely that salivary cues of identity, and possible status, are communicated. This has been noted in the horse and sheep and it seems that salivary pheromones may affect social behaviour in animals to extents not appreciated before. Many so-called 'naso-nasal' contacts are in fact found to be nose-to-mouth contacts when carefully studied.

The production and the effects of pheromones have been noted in the pig (Signoret *et al.*, 1975; Booth and Signoret, 1992). The boar produces a chemical substance called androstenedione, which appears to have a releaser effect on sows during oestrus so that they show the rigid mating posture more readily after they have been exposed to the male pheromone. The boar's pheromone is apparently produced in the submaxillary salivary gland. There is copious production of saliva, in frothy form, during the natural pre-mating activities between boar and sow. During these activities, nose-to-mouth contacts are clearly shown. Pheromone excretion in the boar also occurs via the prepuce. Proof of this has been shown in large-scale, experimental studies in pig artificial insemination work. Small quantities of boar seminal fluid dropped onto the snout of a sow or gilt showing dubious or incomplete signs of oestrus causes the complete oestrous display to be shown. The boar odour may also have effects on other species such as humans.

Many animals leave olfactory messages by the use of urine or faeces. For example, cats mark defended territories, principally their peripheries, by squatting and depositing urine, depositing faeces in prominent places and scratching objects to leave a visual and olfactory message (Borchelt, 1991; Bradshaw, 1992; Simpson, 1998). Male cats also spray urine on to objects, especially when there are other cats in the close vicinity and they are anxious (Beaver, 2003). In houses, furniture, walls and windows at the periphery where cats have visual access to the outside are often the targets for spray-marking (Pryor *et al.*, 2001).

Further Reading: Fraser, D., Duncan, I.J.H., Edwards, S.A., Grandin, T., Gregory, N.G., Guyonnet, V., Hemsworth, P.H., Huertas, S.M., Huzzey, J.M., Mellor, D.J., Mench, J.A., Spinka, M. and Whay, H.R. (2013) General principles for the welfare of animals in production systems: the underlying science and its application. *The Veterinary Journal* 198, 19–27.

2 Describing, Recording and Measuring Behaviour

Some general principles of behaviour organization are explained in this chapter. Recording and measuring are also discussed here and by Martin and Bateson (2007).

Levels of Description of Behaviour

The words used to describe behaviour are a consequence of how people think about what they see or hear. Those words in their turn, however, may change the way of thinking of the person who uses them or of the person who hears or reads them. This occurs especially when the word used to describe the behaviour implies something about the emotional state or the intentions of the animal whose behaviour is described. Hence it is important to be accurate and cautious when describing. As an example of the problem, suppose that a hen is seen to move rapidly, flapping her wings, over a distance of 3 m starting at the edge of one group of birds and finishing next to a wall where a single bird standing there moves away. An observer might describe this sequence of behaviour by saying that the hen is frightened, angry or aggressive. This may be true, but does not inform others about what has been observed. In order to communicate effectively with the listener or reader when describing behaviour, it is best to state what is seen in the manner of the descriptive sentence above. The interpretation in relation to emotions requires further evidence (see Chapters 4 and 6).

Objective description of behaviour can be at various levels of detail. When considering the hen mentioned already, her behaviour could be described in terms of:

- The contraction of each muscle.
- The movement of each group of muscles.
- The movement of one part of the body relative to another, e.g. the wing is flapped or the legs are moved with a running gait.
- The movement of the animal, or part of it, in relation to the environment, e.g. moving from point A to point B 3 m away or touching the wall.
- An effect on the physical environment, e.g. flattening straw by stepping on it, knocking over a food container or pecking a key.
- An effect on another individual, e.g. causing another bird to move, eliciting a submissive posture or a courtship display.

The decision as to whether to use much or little detail, or to describe the consequence of the behaviour, depends upon the aims of the observation.

In selecting measures for a particular study it is useful to know the array of behaviours the animal is capable of showing. A largely complete description of such an array is called an ethogram, and papers have been published that present an ethogram for a species. These papers are necessarily based on an extensive study of that species and they can be very useful if the behaviour description is precise enough. It is still necessary, however, for the observer to spend some time becoming familiar with the behavioural repertoire of the animal. It is likely that any detailed behavioural study will add to our knowledge of the repertoire and organization of that animal's behaviour, so no ethogram is ever complete. The actual selection of measures should take account of whether or not the measures are independent of one another; for example, one activity may necessarily be preceded by another or may prevent the occurrence of another.

When considering the level of detail of description it becomes apparent that some behaviours like sleep are continuous, while others like walking are a series of repeated movement sequences and others like displays or grooming are made up of sets of recognizable units.

Rowell (1961) distinguished between acts, such as a finch bill-wiping: 'bending forward, wiping the bill on the perch and resuming an upright posture' or a step during walking, and bouts such as a sequence of walking with a gap before the next sequence. The question of how to decide, for activities such as walking, preening, eating or displaying, when a bout ends is discussed by Broom (1981) and Martin and Bateson (2007). When acts can be grouped together into longer units, because they are frequently combined in a particular way, this probably reflects some neural control circuit and it is convenient to use a word for that unit. Many of our measures of behaviour are of this kind. The units might be movements in: prey-catching, other food acquisition, grooming, courtship display, mating or parental behaviour and these are often referred to as action patterns. As discussed by Broom (1981), the term 'fixed action pattern' has been used in the past, but careful studies show that the action pattern is fixed neither in detail of sequence nor in a genetic sense, so the term is inappropriate: 'action pattern' is sufficient.

More prolonged sequences of individual behaviour by domestic animals requiring description on occasion include stereotypies (see Chapter 23) and rhythms. *A rhythm is a series of events repeated in time at intervals whose distribution is approximately regular* and the more precise term *periodicity means a series of events separated by equal periods in a time series* (Broom, 1980). Rhythmic activities include heart-beating, breathing, walking, flying, chewing, being active rather than resting, diurnal activity, oestrous behaviour and breeding. Some methods of detecting and analysing activity rhythms are reviewed by Broom (1979). Investigations of sequences of behaviour show that rhythms can be important variables affecting behaviour, so they must be taken into account in certain sorts of behaviour investigation. For example, young domestic chicks showed periodicities in walking behaviour and heart rate with wavelengths of 24 h, 20–30 min, 13–15 s and 1–2 s (Forrester, 1979; Broom, 1980).

The basic mechanism for biological clocks is present in a wide range of animals. Fruit flies, honeybees, hamsters and humans all have a gene that has the information to produce the PER protein (Young, 2000; Toh *et al.*, 2001). This protein acts within the suprachiasmatic nucleus (SCN) in mammals to affect circadian rhythms. If the base sequence in the gene – and hence the amino acid sequence in the protein – is altered by as little as one base pair, the circadian rhythm is changed. The signals from the SCN activate a gene (CREM) in the pineal gland coding for a protein (ICER) that is involved in the production of melatonin (Stehle *et al.*, 1993; Foulkes *et al.*, 1996). Melatonin levels change with photoperiod length and lead to adjustments in daily rhythms of behaviour.

In addition to the mechanism leading to the occurrence of daily cycles of activity, there are mechanisms related to annual cycles of activity (Gwinner, 1996). Seasonally adaptive changes in behaviour may be mediated by hormones, for example an increase in the concentration of testosterone in the blood precedes the major changes in the behaviour of the red deer stag at the beginning of the rut in September and October in the UK (Lincoln *et al.*, 1972). Such effects are discussed further by Wingfield *et al.* (1997).

The actual measurements of behaviour used will depend upon the objective of the study. Careful observation allows more sophisticated description. For example (Fig. 2.1), in horses the posture, eyes, facial muscles and ear position indicate attentive individuals.

Behaviour Measures During Veterinary Examination

Behaviour is used during clinical appraisal and is generally qualitative, in that the presence or absence of a certain kind of behaviour is noted, rather than quantitative. The examination of the animal is more reliable if some

Fig. 2.1. Three horses showing an alert, attentive response (photograph R. Harnum).

previous information about that individual is available and if the reactions, as well as the posture of animals or groups of animals, can be assessed. The animal's attitude, disposition and temperament should be assessed before any handling is performed. Its alertness and apparent awareness of its general environment should be noted. In particular, efforts should be made to record the subject's appreciation of visual, auditory and positional stimuli. Eye movement, involving the lids and orbit, is an important feature. A high degree of exposure and mobility of the orbit suggests anxiety, while a very fixed orbital position may indicate some distress.

The animal's willingness to move and the nature of its gait are both important considerations. Does the animal show reflex responses, either general responses such as those to a sudden sound, or specific responses such as those to pressure on a site on the body? Reflex responses to localized pain such as skin-pricking or pinching may also be noted if circumstances indicate a need to determine nerve function. Is there normal behaviour? This might include self-maintenance such as feeding, response to offering food or body care. The significance of acts of body care in a long behavioural examination is considerable, since this behaviour often ceases as a first sign of sickness.

Common actions, or their absence, should be noted. For example, self-grooming and stretching commonly occur after rising in healthy animals but several factors, including illness in general, can inhibit grooming and stretching reflexes. In cattle, tonguing of the nostrils may be inhibited during illness. It has also been suggested that the eructation reflex in ruminants becomes inhibited in many illnesses and, as a consequence of this, distension of the rumen develops and leads to the condition of bloat that is associated with various illnesses in ruminants. Recognition of such reflexes in normal behaviour is an indication of good health and, consequently, their absence suggests poor health.

In a behavioural appraisal, many postural abnormalities are not shown unless the animal is at rest in its usual environment. For this reason, patient and quiet observation of the animal may be necessary before abnormalities of posture can be detected and appreciated.

Behavioural examinations are best performed in a quiet space or enclosure with limited light, where distractions will be minimal. Tranquillization should be avoided. Parts of the examination that may tend to excite the animal should be postponed until the end. At the conclusion of the examination it may be apparent that further, specialized clinical tests and specific medical examinations are required. In the course of convalescence, the behaviours of self-maintenance return to the animal's behavioural repertoire. An extended period of behavioural examination of the animal in relaxing conditions is helpful, perhaps using video recording.

Behavioural signs of pathological state are reviewed in Chapter 22. In general, the use of behaviour in clinical examinations has not yet utilized modern techniques of behaviour assessment, including sophisticated behavioural tests and aids such as video recorders and telemetric devices, which can tell the clinician much about the behaviour of the animal when it is not being examined.

Design of Experiments and Observation Procedures

Before commencing a behaviour study it is important to consider whether the design of the procedures to be used is adequate to allow reliable conclusions to be drawn from the results of the work. The first precaution concerns the effects of the presence of the observer on the behaviour of the animals. As mentioned above, an animal that is being examined may behave in different ways in the presence and absence of an observer. Small animals like chickens, unless handled frequently and gently from an early age, treat man as a dangerous predator. Hence, their behaviour can be affected very substantially by the proximity of a person watching them. Other animals are also affected by human presence so it is advisable, when watching any of these animals, either to observe from a hide or to carry out checks to ascertain how much behaviour is changed by the observer.

Behaviour observation can be accurately replicable if the definitions of measures and precision of recording are sufficiently clear. It is desirable, if more than one observer is involved, however, for studies of inter-observer reliability to be carried out. The possibility of bias, deliberate or unintentional, should also be considered when designing observation procedures. If two treatments are being compared, wherever possible the observer should be 'blind' in the sense that the treatment category to which each animal belongs is not known at the time of observation.

Wherever experiments are carried out, one or more control situations should also be studied. For example, in a study of the effect of a hormone treatment on behaviour, a control group whose conditions are exactly the same as the experimental group, but with an inert substance given to the animal in the same way as the hormone, should be used. Studies of behaviour often require replication, since unknown variables can sometimes lead to spurious results. An illustration of such necessity is the study of orders of movement of a group of animals from one place to another. The order of animals on one occasion, or in one situation, will be affected by chance and may be substantially changed by local conditions, so orders should be recorded on several occasions and in several different situations before any conclusion about social relationships can be reached. Whenever sets of observations are replicated, the experimenter must be aware of any possible effects of learning on the results. No animal that has been exposed to experimental conditions can be assumed to be unaffected by them, so its behaviour may be different during any repetition of these conditions. In some studies these very changes are under investigation or, as in the case of regular movement orders, the situation is a very frequent one in the animals' lives so behaviour is not likely to change rapidly, because of previous experience of that situation. In other studies, however, an unusual stimulus is presented to the animal and a subsequent response to the same stimulus may be either much less, due to habituation, or much greater, due to sensitization. The design of experiments is explained in more detail by Lehner (1996) and Hawkins (2005).

Marking of Animals for Behaviour Studies

One practical point about setting up experimental studies concerns the marking of animals (Martin and Bateson, 2007). Much more information can be obtained from studies of behaviour where the identity of individuals is known. The idea that animals behave in a 'species-typical' way has been shown to be a gross oversimplification by observations of individuals. There is often considerable variation among individuals in how they respond to a particular situation and how they attempt to cope with difficult conditions. If the animals are in groups, then some form of marking is needed.

Sometimes, animals are sufficiently variable without additional marking: dogs and cats may be readily identifiable, and Friesian or Holstein cows can often be distinguished individually by coat pattern when there are not too many animals in the group. Horses are commonly described according to an internationally agreed procedure (Fédération Équestre Internationale, 1981), and some of these markings are sufficient to allow recognition during behaviour observation. Animals such as dairy cows are often marked with a collar, an ear-tag or a freeze-brand. They might also be marked using methods that cause a certain amount of suffering, like an ear notch, or a great deal of suffering, like a hot-iron brand.

Whenever animals are marked, care must be taken to discover whether the mark itself affects the outcome of the study or the welfare of the animal. In a study by Burley *et al.* (1982) with zebra finches, it was found that coloured plastic leg bands (or rings), put on to allow individual recognition, altered the attractiveness of the birds to the opposite sex. A red leg band made a male more attractive to females while a black leg band made a female more attractive to males. Birds wearing green or blue bands were avoided by members of the opposite sex. Other effects of marks on animals include the possibilities that the marking method or the mark may be painful, that the social status of the animal may be changed or that the vulnerability of the animal to predators may be affected. In order to be aware of the possible effects of marks on the study that is being carried out, it is advisable to check such effects separately. If marks are used, either all or none of the animals should be marked.

The nature of the mark used will depend on the requirements of the observer. Animals that are watched only at feeding time may be identifiable from a small ear-tag, but those ranging over a large area need a large mark on the sides and back. Where video recording is used, a larger and clearer mark is needed than when direct observation only is carried out.

Electronic methods of marking animals were at one time too expensive or too bulky for widespread use. However, very small attachable or implantable capsules are now available, even for small animals. These can produce a signal that identifies the animal if the appropriate receiver unit is available and can also provide data on variables such as body temperature, chemical concentration and physiological change. The advantages to animal welfare of the traceability of animals that are adequately marked are described in Chapter 21.

In some studies, a mark that persists throughout the life of the animal is needed. This can be a tattoo for dogs, cats, pigs or horses; an ear-tag for pigs, sheep, goats, buffalo or cattle; a freeze-brand for cattle, buffalo or horses; or a leg-ring or wing-tag for poultry. Some electronic tags or transponders may function for a significant part of the animal's life, while some transponders are useful only when the animal is close to the identifying equipment. However, this equipment can be located in a human home for companion animals or, for farm animals, in animal accommodation and can operate gates, for example at the entrances to an electronic sow feeder, a boar pen or a farrowing area.

Temporary marking of animals can be carried out using leg-rings or feather dyes for poultry and hair dyes, paint, collars or coat-clipping for mammals. Paints and dyes are available commercially, as many farmers need to mark animals. Some dyes or paints last for only a short time because the mark itself fades quickly or because the animals rub or lick it off themselves – or one another. The life of the mark in the experimental situation should be checked before the investigation proper is initiated. It is possible to use numbers or letters as marks on farm animals but it is easy to confuse some of these if they become indistinct. Letters are better than numbers as there are more of them, even after confusingly similar letters have been excluded. An example of pigs marked in this way is shown in Fig. 2.2. If wear or other effects might reduce clarity it is better to use combinations of simple marks. Figure 2.3 shows an example of marks on poultry that are distinguishable even if feathers are ruffled or some loss of mark occurs. The presence or absence of each of these marks in four positions gives 15 possible combinations, excluding complete absence of marks. Spots or bars can also be used on pigs or other mammals. Collars are suitable for dogs, cats, cattle, buffalo, goats and some sheep, but there is a risk that they will come off so a permanent mark is also necessary.

Sampling and Measuring

Several decisions have to be taken when behaviour is to be measured and these are interrelated in that they are limited by the capabilities of the observer, and greater detail in one aspect means potentially less detail in another. The first decision concerns which animals to observe. If much detail from direct observation is required, then it will be possible to observe only one animal at a time. This may be an individual in its own pen or home, or it may be a focal animal that can move around within a group. With appropriate sampling methods, data on several or many animals at once can be collected by scanning them, but information about each individual is lost by sampling.

Fig. 2.2. Each pig in this photograph is marked to facilitate individual recognition in a study of social interactions (photograph D.M. Broom).

The information about one kind of behaviour that can be obtained from observation and recording might include:

- The presence or absence of the particular activity.
- The frequency of occurrence of each activity during the observation period.
- The duration of each bout of each activity.
- The intensity of the activity at each occurrence.
- The latency of occurrence of the activity.
- The timing and nature of subsequent activities.
- The timing and nature of behaviour changes in relation to physiological changes.

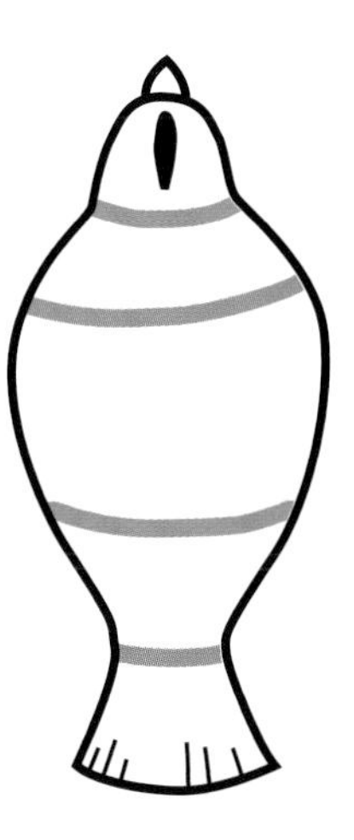

Fig. 2.3. One example of how poultry may be marked with spray colour. By painting lines across the backs of the birds, one or more of the four positions may be used, giving a large number of possible combinations. The symbols are easily recognized, even when the plumage is ruffled (after Jensen *et al.*, 1986).

Continuous recording of behaviour

This technique can be difficult if many measures are used, and recording aids (see below) are often needed, but it offers opportunities for all the different methods of analysis. Sampling behaviour makes possible the collection of data on more than one individual and it allows an estimate of the duration of activities in situations where continuous recording is not possible, but some information is lost. There are three sorts of sampling that can be used: two types of time sampling and behaviour sampling (see Fig. 2.4).

Behaviour sampling

Also known as 'conspicuous behaviour recording', this involves continuous observation of animals but recording only certain kinds of behaviour. For example, a group of dogs may be watched and all occasions where one animal sniffs another recorded in detail. Behaviour sampling may also occur automatically in that a single action, such as pecking a key by a chicken, may be automatically recorded but all other actions are ignored. This method is particularly useful for rare behaviour patterns that might otherwise be missed.

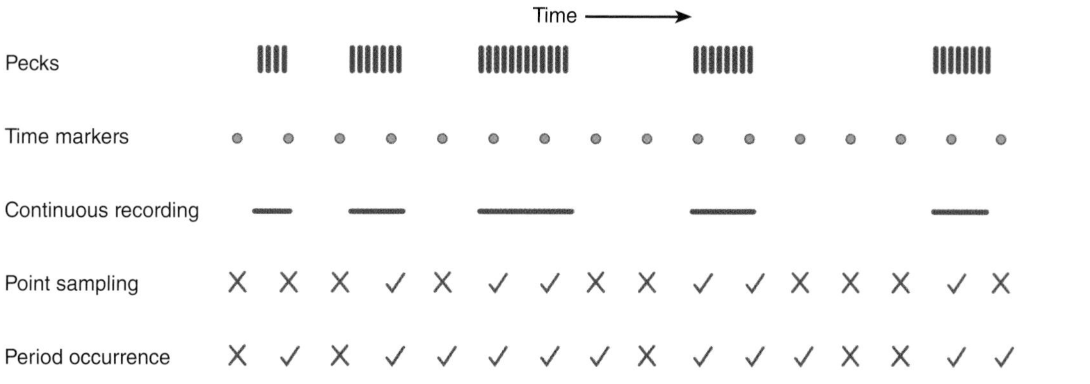

Time →

Pecks

Time markers

Continuous recording

Point sampling X X X ✓ X ✓ ✓ X X ✓ ✓ X X X ✓ X

Period occurrence X ✓ X ✓ ✓ ✓ ✓ ✓ X ✓ ✓ ✓ X X ✓ ✓

Fig. 2.4. Comparison of behaviour recording methods. A series of pecks by a chick is shown as if produced by an event-recorder moving at a constant speed. If continuous recording was used, lines like those shown – or precise times of stopping and starting each bout of pecking – would be produced. Point sampling and period occurrence would produce yes or no answers at each time mark, as shown.

Point sampling

Also known as 'instantaneous sampling', this involves observing animals at regular, predetermined points in time and recording whether or not each of a range of behaviours is being shown at that instant. As shown in Fig. 2.4, a useful estimate of duration of the more common activities is obtained if the observation period lasts long enough and if the interval between the samples is not too long. Rare activities might be missed altogether, however. It is a problem of the method that observers tend to try to include activities that do not actually occur at the moment of sampling. A further problem is that some activities take some time to recognize. For example, when a cow is ruminating it takes a few seconds to be sure of this since the characteristic jaw movement takes time to identify and the animal might be swallowing just at the moment of sampling. The major advantage of this method is that it can be used when many individuals are scanned, so one person can collect much information.

Period occurrence

This type of recording (Broom, 1968b), often rather confusingly called 'one-zero sampling', is another form of time sampling in which the events that have occurred during a predetermined time period are recorded at the end of the period. Several animals can be observed simultaneously because the data do not have to be recorded continuously. As apparent from Fig. 2.4, this method has the advantage that even rare events are not missed, but its important disadvantage is that the figure obtained is not a true representation of the actual duration of each behaviour. If the period between samples is short in relation to the activity bout length, however, then the figures obtained are quite good estimates of activity duration. Period occurrence recording is much easier than continuous recording, except in some studies and where computer-linked systems are available.

An important final step, after behaviour recording, is the statistical analysis of data. A survey of common statistical tests useful in behaviour studies, modified after Martin and Bateson (2007), is summarized below. These tests are non-parametric if unmarked or are parametric if marked with an asterisk.

1. Question 1: Does the sample come from a specified population? (Tests of goodness of fit for single examples)
Chi square test for one sample (nominal data)
Binomial test (nominal data)
Kolmogorov–Smirnov one-sample test
2. Question 2: Is there a significant difference between the scores of two unrelated samples; for example, between the scores of two different groups of subjects? (Tests of difference between two unmatched samples)
Chi square test for two independent samples (nominal data)
Fisher exact probability test (nominal data)
Mann–Whitney U test
Student's *t* test for unmatched samples ('*t*-test')*
3. Question 3: Is there a significant difference between the scores of two related samples; for example, between the scores of the same subjects under two different conditions, or between siblings? (Tests of difference between two matched samples)
Wilcoxon matched-pairs signed ranks test
Student's *t* test for unmatched samples ('*t*-test')*
4. Question 4: Are there significant differences between the scores of several related samples? (Tests of difference between *k* unmatched samples)
Chi square test for *k* independent samples (nominal data)
Kruskall–Wallis one-way analysis of variance
Analysis of variance*
5. Question 5: Are there significant differences between the scores of several related samples; for example, between the scores of the same subjects measured under several different conditions? (Tests of difference between *k* matched samples)
Friedman two-way analysis of variance
Repeated measures analysis of variance*
6. Question 6: Are two sets of scores associated? (Measures of correlation between two samples)
Spearman rank correlation coefficient
Kendall rank correlation coefficient
Pearson product–moment correlation coefficient ('correlation')*
7. Question 7: Are several sets of scores associated; for example, are the scores of one group of subjects consistent when measured several times, or is there an overall association between several different measures for the same set of subjects? (Test of concordance between *k* rankings of the same subject)
Kendall coefficient of concordance

Recording Social Behaviour

Much of the methodology for recording behaviour in social situations, in which animals may interact, is the same as for the description of individual behaviour. When social behaviour is described, it is often desirable to produce data on general activity and on specific kinds of interactions. In these circumstances more than one method of behaviour measurement can be used simultaneously. In a herd of cows, general activity can be recorded by point sampling while rare events such as fights or mutual grooming can be recorded by behaviour sampling. The data from such behaviour sampling are produced as a list of initiators and targets of attacks, as winners and losers of fights, as groomers and groomed or as pairs of individuals associating. These data can be analysed further in several ways. One method is to describe them using social networks (see Chapter 14). The relationships are then depicted, as in Fig. 2.5 showing them for associations in four groups of salmon, with lines connecting all animals that have shown the interaction and thicker lines where there are more interactions. The more central individuals are those that were involved in more of such interactions.

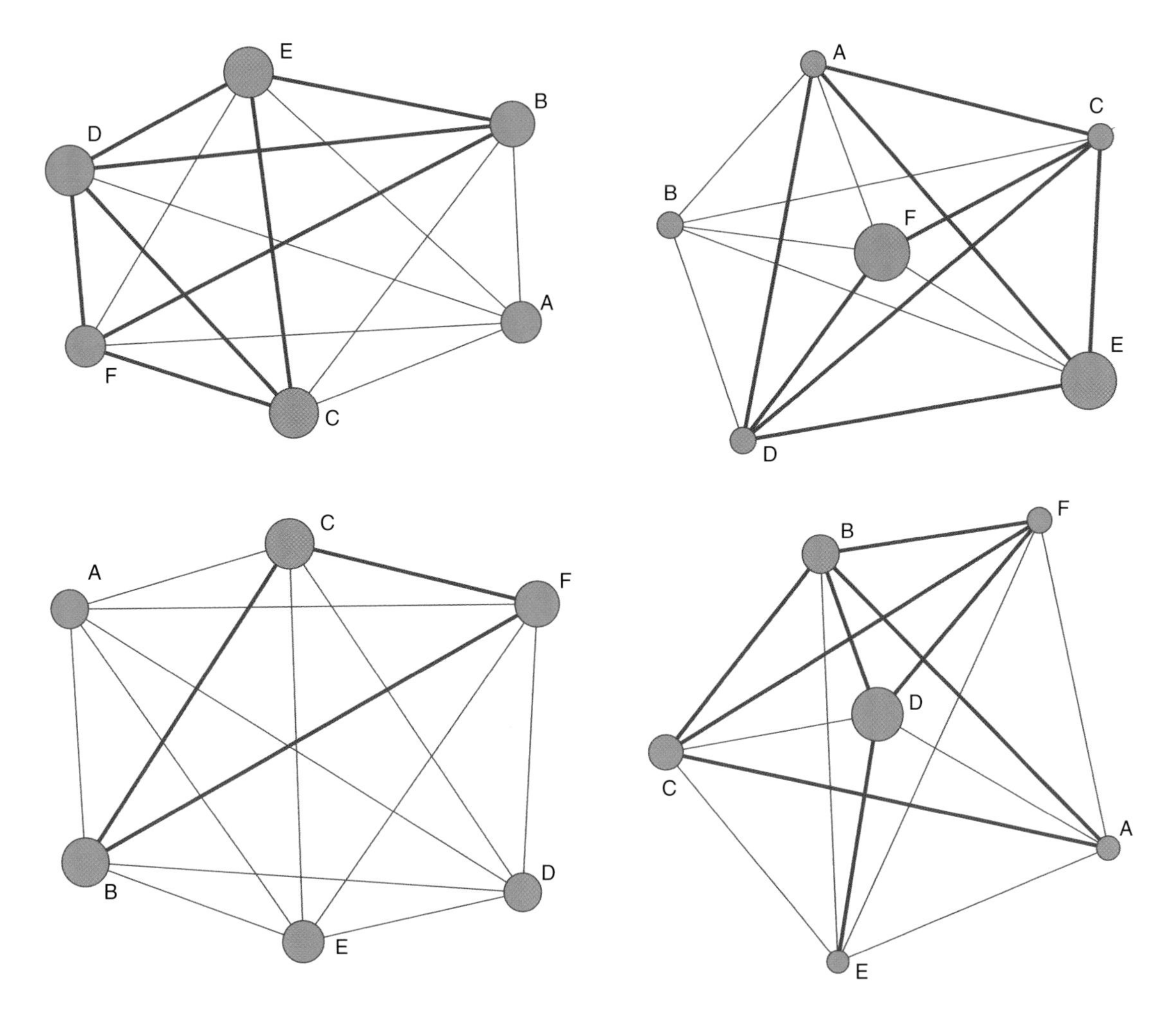

Fig. 2.5. Social networks for association in four groups of salmon. Circles represent individual fish and their diameter indicates centrality in the group of fish. The thickness of the lines between fish is proportional to the strength of association for that pair (from Cañon Jones *et al.*, 2010).

Recording Aids

Behaviour recording is often difficult because events occur too quickly to write down both the nature of the event and the time at which it occurs. When making a written record of behaviour, much time is saved by using a single-symbol abbreviation of the title of a measure. If a sampling procedure is used, then a recording sheet can be prepared with spaces for writing symbols or ticks at the predetermined time intervals. For continuous recording, the simplest method is to use a stopwatch on a board with squared paper on which a line across the page represents time – for example, 30 or 60 s. The hand is moved across the page and symbols are written at a point on the line that represents a certain time. The duration of each activity can then be measured as the total distance on all the lines on the page where that activity was recorded. The next step is to use a computer-linked recording system. Keyboards adapted for rapid use facilitate efficient recording.

Where behaviour changes are very rapid, audio recording and video recording are of particular value. The study of birdsong and many other detailed sequences of behaviour have been facilitated by the possibility that behaviour could be recorded and then slowed down. The data can also be played back repeatedly. A representation of a sound sequence can be produced electronically. Video recording also has the advantage that animals can be observed by a camera rather than by a person, so observer disturbance can be minimized and there is no necessity for an observer to be present. For some sorts of behaviour recording, automatic systems that monitor movement or output from various forms of sensors can be used. These may either locate an individual on the video and track it or produce an indication of total movement by each individual monitored. The output, one type of which is known as an actigraph, can be of use in evaluating individual activity without time-consuming video analysis. However, a check of what is measured with the actigraph is necessary and many behaviours cannot be measured using automatic recording methods.

Field Studies

Studies of animals in a natural or semi-natural environment provide valuable information about their range of behaviour and how they allocate resources, but there is a considerable likelihood that behaviour will be altered by the presence of an observer. Hence if wild, feral or free-ranging animals are to be watched, then aids to distant observation are needed. Close observation requires the use of a hide unless the animals are fully habituated to human presence. Experiments can be very valuable in field situations: for example, animals can be presented with food items, sounds or scents.

Test Situations

One type of test that has provided information about the needs of animals is the preference test. Any investigation of animals in a varied environment offers the opportunity of finding out what animals choose to do, but specific choice or operant tests are also possible. Different foods, flooring, housing design, companions, temperatures, light levels or air flow conditions can be presented. Tests of strength of preference are important in animal welfare research (see Chapter 6).

Another procedure, useful in behavioural research, is to deprive the animal of some resource or ability in a controlled way and then to monitor behaviour when the deprivation period ends. The effects of deprivation of, for example, particular foods, social contact or space to flap wings for a long or a short period can be assessed by recording immediate and long-term changes in behaviour. Deprivation is often used as a prelude to studies of learning. Domestic animals perform very well in learning tasks provided that they are given adequate cues and the responses required are appropriate.

Other test situations include exposure to a novel environment and tests associated with reproduction. If an animal is moved to a new pen, sometimes misnamed an 'open field', it shows exploratory behaviour and may also be disturbed by the conditions, so that responses associated with adrenal activity are shown. Mating behaviour can be tested by presenting an individual with a potential partner or a model, and parental behaviour can be tested by exposing the animal to young animals, sounds or models.

Further Reading: Martin, P. and Bateson, P. (2007) *Measuring Behaviour*, 2nd edn. Cambridge University Press, Cambridge, UK.

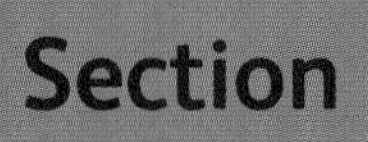

Section 2

Fundamental Topics

Learning, Cognition and Behaviour Development

Experience

During development, the expression of genes and the synthetic processes that lead to the growth of cells and organs into a particular form are dependent upon environmental factors. In adulthood, many genes are still active and a wide range of bodily processes are modified by input from within the rest of the body and from the impact of the outside world on the body. Behaviour is controlled by the nervous system and brought about by muscles, bones, etc., within the body. The environment affects the development and continuing functioning of all of these systems (Bateson *et al.*, 2004; Gluckman *et al.*, 2005; Bateson and Gluckman, 2012). One consequence of this is, as mentioned in Chapter 1, that no behaviour is independent of the genetic information in the animal and no behaviour is independent of all environmental factors. When dog pups are reared, the treatment of the dogs during this rearing and their consequent experiences have great effects on the later behaviour of the dog. However, when dog pups and wolf pups were hand-raised and intensively socialized in the same ways, at 5 weeks of age the wolf pups showed less distress vocalization, less tail-wagging and less gazing at the human face but more aggressive behaviour than the dog pups (Gácsi *et al.*, 2005). The behaviour always depends on both genetic information and environment. As indicated by the wide range of examples in this and other chapters, every interaction between an individual and its environment has a potential for modifying that individual, including its brain and behaviour.

When people say that something is experienced, they often imply that some sensory perception has occurred. An experience is thought of as a mental construct that results from an event in the environment of the body and brain. However, an effect of the environment on behaviour may be mediated via sense organs or via other cells in the body. For example, a change in environmental oxygen concentration might result in a change in behaviour without a sense organ being involved. The animal might discover that, if it goes to a certain place where there is a lack of oxygen, a wide variety of consequences follow, some of which are sensory and some are not. If it goes to a second place a dangerous predator might be detected. Each of these would add to the experience of the animal. There is no fundamental reason for distinguishing between a direct environmental effect and one that reaches the brain as a perceived stimulus.

Experiences may be a consequence of: sensory input, changes in hormone levels or changes in other aspects of the physical and chemical environment of the brain. The input to the brain, however mediated, will usually result from some change outside the body. However, it will sometimes result from physiological changes consequent upon previous changes outside the body, or from changes that are entirely internal. An imaginary event might lead to adrenal activity that will result in a bodily change being experienced. *An experience is a change in the brain that results from information acquired from outside the brain.* Some experiences are very brief while others are very long-lasting. The information that already exists in the brain affects whether or not the effects of an experience are long-lasting.

Learning

When learning occurs, an experience of some kind has led to a change in behaviour. This change must itself be a result of a process within the brain. *Learning is a change in the brain, which results in behaviour being modified for longer than a few seconds, as a consequence of information from outside the brain.* The reference to 'a few seconds' excludes simple responses. The experience and the consequence might range from the effects of oxygen concentration on enzyme action

during the development of a motor control mechanism to the effects of scarcely perceptible social signs on the behaviour of an individual in a complex society. There is thus a wide range of situations where learning occurs. This means that it is frequent and involved in almost all aspects of behaviour.

The brain mechanisms that make learning possible include very complex processes in which there is awareness of what is occurring, what has occurred and what is likely to occur. Some authors always attempt to provide simple explanations for apparent learning rather than allowing for the possibility that complex learning has occurred. Simple explanations should be considered, but it is not a good scientific approach to assume that these simple explanations are always more likely to explain what is observed (Roberts, 2000; Forkman, 2002). If an animal has complex and efficient abilities, it will use them where possible.

Learning experiments are important in relation to a wide range of objectives in research and no one studying behaviour can ignore the effects of learning. Having said this, however, it is apparent that many events in the environment of an animal do not lead to any change in future behaviour. Some events are not detected by the animal; some do not reach decision-making centres as a consequence of sensory filtering mechanisms. One question of fundamental importance is how animals learn to ignore irrelevant cues (Mackintosh, 1973). Such learning occurs partly because animals have a predisposition to respond to certain cues, associating them with particular actions (Lorenz, 1965; Shettleworth, 2009), and partly because responsiveness changes with the repetition of cues.

Predispositions to learn

When there is an array of detectable cues, animals are much more likely to learn to associate some of them than others with an action or another cue. Bolles (1975) points out that a rat is more likely to learn to show an avoidance reaction to an aversive stimulus, such as an electric shock, and learns this faster than it learns to press a lever in response to a signal indicating imminent food arrival. Rats that modify their behaviour quickly when a cue indicating danger is recognized will produce more offspring than rats which are less good at doing so. An animal may also learn, because of negative reinforcement, when a harmful or unpleasant consequence is prevented by an action of the animal. Since novel objects that might be poisonous or otherwise hazardous are best avoided, it may be adaptive to take time to learn about new food.

The work of Shettleworth (1972, 1975, 2009) shows that some sorts of responses are very difficult to associate with food while others are associated easily. Food was presented to hamsters on a series of occasions when they were carrying out some activity. The hamsters learned readily to associate some activities with food but did not learn as readily with other activities. Pressing a bar in the cage, rearing on their hind legs in the centre of the cage, scrabbling or digging were all increased in frequency if food was presented at the instant that they occurred. Washing their faces, scratching themselves with the hind leg or scent-marking on the cage were not increased in frequency by food presentation. The activities that the hamsters did learn to associate with food were more like those involved in food-finding and acquisition.

The examples given above are of learning to associate a cue with an action. Similar examples of predispositions to learn could also be given for other learning situations, for example habituation to cues that might be associated with a dangerous predator is less likely than habituation to minor disturbances or social signals of minor importance (see next section). Predispositions to learn develop as a consequence of genetic and environmental factors, just like other behaviour-controlling systems. Animals that had more effective predispositions would have produced more offspring, so genes that promoted effective learning about predators would have spread in the population.

Habituation and sensitization

If a flock of sheep is moved from a quiet field to one near a road, they will show an escape response on the first occasion that they see or hear a motor vehicle pass along the road. Subsequent vehicles elicit less and less response until each member of the flock ceases to show any behavioural response: it habituates. *Habituation is the waning of a response that could still be shown, to a repeated stimulus.* The repetition might be very frequent or as infrequent as once per day, but habituation could still occur. The likelihood of habituation and its rate would depend upon the nature of the stimulus, its rate, its regularity and the state of the animal. A stimulus like the sound of a falling pine cone might elicit a startle response initially from a sheep, but habituation would occur rapidly if pine cones fell every minute and less rapidly if they fell irregularly at a rate of five per day.

The sheep would not be likely to habituate to the sight of a hunting wolf, however. Indeed, the reverse might occur in that they would become sensitized. *Sensitization is the increasing of a response to a repeated stimulus.* The wolf might elicit a greater response the second time it is seen than the first. A sound like a cracking stick might also elicit little response the first time, but more and more response when it recurred. In each of these examples the repetition might mean greater danger than a single stimulus, so sensitization is advantageous. Habituation is also an adaptive process for it saves energy that would be wasted on repeated response to a trivial stimulus, and it may also prevent the animal from being detected by a predator during a response to a trivial event.

Habituation could occur as a consequence of receptor fatigue or adaptation of a neuron in the brain pathway and it may be that such simple habituation does occur. It seems likely, however, that much habituation is just as complex as the various forms of associative learning because it is so specific. Sokolov (1960) describes the habituation of the startle response in dogs to a tone, but the reappearance of the full response when these habituated dogs are presented with a tone differing only slightly in frequency. Sokolov also demonstrated very specific habituation of the physiological startle response of dogs to the duration of a sound. Broom (1968a) carried out a similar experiment in which the behavioural response of young domestic chicks, to a light repeatedly switched on for 10 s and off for 20 s, habituated. When the light was switched on for only 5 s, so that it went off early, the chicks responded and when it was left on for 15 s they responded at 10 s when it should have gone off. Such studies demonstrate that the animal must be establishing a model of the environmental change in its brain and comparing each input with that model. This is much more elaborate a form of learning than simple neuronal adaptation.

Important practical aspects of habituation concern the waning of the responses of domestic animals to handling procedures and housing conditions. If much handling is likely to be needed during an animal's lifetime, as it is for most pets, milking cows, breeding stock or show animals, then careful habituation of animals to handling procedures is necessary. Any new procedure or item of equipment may need further training. The sudden introduction of a different kind of food container or vehicle for distributing food may result in food refusal or escape responses, but preliminary exposure and slow approach will allow rapid habituation. Similarly, a different item of clothing like a suit or white coat may elicit a response in an animal that has habituated to a person in their normal clothes. This last point is especially important for animals taken to shows, markets or other situations where there are many people and other novel stimuli. Farm and companion animals that are transported are often not adequately habituated to handling, etc., and are disturbed by the procedure (Grandin, 2000).

Experimental Learning Studies

Much of what is written about learning refers to learning tests in laboratory situations, perhaps misleading the reader into thinking that learning is scarcely relevant to real life. Nothing could be further from the truth, as learning is involved in all of the functional systems. It is necessarily mentioned in every chapter of this book. The animal that is efficient in its ability to learn about its physical environment, individuals of its own species, sources of danger and resources of various kinds will survive and reproduce. Poor ability to learn such things means reduced fitness.

Experimental studies are of value, for it is difficult to carry out controlled studies in the field. Experimental results are vulnerable to errors of interpretation, however, because in some studies animals fail to learn because they are frightened by the experimental situation or because either the stimulus presented or the action required of them is inappropriate. A sheep will not work in a learning test, if what it views as a dangerous predator is present, or if its long-term companions are absent. Neither will a cat work readily for food that it does not want to eat or a horse perform an action such as putting its foot on a lever.

Examples of experimental situations that have been used for domestic animals include: (i) classical conditioning, in which an animal learns to show an existing response to a new stimulus; (ii) operant conditioning, in which the animal learns to perform an 'operant' response in order to obtain a reward or avoid an aversive experience; and (iii) maze learning, in which the animal learns to take a particular path in order to obtain a reward.

These terms are consistent descriptions of procedures, but it is likely that the changes that occur in the brain during learning are similar in different situations. In each situation, an environmental change may act as a positive reinforcer, a reward increasing the likelihood of a response; or a negative reinforcer, which punishes and decreases the response likelihood. Positive reinforcers include various correctors of homeostatic imbalance (see Chapter 4) such as food, water, temperature change, etc., and opportunities for social behaviour, sexual behaviour or exploration. Negative reinforcers include painful events, frightening stimuli or extensions of homeostatic imbalance.

The classical conditioning studies of I.P. Pavlov, in which the unconditioned response (UR, salivation), normally shown to the unconditioned stimulus (US, food), is associated with the conditioned stimulus (CS, a bell) as a conditioned response (CR) after pairing of CS and US (Pavlov, 1927), may be interpreted in more than one way (Forkman, 2002). It could be that the dog reacts to the CS with the CR because it cannot tell the difference between the CS and the US. However, an alternative explanation is that the idea of food is evoked by the bell and then salivation results (CS→US→UR). Dickinson *et al.* (1995) and Balleine and Dickinson (1998a) explain how devaluation experiments can help to discover which of the above is occurring. If an animal is conditioned to a sound (CS) with a new food (US), the UR is to approach the food. After learning, the animals start to approach the sound. If the new food is given and a nausea-inducing drug is given, the animals learn not to eat the food: i.e., it is devalued. If the original learning was CS→CR, the devaluation of the food should not affect the approach (CR) response to the sound (CS). If, however, the original learning was CS→US→UR, the tone (US) should evoke the approved response (CR). In rats and hens it is the latter, more complex, learning that occurs (Holland and Straub, 1979; Regolin *et al.*, 1995; Forkman, 2001). This suggests a high level of awareness in animal subjects as it would appear that the individuals are aware of the significance of the CS, i.e. the sound, in relation to the food (US). It has also been shown (Haskell *et al.*, 2000) that frustrating chickens by depriving them of expected food leads to behavioural changes that indicate a CS→US→UR sequence (Forkman, 2002).

A well-known consequence of classical conditioning on farms is milk let-down by dairy cows in response to the typical sounds of a milking parlour. Milk let-down is initiated by oxytocin release following stimulation of the teat by a calf attempting to suckle. Cows with calves soon start to release oxytocin when other stimuli from the calf are detected, and cows milked using a milking machine may respond to other cues, such as the sounds, in the same way. There is variation among breeds of cattle with respect to how readily such conditioning can occur, and old breeds such as the Salers in France are much less ready to let down milk to stimuli that are not those emanating from real calves than are Friesians or Holsteins.

Pavlov's original studies of classical conditioning involved dogs, which show the UR of salivation to the unconditioned stimulus of detecting food. They started to salivate at the sound of a bell if the bell had been paired with food presentation on a number of occasions. The bell is referred to as a CS and the salivation as a CR. Using these terms, milk let-down becomes a CR to the CS of clanking noises, etc., in the milking parlour. Farmers need to be aware of the fact that milk let-down in a parlour is a CR and that such learning depends upon adequate training. If disturbing stimuli are present in the milking parlour, the young animal may not learn and the older animal that is conditioned may be inhibited from showing the response. As Kilgour (1987) points out, any veterinary work involving discomfort for the cow should not be carried out in the milking parlour but in a separate facility.

Another form of associative learning is operant conditioning. The action of carrying out the operant, such as switch-pressing, is increased or decreased according to the outcome of the behaviour (Domjan, 1998). Sheep and pigs, studied by Baldwin (1972, 1979), learned to operate a switch for food, light or heat. A sheep was able to switch on a heater over its pen by putting its nose in a slot and breaking a beam monitored by a photocell; it learned that, when the ambient temperature was low, it could warm itself by the operant behaviour of putting its nose in the slot. Sheep did not do this when they had a full fleece but only when they had been shorn. Substantial innovation may be shown by animals learning to operate a food delivery system. For example, when Atlantic cod *Gadus morhua* were in a tank where pulling a string would cause food delivery, they learned to do this, as several species of farmed fish have been shown to do. Three of the cod, which had tags in their dorsal fins, learned to pull the string more efficiently by hooking the tag on to the string and thus pulling it. These cod were using the tag as a tool (Millot *et al.*, 2013). Tool use has also been demonstrated in sting rays *Potamotrygon castexi* (Kuba *et al.*, 2010).

Operant conditioning, which is also called instrumental conditioning, may involve a reward or positive reinforcement. For example, when a dog sits, this operant response is followed by the receipt of a biscuit. This reward positively reinforces the action of sitting and increases the likelihood that the dog will show the behaviour in the future. If, on the other hand, a response stops or prevents a negative consequence of that action, the animal may learn because of negative reinforcement. For example, a dog showing submissive behaviour is less likely to be bitten by another, stronger dog. The negative reinforcer of biting increases the probability that the dog will learn to show the action of showing submissive behaviour. Very many studies of this kind have been carried out by experimental psychologists using rats pressing a lever for a reinforcer. Much used in experimental psychology has been the 'Skinner box', in which the lever pressing and food delivery are automatically monitored. Some experimentation has involved studying the effects of different schedules of reinforcement. If food is delivered every 5th time that the lever is pressed, this is referred to as a fixed ratio of reinforcement (FR5 in this case). Animals still learn when the ratio is very high and this is not surprising because, in wild conditions, they might often need to carry out a food-searching behaviour many times in order to obtain a food item. As a consequence, animals may repeat many times a movement that has resulted in reinforcement. A dog was readily trained to bark 33 times for small food rewards (Salzinger and Waller, 1962). Dogs that are occasionally fed scraps when their owners are eating, and which sometimes bark in this situation, may associate barking with feeding and bark even more. The operant response of barking is effectively reinforced on a large, rather variable, ratio of responses to rewards, but the dog still learns the association and the barking may be perceived as a problem by humans. When a stimulus or behaviour is no longer followed by a reinforcer, the learned response will rapidly or gradually disappear: this is extinction of learning. If the owner of a dog ceases to give a food scrap when barking occurs, extinction of the barking response occurs. However, occasional reinforcement with a food scrap (partial reinforcement) makes the extinction very slow.

A quite different schedule of reinforcement is that where the reward follows the operant behaviour but only after a fixed interval. In a laboratory experiment the animal still has to press a lever in order to obtain food, but lever-pressing has no effect except after the predetermined interval. If the interval is long enough, animals usually learn not to press the lever except when the time of reinforcement is close. The existence of quite accurate internal clocks is apparent from such studies, as it is from observation of domestic animals. Dogs may prompt their owners when it is the time of a regular walk; cows use the operant response of bellowing when it is time for the farmer to collect them for milking; and sows using an electronic sow feeder enter the feeder more often at the time that the daily feed cycle is about to begin.

The extent to which conditioning and other learning occurs in everyday life was exemplified by Forkman (2002) as follows. A hen may approach (UR) any worm-like object (US) but may also have learned by conditioning to approach (CR) any dark leaves (CS) because it had found that there were often worms (US) under them. It may have learned by operant/instrumental conditioning to peck at the front end of the worm because this stops it moving quickly, so allowing the positive reinforcement of ingesting the food to occur. Some worm-like animals bite or taste bad (negative reinforcement), so pecking at them is reduced or avoided. In an area where there are no worms under leaves, extinction of leaf-turning behaviour may occur.

Cognition

Cognition was defined by Shettleworth (1998, 2009) as the mechanisms by which animals acquire, process, store and act on information from the environment. Many authors, such as Mendl and Paul (2004) and Paul *et al.* (2005) refer to Shettleworth's definition of cognition. However, a high proportion of all brain function is included within this definition, for example some perceptions and motor system outputs, and these would not normally be considered as part of cognition. The basis of Shettleworth's definition is that during cognition something is formed in the brain that represents what is acquired and processed and can be retained and subsequently used. In a summary of what is included in cognition, Heyes (2000) included: (i) an entity in the brain; (ii) 'input from other cognitive states and processes and from perception'; and (iii) 'outputs to other cognitive states and processes and to behaviour'. A key aspect of cognition is that the entity in the brain does not disappear when perception ceases. Hence Terrace

(1984) argued that cognition involves explicit representations of absent stimuli. This is not a sufficient definition by itself because the representation would often be of more than just stimuli and would itself be a result of processing inputs in relation to other relevant information. A definition proposed by Broom (2014b) is: *cognition is having a representation in the brain of an object, event or process, in relation to its context, where the representation can exist whether or not the object, event or process is directly detectable or actually occurring at the time. The representation of something absent is an abstraction.*

Do domestic animals have cognitive representations of objects or other resources? A dog that searches for a thrown stick after it has lost sight or smell of it must have some representation of that stick in its brain while it is looking. A cow whose calf has been removed has a brain representation of that calf during the period when she is showing distress and, perhaps, thereafter. Any animal that is working towards a goal is utilizing cognitive processes in its behaviour control. At one time it was thought that a chicken would lose any concept of an object if it moved out of sight and throughout any period when it was not visible. When this was investigated by Vallortigara and colleagues they found that not only could young domestic chicks go to objects hidden behind screens but that, when two or three objects were hidden behind screens, the chicks went to the screen with the larger number of objects (Rugani *et al.*, 2009). The domestic cats studied by Witt *et al.* (2009) were able to learn that they could obtain a reward by pulling a string but when different strings resulted in the cat obtaining a reward or no reward, the cats did not learn to differentiate between the strings. In a somewhat similar study, neotropical parrots were able to learn to look at the continuity of strings and to pull the right string to get food (Schuck-Paim *et al.*, 2009). In experimental studies, in the view of the watching animal, food was hidden under cups and the cups were then moved or substituted with a different cup. Apes were successful in obtaining the food in all situations but dogs tracked the food less well, generally solved the simple cup movement problems only, and tended to focus on the place where they last saw the food (Rooijakkers *et al.*, 2009). Corvids, parrots and apes successfully solve multiple displacement tasks but dogs, cats and monkeys that are not apes do not.

Other experimental studies show that domestic animals can use a visual or auditory symbol for objects. Langbein *et al.* (2004) were able to train goats to respond by carrying out an operant, with water as a positive reinforcer, when they saw one particular picture rather than others. Goats also learned shape discrimination successfully when living in a social group (Langbein *et al.*, 2006). A further example will be familiar to those who have trained dogs but was studied in a carefully controlled way by Rossi and Ades (2008). When a dog was given commands that required her to respond to one of several objects, such as a ball, a stick, a bottle, a key or a toy bear, and to carry out one of several actions, such as to point to it or fetch it, she was successful. Similarly, Kaminski *et al.* (2009) found that dogs shown replicas or photographs could use this information and fetch the objects that were thus iconically portrayed. When Rossi and Ades' dog was provided with a keyboard that had symbols on it that indicated 'water', 'food', 'stroke me', 'I go out', 'I get a toy' or 'I urinate', she could indicate what she detected or what she wanted to do next. A further example demonstrating the existence of a concept of an absent object is of pigs studied by Held *et al.* (2000, 2002). They were put in a room and allowed to find hidden food. On the next day they were returned to the room and they went immediately to the place where they had found food. The studies described above show that some of these animals had a concept of an object in the absence of that object, had a concept of a symbol or of a location, and had a concept that pressing the symbol or going to a particular place was linked in a causal way to obtaining the resource. Cognitive ability has been demonstrated in a wide range of animals (Shettleworth, 2009; Broom, 2014b). Some aspects are discussed in Chapter 4.

If an individual has the novel visual experience of viewing images in a mirror, this could be followed by learning about what it sees in the mirror in relation to itself and then using such information at a later time. Human infants, chimpanzees, an elephant, dolphins and magpies with previous experience of mirrors showed their awareness of marks on their body, visible in a mirror, by touching or apparently looking at the marks. Broom *et al.* (2009) exposed 4–6-week-old pigs to a mirror for the first time in such a way that they could see a food bowl otherwise out of view behind a barrier (Fig. 3.1). The young pigs went behind the mirror to the apparent position of the food bowl. However, when given 5 h experience of a mirror, they responded initially to it as if to another pig but later by looking at it as they moved. They moved and then stopped still,

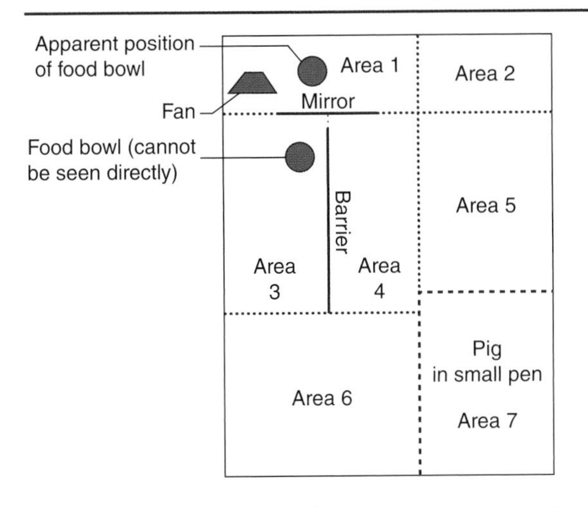

Fig. 3.1. Plan of the pen where the experiments were carried out, showing the small pen with solid walls (Area 7), mirror in a frame, solid wood barrier, fan position (above pig head level in the pen, to prevent detection by odour) and the numbers of floor sections used to describe the position of the pig. Area 3 is where the red food bowl, whose reflection was visible in the mirror when the pig was in Areas 4 or 7, was placed during the mirror test. The food bowl would appear to be in Area 1 to a naive pig that had not had experience with a mirror. The pig had not been given food in this place before the experiment (Broom, 2014b).

Fig. 3.2. The movements of young pigs in front of the mirror, to which the animals were exposed for 5 h, often suggested that they were looking at their image while making body movements (photograph D.M. Broom).

apparently looking at their image and its surroundings, oriented either with nose towards the mirror or with the head parallel to it. As a consequence of the lateral position of the pig's eye, it is not possible to record duration of looks towards the mirror and pigs show little change in facial expression. Some pigs vocalized. As with the movements in front of a novel mirror described for chimpanzees, humans, capuchin monkeys, dolphins and elephants some of the movements of these young pigs suggest that they could have been monitoring the movements in the mirror image when they moved their own head or body (Fig. 3.2). The animals could be comparing the kinaesthetic information and the external visual effects. After this experience with the mirror, seven out of eight pigs tested in the pen in Fig. 3.1 moved away from the mirror and around the barrier to the food bowl. Location by odour was prevented by fans and the naive controls that went behind the mirror had exactly the same olfactory situation. To use information from a mirror and find a food bowl, each pig must have observed features of its surroundings, remembered these and its own actions, deduced relationships among observed and remembered features and acted accordingly. The significance of mirror studies in relation to awareness and the issue of self-awareness is discussed further by Broom (2014b).

Learning Ability of Domestic Animals

What learning is possible for domestic animals? Observations of behaviour and experimental studies of learning show that they can, for example, learn to navigate in their environment, distinguish the qualities of food, return to food sources, avoid physical dangers, minimize predation risk, discriminate individual animals and respond differentially to individuals according to previously acquired information. Some studies indicating such abilities are described here.

A form of associative learning, in which several successive responses are associated with a reinforcer, is involved in learning how to get from one place to another. Domestic animals readily learn their way around the area available to them in human habitations or on farms, so it is to be expected that they can learn to run mazes. Kilgour (1987) compared the performance of several species in a variable (Hebb–Williams) maze, which involved a set of six different simple detours to reach an objective by walking. A score of 100% could be obtained if the animal solved the problem in the first four of eight runs given daily, and slower learning resulted in a lower score. He used various farm animals, dogs, cats and man. The learning scores were 99 for children, 90–93 for dogs, cows, goats and pigs, 85 for sheep, 81 for cats and rats, 61–66 for hens and pigeons and 48–53 for mice and guinea pigs. Using a score based upon numbers of errors, sheep were as good as cows and dogs but pigs were less good.

Some results from such experiments might be affected by variation in motivational state, especially that of being frightened, but it is clear that the farm animals performed very well. A wide range of further studies shows that farm animal species and dogs can learn simple and complex tasks with great rapidity. The maximum possible performance in any one of a wide range of learning situations is probably the best indicator of learning ability. Using such a criterion, cattle, sheep, goats and pigs learn at least as well as dogs. Horses may be slightly less good at learning but there are few well-controlled studies on them. Domestic fowl perform somewhat less well, but all of these species are very competent at various tasks. Measurements of brain size, or brain size in relation to body size, offer little additional information (Broom, 2003). For mammals, the degree of folding of the cerebral cortex may be related to intellectual ability. The ungulates such as sheep, cattle, goats, pigs and horses have more folding than most mammals, with only the primates and whales clearly showing more folding.

Observations of learning in the real world that domestic animals encounter offer the most impressive evidence of their ability. Pet-owners are familiar with the abilities of their animals to learn how to get food and other resources in their daily lives. Grazing animals are often thought to lead an uncomplicated life, but recent research shows that this is certainly not so. As explained in Chapter 8, sheep and cattle are very selective about what they eat and they have to learn about all the different plants they encounter. They also have to learn to identify patches of good grazing and to return to them after intervals so that they obtain adequate quantities of a good mix of plant material without wasting energy going to places that have not regrown after the last grazing.

These animals live in groups with an elaborate social structure and they have to learn a lot about other individuals. The most complex tasks in the lives of animals are those associated with establishing and maintaining social relationships. Hence, pack-living dogs or wolves, and flock- and herd-living farm animals have to have a considerable intellect for this purpose alone. Farmers are accustomed to rapid learning by farm animals to the extent that they may not appreciate that much is being demanded of the animals. A simple form of conditioning with a negative reinforcer is learning to avoid an electric fence. Some individuals explore the fence and learn to avoid it after receiving a shock. Some learn, by watching others, that the fence has some unpleasant characteristics. A few discover that, while a moist nose applied to the fence results in a substantial shock, a touch with a better-insulated area has less effect, so they monitor the fence at intervals to check that it is still activated.

The provision of food in troughs whose lids have to be lifted requires quite sophisticated operant conditioning, as does the use of Callan–Broadbent doors (see Figs 3.3 and 3.4). One of these doors (or gates) opens only when the cow wearing the correct transponder comes close to it. Hence, a cow that is newly equipped with such a transponder and is then faced with a row of doors has to learn that food is available to it when it pushes down with its head on one of the doors in this row. This very complex task is learned very quickly, with little training, by most cows. In experimental studies, Langbein *et al.* (2004) were easily able to train goats to respond to a picture in order to obtain water.

Another complex automatic feeder whose operation is readily learned is the electronic feeding-stall for cattle or sows, which again depends upon the wearing of a transponder but in which the operation is more complex. Problems associated with training animals to use these stalls are more to do with social contact with

Fig. 3.3. Holstein cow with transponder that could be used for opening a Callan–Broadbent gate or for entry to a feeding stall. The transponder is recognized individually and an appropriate electrical response initiated (photograph D.M. Broom).

Fig. 3.4. Callan–Broadbent gates, each of which can be opened by an individual cow's transponder (photograph D.M. Broom).

other cows or sows near the feeder than with the operation of the system. Not only do the animals learn to operate these feeders, but they learn how to beat the system and obtain extra food by chasing other animals out, by coming back in through the exit gate or by banging the food dispenser so that it delivers a few extra pellets of food. One sow in a group wearing transponders on their collars and fed in an electronic feeder learned that, if she picked up a collar that had fallen off another sow, she could receive an extra ration of food. In an experimental study, pigs have been found to discriminate between food sites, choosing the one with more food or less difficult access (Held *et al.*, 2005). Food selection is also affected by social factors as they avoid the risk of being robbed of food by another pig (Held *et al.*, 2010; Chapter 8).

The ability of sheep to distinguish other individual sheep and individual humans has been demonstrated by choice experiments and by recording from cells in the medial temporal and prefrontal cortex (Kendrick *et al.*, 2001). Sheep could discriminate between 25 pairs of photographs of other sheep, could remember this for more than 1.5 years and had the same neuronal responses in the brain at the beginning and end of this period (Broad *et al.*, 2002). In the same study it was shown that, when a ewe recognized a lamb, there was an increase in the brain-derived neurotrophic factor, in the density of receptors trk-B and in the expression of messenger RNA in eight regions of the brain. Other studies have shown that young cattle can learn a task that requires them to discriminate between two herd members (Hagen and Broom, 2003), hens can discriminate between other individual hens (Bradshaw, 1991; Abeyesinghe *et al.*, 2009) and sheep can discriminate between photographs of calm or disturbed conspecifics (Elliker, 2007).

The ability of pet animals to acquire and use concepts of objects and actions described by words has long been assumed to exist by their owners, and experimental studies such as those described above demonstrate some of these abilities (Rossi and Ades, 2008). It is clear that a dog can respond to objects and to a desire for an action or resource. However, a question that has been asked is whether dogs or other animals have a concept of an object when they cannot see it or otherwise directly detect it. Young (2000) addressed the question by putting a dog in a situation where it was accustomed to being allowed to feed from one of three bowls. The dog observed person A enter the room and place food in one of the covered bowls, but could not see which bowl had food put in it. The dog could also see that person B was present in the room (who could see into which bowl the food had been put). A third person (C) was not present in the room. Later, the dog was again allowed to see the three covered bowls but with no olfactory cues to distinguish them. One of the bowls was pointed to by person B and another by person C. When the dog was able to go to one of the bowls, it went to the bowl indicated by person B. This study shows, first, that the dog had a concept of the food that it could not see; second, that information provided by a person perceived by the dog to have accurate information was used in preference to that from another person. Persons B and C did not differ in their history of providing accurate information. Both dogs and cats can respond to human pointing while, in the reverse situation, dogs observed where a toy had been hidden by a person and then indicated the position of the toy to another person who might retrieve it (Miklósi *et al.*, 2005; Virányi *et al.*, 2006). The same group reported that dogs could watch a person and learn to make a detour around a fence to reach a reward (Pongrácz *et al.*, 2001, 2005). Dogs that had learned a detour, involving going through a gap in a fence, showed perseveration of this learning and did not readily learn to use a new gap in the same fence (Osthaus *et al.*, 2010). Horses, donkeys and mules also learned the detour, mules being fastest to learn, but had much difficulty in learning when the gap position was changed (Osthaus *et al.*, 2013). Mules were also better than donkeys and ponies in a discrimination learning task (Proops *et al.*, 2009). Studies by Held *et al.* (2001) showed that pigs could watch another pig and learn from its actions how to choose correctly which of four corridors to go down in order to receive a food reward.

The most impressive cognitive abilities demonstrated by a domestic animal are those shown by parrots. Indeed, their abilities are at least as good as those of apes. In the studies by Pepperberg (2000) the subjects were African grey parrots (*Psittacus erithacus*). One parrot learned to respond to and say words referring to 50 different objects, seven colours, five shapes and quantities from one to six. The parrot could use the words 'no', 'come here', 'I want to go out' and 'I want' (a particular object). Combinations of words could be used. For example, when asked 'What is in the box?', the parrot could answer 'five red squares'. If asked 'How many blue circles are in the box?' when there were 15 objects of various shapes and colours, the parrot could answer correctly, 'four'. The parrot could also ask for items, for example, 'want blue bell'. In the course of Pepperberg's studies, parrots were observed to ask for absent objects, react to hidden objects, use a tool to discover hidden objects and, on seeing a person discovering an object, to say after the object could no longer be seen what it was and what colour it was.

Another kind of study that shows awareness of learned actions and situations involves assessing emotional and other responses to them. Mendl *et al.* (1997) trained pigs to find hidden food and to remember the task until the following day. However, the pigs had to work hard to obtain the food and thus remembered better how to find it on the following day (Laughlin and Mendl, 2004). If the pig was subjected to disturbing treatment during the day after learning, the efficiency of remembering was impaired (Laughlin *et al.*, 1999). It appears that the emotional disturbance affected memory consolidation. Other examples of emotions influencing cognition and of cognition inducing emotions are described by Boissy and Lee (2014) and Broom (2014b); see also Chapter 4.

A different approach is to study the response of an animal that learns. Langbein *et al.* (2004), studying goats that learned to respond to a picture, found that the heart rate frequency of the goats was more variable after they had learned than before. Hagen and Broom (2004) put young cattle (heifers) in a pen where they had to put their noses in a hole in the wall of the pen in order to make a gate open, thus allowing access to a bucket of food 15 m away in a field. The heifers learned to do this and were matched with control heifers that spent the same time in the pen and then gained access to the food bucket. The actions of the control heifers did not cause the gate to open. When compared with the controls, and with heifers that had already learned to open the gate, the heifers that were just learning the task had higher heart rates and were more likely to show excited behaviour such as running and jumping. It seemed that the excitation was associated with the time of learning, and so might be a response to it. This was called a 'eureka effect'. Broom and Barone (in preparation) obtained a similar result with sheep.

Behaviour Development

There are two kinds of problem for the young developing animal. The immediate problem is how to survive during the first period of life when it is very vulnerable to predation, to physical conditions and to the risk of not obtaining adequate nutrients. This is often quite a different problem from that of the adult because the young animal is smaller, less able to defend itself than are adults, subject to attack by predators that might not attack its parents, requires different physical conditions and requires a different diet from adults. The other problem for the developing animal is how to change in such a way that it becomes an effective adult. There is often an assumption that most behaviour during development is directed towards the adult objectives, but the very high incidence of early mortality in most species means that there is a high selection pressure promoting efficient survival mechanisms at this time.

Development of domestic chick behaviour

What does a young domestic chick have to do in order to survive, grow and eventually become a successful adult? In the early stages, especially, it is not possible to understand how behaviour changes without knowing about the development of body anatomy and the biochemistry and physiology of the brain. Developmental changes in behaviour start before hatching, but increase dramatically in number and complexity after hatching. A summary of pre-hatching and early post-hatching changes is shown in Table 3.1. Some behavioural contact with the mother occurs before hatching by the chick calling and reacting to parental calls. Just before hatching, the embryo chicks commence making clicking noises. Vince (1964, 1966, 1973) demonstrated that the chicks in a clutch of eggs are communicating with one another by this means. The clicks are perceived and have the effect of accelerating hatching by some chicks and hence synchronizing the hatching of the eggs in a clutch. A chick that hatches early or late is more vulnerable to predation, so synchronization is advantageous to all.

After hatching, the young chick needs to recognize the mother, keep the mother close or follow her, make the mother do things that benefit the chick, recognize anything dangerous and respond in a way that minimizes the danger. Within a short time after hatching, the chick needs to develop an ability to look after itself: it must learn about feeding, body temperature regulation and many other aspects of life. The motor and sensory ability of the chick is quite good at hatching, but both improve during the first few days of life (Kruijt, 1964). Pecking is an important way of investigating the environment and it improves in accuracy during the first week after hatching (Padilla, 1935). Dark-rearing only slightly impairs this improvement and corrections for distorted vision are possible. The chick prefers to peck at objects that are small, shiny, three-dimensional and of certain colours. These preferences are modified by experience during the first few days of life. The chick has the sensory ability to see and hear the mother in the few hours after hatching.

Newly hatched chicks, ducklings and goslings approach objects larger than themselves and which move at approximately walking speed (Lorenz, 1935; Hinde *et al.*, 1956). These do not have to be the mother, and

Table 3.1. Some changes in the domestic chick's anatomy, physiology and behaviour during development from fertilization (from Broom, 1981).

Change in	Incubation (days – hatch at 21)	First 7 days post-hatching
Behaviour	Moves (4)	Walk/preen/peck (1)
	Reacts to light/sound (17/18)	Copulate (2)
	Pipping/clicking (18/19)	Pecking accuracy improves (1–5)
Brain recording	Cerebral activity (13)	
	Response to light (17)	Latency drops (1)
	Sleep waves amplitude lower (17–20)	Spindle frequency less (1–5)
Brain biochemistry	Spinal cord cholinesterase lower (6–10)	
	Optic lobe cholinesterase lower (18–21)	Peak (2)
	Alkaline phosphatase lower (17–20)	
Anatomy	Basic structure established (8–12)	Cold tolerance less (2–6)
	No further gross change (17)	
	Albumen all utilized (18)	Yolk all utilized (modern breeds) (2)

particular visual or auditory patterns are attractive. A flashing light or a rotating disc could also be maximally effective visual stimuli in attracting chicks. In normal circumstances, the object that the chick sees with these characteristics is the mother. The chick rapidly learns her precise characteristics and is subsequently more likely to follow something with these characteristics. This period of rapid learning is associated with particular structural and biochemical changes in the chick's brain.

At the same time as learning about its mother, the chick is learning the characteristics of other aspects of its environment and starting to avoid the unfamiliar. The chick must form a neural model of this familiar world in order that discrepancies can be recognized and avoided in case they are dangerous. The manipulation of maternal behaviour by the chick is important for early survival. A cold chick calls loudly and thus encourages the mother to come and brood it. Mother hens respond to this call (Edgar *et al.*, 2011). Chicks make twitter calls that encourage the mother to stay with them and they are able to copy her pecking movements. Hence, they can learn from her as well as deriving protection from her.

Development in each functional system

The chick whose development is described above is very different from a sparrow or rabbit in its stage of development when it emerges into the complex and dangerous outside world. *Animals that are well developed when they are born or when they hatch are precocial, whereas those that are helpless are altricial* (see Fig. 3.5). There is a continuum between the two extremes, man being more altricial than precocial, but some farm animal species are at the precocial end of the scale.

Just as the experience before hatching can affect chick functioning and behaviour, mammalian experience *in utero* can have effects. For example, Rutherford *et al.* (2009) compared the piglets of sows that had been subjected to stressful social mixing three times during pregnancy with those of sows that had not. Piglets of stressed sows showed greater responses to pain during tail-docking.

After birth or hatching, the development of sensory systems is affected by experience. For example, animals reared in darkness have fewer cells and fewer synapses between neurons in the visual pathway. Studies of the

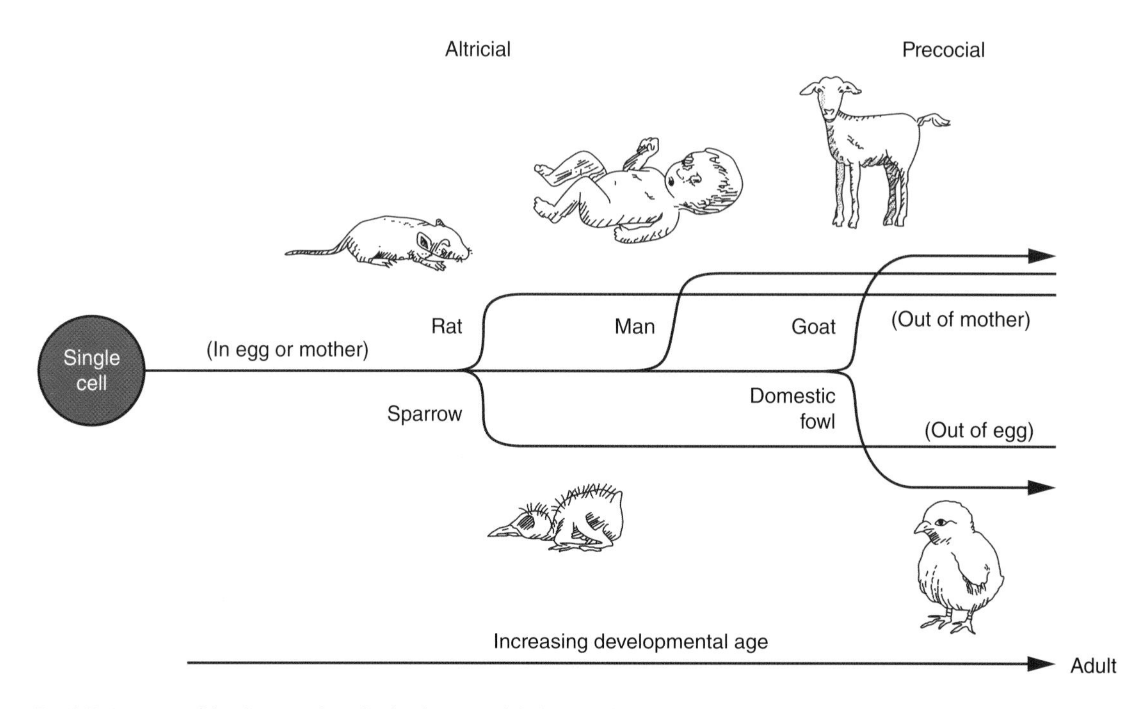

Fig. 3.5. Diagram of development from fertilized egg to adult showing that altricial animals emerge into the outside world at an earlier stage of development than do precocial animals. The origins of arrows show the point of hatching or birth (modified after Broom, 1981).

functioning of the visual system show that various sophisticated analysers do not develop if the eyes are alternately covered or if the environment is limited to vertical stripes during early rearing (Blakemore and Cooper, 1970). Such developmental changes will affect all aspects of behaviour, as will the development of ability to make certain movements. Altricial mammals do not avoid a cliff in the early stages of life but, when perceptual and motor abilities are sufficiently developed, they do avoid it. Young domestic chicks, on the other hand, avoid cliffs as early as 3 h after hatching. Their depth perception and awareness of danger are adequately developed by that age.

The development of predator avoidance may be very unspecific initially; for example, there is an increase with age in avoidance by young domestic chicks of anything unfamiliar (Broom, 1969a). Such general effects become more specific as the complexity of experience increases. Studies with rhesus monkeys, chimpanzees and domestic chicks have shown that responses to relatively harmless novel stimuli were less if the animals had been reared in more complex early environments than if they had been kept in more barren environments (Broom, 1969b).

The recognition of specific predators also depends upon experience. Lorenz (1939) reported that young precocial birds showed a flight response to hawks or falcons but not to flying geese. However, experiments by Schleidt (1961) showed that this difference could have been due to habituation to geese but not to hawks. Models of hawks and geese can be distinguished by ducklings, and both elicit responses (Mueller and Parker, 1980), but extreme escape is more likely to be shown if other birds have been observed to show it. If an actual predator attack occurs, surviving individuals usually improve their ability to deal with such an attack as a result of this experience. This is apparent when deer are chased by wolves or larks are chased by falcons. The more experienced individuals have learned tricks that help them to escape.

During the neonatal period, a puppy is comparatively helpless and dependent on the mother. Suckling and care-soliciting are the main behaviours (Serpell and Jagoe, 1995). Before 2 weeks of age, neither their eyes not their ear canals are open or functional. However, human handling and noxious stimuli can elicit responses and learning. In the 3rd week of life, puppies develop the abilities to crawl backwards as well as forwards, to stand, to walk, to defecate outside the nest, to take an interest in solid food, to play-fight, to growl and to wag the tail. These dramatic behavioural changes are accompanied by an increase in alpha waves in the electroencephalogram (EEG). A more adult EEG pattern is not seen until about 8 weeks of age. The socialization period in puppies starts at 3–5 weeks and continues until about 13 weeks. If the dog does not have experience of a range of members of its own species during this period, it is likely to have difficulty adjusting to normal canine social life. Serpell and Jagoe (1996) explain how various behaviour problems in dogs may be exacerbated or ameliorated by experiences during development.

Behavioural efficiency in other functional systems also develops with age and experience. Grazing time increases during the first 4 months in calves as ruminal function develops. Young animals are, however, less efficient as grazers than are older animals. Arnold and Maller (1977) and Arnold and Dudzinski (1978) found that sheep reared with no grazing experience for 3 years grazed much less efficiently than experienced sheep (see Fig. 3.6). Feeding is also affected in some ways by early experience in predatory animals; for example, the opportunity to play with objects by dogs and cats can lead to some refinement in sophisticated movements but no change in certain killing movements (Hall, 1998).

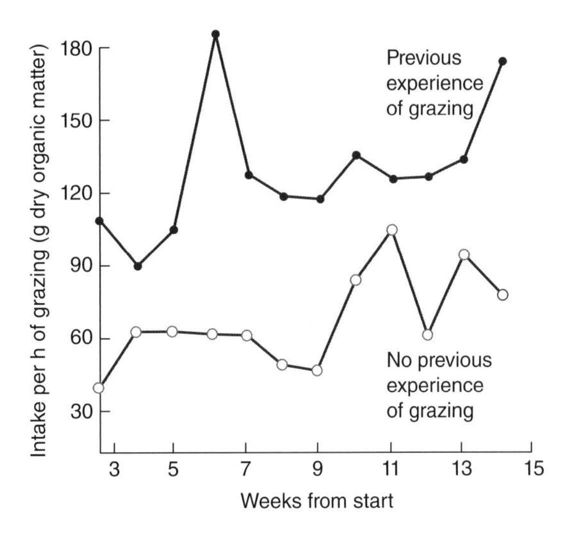

Fig. 3.6. 3-year-old sheep that have had no experience of grazing (○) are less efficient grazers than are sheep that have grazed (•). Their performance improves after 10 weeks' practice (after Arnold and Dudzinski, 1978).

Communication and courtship behaviour are sometimes not very variable within a species, but detailed studies show how much the final form depends upon experience during development. Although most bird calls develop normally when birds are reared in isolation, they depend upon the ability of birds to listen to themselves, for early-deafened birds are abnormal (Nottebohm, 1967). Songs of birds like chaffinches (*Fringilla coelebs*) and white-crowned sparrows (*Zonotrichia leucophrys*) vary according to the sounds of singing birds heard during the juvenile period (Thorpe, 1958), and the songs of territorial rivals heard during adulthood (Thielcke and Krome, 1991).

The mechanisms in the brain of a songbird that are involved in song learning differ from one group of birds to another and may include four 'modules': (i) brain stem nuclei controlling song production; (ii) telencephalic nuclei for producing learned sounds; (iii) telencephalic and thalamic nuclei for song learning; and (iv) telencephalic nuclei in the ascending auditory pathway (Nottebohm, 1999). Song development depends upon the nature of the development of one or more of these modules as a consequence of environmental input. For example, when other birds sing, the bird itself sings or hormonal changes occur, song may be modified. Although most changes during the development of birdsong occur quite early in the bird's life, there are circumstances when later innovation in song occurs and is adaptive (Lachlan and Slater, 2003).

Developmental studies of sexual behaviour in the jungle fowl, the wild ancestor of the domestic fowl, show that movements involved in courtship and mating appear in the behavioural repertoire at an early age but become organized into sequences and used at appropriate times as the bird gets older and more experienced. Jungle fowl, guinea pigs, rats, cats and rhesus monkeys reared with insufficient contact with social companions show abnormal courtship and mating behaviour.

Mate selection is substantially affected by early experience. The discriminations between potential mates are very subtle, and many animals devote much energy to behaviour that maximizes the chances of obtaining a high-quality mate so that offspring are likely to be successful. Many studies of waterfowl have shown that birds reared by foster-parents often directed courtship to that species when they became adult. For example, Schutz (1965) found that male mallard ducks, the most common domestic duck species, reared by their parents for 21 days but by another species from 21–49 days or later, showed most sexual behaviour when adult towards the other species.

Studies of zebra finches (*Poephila castanotis*), exposed to their own and another finch species by Immelmann (1972) and then by ten Cate (1984), showed that the later mating preferences of males depend upon the species that is active as a parent during the rearing period, the species of the other young reared with them, the species with which they interact after fledging, their age at each of these contacts and the duration of the contacts. Cockerels, turkeys and doves reared alone by hand will court and attempt to mate with a human hand, but prolonged contact with females of their own species usually reverses this (Schein, 1963; Klinghammer, 1967; Schleidt, 1970). Within a species, work on quail has demonstrated that the sexual partners chosen are those that are similar to but not identical to the birds that were present during the first 35 days of rearing (Bateson, 1978).

Much social behaviour, as mentioned already, is complex and difficult to learn, so it is not surprising that its development is slow, prolonged and greatly affected by experience. Isolation rearing does not prevent the development of the motor patterns involved in fighting, threatening or caring for mate and offspring. It does alter the production of social signals and the timing of all behaviour during social interactions, so that isolation-reared animals are socially incompetent. Mason (1960) and Harlow (1969), working with rhesus monkeys, found that their social behaviour after 6 months of social isolation was drastically altered and they tended to avoid other members of their species when exposed to them. Monkeys, rodents and cattle did badly in social competition after isolation rearing and failed to 'acquire effective elementary communicative skills which serve to coordinate and control the form and direction of social interactions' (Mason, 1961; Broom and Leaver, 1978). Gaillard *et al.* (2014) described how isolation-reared calves performed less well at 7 weeks of age than pair-housed calves in two cognitive tests.

Broom (1981, 1982) describes how heifers that had been kept in individual pens over the 8 months from birth to turnout to pasture in spring were quite inadequate in their social responses. In encounters with heifers experienced in social encounters they did not return the gaze, kept their ears back much more when approached, failed to retaliate if attacked and lost most

Fig. 3.7. Young animals learn by following the behavioural examples of their mothers, for example, in selection of pasture plants and avoidance of danger. The mother goat is a Saanen–Toggenburg cross and the kid is three-quarters Saanen (photograph D.M. Broom).

competitive encounters. As a consequence, they procured less food in competitive feeding situations. These social inadequacies persisted for at least 1 year. Gygax *et al.* (2009) found that heifers reared in pairs adapted better to the dairy herd situation than heifers reared individually.

Social encounters during development help individuals to improve their performance in such encounters. For example, in the development of feeding, predator avoidance and social skills in monkeys, innovative behaviour results when individuals learn from one another in a social group (Box, 2003). Domestic animals learn from one another, especially from their mothers (see Fig. 3.7). Some of the improvement in ability to forage effectively and manage other resources comes from observing efficient individuals and doing what they do. Predator avoidance improves with age, and some of this is a consequence of being with and copying experienced individuals. The development of responses to humans by companion animals is much affected by early exposure to people. For example, if kittens were given extra social contact with people from 2 to 9 weeks of age, as adult cats they showed less fear of humans and were reported by their owners to give more emotional support than cats that did not have this experience (Casey and Bradshaw, 2008).

Farm animals develop their behaviour in various ways relevant to management and welfare. For example, farmers know that, once an animal in a group learns to open a gate, to manipulate a piece of equipment so as to get more food or to intimidate a stockman, others in the group are likely to learn to do it too. We have much to learn about the role of social factors in the development of behaviour and, in particular, of the animal's ability to control its environment.

Further Reading: Shettleworth, S.J. (2009) *Cognition, Evolution and Behavior*, 2nd edn. Oxford University Press, Oxford, UK.

Motivation

Introduction

When a dog steps out of the house or kennel in which it lives, or a pig awakens after lying asleep in the corner of a field or yard, what determines which movements it will then make and which functional system will be served by its behaviour? If a young domestic chick is observed and its behaviour categorized as shown in Fig. 4.1, what determines the nature and timing of each transition from one behaviour to another? What causes the chick to start or stop pecking, preening or walking when it does? These are questions about the motivation of the animal. *Motivation is the process within the brain controlling which behaviours and physiological changes occur and when.* An understanding of motivation is fundamental to all studies of behaviour, and is especially relevant to most of the questions asked about domestic animal behaviour by those managing animals, e.g. concerning feeding, reproduction and handling. An appreciation of the subtleties of motivational systems is also necessary in order that behaviour can be used as an indicator of animal welfare. The explanation of causal factors and motivation below is based on that by Broom (1981), and a useful discussion of the issues is provided by Toates (2002).

Causal Factors

The dog, pig or chick might initiate an activity that is likely to result in obtaining food, such as standing and walking to a place where food could be present or starting to peck at the ground in a particular place. A number of factors could affect whether or not these behaviours are shown. There might be sensory input to the brain about the body's environment, e.g. when a food odour is detected or a possible food item is seen. There will be internal input from body monitors, such as those affected by gut distension or blood nutrient levels, which provide information about general or specific body deficiencies. There could be internal input from oscillators within the body that produce an output after a particular time and can indicate normal feeding time or interval since the last feed. Each of these factors has some direct relevance to the feeding functional system, but the likelihood of food-searching will also be affected by inputs to the brain about other aspects of the animal's life. Possibilities include: (i) input about a skin irritation that results in scratching and rubbing rather than food-searching; (ii) input about the presence of a potential mate, rival or predator that again leads to some other activity being given priority over food-searching; or (iii) various aspects of hormonal status that change the likelihood of occurrence of the various behaviours.

All of the factors mentioned above, and the behaviour that they initiate, will be altered by the previous experience of that animal. Many kinds of previous experience might make a dog less likely to start food-searching behaviour: (i) if an odour of food is detected by the dog but it comes from a shop and, in the dog's previous experience, it has never been followed by food being made available to the dog; (ii) if the gut is empty, but experience shows that the gut has to be empty for several hours before food is forthcoming; (iii) if an oscillator indicates feeding time, but recent experience is that a more dominant dog always gets to the food first; (iv) if the odour of a potential mate is detected and there is experience that the potential mate is likely to be accessible; or (v) if the sound of people talking means that food provision is imminent.

Each input to the brain must be interpreted in relation to previous experience. Some inputs will never reach the decision-making centre in the brain because the interpretation results in their relevance being assessed as zero. It seems likely that most inputs will reach the centre after modification. *The actual inputs to the decision-making centre, which are interpretations of a wide variety of external changes and internal states of the body,*

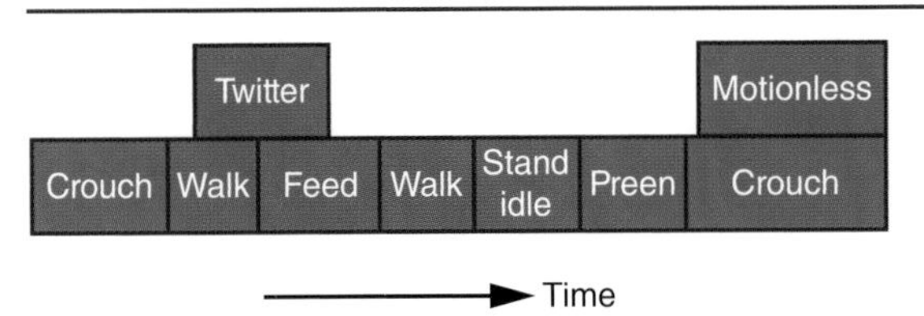

Fig. 4.1. Sequence of behaviour shown by young domestic chick.

are called causal factors. At any moment there will be very many different causal factors and the levels of these will determine what the individual actually does. Some causal factor levels will change very rapidly because they are altered by rapidly changing environmental events. Others, such as those that depend upon the levels of certain steroid hormones in the blood, change slowly.

All changes in behaviour are a manifestation of the animal's response to changes in causal factors. The experimental investigation of the relationships between causal factors and behaviour may involve either attempting to find out all the effects on behaviour of a single causal factor, or assessing the effects of variation in a wide range of causal factors on a single behaviour (Hinde, 1970). One difficulty of research in this area is that causal factors cannot be measured directly and some are very hard to estimate at all. Many valuable studies have been carried out, however, in which one activity or one experimental manipulator of causal factors has been studied in detail: for example, drinking or effects of water deprivation. In order to understand motivation in real life, though, work investigating situations where more than one experimentally modifiable set of causal factors is acting needs to be considered. This approach was pioneered by McFarland (1965, 1971), who worked initially on feeding and drinking in doves.

Motivational State

If a pig is deprived of water, after some time there will be input to the brain from: (i) monitors of body fluids; (ii) sensory receptors indicating a dry mouth; (iii) oscillators that would normally prompt drinking; and (iv) other brain centres, indicating that the animal is aware of the fact that drinking has not been possible for some time. The change in the state of the animal with respect to this group of causal factors is shown in Fig. 4.2. As the levels of these causal factors rise, there will also be increases in the likelihood of drinking if the opportunity arises and in the extent of activities that should result in water acquisition. If the pig were deprived of food as well, then its state with respect to another group of causal factors could be described and combined in a plot indicating the consequences of both food and water deprivation. In Fig. 4.3, a pig whose state has reached B is more likely to eat and to work in some way to get food than one whose state is at O, while an animal whose state is at A is more likely to drink than one whose state is at O. Indicating the state of the animal in this two-dimensional state space plot allows interactions between the two sets of causal factors to become clear.

When a pig is deprived of water its state moves up towards A on the state space plot, but it also moves to the right because pigs given no water cannot eat as much. The change in state of the animal as a consequence of water deprivation is shown as a trajectory from O to C_1. If this animal were then to be deprived of food as well as water, its state would move sharply to the right and up further to C_2. The behaviour of an animal that is given the opportunity to either eat or drink will depend upon its state as represented in Fig. 4.3. A pig whose state is at C_2 is a little higher on the water deficit side, so it might drink. This would bring its state down across the boundary line to C_3, at which time it might switch to eating, thus lowering the causal factors resulting from food deprivation. An example of a possible course of the animal back to O is shown. Various actual paths may be chosen by animals whose state is manipulated in this way, and the decision-making mechanism for switching from feeding to drinking can be affected by whether or not the animal has to search harder for the food or use more energy to get the water (McFarland and Sibly, 1975; Sibly, 1975; Toates, 2002).

In the state space plot shown, only two sets of causal factors are considered. In reality, the likelihood that the animal would eat or drink at any moment would be affected by many other causal factors as well. Each of these could be plotted in the same way, so the motivational state of the animal is its position in a multidimensional causal factor space. Each of these causal factor levels might interact with others in the same way that those relating to food and water deficit interact. A simpler definition, however, is that *the motivational state of an animal is a combination of the levels of all causal factors*.

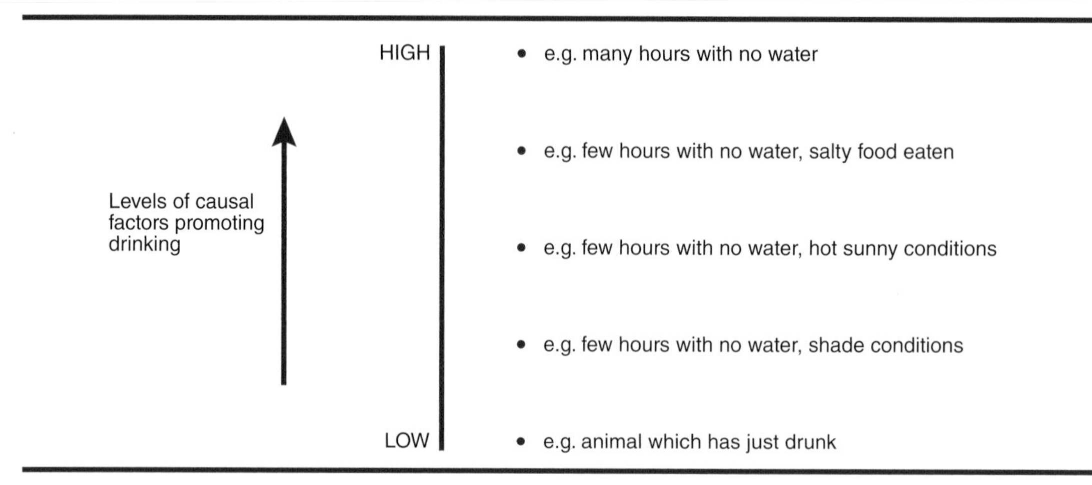

Fig. 4.2. Levels of causal factors that promote a particular action vary over a range, and the state of the animal can be described in terms of these.

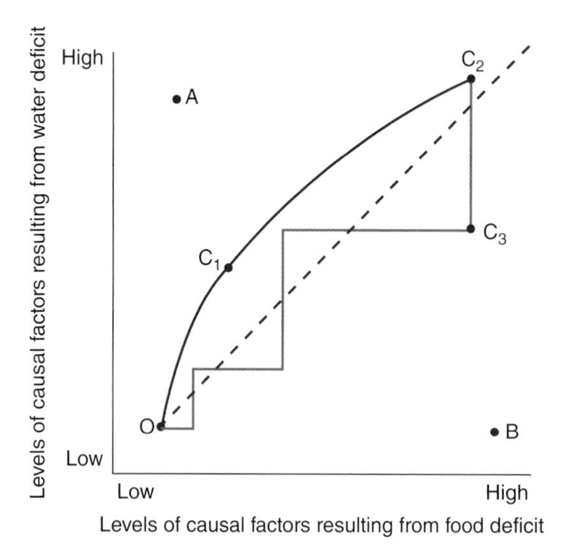

Fig. 4.3. Motivational state of animals A, B and C in two-dimensional causal factor space. Animal A is most likely to drink whereas animal B is most likely to eat. The changes in state of animal C are explained in the text.

Motivation Concepts

Early attempts to explain how it came about that animals showed a behaviour when they did so referred to 'instincts'. These were thought of as some inherited property of an animal that made it act in an automatic way in certain circumstances. The term implied development without environmental influence, an idea now discredited, and detailed studies of behaviour showed that animals, especially vertebrates, are far from being automata, so the term is no longer used. The term 'drive' was thought of by some people as a component of a homeostatic control system and by others as the agent causing a particular behaviour to occur. The idea of a thirst drive that caused drinking and an exploration drive that caused exploration was criticized by Hinde (1970), who said that 'Drive concepts can be useful if defined independently of the variations in behaviour which they are supposed to explain'. Hinde follows Miller (1959) in suggesting that it could be useful to think of 'thirst' as an intervening variable between effects on animals (independent variables) – such as water deprivation and amount of dry food eaten – and behavioural responses (dependent variables) such as amount of water drunk and rate of pressing a bar for a water reward.

Figure 4.4 shows relationships of 'thirst' to six variables. When these relationships were measured, both bar-pressing rate and amount of bitter quinine required to stop drinking had a linear relationship with the level of a dependent variable supposedly manipulating thirst, but the amount of water drunk had quite a different relationship. Toates (2002) emphasizes the complexity of motivational systems that is demonstrated by such experiments but affirms that it is helpful to think of motivation as involving incentive objects or goals. He explains that the defence of body fluid volume depends upon detection of interleukin concentration by neurons in the brain, release of the hormone vasopressin, action

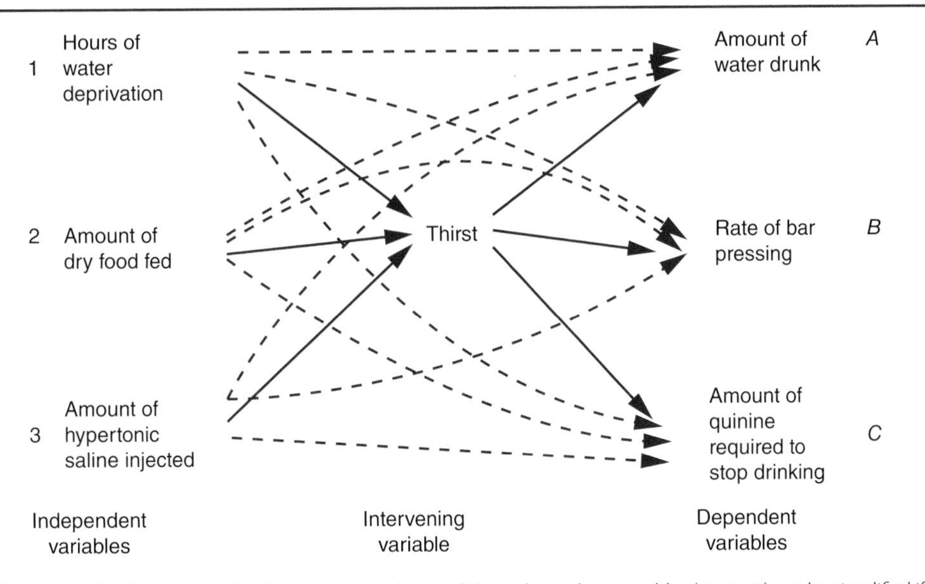

Fig. 4.4. The relationship between each of three independent and three dependent variables (see text) can be simplified if an intervening variable is considered. This diagram refers to experiments on the maintenance of water balance in a rat (from Broom, 1981, modified after Miller, 1959).

of this hormone in the kidney to conserve water, secretion of renin from the kidney and hence production of angiotensin, which acts in the hypothalamus to initiate drinking.

Another view of motivation is that of Lorenz (1966), who thought of motivation as an accumulation of action-specific energy that is released when the action occurs. However, energy is clearly an inappropriate term for the accumulated potential to perform an action and, although the potential to carry out some actions may accumulate, this does not happen for other actions. Hence, the concept might be useful in certain situations but it is not a general model. Similarly, ideas about general arousal or activation (Berlyne, 1967) are clearly important, for animals sometimes increase their responsiveness to a whole range of inputs, but they do not explain a high proportion of changes in behaviour. Levels of arousal, referring to a defined range of effects, are best thought of as causal factors that are combined with others in decision making.

Observations of the behaviour of rats learning to press a lever for a food reward led to ideas of motivation as merely the link between a stimulus and a response. Many people assumed that behaviour could be explained as being the largely automatic response to a series of stimuli from the environment. Toates (1987) explains that the stimulus–response model is inadequate in important respects: 'Contemporary theory sees the animal as being (i) intrinsically active rather than passive, even in the absence of impinging stimuli; (ii) goal-seeking, or, in other words, purposive; (iii) flexible; (iv) able to learn cognitions; and (v) exploratory.' An animal deprived of an important resource is seldom just awaiting an appropriate stimulus so that it can respond. Where the animal has the capacity for cognition, many of the mechanisms for preventing the continuation of a deprivation involve reward and incentive learning (Dickinson and Balleine, 2002). For example, an animal deprived of food or water may have lower perceptual thresholds for relevant stimuli (Aarts *et al.*, 2001). The cluster of neuronal representations associated with particular food objects, positive tastes and the joy of eating may be activated by food deprivation (Ferguson and Borgh, 2004).

Two other terms that have been used when referring to motivation are 'conflict' and 'displacement activity'. The idea of motivational conflict arose when people were trying to explain situations where an animal was thought to have two important drives, each of which would make it carry out a different behaviour. Although the term conflict appeared to be necessary when it was thought that one drive operated at a time, it is of much less value in modern thinking. Where there are very many different causal factors, in a sense there is conflict all of the time. Several different causal factors will always be competing for the animal's time and energy. The situation where two activities are both very likely because of the levels of the causal factors that promote them is of great interest, but it is different only in degree from all other motivational states. The term 'displacement activity' has been used for behaviour that appeared irrelevant to a human observer, but seems to be of little value.

As emphasized in several chapters of this book, the feelings and emotional states of animals have much influence on their behaviour. Many decisions are influenced by past and present feelings. Stimuli remembered by an animal as being associated with positive feelings will lead to quite different responses from those associated with pain, fear or other negative feelings. If a stimulus has been associated with fear, it will tend to induce avoidance and also a complex of behaviours whose effect is to reduce the likelihood that there will be a close encounter with the stimulus (Lang, 1995).

Monitoring Motivation

Although causal factor levels cannot be measured directly, some estimate of the likely levels of certain causal factors can be made by direct physiological measurement. Blood sugar and hormones can be assayed, for example. Most estimates of motivational state, however, come from behaviour observation, especially where the change in state is rapid. Brain recording and brain chemistry assessment can also be used. However, both sorts of measurement can be misleading because, first, many activities and brain states are common to a wide variety of motivational states; and, second, many causal factor levels may be high without there being any evidence of this from current activity. Additional information about motivational state can be obtained from experiments in which behaviour is interrupted, stimuli are presented and brain activity is artificially modified.

There is a large literature on a wide variety of animals in which sequences of activities are studied while various attempts are made to manipulate motivational state. It is clear from such work that changes from one behaviour to another may depend much more on some causal factors than on others. Sometimes, a particular causal factor has much urgency, for example when a predator is suddenly detected, so the fact that this input overrides any others in determining behaviour is not surprising because of the survival advantage. Experimental studies have shown that activities such as feeding and courtship sometimes seem to have a higher priority than other activities such as grooming or nest repair. Hence, it seems that the input to the decision-making centre must include a weighting that evaluates the importance and urgency of that input.

These importance or urgency ratings should therefore be regarded as causal factors. They will certainly change according to the conditions of the individual's life. For example, grooming could be very important if it is the eyes that require grooming and the situation is dangerous.

Motivation can be investigated by experiments in which an animal is trained to carry out an operant response, and the amount of work that it will do is related to particular positive or negative reinforcers (see Chapter 3). Using such experiments, an animal can be asked about its own criteria of what is important at that time. Animals will work for food, water and comfortable physical conditions. They will also work for access to social companions, for opportunities to manipulate bedding material and for certain kinds of novel stimulation. In some situations it is clear that positive reinforcers can be behaviours as well as objects or physical changes. For example, young calves try hard to suck at teats as well as to obtain milk, even after milk has been drunk to satiation: the sucking behaviour itself is a reinforcer. Herrnstein (1977) suggests that stalking and capturing prey is a reinforcer for a predator, this reinforcer being additional to the reinforcers resulting from the ingestion of the prey. Ideas about what constitutes a reinforcer are important for our general understanding of animal behaviour and, since absence of important positive reinforcers can cause difficulties for an animal, for appreciation of welfare (Hogan and Roper, 1978; Dickinson and Balleine, 2002).

Motivational Control Systems

Body state is maintained within a tolerable range of temperature, osmotic state, nutrient level, etc., by a set of homeostatic control systems. The concept of the tolerable range as that within which the animal seeks to maintain itself is fundamental in biology. It implies that there are tolerable ranges outside which remedial action is taken. Some of the regulatory actions are physiological, such as sweating or changing blood vessel dilation, but many are behavioural. Some of the variations in state can easily be described in terms of body physics and chemistry, as in temperature or blood sodium level. Others, which might be just as important to the animal, are not easy to describe in that way: for example, level of total sensory input or degree of reassurance from parental contact by a young animal.

One sort of control mechanism works by negative feedback (see Fig. 4.5). As displacement from an initial state within the tolerable range occurs, this change is monitored and, as soon as the edge of the tolerable range is reached, some corrective action is taken. Another form of control that is similar but needs no sensory feedback is that where a body variable such as blood glucose is automatically prevented from passing a certain level by a mechanism such as storage.

The major alternative to negative feedback is feedforward control (see Fig. 4.6), in which a displacement from the tolerable range is predicted and a correction is made before the state changes. As a consequence of many detailed studies of body biochemistry and physiology, the importance of negative feedback control has been apparent for some time. A decline or an increase in causal factors can often result in a sufficient change in motivational state for a corrective behaviour or physiological response to be made. Research on animal behaviour is providing more and more evidence of feedforward control in operation (e.g. Tolkamp *et al.*, 2012). Animals use a variety of cues and previous experience to predict that the state will depart from the tolerable range, and so they act in a way that prevents this from happening. When feedforward control is very efficient an observer may be unaware that any change in state would have occurred, because the action compensates for it exactly.

The realization that animals often predict likely changes in body temperature, body nutrient levels or social actions has resulted in our view of animals changing to one of cognitive beings aware of the complexities of their environment. Animals are information-processing systems that utilize information about how their environment is in relation to how it should be. The idea that animals have a 'should-be value' or Sollwert for each important aspect of their environment and that they compare this with an 'actual value' or Istwert was presented by Wiepkema (1985, 1987). The Sollwert is the animal's neural construct of the tolerable range. Related to this is the animal's expectation of what input it will receive when it performs a certain action. It is essential for simple movement control for animals to have a model of the expected input following actions with which they can compare actual input (Broom, 1981). With more complex actions, too, the animal is continually predicting changes in input and comparing actual and expected input.

Many responses to perceived environmental changes are not related just to that change, but also involve predictions about what will happen next. Studies of animals in learning situations show that they not only associate successive events, but also assess the probability that

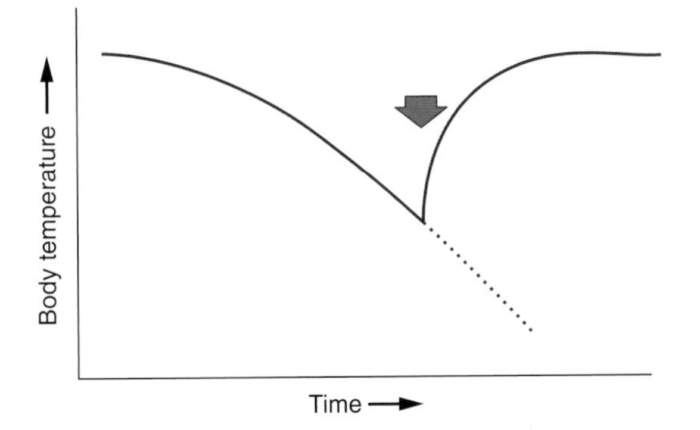

Fig. 4.5. In negative feedback control, after the state of the animal has changed, a correction is made that restores the state to the former condition. In the example shown here, a drop in body temperature is detected and corrective behavioural or physiological action (marked by arrow) is taken. The dotted line shows how the state would change if no correction occurred.

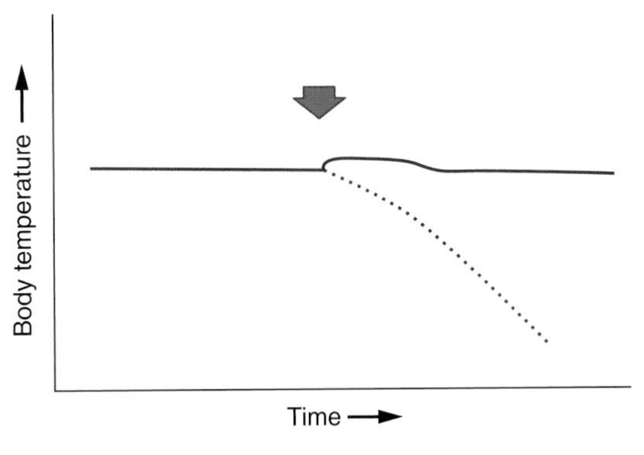

Fig. 4.6. In feedforward control, a change in state is predicted, and corrective action taken before it can occur so that the state changes little from its former condition. In the example shown, a drop in body temperature is predicted and behaviour or physiological action (marked by arrow) is taken. The dotted line shows how the state would change if no correction occurred.

events will occur (Dickinson, 1985; Dickinson and Balleine, 2002; Raby and Clayton, 2009). As Forkman (2002) says: 'Any animal that can predict the future has a tremendous advantage over one that cannot. Predicting and to some extent controlling the future, is really what learning is all about.' Rats running in a maze show clearly their expectation of a food reward at the end if that reward is not present or inadequate. Pigs fed at a particular time of day change their behaviour in the hour before feeding and cattle show responses if their feed gate does not work.

Previous unpleasant experiences also result in expectation, so that a cow that has experienced unpleasant veterinary treatment in a crush may be unwilling to enter it later (Broom, 1987b). Rushen (1986) showed that a sheep which had been roughly or painfully treated at the end of a race was difficult to drive into and along that race on subsequent occasions. Previous experience with stockmen can substantially alter later farm animal behaviour and ease of management by people (Hemsworth and Coleman, 1998).

If animals live in a world that they organize so that many of the events in it are predictable and the state of the animals is closely regulated, then it is logical to ask whether unpredictability is especially aversive to those animals. Work by Overmier *et al.* (1980) showed that rats and dogs show a clear preference for prediction and control over unpredictability and lack of control. Predictable shocks cause fewer ulcers in rats than do unpredictable shocks (Weiss, 1971), and unpredictability in feeding after previous regular feeding leads to increases in adrenal cortical activity (Levine *et al.*, 1972). If aversive events are predictable, animals can prepare for them behaviourally or by brain change. They can also prepare for events that are not aversive. Inability to prepare makes body regulation more difficult. Unpredictability of a wide range of events is hard for animals to cope with and can lead to adverse effects.

One kind of situation where there is no match between expected and actual input leads to frustration. *If the levels of most of the causal factors that promote a behaviour are high enough for the occurrence of the behaviour to be very likely, but because of the absence of a key stimulus or the presence of some physical or social barrier the behaviour cannot occur, the animal is said to be frustrated* (Broom, 1985). For example, Duncan and Wood-Gush (1971, 1972) thwarted hens about to feed by covering their food dish with a transparent perspex cover. The hens showed stereotyped pacing and increased aggression. Feeding is often frustrated by the presence of stronger rivals in group-housing situations. Some frustration must be of trivial importance in the life of the animal, but frustration can be so frequent and involve so fundamental an activity that the fitness of that animal is impaired. For a review of the various circumstances in which frustration can occur see Amsel (1992).

Feelings and Emotion

Some decision making is the result of modification by feelings and emotions. That is to say that feelings and emotions are causal factors that can have a substantial effect on decision making. The more that we know about brain processing in mammals, the more that we find that the limbic system has a modifying or directly causal effect on much behaviour. The limbic system is active when an individual has feelings.

While the words emotion and feelings can sometimes be used interchangeably, feelings are thought of as stemming from the brain and involving sophisticated processing, while emotions can be described more readily in physiological terms. For example, fear is clearly a feeling but there are bodily changes often associated with fear such as increased heart-rate, increased sweating, facial movements and a greater tendency to flee. The term feeling encompasses the emotion.

The meaning of the term feeling is of key importance in this discussion. *A feeling is a brain construct involving at least perceptual awareness which is associated with a life-regulating system, is recognizable by the individual when it recurs and may change behaviour or act as a reinforcer in learning* (Broom, 1998). In relation to this definition, emotions are considered to be similar but physiologically describable (Broom, 2007). Rolls (1999) considered feelings to be the subjective consequences of emotions involving consciousness or awareness. If the concepts of emotions and feelings are so closely overlapping, perhaps emotion should be defined referring to feelings, i.e. *an emotion is a physiologically describable component of a feeling characterized by electrical and neurochemical activity in particular regions of the brain, autonomic nervous system activity, hormone release and peripheral consequences including behaviour.*

During emotions there are physiological changes in the body, associated with brain activity, and often also with behaviour modification. Heart-rate variability is reduced during negative and increased during positive emotional changes. This is a result of sympathetic nervous system–adrenal–medullary activity. Several different negative emotions and feelings, such as pain, are associated with increases in adrenal cortex production of cortisol or corticosterone, depending on the species. The assessment of pain is described in Chapter 6. *Pain is an aversive sensation and feeling associated with actual or potential tissue damage.* Negative experiences may lead to immunosuppression and positive emotions may occur at the same time as increases in oxytocin production. Positive experiences may occur at the same time as increases in cardiac vagal tone. Gygax *et al.* (2013) made a comparison of the responses of goats to two situations expected to produce positive or negative feelings. The goats were repeatedly confronted with a food bowl, associated with reward, and a covered food bowl, expected to be frustrating. The covered food bowl was associated with time spent away from the bowl, increased locomotion, little indication of autonomic changes and increased prefrontal cortical activity. This last was measured by functional near-infrared spectroscopy which indicates haemodynamic changes in the brain. When feed was available, the goats reduced locomotion, stayed near the food bowl, showed sympathetically mediated arousal indicating anticipation and on some occasions showed cortical activity in the left hemisphere. This correlational evidence suggests that there are negative and positive feelings in these two situations.

Feelings are associated with some situations that occur in many species of animals, for example a mother responding to her offspring (Chapter 19). Mothers defend and care for their young and the associated brain mechanisms can result in affection, empathy and distress. For example, Edgar *et al.* (2011) described empathic responses of mother hens when their chicks were somewhat distressed. However, hens did not show such responses when other adults were distressed (Edgar, 2012). Feelings that are discussed in other chapters of this book include pain, fear, anxiety, various forms of pleasure, social affection, guilt, anger and rage. In each case, evidence for the feeling is from observations of behaviour, physiological change or change in the brain. Such data are more reliable than the self-report that is commonly accepted as evidence for feelings in humans (Broom, 2014a). However, much care must be taken in the interpretation of the information as some physiological and behavioural changes are associated with more than one feeling. For example, the behaviour of a dog that has done something that the owner told it not to do is often interpreted as indicating a feeling of guilt but it may be entirely a response to the current behaviour of the owner (Horowitz, 2009). Torres-Pereira and Broom (2010) observed pet dogs that had been told by their owners not to take a food object. When the owner then left the room, some dogs ate the food object. While the owner was absent, these dogs had a higher heart-rate than dogs that obeyed. When the owner returned after 30 s and behaved exactly the same whatever the dog had done, carefully avoiding looking at the dog, there was a difference in behaviour between dogs that had taken the food and dogs that had not. The existence of the concept of wrongness, which cannot be separated from a prediction of negative consequences, is implied by these results.

Another term, often used to describe an affective state, is mood. Factors affecting moods tend to persist so moods endure over time (Nettle and Bateson, 2012). Rolls (2005) says that a mood state could be produced as a result of the initiation of an emotion but that the mood is not directed towards an object, as the emotion is. However, a mood could also be produced without the emotion, or at least without continuing emotion, and continues for longer than an emotion. Feelings involve some complex analysis but a mood might occur with lower level brain activity such as a high degree of perceptual processing. Some moods are referred to using words that imply that a feeling is involved, for example anxious, fearful, happy or contented. A second category may or may not involve feelings, e.g. tranquil or excited, while a third category would not normally include feelings, e.g. optimistic, pessimistic or analytical (Broom, 2014a). The measures of mood are a combination of behaviour and physiology. For example, Reefmann *et al.* (2012) kept sheep for 3 weeks in a barren, unpredictable environment or in an enriched, calm environment. When tested in more positive or more negative situations, the 3 weeks of negative experience resulted in a more negative mood in the negative test, indicated by more frequent ear-position changes, more time with ears in asymmetric positions, shorter inter-beat intervals in heart recording and higher breathing rates.

Many studies show how activity in particular regions of the brains of mammals, for example the amygdala, prefrontal and orbitofrontal cortex, anterior cingulate, insula, nucleus accumbens, ventral tegmental area and peri-aqueductal grey, is associated with emotion and feelings (Panksepp, 1998; Rolls, 2005; Murray, 2007). Cognition in relation to emotion is discussed by Rolls (2005) and Broom (2014a). Most changes in the concentrations of hormones that occur in humans during emotional responses also occur in other mammals, birds, fish and, in some instances, invertebrate animals. However, there is variation in the region of the brain that is involved in the control of the hormone release.

Motivation, emotion and cognitive bias

As discussed in Chapter 3 and by Boissy and Lee (2014) and Broom (2014b), emotions can influence cognition and cognition can induce emotions. The concept of cognitive bias is more to do with motivation and emotion than cognition and the results of cognitive bias studies are used in welfare assessment. When a situation is evaluated, there will often be some degree of ambiguity in the information available. In some circumstances the same sensory input could indicate that there is a likely consequence that could be either positive or negative. As Mendl *et al.* (2009) put it, should an individual interpret a rustle in the grass as danger or food? This depends on the overall set of information available. In a series of studies, Mendl and collaborators have investigated the possibility that an animal's interpretation of an ambiguous situation may be altered by its emotional state (Mendl *et al.*, 2010a). This has been called cognitive bias or cognitive affective bias. Paul *et al.* (2005) and Mendl *et al.* (2009) used cognitive bias to refer to 'the influence of affect on a range of cognitive processes including attention, memory and judgement'. The bias referred to here is not a bias in the cognitive mechanism but rather in the direction of the decision reached (Broom, 2014a). While judgement is necessarily cognitive, the term has also been used for effects on attention, motivation and memory that may not be cognitive. Hence the definition used here is: *cognitive bias is the influence of affect on a range of processes, some of which are cognitive, e.g. judgement.*

Harding *et al.* (2004) presented rats with one tone followed by positive consequences and another followed by negative consequences and then tested them with an intermediate tone. Rats which had been living in a relatively rich environment were more likely to respond to an ambiguous tone as positive. Similarly, Burman *et al.* (2008a) found that rats from a better environment treated an ambiguous food bowl position as positive; and Matheson *et al.* (2008) showed that starlings living in a more complex environment were more likely to treat as positive a stimulus of duration intermediate between the positive and negative, than starlings living in a more impoverished environment. Mendl *et al.* (2010b) found that rescue shelter dogs with a higher separation anxiety score were more likely to react to an ambiguous position as negative but Titulaer *et al.* (2013) did not find a difference in judgement bias between dogs that had been living in rehoming kennels for more than 6 months rather than 1–12 weeks. Doyle *et al.* (2011) trained sheep that there was a positive and a negative bucket position. If they were shown a bucket in an ambiguous position between the positive and negative positions, they did not go to it after being stressed. Studies using such a paradigm, with three-quarters of the results indicating cognitive bias and some that do not, are reviewed by Mendl *et al.* (2009).

A negative cognitive bias may be an indicator of depression. Indeed, the likelihood that depressed people will judge ambiguous situations as negative encouraged the use of the experimental paradigm. Hymel and Sufka (2012) measured approach or avoidance by young domestic chicks to a picture that was manipulated to range from a chick, more likely to be approached, to an owl, less likely to be approached. Chicks previously isolated for 5 min, intended to make them anxious, or 60 min, intended to make them depressed, were less likely to approach intermediate images. After the 5 min of isolation, the anxiolytic drug clonidine produced some sedation and made the chicks less likely to avoid intermediate images. After 60 min of isolation, the anti-depressive drug imipramine increased approach to intermediate images. While such results are not easy to interpret, the impacts of the drugs suggests that they are altering something that is varying in the animals showing cognitive bias.

The existence of cognitive bias may well give information about the emotional state of the animals and about their welfare but in order to understand cognitive bias there is a need to consider the strategies adopted by individuals in life as a whole and in the test situation. What are the possible strategies, which of these are shown, and what will be the consequences of showing one or other strategy? Once this is ascertained, the probability that the supposed optimistic or pessimistic response will be shown can be calculated and compared with the data obtained (Broom, 2010). For further discussion about how to interpret cognitive bias results see Mendl *et al.* (2009) and Broom (2014a). Cognitive bias is a potentially valuable indicator of affect and of welfare. However, it has not yet been demonstrated that either the affect or the welfare will be reliably indicated by cognitive bias studies alone. A combination of studies is needed to increase the accuracy with which cognitive bias reveals feelings or allows assessment of welfare.

Needs

The importance for understanding animal welfare of considering the needs of animals, as proposed by Thorpe (1965), has been described in Chapter 1. *A need is a requirement, which is part of the basic biology of an animal, to obtain a particular resource or respond to a particular environmental or bodily stimulus.* The need itself is in the brain. It allows effective functioning of the animal. It may be fulfilled by physiology or behaviour but the need itself is not physiological or behavioural (Toates and Jensen, 1991; Broom, 1996, 2008). Needs can be identified by studies of motivation and by assessing the welfare of individuals whose needs are not satisfied (Hughes and Duncan, 1988a,b; M. Dawkins, 1990; Broom and Johnson, 2000). The studies of motivation involve starting from either a known resource, or a known action by the animal, and finding out what are the causal factors influencing decision making. The animal's need is a consequence of the motivational state that arises in given circumstances. This will depend on the biological functioning of such an animal. Most of the needs of animals of a certain species will be the same but there will be some inter-individual variation depending upon the experience during the development of each individual. However, the means of fulfilling some needs will vary greatly depending on the environment at the time and the capabilities and experience of the individual. Individuals often develop one or more particular strategies for achieving objectives.

Some needs are for particular resources, such as water or heat, but control systems have evolved in animals in such a way that the means of obtaining a particular objective have become important to the individual animal (Toates and Jensen, 1991; Broom, 1996). The animal may need to perform a certain behaviour and may be seriously affected if unable to carry out the activity, even in the presence of the ultimate objective of the activity. For example, rats and ostriches will work, in the sense of carrying out actions that result in food acquisition, even in the presence of food. In the same way, pigs need to root in soil or some similar substratum (Hutson, 1989), hens need to dust-bathe (Vestergaard, 1980) and both of these species need to build a nest before giving birth or laying eggs (Brantas, 1980; Arey, 1992). As explained further in Chapter 6, sophisticated strength of preference studies depend upon the use of operant and other techniques that exploit the abilities of animals to learn to carry out new procedures. The terminology used in motivational strength estimation is that developed for micro-economics (Matthews and Ladewig, 1994; D. Fraser and Matthews, 1997; Kirkden *et al.*, 2003).

The idea of providing for 'the five freedoms' was first suggested in the Brambell Report in 1965 but it was not quite in line with Thorpe's concept of needs. The list of freedoms provides a general guideline for non-specialists. Animals have many needs and these have been investigated for many species. Hence the rather general idea of freedoms is now replaced by the more scientific concept of needs. This is the starting point for reviews of the welfare of a species. A list of needs has been the first step in Council of Europe recommendations and EU scientific reports on animal welfare for over 20 years. The freedoms are not precise enough to be used as a basis for assessment of the welfare of a particular species or group of closely related species.

The 'five freedoms' are illogical in places if the wording is considered exactly. For example, 'freedom from pain, injury and disease' for domestic animals is a desirable state and those responsible for animals should aim at it but it is not achievable. Any animal might slip and fall or collide with something and be caused pain and injury. Freedom from disease cannot be achieved because pathogens may result in disease that could not have been prevented. Similarly, 'freedom from fear and distress' could not be achieved in some individuals because they have to encounter humans and are disturbed by such contact. Also, 'freedom from hunger and thirst' would not be possible unless food and water were available at all times. 'Freedom to express normal behaviour' would include giving the animals the possibility to show aggression to others and other antisocial behaviours that are normal for the members of the species. Just as unlimited freedom is harmful for humans (Broom, 2003, 2014a), the freedom for the animals should have social limitations. A consequence of these logical inconsistencies is that some of those who use animals may say that they follow the five freedoms approach, knowing that they are not fully achievable and, as a consequence, failing to follow the general guidelines that the five freedoms list.

The four welfare principles and 12 criteria proposed, as a development of the five freedoms concept, by the Welfare Quality project (Blokhuis *et al.*, 2010) are a more useful general guideline. They have some of the same disadvantages, for example they also include: 'no disease', 'no injuries', 'expressing normal behaviour', although normal behaviour is qualified by 'non-harmful'. A limitation of these 12 criteria is that they are aimed at certain, housed land animals so some wording is difficult to apply to extensively kept or aquatic animals, e.g. 'good housing' and 'comfort around resting'. A general guide such as this should be used only to help in general planning. It should not be used when preparing guidelines for a particular species. For any animal, the best approach to a scheme for ensuring good welfare is to start with details of the needs of animals of that species, as determined from knowledge of their biological functioning and scientific studies of their preferences. Laws and guidelines for animal care, whose aim is to ensure good welfare, should refer to needs rather than to freedoms.

Brain Structure and Mechanisms in Relation to Motivation, Emotion and Other Behaviour Control

Aspects of brain structure and function involved in behaviour control are listed briefly here. For a review of neuroanatomy see Dyce *et al.* (1996) and, for nervous system structure in relation to its potential malfunction, see Bagley (2005). Some of the central nervous system (CNS) involvement in the bull's threat display is shown in Fig. 4.7. The vertebrate CNS consists of the following principal items:

- A spinal cord and brainstem.
- A highly organized cerebellum connected with the spinal cord.
- Basal ganglia and their links with the brainstem.
- A diencephalon, including the hypothalamus, linked to the pituitary and other endocrine glands.

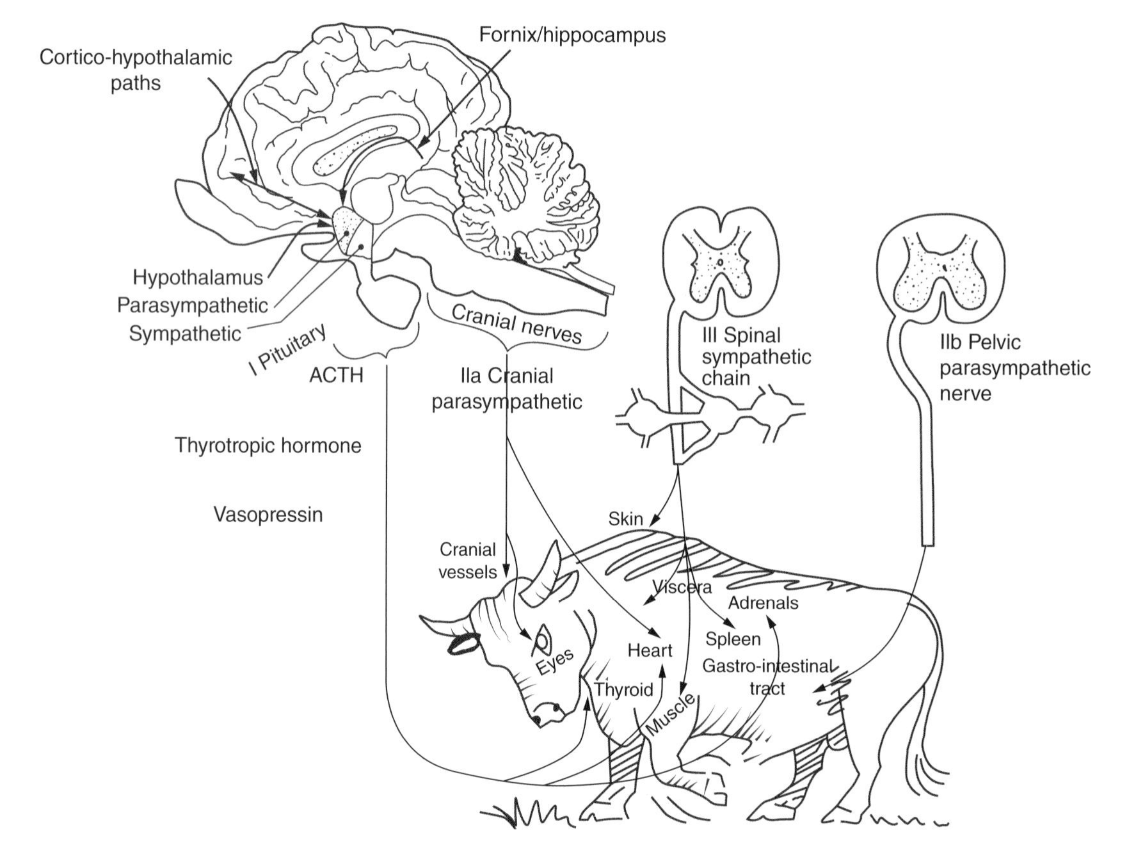

Fig. 4.7. Diagram showing some of the central nervous system involved in the threat display of a bull.

- A limbic system consisting of parts of several brain structures such as hypothalamus, hippocampus, amygdala, mammillary bodies, thalamus and cortex.
- A thalamo-cortical system mediating both specific and non-specific sensory influences.

The functions of the reticular formation, an important part of the brain stem, include the production of general arousal in the CNS. States of the CNS are regulated: for example, sleep, wakefulness and degrees of alertness and inattentiveness.

Three large subcortical nuclear groups – the caudate nucleus, the putamen and the pallidum – are collectively called the basal ganglia. Together, they account for 5% of the brain mass. The basal ganglia and several associated subthalamic and midbrain structures are referred to as the extra-pyramidal system. These nuclei participate in the control of movements, together with the cerebellum, the corticospinal system and other descending motor systems. Although part of the motor system, the basal ganglia do not project directly to the spinal cord.

In the hypothalamus, patterns of nervous activities become integrated so as to establish the adaptive reactions of the animal. Much of the influence of hypothalamic activity is directed at the production of hormones in the subjacent pituitary gland. The pituitary is the principal endocrine gland in the body and its hormonal production is all-important in the maintenance of the bulk of the body's activities, including behaviour.

The limbic system is a determinant and integrator of strategic and tactical functions, is involved in emotional behaviour and is a central representative of the autonomic system. It includes portions of the frontal lobe of the cortex, temporal lobe, thalamus and hypothalamus, together with certain midbrain parts (see Fig. 4.8). The limbic system contains neural centres, such as the amygdala, that control aggressive behaviour in its various forms. The limbic system, or the paleo-mammalian brain, represents a device for helping to cope with the environment. Parts of the limbic system are concerned with activities related to food and sex; others are related to emotions and feelings and still others combine messages from the external world. On the basis of behavioural analysis, its regulation seems often to be inhibitory in nature. There have been too many studies of brain function during cognition, emotional responses and decision making to review them here. Some studies make it clear that there is activity in many parts of the brain during some responses. For example, Muehlemann *et al.* (2011) measured oxy-haemoglobin concentration in the cerebral hemispheres of sheep (i.e. in a large brain area) using functional near infra-red spectroscopy, and reported different changes during a positive affective state associated with grooming and a negative affective state resulting from spending time in a barren environment.

Within the brain and spinal cord, many neurotransmitters are involved in the various functions. Acetylcholine (ACh) is the transmitter used by the motor neurons of the spinal cord and operates, therefore, at all nerve–skeletal muscle junctions in vertebrates. ACh is diffusely localized throughout the brain,

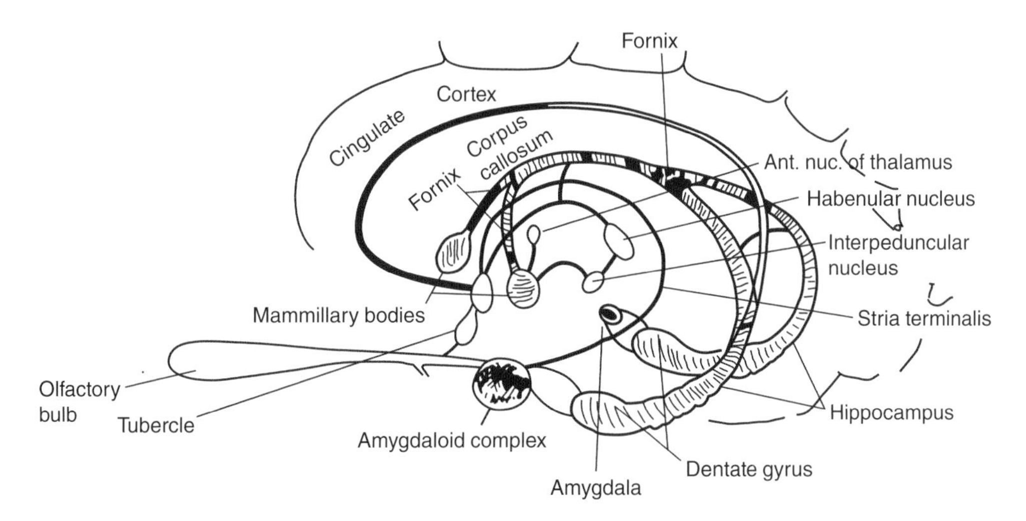

Fig. 4.8. Diagram showing the parts of the limbic system in mammals (Broom, 2007).

but is highly concentrated in neurons of the basal ganglia. It is important in the autonomic nervous system, being the transmitter for the parasympathetic neurons. In studies of decision making in mammals, much of which has focused on primates (Damasio *et al.*, 1996), somatic states can influence decisions via, for example, pathways to the striatum and anterior cingulate, serotoninergic pathways to the anterior cingulate and working memory in the prefrontal cortex (Bechara *et al.*, 2006). Dopamine, noradrenaline (norepinephrine) and serotonin (5-hydroxytryptamine) are important transmitter amines. In the CNS, noradrenaline-containing nerve cell bodies are prominent in the *locus coeruleus*, a nucleus of the brain stem concerned with arousal. These neurons project diffusely throughout the cortex, cerebellum and spinal cord. In the peripheral nervous system, noradrenaline is the transmitter in the post-ganglionic neurons and is thus the transmitter for the sympathetic nervous system.

Dopamine-containing cells are located in, for example, the *substantia nigra*, where the cells project to the striatum; the midbrain, where they project to the limbic cortex; and the hypothalamus, where they project to the pituitary stalk.

Serotonergic cell bodies are found in the midline of the brain stem and send fibres throughout the brain and spinal cord. Histamine is concentrated in the hypothalamus. Noradrenaline and dopamine are catecholamines and serotonin is an indoleamine. Catecholamines, indoleamines and histamine are all referred to as 'biogenic amines' and have considerable effect on behaviour.

Several amino acids are transmitter substances. Glycine and glutamate are two of the 20 common amino acids that are incorporated into the proteins of all cells; glutamate and 7-aminobutyric acid (GABA) also serve as substrates of intermediary metabolism. Glutamate is a transmitter in the cerebellum and the spinal cord, while glycine is an inhibitory transmitter in spinal cord interneurons. GABA is present in neurons in the basal ganglia which project to the *substantia nigra*; cells of the cerebellum are GABA-minergic, as are certain inhibitory interneurons in the spinal cord.

Many peptides have been found to be localized in neurons (see Table 4.1) and to be pharmacologically very active, causing inhibition, excitation or both. Some of these peptides were previously identified as hormones with known targets outside the brain, for example angiotensin – or as products of neurosecretion, for example oxytocin, vasopressin, somatostatin, luteinizing hormone (LH) and thyrotrophin-releasing hormone (TRH). In addition to being hormones in some tissues, these peptides may act as neurotransmitters in other tissues. These neuroactive peptides are localized in regions of the brain thought to be involved in the perception of pain, pleasure and emotion. Peptides with opiate-like actions include the endorphins and enkephalins. These opioid peptides are involved in a variety of functions, including the modulation of pain. An animal deprived of an important resource is seldom just awaiting an appropriate stimulus so that it can respond. Where the animal has the capacity for cognition, many of α-, β- and γ-endorphins, together with adrenocorticotrophic hormone (ACTH), melanocyte-stimulating hormone (MSH) and the enkephalins, are derived from a large peptide with 91 amino acids, pro-opiomelanocortin (POMC). Particularly active is β-endorphin, which is synthesized in the hypothalamus as well as in the pituitary. Two pentapeptide enkephalins are metenkephalin and leu-enkephalin. The enkephalins are also derived from the precursor molecule pre-enkephalin and are synthesized in ribosomes and transported within secretory vesicles to nerve terminals. Unlike β-endorphin, the enkephalins are widely distributed in the brain. The distribution matches that of the opiate receptors.

Substance P is a peptide concentrated in certain neurons of the dorsal root ganglia, basal ganglia, hypothalamus and cerebral cortex. It has been proposed as the transmitter for sensory fibres involved in mediating pain and is a transmitter involved in the modulation of motor movement. Transportation of transmitters may be fast, e.g. acetylcholine, or slow, e.g. the catecholamines, noradrenaline and dopamine.

Hormones that Affect Behaviour

The nervous system and the endocrine system are adapted for different roles, but contact between them is essential since they are interdependent. The two systems interact through neural secretion and the priming effects of hormones on the brain. Hormone secretion is subject to the influence of many forms of stimulation. Endocrinologists now recognize an elaborate system of interactions between the animal's own activity, the external stimuli that it receives and its internal physiological state. Any of these can alter so as to cause a change in others.

Table 4.1. Hormones and some of their sources (adapted from Schulkin, 1999).

Source	Hormone(s)
Endocrine gland	Some hormones secreted
Anterior pituitary/adenohypophysis	Growth hormone (GH), prolactin, pro-opio-melanocortin (POMC), melanocyte-stimulating hormone (MSH), adrenocorticortrophin (ACTH), β-endorphin, luteinizing hormone (LH), thyroid-stimulating hormone (TSH), follicle-stimulating hormone (FSH)
Posterior intermediate pituitary/ neurohypophysis	Arginine or lysine vasopressin, oxytocin, dynorphin, POMC, enkephalins, β-endorphin
Pineal gland	Melatonin
Thyroid gland	Thyroxine, calcitonin
Parathyroid gland	Parathyroid hormone
Heart	Atrial natriuretic factor
Adrenal cortex	Glucocorticoids, mineralocortocoids, androgens
Adrenal medulla	Adrenaline (= epinephrine), noradrenaline (= norepinephrine)
Kidney	Renin
Skin, kidney	Vitamin D
Liver, lung	Pre-angiotensin/angiotensin
Pancreas	Insulin, glucagon
Stomach, intestines	Cholecystokinin, vasoactive intestinal peptide, bombesin, somatostatin
Ovary	Oestrogen, progesterone
Testis	Testosterone
Macrophages, lymphocytes	Various cytokines

The generation of cyclic adenosine monophosphate (AMP) is the common biochemical action of most hormones. In spite of this common denominator, different target organ cells vary in their abilities to be activated by different hormones. This is a result of qualitative differences in membrane receptor sites in different tissues. Consideration of hormone physiology can be based on the roles and effects of the endocrine system by hormone production sites.

The actions of many hormones are described in this book, largely where they affect behaviour or are affected by behaviour and stimulation resulting from it. They are involved in sexual behaviour, regulation of water and salts, regulation of feeding, biological clocks, parental care, attachment behaviour, fear, stress and pleasure (Schulkin, 1999; Berne and Levy, 2000; Feldman and Nelson, 2004). The sources of some hormones that effect behaviour are shown in Table 4.1.

Corticotrophin-releasing hormone (CRH) leads to ACTH being secreted into the blood from the anterior pituitary (anatomically the adenohypophysis) and being transported via the blood to the adrenal gland, where it initiates the output of glucocorticoids. The most active glucocorticoid is cortisol in cattle, sheep, pigs, cats, dogs and man but, in rodents and poultry, corticosterone is the most active. The role of the hypothalamus includes the secretions of both anterior and posterior pituitary lobes, the production of releasing and inhibiting factors, the operation of feedback mechanisms and the control of rhythmic phenomena and sexual behaviour. Among releasing factors is gonadotrophin-releasing hormone (GRH), which acts on the anterior pituitary gland causing it to produce and release luteinizing hormone (LH) and follicle-stimulating hormone (FSH).

The adrenal medulla is activated from the sympathetic nervous system to produce adrenaline (epinephrine) and noradrenaline (norepinephrine). The effects on the body of adrenaline are complex and various. In principle, this hormone prepares the animal for gross physical activity in situations of emergency. It is particularly responsible for temporary increase in blood pressure in response to stressful stimulation. Sudden increases in adrenaline output are associated with events such as male fighting and maternal protection. Adrenaline also provides the basis for alarm responses in general.

Thyroid-stimulating hormone (TSH) is produced by the anterior pituitary, which is anatomically equivalent to the adenohypophysis. This hormone initiates thyroxine output, which determines metabolic rates, energy provisions and thermal homeostasis. It is also concerned with reproductive behaviour in general, since many reproductive acts require suddenly increased energy output.

The posterior pituitary, which corresponds anatomically to the neurohypophysis, produces the hormones oxytocin and vasopressin. Oxytocin encourages the outflow (let-down) of milk at suckling and also plays a predominant part in parturition. It is produced in various situations associated with pleasure. Vasopressin is associated with ACTH and its output and also functions as a neurotransmitter. Vasopressin also facilitates social recognition via its action on the mitral cells of the olfactory bulb in rats; social recognition is blocked when vasopressin action is prevented by injection of a vasopressin receptor agonist (Tobin *et al.*, 2010).

Further Reading: Hogan, J.A. (2005) Motivation. In: Bolhuis, J.J. and Giraldeau, L.A. (eds) *The Behaviour of Animals*. Blackwell, Malden, Massachusetts, pp. 41–70.

Evolution and Optimality

Introduction

The mechanisms that control behaviour have evolved by natural selection like any other characteristic of living organisms. Domestic animals still have the systems of their ancestors, for defence against predation for example, and these have important effects on a wide range of their activities. Domestication has changed these animals anatomically and, to some extent, physiologically, but it has changed a relatively small proportion of their behaviour. Evolution has continued during domestication; new environments and active selection by man have been important factors.

Many evolutionary adaptations of animals have not occurred because of active human selection during breeding. People selecting animals for breeding have looked for certain qualities, but very many other qualities have changed during the generations of association with humans and some of these, such as reduced maternal ability in some species, are not beneficial to those who wish to use the animals. The species that were domesticated had to be suitable for assisting in hunting, providing companionship or providing meat, milk, eggs, wool, etc., but they also had to have certain behavioural characteristics for domestication to be possible. The potential for showing reduced antipredator responses to man, low aggression to man and effective reproduction in captivity were essential if these animals were to be kept for human use. Given this potential in their animals, the early domesticators and all those involved with animal husbandry ever since have had to be able to assess and interpret animal behaviour.

The way in which behaviour control mechanisms must have evolved will be emphasized in many of the chapters in this book, but it is useful to refer first to some general principles concerning the evolution of behaviour.

Variation, Heritability and Selection

There is much variation in behaviour among the individuals of a species, and some of this variation is a consequence of genetic differences (Jensen, 2009). Much of the research that has investigated the genetic aspects of behaviour has been carried out on fruit flies, other insects, fish or birds. Variation among species of ducks in the displays they perform during courtship were described by Lorenz (1941) and used to classify the duck species. Certain displays of ducks are common to several closely related species but absent from others, so it is likely that the mechanism for showing the behaviour has been inherited from a common ancestor.

One study of variation and heritability of behaviour within a species (Scott and Fuller, 1965) compared the behaviour of two breeds of dogs: attempts to restrain Cocker Spaniel puppies resulted in little struggling, but similar attempts to restrain Basenji puppies led to much struggling, avoidance and vocalization. Similarly, Jensen (2009) describes Retriever puppies as being generally more inclined to retrieve, while Border Collie puppies have a strong tendency to herd. Dogs of these different breeds were selected for particular characters. The genes function by interacting with their environment, but the average function following that interaction becomes substantially different in the different breeds. When Scott and Fuller crossed Cocker Spaniels with Basenjis, the F1 hybrids behaved like the Basenji parent and the back-crosses indicated that a single dominant gene influences this behaviour in Basenjis. Other behavioural characteristics are clearly affected by more than one gene.

Behavioural differences among breeds are well known. In farm animals, for example, Le Neindre (1989) showed various consistent differences in behaviour between Salers and Friesian cattle. These differences are important when recommendations about management procedures are being made, as the best method for one breed may not be suitable for another. Some of the variation in behaviour is relevant to welfare; for example, Håkansson *et al.* (2007) compared two strains of jungle-fowl kept in identical conditions and found that they varied in fear behaviour but not in exploratory or social behaviour. A further example is the different amounts of aggression in strains of pigs and the successful incorporation in selection indices of behaviour that does not result in injuries to others (Canario *et al.*, 2012).

To what extent are behavioural characteristics heritable? Heritability is an indication of how much a character or trait in an individual is likely to be passed on to its offspring. Willis (1995) explains that the mean heritability of those characteristics of Labrador Retrievers that results in them being successful guide dogs (Goddard and Beilharz, 1982) is 44%, while the heritability of German Shepherd dogs for retrieving was 19% for sires but 51% for dams. Most characters of dogs showed much lower heritabilities, so more information is now being taken into account in dog breeding. The Best Linear Unbiased Prediction (BLUP) technique takes account of information on the dog itself, its parents, siblings, half-siblings and progeny.

Natural selection acts on animals so as to increase the proportion of some genes in the population at the expense of others. If a gene's action is such that an animal shows a particular kind of display during courtship and this display is more effective at attracting a mate – and hence producing offspring – than that shown if a different gene is present, then the bearers of the first gene will be more common in the next generation. There are many examples of genetic lines with behaviour less successful than the norm; one such example comes from the work of Bentley and Hoy (1972), who studied the songs of the male cricket, *Teleogryllus*. These songs are produced by scraping one wing over another and are a consequence of nerve impulse output from central ganglia. If there is a genetic difference that is such that the song has too many or too few pulses in it, females are much less likely to approach the singing male. Hybrids between two cricket species were produced: the males were intermediate between the parents in normal output and song characteristics and less attractive to females of the parent species.

In all aspects of life, some characteristics will result in more offspring being produced than others. Genetic variation can lead to more or less efficient food-finding, predator avoidance, poison avoidance, etc., and those genes that on average result in better survival and offspring production will become more common in succeeding generations. A gene is a length of deoxyribose nucleic acid (DNA) with its particular arrangement of base pairs. Most DNA is in the nucleus of a cell. The gene codes for a protein that could be structural or could be an enzyme that promotes a key biochemical reaction in a cell. A very large variety of base pair combinations, and hence proteins, is possible. The product may be released from that cell, sometimes influencing hormones or nervous system function, including sensory and analytical ability. Many gene mutations result in gross biochemical changes. For example, mutations leading to improvement in production of cyclic adenosine monophosphates (cyclic AMP) significantly alter cell metabolism while changes in myosin production result in modification of efficiency of muscle contraction.

To say that a gene affects a trait does not mean that it controls it, as most traits are affected by many genes and each depends on interaction with the environment, but one gene failure can mean that a biological feature cannot develop. Genes associated with particular behavioural traits may be inherited very frequently together with some other trait such as an anatomical feature; that is, the gene affecting the behavioural trait and that affecting the anatomical trait are linked. This means that they are close together on a chromosome so relatively unlikely to be separated by crossover during meiosis. Since non-coding sequences of base pairs on chromosomes can be identified, it is now possible to use these quantitative trait loci (QTL) to map genes. QTL analyses have allowed the identification of chromosomal regions influencing behaviour, for example: (i) the tendency of honeybees to sting; (ii) the cyclicity of activity in mice; (iii) preference for alcohol in mice; and (iv) hypersensitivity in rats (Jensen, 2009). Genetic analysis can improve understanding of practical problems. For example, genes have been found in pigs that link adrenal weight, cortisol concentration in blood and aggressive behaviour (Muráni *et al.*, 2010). Animal welfare problems may arise in farmed animals if animals selected for particular characteristics in one environment are then kept in another environment where the gene–environment interaction results in a less well-adapted phenotype (Lawrence and Wall, 2014).

Another technique that provides some information about the genetic basis of behaviour is the production of strains of animals in which genes are 'knocked out' or prevented from expression. 'Knock-out' animals, which are normal except that they lack a particular gene, can be studied to help in elucidating the function of that gene. For example, Crawley (1999) showed that mice lacking a gene for oxytocin production, and which as a consequence failed to eject milk, also showed reduced aggressive behaviour. A gene will survive and spread if its effects promote that survival and spreading.

Sometimes behaviour can affect the survival of other, related individuals, and it is the overall spread of the gene, in whatever individual, that is important. This point, first made clearly by Hamilton (1964a, 1964b) and followed up by Wilson (1975), R. Dawkins (1976, 1986) and Maynard-Smith (1982), has explained how natural selection has led to many aspects of social and other behaviour. A behaviour that promotes the survival of a close relative, including brothers, nieces, etc., as well as offspring, can be selected for in that genes which promote it can spread in the population.

Since it is easier to consider individuals bearing genes than the genes themselves, Hamilton (1964a) introduced the term 'inclusive fitness' to refer to gene frequency in terms of the effects of that gene on individuals. Some genes affect the survival of the bearer only, so individual fitness refers to the number of offspring of that individual that themselves survive to breed. Other genes affect relatives, so the inclusive fitness must take account of this. Close relatives, such as offspring and siblings whose coefficient of relatedness is 0.5, must count for more than distant relatives, such as cousins, for which the coefficient in this calculation is 0.125. However, as Grafen (1984) points out, when considering the effects of a particular gene that results in helping relatives, only those individuals that are actually helped should be included in the calculation of inclusive fitness. In practice, the number of offspring that survive to adulthood is generally the best estimate of inclusive fitness.

All evolutionary changes can be explained in terms of gene survival. The idea that a characteristic might be present in an individual solely for 'the good of the species' is now shown to be incorrect. The problems of how social behaviour might have evolved are explicable using Hamilton's ideas, and group selection ideas are unnecessary.

Ideas About Optimality and Efficiency

Animals are likely to have mechanisms for the optimal allocation of time and energy expenditures (MacArthur and Pianka 1966). Early attempts to assess what was optimal involved measuring energy usage. For example, for feeding behaviour this would be the energy obtained from food or utilized during attempts to obtain food. Later studies have made it clear that energy measurements are relevant in certain circumstances but not in others. Energetically efficient food acquisition would be of no use if the individual concerned was then much more likely to be eaten by a predator or much less likely ever to obtain a mate. Where optimal refers to the whole life of an animal, it should be measured in terms of the fitness of the animal and, for behaviours that affect relatives, in inclusive fitness. For behaviours such as foraging for food (see Chapter 8), energetic efficiency is of particular interest provided that it is remembered that being able to achieve a good energy balance is only one of the things that an animal has to do.

As emphasized in Chapter 4, at all times individual animals have to decide on the allocation of time and energy to functional systems. For example, they decide to try to obtain water and then decide how to do so. The initiation of behaviour that serves a new function will also involve the termination of a previous behaviour, even if this is only resting. A cow has to decide when to stop tending her newborn calf, by licking it and staying near it, and go off to find a patch of pasture and graze. Genes that increased the chances that a cow would leave her calf too early, before adequate licking and a meal of colostrum, or too late, so that the cow lost weight and could not lactate adequately, would be less likely to survive in the population than those that facilitated an accurate assessment of biological priorities.

Another kind of decision is whether to forage for food in a risky place where there is much food, or in a much less risky place where there is less food. For an animal that behaves in a rational way, the decision should depend upon the actual risk involved and the advantages to survival of obtaining more food. Again, a gene that promotes good decision making in this situation should survive better in the population than one that leads to poor decisions and, as a consequence, motivational systems evolve (Broom, 1981). There will, of course, be much individual variation in decision making, and every individual will depend upon its own experience affecting the development of its motivational systems.

The Evolution of Social Behaviour

Animals free to move about are almost always clumped rather than spaced out when their distribution is compared with random spacing. Sometimes, clumping is a consequence of individuals choosing the same resting or living place. However, it is often the case that one individual remains with another because it chooses to be near it and not just because of the place where it is. Once associated, animals of social species show sophisticated interactions and usually establish a complex social structure (see Chapter 14). Most domestic animals show social behaviour, but how might this have evolved? There are disadvantages associated with being close to others with the same requirements. As a consequence of this there might be competition for food, resting places or a mate, and predators might be attracted by an aggregation of animals. Presumably any advantages outweigh these disadvantages, otherwise social behaviour would not have evolved in so many species. The arguments summarized here are discussed in detail by Broom (1981, 2003). Individuals might benefit from being in a group in that their local environment is modified by the others of their own species. Small grazing animals such as rabbits and prairie dogs have difficulty feeding in long grass but can graze readily in areas kept short by others. Termites, ants and bees can collectively change their physical environment by building nests, and social animals of many species can huddle together to reduce heat loss.

Food finding can be facilitated by watching others, and this may be very important to individuals, especially when food is scarce. A hungry individual bird in a roost or a mammal in a resting place may be able to find food by following others more knowledgeable than itself. This would be a major advantage at a time when an isolated individual might find little or no food. Feeding methods can be learned by watching others, so young birds derive advantage from flocking with others who know more about how to feed efficiently, and older animals may benefit from others if new foods become available. Once found, food may be acquired more readily if others are present. Packs of wolves can catch prey that single animals could not catch, and pelicans synchronizing scoops into the water for fish catch more than do single pelicans.

One cost of foraging in a group is that the individuals may have to share some food (Giraldeau and Caraco, 2000), but it is likely that there is a net benefit of being in a group or otherwise individuals would leave. Groups of animals are less likely to return to a depleted food source. Favre (1975) found that sheep in alpine pastures did not return too early to areas that they had grazed, and this was probably facilitated by the presence of experienced ewes that controlled flock movements. Groups of animals may also be able to defend a food source.

Predator attack is a major selective factor in the evolution of group-living. Individuals can reduce the risk to themselves simply by ensuring that another individual is between themselves and any predator. Animals can hide within a group when no predator is present and can move into the centre when danger threatens (Treves, 2000). Colonially nesting birds do better if they position their nest in the central part of a colony. Living in a group also allows the possibility of responding to alarm signals given by others. For some species, collaboration in defence is possible.

Reproduction may be facilitated by group-living. Mates are much more readily found in a group, but this must be balanced against the necessity to compete for them. For females, males can be more readily tested if they are forced to compete with others before they will be accepted. In some species that live in groups there is collaboration in rearing young, the most extreme examples coming from the social insects.

The relative importance to group-living of the various advantages and disadvantages will vary from species to species. The first step in the origins of social behaviour might have been either aggregation in localities where there was abundant food or good shelter, or when parents and offspring did not separate. The possible sequences of events are illustrated in Fig. 5.1. If aggregation was at a food source, individuals might subsequently stay together in order to reduce predation or to find other food more effectively. Individual offspring benefit in various ways by staying longer with the parents and, in some species, the parents might tolerate this if the presence of the older offspring increases the survival chances of the next set of offspring.

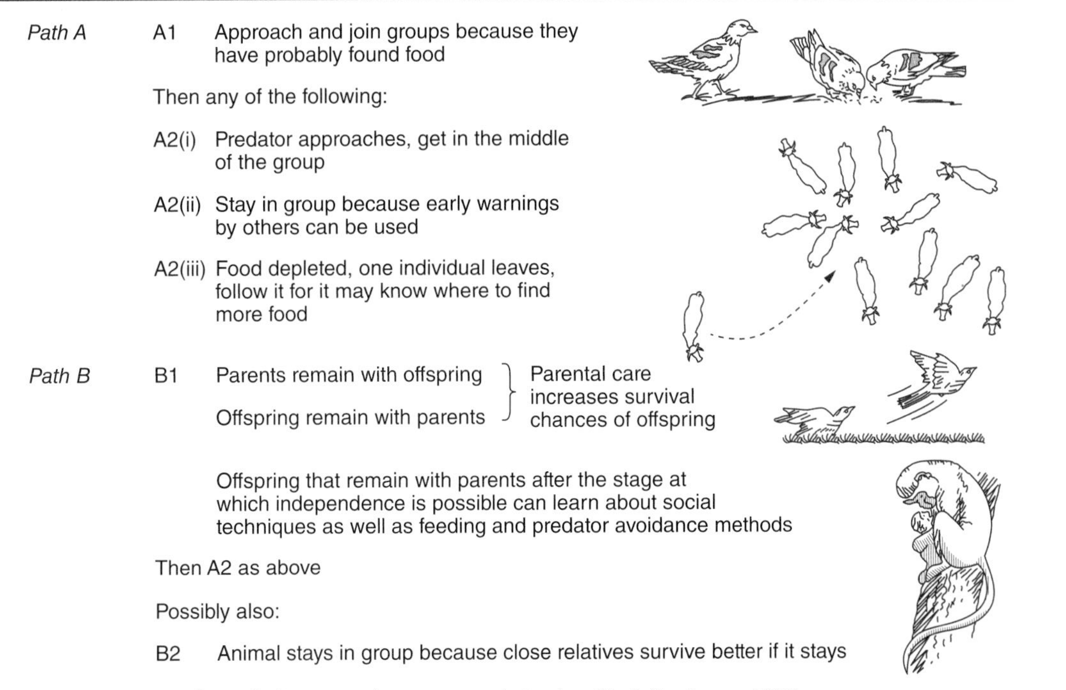

Fig. 5.1. Possible origins of social behaviour and steps in its evolution (modified after Broom, 1981).

The possibility that altruism might be shown to relatives other than offspring or parents is explicable following Hamilton's (1964a) and R. Dawkins' (1976) arguments that selection acts on the replicators and that a gene that promotes a kin-helping action could survive if enough kin bearing the same gene are helped. Altruism can also be directed at individuals that are not relatives. *An altruistic act by an individual is one that involves some cost to that individual, in terms of reduced fitness, but increases the fitness of one or more other individuals.* Trivers (1985) said: 'There can hardly be any doubt that reciprocal altruism has been an important force in human evolution.' *Reciprocal altruism occurs when an altruistic act by A directed towards B is followed by some equivalent act by B directed towards A or by an act directed towards A whose occurrence is made more likely by the presence or behaviour of B.*

There are many examples of reciprocal altruism in human society and some in other species. Packer (1977) reported that two sub-dominant baboons took it in turn to engage the dominant male in fighting or chasing while the other mated with females. There are also many examples of allogrooming, first by animal A on animal B and then the reverse, among primates and ungulates. Benham (1984) found that cows in a suckler spent much time close to another individual and took turns to groom one another. The sharing of food by vampire bats, ravens, wolves, dogs and chimpanzees (e.g. Wilkinson, 1984; Heinrich, 1989; Savage-Rumbaugh and Lewin, 1994) is clearly reciprocal altruism.

In addition to the more obvious kinds of cooperation, the commonest kind of altruistic behaviour in social groups, which is often reciprocated, is to avoid injuring other individuals (Broom, 2003). Great care is usually taken by individuals to avoid collisions. Such avoidance would benefit the avoider as well as the avoided, but efforts are also made not to step on others, to injure them with horns or teeth or to push others out of trees, over cliffs or into places of danger from predators. If any accidental and perhaps avoidable harm to another does occur, this can be followed by changed behaviour on the part of the harmed individual and on the part of the one who has harmed. Harm may be followed by some form of retribution, but either accidental or deliberate harm may also be followed by reconciliation, at least in primates (de Waal, 1996). The individuals that take part in reconciliation may form alliances in order to achieve social and other objectives.

Once altruism occurs and is reciprocated, the possibility of cheating becomes important. A variety of characteristics of individuals, any of which would tend to promote altruistic or moral behaviour, is listed in Box 5.1 (Broom, 2006b). Among these are ways of detecting and responding to individuals who cheat, in that they fail to avoid harming others or make no effort to reciprocate to an individual or contribute in a more general way within a group if benefit is received. *Something is moral if it pertains to right rather than wrong* (Broom, 2003). True morality does not include customs or attitudes to sexual behaviour stemming from mate guarding, etc., except indirectly by effect.

Box 5.1. Characteristics of individuals, any of which would tend to promote altruistic or moral behaviour (after Broom, 2006b).

- Affection for certain types of individuals, perhaps those that are close relatives or group members, or are likely to be, which reduces the chances that harm will be done to them.
- Affection for those same individuals that increases the likelihood of carrying out behaviour that is beneficial to them.
- Ability to recognize individuals that might be beneficiaries or benefactors.
- Ability to remember the actions of others that resulted in benefit to oneself or to others in the group.
- Ability to remember one's own actions that resulted in benefit to another individual.
- Ability to assess risk or benefit of own and other actions and either to compare these or to avoid high risk and try to attain high benefit.
- Ability to detect and evaluate cheating.
- Ability to punish or facilitate the punishment of those that cheat.
- Ability to support a social structure that encourages cooperation and discourages cheating.
- Having a desire to conform.

One key question in relation to morality and its evolution is whether or not genes that promoted co-operative, altruistic behaviour would be out-competed by those that promoted subject benefit at the expense of others. The question of whether a gene that promoted altruistic behaviour would spread in a population of a social species is discussed by Riolo *et al.* (2001) and Broom (2006b). Reciprocal altruism is important in the evolution of morality, but does not comprise all of the biological basis. Some actions that do not harm, or that directly benefit others, are not reciprocal but are directed towards individuals that need help and that have not previously provided benefit to the actor. Such actions may make a contribution to the stability of the social group.

Domestication

Price (1984, 2002) defined domestication as 'that process by which a population of animals becomes adapted to man and to the captive environment by some combination of genetic changes occurring over generations and environmentally induced developmental events recurring during each generation'. However, the developmental events may not be the same during each generation. Wild animals vary in their potential to adapt, some being unable to do so and others having more or less ability to adapt, so there is likely to be much variation in chances of surviving and breeding when animals are first brought into captivity. In some species all will die, while in others a few will survive and breed in captivity. Some studies have been carried out on the effects of genes that help or hinder such adaptation (Price, 2002). A clearer definition of *domestication* is: *the process, occurring over generations, by which a population of animals becomes adapted to man and to the captive environment by some combination of genetic changes and environmentally induced developmental events.*

The animals that have been domesticated – or, one might say, the animals that have allowed themselves to be domesticated or which have domesticated humans – have certain characteristics. The most obvious of these is that they are social species, living in groups irrespective of human action. These are generally the animals with the highest levels of cognitive ability (see Chapter 3). Another characteristic, described in Chapter 19, is that the father plays an unusually small part in parental care. This fact has led to some behavioural scientists having the unbalanced view that most animals with any parental care have little paternal care. In reality, in many animals the father does have a role in caring for the young. Why should it be that the species that have formed the link with humans that we, anthropocentrically, call domestication have maternal behaviour but little paternal behaviour?

Further Reading: Jensen, P. (2009) Behaviour genetics, evolution and domestication. In: Jensen, P. (ed.) *The Ethology of Domestic Animals*, 2nd edn. CAB International, Wallingford, UK, pp. 10–24.

---

6 Welfare Assessment

The Range of Measures

As explained in Chapter 1, coping with the environment, including its negative impacts, is an important fundamental process in the biology of animals. Since welfare concerns the state of the various coping mechanisms, welfare assessment is measurement of that fundamental process.

The general methods for assessing welfare, including measures of poor welfare, are summarized in Box 6.1. and by D. Fraser (2008). Most indicators will help to pinpoint the state of the animal wherever it is on the scale, from very good to very poor welfare. Some measures are most relevant to short-term problems, such as those associated with human handling or a brief period of adverse physical conditions, whereas others are more appropriate to long-term problems (Broom, 1988c, 2014a,b; Broom and Johnson, 2000; Keeling and Jensen, 2009). The information in most of this chapter refers to methods that can be used by animal welfare scientists. Some measures can also be used by those who are inspectors who have to evaluate whether or not standards have been met and laws obeyed. The welfare outcome indicators that such inspectors can use are discussed by Broom (2014b) and mentioned briefly in the last section of this chapter.

Using many of these measures, some indication of the positive or negative feelings of the animals may be obtained. The feelings of another human can never be known with certainty and, similarly, only an estimate of the feelings of individuals of other species can be obtained. In welfare assessment, we use a variety of direct measurements of welfare in the categories listed in Box 6.1, some of which are measures of pain, fear, pleasure and other feelings. Measures of strength of preference are used to understand what is likely to lead to good or poor welfare and to develop better housing and management methods (M. Dawkins, 1983, 1990; Duncan, 1992; Kirkden *et al.*, 2003).

Direct Measures of Poor Welfare: Physiological Measures

Some signs of poor welfare arise from physiological measurements. For instance, increased heart rate, adrenal activity, adrenal activity following challenge with adrenocorticotrophic hormone (ACTH) or reduced immunological response following a challenge can all indicate that welfare is poorer than in individuals not showing such changes. Care must be taken when interpreting such results, as with many other measures described here. In the case of hypothalamic–pituitary–adrenal cortex (HPA) activity, which leads to increased production of cortisol or corticosterone, the response may occur because of increased activity level, or courtship or mating excitement. If the objective is to identify the extent of any emergency response, the context of the HPA response must be taken into account. It is usually obvious whether a response is to potential danger or to a sexual partner and it is usually possible to take account of activity levels. If a treatment results in more walking or running this can be measured, and control glucocorticoid responses with such an activity level can be used to assess the component of the response that is emergency response.

The glucocorticoid cortisol is produced by the HPA axis in primates, Carnivora, Ungulata and many fish and other animals. Corticosterone has the same function, in particular making more energy available from glycogen reserves, in rodents, poultry and other birds. Glucocorticoids have other important functions. There is a daily fluctuation in plasma cortisol concentration and this may well be associated with the facilitation of effective learning via hippocampal function. Hippocampal cells metabolize cortisol (see Fig. 6.1) for, when incubated with cortisol, pig hippocampal cells took it up actively while other control tissues did not. There are cortisol receptors in many parts of the mammalian

Box 6.1. Measures of welfare include the following (modified after Broom, 2000):

- Physiological indicators of pleasure.
- Behavioural indicators of pleasure.
- Extent to which strongly preferred behaviours can be shown.
- Variety of normal behaviours shown or suppressed.
- Extent to which normal physiological processes and anatomical development are possible.
- Extent of behavioural aversion shown.
- Physiological attempts to cope.
- Immunosuppression.
- Disease prevalence.
- Behavioural attempts to cope.
- Behaviour pathology.
- Brain changes.
- Body damage prevalence.
- Reduced ability to grow or breed.
- Cellular changes indicating system failure and ageing.
- Reduced life expectancy.

brain including the hippocampus, amygdala and frontal cortex (Poletto *et al.*, 2003, 2006; Broom and Zanella, 2004). The diurnal rhythm with a morning peak in cortisol concentration was suppressed or modified in women who were severely stressed (Jones *et al.*, 2006; Kivlighan *et al.*, 2008).

Cortisol is produced as a consequence of, first, the production in the hypothalamus of interleukin 1-beta and then corticotrophin-releasing hormone (CRH), also called corticotrophin-releasing factor (CRF), which leads to the release of ACTH from the adenohypophysis, or anterior pituitary. The ACTH travels in the blood to the adrenal gland where the outer part of this gland, the adrenal cortex, produces the glucocorticoid and releases it into the blood. The ACTH is part of a larger peptide molecule called pro-opiomelanocortin (POMC) that breaks up to produce beta-endorphin, dynorphin, met-enkephalin and leu-enkephalin as well as ACTH. When ACTH is injected into a pig, the plasma cortisol concentration increases (see Fig. 6.2). Since cortisol in the form that is not bound to protein in the blood diffuses into saliva in the salivary gland after a short delay, it is also possible to measure cortisol in saliva, although the concentrations are lower than in plasma. Measurements of glucocorticoids in plasma and saliva are of particular use in studies of the welfare of animals during short-term management practices. If a dog or cat is handled or treated by a veterinary

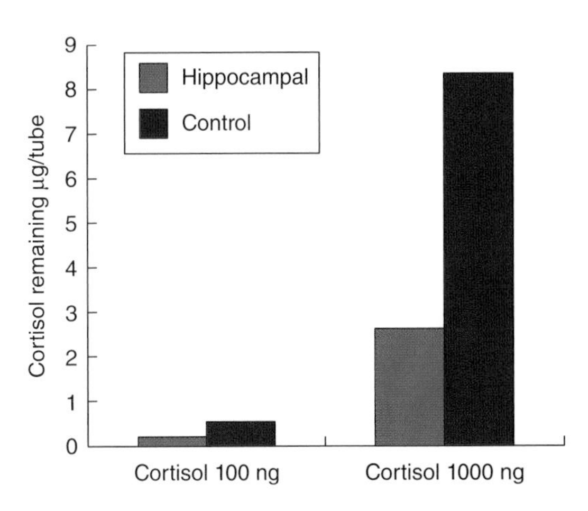

Fig. 6.1. Hippocampal cells and other (control) mammalian tissue cells were compared when they were put in a medium containing 100 or 1000 ng of cortisol. The hippocampal cells actively took up cortisol so less remained (from Poletto *et al.*, 2003).

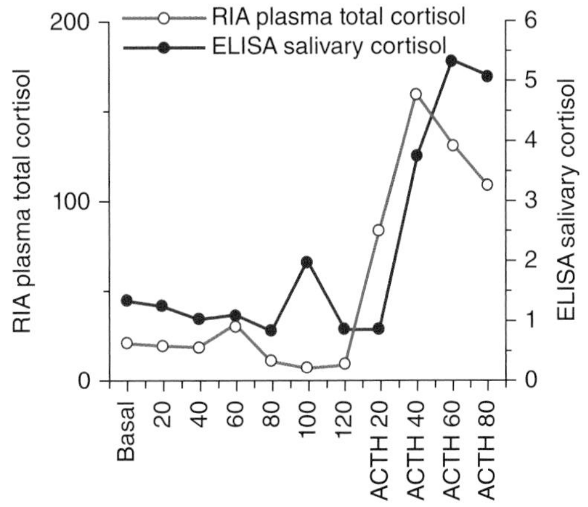

Fig. 6.2. The concentration of cortisol in pig plasma and saliva is shown for 120 min at the basal level and for 80 min after ACTH is injected into the animal. The increase in cortisol following injection is delayed for a few minutes in saliva because of the time taken to diffuse from blood.

surgeon, the magnitude of the coping response of the animal can be usefully estimated by comparing cortisol concentration in handled and control individuals. When animals are transported, the effects of the various components of the transport process can be assessed by monitoring glucocorticoid concentrations (Fig. 6.3; see also Broom, 2014a and Chapter 21, this volume). In all of such studies, the effects of the sampling procedure itself on the animal must be assessed. Since the time to the increase in cortisol concentration is usually 1.5–3.0 min, if samples can be taken quickly the true response is seen. If samples can be taken with minimal disturbance of the animal, the coping response can be evaluated effectively. In a study of the effects on rabbits of various adverse conditions that might be encountered during transport, a period of 4.5 h at 42°C caused greater elevation of cortisol and other physiological measures than a temperature of –5°C, a 96 dB noise or social mixing (De la Fuente *et al.*, 2007).

During the period of monitoring of sheep during a road journey shown in Fig. 6.4, the sheep showed a very marked increase in plasma cortisol when they were loaded on to the vehicle. This occurred despite the fact that the staff concerned were experienced animal handlers and did not treat the animals roughly. The sheep were clearly very disturbed by the loading, and the response lasted for 6 h. The cortisol concentration then dropped to close to the basal level as the sheep became accustomed to their new environment. During the last 3 h of the journey, cornering and acceleration caused problems for the sheep, so cortisol concentration increased. It is clear from studies like this that measurement of cortisol concentration can provide information about the welfare of animals over relatively short periods.

Where it is not easy to take a sample of blood or saliva, HPA axis activity can be assessed by measuring glucocorticoids, or their metabolites, in urine or faeces. The time for the excreted substances to get into the urine or faeces must be taken into account. Indeed, careful validation for each species is necessary in order to understand the information available from measurements of faecal cortisol (Touma and Palme, 2005; Palme, 2012).

Glucocorticoid measurements are of less use when housing or other long-term treatment is being evaluated, because the response adapts after some minutes or hours. Multiple activation of the HPA axis can sometimes lead to measurements of elevation in cortisol over a period of some days. However, lack of increase in glucocorticoids over long periods in a particular housing condition or other treatment does not mean that the welfare is good. For example, severe chronic pain, an ambient temperature above the readily tolerable range or close confinement that allows little movement are welfare issues not usually associated with elevated cortisol: other indicators of welfare are needed in these circumstances. Increased

Fig. 6.3. Sampling of blood from a pre-implanted catheter in sheep during a transport journey. The sample can be taken with little disturbance to the sheep but has to be taken in less than 2 min in order that the concentration of cortisol does not increase because of the sheep's response to the sampling procedure (photograph S. Hall).

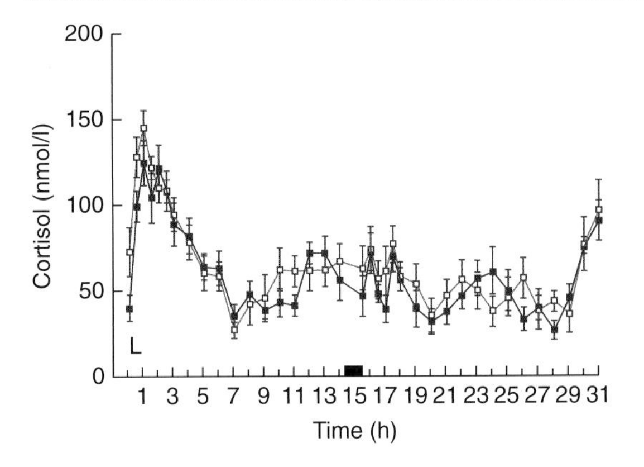

Fig. 6.4. The mean plasma cortisol concentration in two groups of sheep during a road journey are shown. The basal concentration of cortisol is about 40 nmol/l. The changes at loading (L) and during a 14 h motorway journey, a 1 h rest period, a further 13 h on a motorway and 3 h on rural roads can be seen (from Parrott *et al.*, 1998).

glucocorticoid production may lead to immunosuppression. The impaired immune system function and some of the physiological changes can indicate what has been termed a pre-pathological state (Moberg, 1985).

The heart rate of animals changes according to activity and perceived need for activity. The response is relatively rapid and brief, often adapting in a minute or two. If a cat stands up from a lying position, starts walking, then starts running, its heart rate will increase with each of these activity changes. Were the cat to detect imminent danger at any stage during these changes, a further increase in heart rate would be superimposed. Heart rate can be a useful indicator of short-term welfare problems, provided that the component of the heart rate change that is a coping response can be evaluated. The data in Table 6.1 (Baldock and Sibly, 1990) show ovine heart rate in excess of that expected for each animal at that level of activity, e.g. lying, standing, walking or running. Thus this measures the magnitude of responses to stimuli presented to the animals. In these animals, which were accustomed to frequent human contact, one very striking response was the almost maximal response to the approach of a dog.

A variety of other measurements can be used when attempting to assess the welfare of animals during transport or other relatively short-term treatments, and a summary of such measurements is shown in Table 6.2. If transport is prolonged, food deprivation and rapid metabolism can lead to the metabolism of, first, food reserves and the functional body tissues, and the metabolites of each can be identified in blood. Dehydration, bruising,

Table 6.1. The heart rate of sheep subjected to management procedures (from Baldock and Sibly, 1990).

Treatment	Heart rate[a] (beats/min)
Spatial isolation	0
Standing in stationary trailer	0
Visual isolation	+20
Introduction to new flock (0–30 min)	+30
Introduction to new flock (30–120 min)	+14
Transport	+14
Approach of man	+50
Approach of man with dog	+84

[a]Taking account of activity performed.

Table 6.2. Physiological indicators of welfare: short-term problems.

Stressor	Physiological variable(s)
Food deprivation	↑ FFA, ↑ β-OHB, ↓ glucose, ↑ urea
Dehydration	↑ osmolality, ↑ total protein, ↑ albumin, ↑ PCV
Physical exertion, bruising	↑ CK, ↑ LDH5, ↑ lactate
Fear/arousal	↑ cortisol, ↑ PCV ↑ heart rate, ↑ heart rate variability, ↑ respiration rate, ↑ LDH5
Motion sickness	↑ vasopressin
Inflammation, large immunological responses	Acute phase proteins, e.g. haptoglobin, C-reactive protein, serum amyloid-A
Hypothermia/hyperthermia	Change in body and skin temperature, prolactin

FFA, free fatty acids; β-OHB, beta-hydroxy butyrate; PCV, packed cell volume; CK, creatine kinase; LDH5, lactate dehydrogenase isoenzyme 5.

fear, distress, motion sickness and attempts to combat pathogens also lead to recognizable changes in blood.

Behavioural measures

Behavioural measures are also of particular value in welfare assessment. The fact that an animal avoids strongly an object or event gives information about the feelings and hence about the welfare of the individual. The stronger the avoidance, the worse the welfare while the object is present or the event is occurring. An individual that is completely unable to adopt a preferred lying posture despite repeated attempts will be assessed as having poorer welfare than one that can adopt the preferred posture. Other abnormal behaviour such as stereotypies, self-mutilation, tail-biting in pigs, feather-pecking in hens or excessively aggressive behaviour indicates that the perpetrator's welfare is poor.

Stereotypies and other abnormal behaviours that can be used as welfare indicators are described in detail in Chapters 24–28. Individually housed sows may show bar-biting, drinker-pressing or sham-chewing. *A stereotypy is a repeated, relatively invariate sequence of movements that has no obvious function.* Some sows confined in stalls (Fig. 6.5) or on tethers may spend many hours showing such behaviour. The sequence of events recorded using video for such a sow is shown below.

- Sow standing.
- 5–8 s: press drinker with snout.
- 1–2 s: pause.
- 5–8 s: press drinker (water pouring on to floor).
- 1–2 s: pause.
- Above pattern repeated 7–15 times.
- 5–8 s: press drinker, swing head to left and put snout into neighbour's pen.

Other examples of stereotypies include tail-chasing in dogs and crib-biting and tongue-drawing in horses. All stereotypies tend to occur in circumstances where the individual lacks control over its interactions with its environment, and indicate poor welfare. McBride and Hemmings (2009) conclude from a review of equine stereotypies that restricted feeding, reduced social contact and diminished possibilities for locomotor activities are critical in causing increased risk of stereotypies. Other abnormal behaviours that can be quantified and used as indicators of long-term welfare problems include excessively aggressive behaviour and inactive unresponsive behaviour. Meagher *et al.* (2013) found that mink in impoverished environments sometimes showed prolonged lying in the nest-box, interpreted as a measure of anxiety, and sometimes showed lying prone with eyes open, interpreted as an indicator of boredom.

Fig. 6.5. Sows confined in stalls show much stereotypy, other abnormal behaviour and other signs of poor welfare (photograph D.M. Broom).

In some of these physiological and behavioural measures it is clear that the individual is trying to cope with adversity, and the extent of the attempts to cope can be measured. In other cases, however, some responses are solely pathological and the individual is failing to cope. In either case, the measure indicates poor welfare. Mason *et al.* (2007) review methods of using environmental enrichment as a means of reducing the extent of poor welfare associated with the occurrence of stereotypies.

Measures of pain

One type of poor welfare is pain. Although some people have thought of pain as limited to humans or mammals, such ideas have long been thought improbable by many of those involved in pain research. Melzack and Dennis (1980) made these statements: 'The nervous systems of all vertebrates are organized in fundamentally the same way'; and 'the experience of pain is often inferred from the behaviours of mammals, and it is not unreasonable to attribute pain experience to birds, amphibia and fish' (and presumably, reptiles).

The problem often expressed in relation to pain in species other than man is that the animals cannot tell you when they are in pain or how bad it is. The major method used in human pain studies is self-reporting, for example on a scale from no pain to very severe pain. We do not know how reliable this method is. People can lie or deceive themselves in relation to pain. Perhaps measures of observed behaviour or physiological change in people, like those used in non-human studies, will in future be considered more accurate than human reporting (Broom, 2001).

Some methods for recognizing and assessing non-human pain have been used for a long time. For example, the tail-flick response of rats (since 1941), the jaw-opening response (since 1964), limb-withdrawal (since 1975) and self-mutilation (for much longer, Dubner, 1994). Sophisticated behavioural measures are being used more and more in studies of pain. However, there are problems in pain recognition which make comparisons between species difficult. Severe pain can exist without any detectable sign. For example, a major response of rabbits that are in pain is inactivity (Leach *et al.*, 2009). Individuals within a species vary in the thresholds for the elicitation of pain responses and species vary greatly in the kinds of behavioural responses which are elicited by pain (Morton and Griffiths, 1985; Rutherford, 2002). Hence it is important to consider which behavioural pain responses are likely to be adaptive for any species that is being considered.

Humans, like other large primates, dogs and pigs, live socially and can help one another when attacked by a predator. Parents may help offspring and other group members may help individuals who are attacked or otherwise in pain. Hence, distress signals such as loud vocalizations are adaptive when pain resulting from an injury is felt. In species that can very seldom collaborate in defence, the biological situation is quite different. For example, African antelopes that are subject to attack by lions, leopards, hyaenas or hunting dogs, or sheep that are subject to attack by wolves, lynx, leopards or mountain lions survive better if they do not show obvious responses to pain. The predators select apparently weak individuals for attack and vocalizations when injured might well attract predators rather than conferring any benefit. As a result, these animals do not vocalize when injured.

In the mulesing operation, devised by Mr Mules to reduce the likelihood of fly-strike, a sheep is caught by humans, held upside down in a holding frame, has a 15 cm-diameter area of skin around the anogenital apertures cut off with a pair of scissors, and is then turned over and released. The animal often makes no sound and walks away. Farm staff who carry out this procedure may believe that sheep do not feel pain. However, sheep have all of the normal mammalian pain system and they produce high levels of cortisol and β-endorphin after the mutilation (Shutt *et al.*, 1987). Another example concerns monkeys, which, although normally very noisy, are very quiet when giving birth, a time when they are at increased risk from predators. Their silence does not mean that parturition involves no pain. Cattle may give birth with little vocalization but cows with calving difficulties are more likely to show tail-raising behaviour (Barrier *et al.*, 2012).

A knowledge of the selective pressures affecting the species is needed before behavioural responses to pain can be properly interpreted. Having explained the difficulties in using behavioural measures of pain, however, there are many examples of studies in which quantitative measurement of pain has been carried out and these are reported in later chapters. A dog or pig in pain will often vocalize, and the pitch and loudness of the sound can be measured. A rat will change its behaviour in several ways, including changes in the amount of locomotion and adopting recognizable postures, all of which can be quantified (Flecknell, 2001).

Peripheral anatomical and most physiological aspects of the pain system, on the other hand, vary little between species. Most vertebrate animals that have been investigated seem to have very similar pain receptors and associated central nervous pathways. Even some invertebrates have such systems; for example, Kavaliers (1989) reports that gastropod molluscs have nociceptors. Their output following tissue damage indicates that such damage causes sensitization. The most primitive vertebrates are the lampreys and hagfish, which are considerably more different from modern teleost fish than are humans. When Martin and Wickelgren (1971) and Mathews and Wickelgren (1978) made intracellular recordings from sensory neurones in the skin and mouth of a lamprey (*Petromyzon*) during heavy pressure, puncture, pinching or burning, the output was like that which would be recorded in a mammalian pain receptor. The conduction velocity was slow relative to other sensory neurones, and so these are probably of small diameter. There was no fatigue with repeated stimulation, and the receptors were sensitized following local tissue damage.

The neurotransmitter substance P occurs in small fibres in the dorsal horn of the spinal cord in both mammals and fish. In studies of elasmobranch fish, Cameron *et al.* (1990) found substance P, serotonin, calcitonin, neuropeptide Y and bombesin in the outer part of the substantia gelatinosa of the dorsal horn and met-enkephalin in the lateral part. Ritchie and Leonard (1983) found substance P in the afferent neurones of the elasmobranch substantia gelatinosa. These distributions are similar to those in mammals (Gregory, 1998), and substance P occurs more in the regions of the trout brain receiving input from pain receptors, the hypothalamus and forebrain, than in other parts of the brain (Kelly, 1979).

Within the Mammalia, there is considerable uniformity in the areas of the brain having particular functions. However, different vertebrate groups vary considerably in the locations of function. Some analysis that occurs in the neocortex in mammals takes place in the striatum in birds. There is little difference in function between the pain systems of mammals and birds and the analgesics butorphanol (a kappa-receptor agonist) and morphine (a mu-receptor agonist) both stop pain-related behaviour in hens with broken bones, as they do in mammals (Nasr *et al.*, 2012). Within the different groups of fish there is diversity in the localization of complex analysis. It is necessary to look for the site of any particular function rather than to assume that it will be in the same area as in man, and it is not logical to assume that, because an area that has a certain function in man is small or absent in another group of vertebrates, the function itself is missing.

Behavioural responses to stimuli that would be expected to be painful occur in many vertebrates studied. For example, Verheijen and Buwalda (1988) stimulated the mouth of a carp electrically and, while a mild stimulation led to some fin movements and bradycardia, a current three times as strong resulted in freezing or in erratic darting movements in which the glass tank was bumped. When carp were hooked in the mouth using a certain kind of bait, both Beukema (1970) and Verheijen and Buwalda (1988) reported avoidance of such bait afterwards for many weeks or up to a year. This shows that the carp showed learned avoidance as a result of the hooking experience.

Studies of the behaviour of animals after a surgical operation allow behaviour associated with pain to be identified and quantified. Measures found to be useful in rabbits by Farnworth *et al.* (2011) include some measures found to occur only during pain: full-body flexing, tight huddling and hind-leg shuffling. Other measures increased in frequency during pain: staggering, drawing back and eyelids closed. In these studies, carprofen was an effective analgesic but it was considered desirable to look for improved methods for analgesia. In a study of pain associated with ear-tattooing in rabbits, Keating *et al.* (2012) found that a facial grimace scale, called the rabbit grimace scale, was the best of several behavioural and physiological methods tried for evaluating the level of pain. Topical application of a cream made from a eutectic mixture of local anaesthetics (EMLA) was effective in reducing pain as measured in all ways. Facial grimace scales have now been developed for mice, rats, rabbits, horses and sheep (Leach *et al.*, 2012; M. Minero *et al.*, in preparation; K. McLennan *et al.*, in preparation).

The most frequently observed behavioural response to pain is to withdraw, and subsequently to avoid the source of the experience. Avoidance learning is reported for fish by several authors; for example, Brookshire and Hoegnander (1968) administered a shock to paradise fish when they entered a black compartment and found that they avoided the black compartment subsequently and learned to activate an escape hatch to avoid further shocks.

Opioids have many functions, one of which is natural analgesia. Met-enkephalin and leu-enkephalin are present in all vertebrates that have been tested. When goldfish are subjected to difficult conditions, there is an elevation of pro-opiomelanocortin, just as there would be in man (Denzer and Laudien, 1987). Goldfish given an electric shock show agitated swimming, but the threshold for this response is increased if morphine is injected and naloxone blocks the morphine effect (Jansen and Greene, 1970). Work by Ehrensing *et al.* (1982) showed that the endogenous opioid antagonist MIFI down-regulates sensitivity to opioids in both goldfish and rats. In general, it is clear that there are very many similarities among all vertebrates in their pain systems.

Pain receptors are often called nociceptors and a specific term for them that distinguishes their input to the pain pathways from that of other kinds of receptor seems useful. It would seem that the distinction between nociception and pain is a relic of attempts to emphasize differences between humans and other animals, or between 'higher' and 'lower' animals. The visual and auditory systems involve receptors, pathways and high-level analysis in the brain, but the simpler and more complex aspects are not given different names. A perception of pain can exist without the involvement of pain receptors, but so can visual or auditory perceptions exist without their receptors being involved. Wall (1992) said that the problem of pain in man and animals was 'confused by the pseudoscience surrounding the word nociception'. The use of the term nociception, which separates one part of the pain system from other parts when the system should be considered as a whole, should be discontinued (Broom, 2001e, 2014b).

Measures of other negative feelings

Many measures of welfare give information about feelings. In addition to those that indicate level of pain, several measures indicate fear, anxiety or long-lasting depression. In addition to measures of behaviour, physiology, etc., described above, long-term poor welfare can be indicated by mood measures. For example, Burman *et al.* (2008b) describe experimental studies in which sensitivity to reward loss was assessed. When an expected reward was less than the anticipated amount, animals that had been living in deprived conditions showed a greater negative response than those that had been living in enriched conditions. This is interpreted to mean that the extent of the response is an indicator of the extent of depression or other negative feelings in the animal. Stress and other poor welfare during life can have long-term consequences. People who have been greatly stressed in various ways during life are more likely to develop psychoses (Holtzman *et al.*, 2013). Some of the measures of poor welfare in non-humans may also be indicators of psychoses. For example, confined sows may show stereotypies for long periods each day during many months. If a person showed these symptoms, there is little doubt that extreme psychiatric disorder would be diagnosed.

As explained in Chapter 4 (p. 54), studies of cognitive affective bias provide information about the emotional state of animals that can be relevant to the occurrence of mental disorders and good or poor welfare (Paul *et al.*, 2005; Mendl *et al.*, 2009). Studies of a range of species of animals have followed the work of Harding *et al.* (2004). After training to approach a more positive or more negative reinforcer, the response to an intermediate position is assessed. Mendl *et al.* (2010b) found that rescue shelter dogs with a higher separation anxiety score were more likely to react to an ambiguous position as negative. Doyle *et al.* (2011) trained sheep to respond to a positive and a negative bucket position. If they were shown a bucket in an ambiguous position between the positive and negative positions, they did not go to it after being stressed. Pigs from enriched environments showed more positive cognitive bias than those from barren environments (Douglas *et al.*, 2012).

Studies using such a paradigm are reviewed by Mendl *et al.* (2009) and their significance and usefulness for evaluating welfare is discussed by Broom (2014b). Cognitive bias is a potentially valuable indicator of affect and of welfare. However, it has not yet been demonstrated that either the affect, or the welfare, will be reliably indicated by cognitive bias studies alone. A combination of studies is needed to increase the accuracy with which cognitive bias reveals feelings or allows assessment of welfare.

Disease, injury, movement and growth measures

Disease, injury, movement difficulties and growth abnormality all indicate poor welfare. If two housing systems are compared in a carefully controlled experiment and the incidence of any of the above is significantly increased in one of them, the welfare of the animals is worse in that system. The welfare of any diseased animal (see Chapter 23 for definition) is worse than that of an animal that is not diseased. Responses to disease are important parts of coping systems and hence of welfare (Broom, 2006b; Hart, 2010). Our knowledge about how much suffering is associated with different diseases is now starting to increase.

Damage to normal functioning in cells can occur because the individual is not able to cope with its environment, so is stressed. One example of a change that can occur faster in stressed animals is telomere shortening. This is associated with increased risk of cancer and premature aging and has been reported to occur in humans and other species, such as parrots kept in isolation (Aydinonat *et al.*, 2014).

Some injuries are a consequence of attacks by conspecifics so injuries, such as skin lesions in pigs, are useful measures of aggression and fighting (Turner *et al.*, 2009). Other injuries may be inflicted by people, for example those detectable on a horse after the use of spurs or whips.

A specific example of an effect on housing conditions which leads to poor welfare is the consequence of severely reduced exercise for bone strength. In studies of hens (Knowles and Broom, 1990; Nørgaard-Nielsen, 1990), birds that could not sufficiently exercise their wings and legs because they were housed in battery cages had considerably weaker bones than birds in percheries which could exercise. Similarly, Marchant and Broom (1996) found that sows in stalls had leg bones only 65% as strong as sows in group-housing systems. The actual weakness of bones means that the animals are coping less well with their environment, so welfare is poorer in the confined housing. If such an animal's bones are broken there will be considerable pain and the welfare will be worse.

Inhibited behaviour and behaviour strategies

A further general method of welfare assessment involves measuring what behaviour and other functions cannot be carried out in particular living conditions. Hens prefer to flap their wings at intervals but cannot in a battery cage, while veal calves and some caged laboratory animals try hard to groom themselves thoroughly but cannot in a small crate, cage or restraining apparatus.

In all welfare assessment it is necessary to take account of individual variation in attempts to cope with adversity and of the effects that adversity has on the animal. When pigs have been confined in stalls or tethers for some time, a proportion of individuals show high levels of stereotypies, while others are very inactive and unresponsive (Broom, 1987b). There may also be a change, with time spent in that condition, in the amount and type of abnormal behaviour shown (Cronin and Wiepkema, 1984). Unresponsive animals are sometimes described as apathetic, a condition shown by some depressed people and a clear sign of poor welfare. Working equids that have been badly treated over a long period are recorded as being apathetic (Pritchard *et al.*, 2005; Tadich and Stuardo Escobar, 2014).

In rats, mice and tree shrews, it is known that different physiological and behavioural responses are shown by an individual if confined with an aggressor, and these responses have been categorized as active and passive coping (von Holst, 1986; Koolhaas *et al.*, 1999). Active animals fight vigorously whereas passive animals submit. However, this classification of responses is over-simplified. A study of the strategies adopted by gilts in a competitive social situation showed that some sows were aggressive and successful, a second category of animals defended vigorously if attacked, while a third category of sows avoided social confrontation if possible. These categories of animals differed in their adrenal responses and in reproductive success (Mendl *et al.*, 1992). As a result of differences in the extent of different physiological and behavioural responses to problems, it is necessary that any assessment of welfare should include a wide range of measures. Our knowledge of how the various measurements combine to indicate the severity of the problem must also be improved.

Direct Measures of Good Welfare

Most indicators of good welfare are behaviours, but care should be taken in interpreting these. For example, smiling in humans, tail-wagging in dogs and purring in cats are all behaviours that can indicate more than one motivational state in the animals, and that may or may not mean that the welfare of the individual is good at that time. Observations of behaviour with some detail of its context are needed before good welfare can be identified and assessed. In an attempt to identify conditions that result in good welfare for hens, Edgar *et al.* (2013) asked animal welfare scientists to consider the evidence and specify what would lead to good welfare. These scientists would have been using evidence from the whole spectrum of welfare measurements.

Spruijt *et al.* (2001) suggested that successful coping involves a balance between factors with a negative impact and reward systems. They proposed that good welfare would be indicated by showing anticipatory behaviour when a reward is imminent. Such behaviour was found to occur when rats knew from previous experience that they were about to be transferred from a barren cage to an enriched cage or to one where there would be sexual contact (van der Harst *et al.*, 2003a). However, the amount of anticipatory behaviour was greater for rats from a more barren cage than for rats from an enriched cage (van der Harst *et al.*, 2003b). The rats from the more barren cage were not apathetic and, at the time of anticipation, their welfare was better than when they did not anticipate the reward but the anticipatory behaviour did not allow evaluation of welfare in relation to housing condition. The reward was relatively greater after living in more barren conditions than when living in enriched conditions. It is clear that anticipatory behaviour is an indicator of good welfare over a certain timescale but must be interpreted carefully.

Physiological changes in the brain do seem to be associated with good welfare on some occasions (Broom and Zanella, 2004). When people were shown happy pictures there was an increase in magnetic resonance imaging (MRI) activity on one side of the frontal area of the cerebral cortex, and amygdala activity dropped. A set of regions was found in which there was activity during sad, but not during neutral or cheerful, situations.

It is also known that oxytocin concentration in the blood is higher during some pleasurable events. One such event is nursing the young in a female mammal. Oxytocin is not only associated with the let-down of milk but also leads to a feeling of pleasure as well. Oxytocin is synthesized in the paraventricular nucleus (PVN) of the hypothalamus and in the supraoptic nucleus. It binds to receptors that regulate HPA axis activity and its increase is associated with ACTH and glucocorticoid decrease, lymphocyte proliferation, brain GABA increase and cardiac vagal tone increase (Carter and Altemus, 1997; Altemus *et al.*, 2001; Redwine *et al.*, 2001).

Qualitative behavioural assessment

A problem with many of the animal welfare assessment methods described in this chapter is that each of them gives information about an aspect of the attempts to cope with the environment but an understanding of one, or of several, aspects does not necessarily allow an appreciation of welfare in general. In particular, there may be inadequate indications of the positive and negative feelings of the individual. In an attempt to address this problem, Wemelsfelder (2007) advocated the use of human abilities to describe an animal's demeanour in psychological terms that are relevant to welfare. Those who are very familiar with such animals sometimes have considerable ability to do this. This methodology has parallels with scoring of temperament or personality. Observers are asked to describe the whole animal (qualitative behavioural assessment, QBA). In order to do this they observe animal behaviour and try to evaluate the animals' feelings. Their ability to produce a whole animal assessment of welfare is exploited. As Wemelsfelder (2007) and D. Fraser (2008) emphasize, it is important for observers to know enough about the behaviour of the species observed to avoid misinterpretation, e.g. of a chimpanzee's grin as friendly when it actually indicates threat.

A situation where the use of QBA is valuable is where quantitative welfare indicators are difficult to use. This can often be the case where the welfare of the animals is good. Human observers familiar with such animals may observe subtle changes in the animals and incorporate these in their assessment. In one study in which farmers, veterinarians and those in animal protection groups were asked to use QBA to assess the welfare of pigs from video clips, there was substantial agreement between these people from different backgrounds (Wemelsfelder *et al.*, 2012). An example of a study that used QBA is that of Rutherford *et al.* (2012) who found that pigs disturbed by being put in a novel situation were scored as more positive emotionally if treated with the neuroleptic, tranquillizing drug azaperone. The scoring by 12 observers, blind to the treatment, shows that they were making observations of behaviour, especially that involving emotional components. The analysis revealed two expressive dimensions of pig demeanour. This result shows that the observers were able to make a useful evaluation using QBA. While this itself is useful, it would also be of interest to ascertain exactly what they were observing and to develop these as new measures.

During QBA use, there can be a risk of observers not using the same observations, and in particular the same weighting of observations, and hence that there could be poor inter-observer reliability. While all scientific measurement is subject to some degree of lack of inter-observer reliability, this risk can be higher when QBA is used. However, in several published studies, carefully planned training led to good inter-observer reliability during QBA (e.g. Stockman *et al.*, 2011b).

A further possible problem is that the observers have a bias in their evaluation because of their expectations about animal welfare in the living conditions or after the treatment. Wemelsfelder *et al.* (2009) found that some components of QBA of pigs were affected by the perceptions of observers about the environment of the animals. This is not surprising because, in most studies, it is not possible for the observer to be entirely blind to the environment and treatment of the animals. In the study of Tuyttens *et al.*, 2014, veterinary students were shown the same videos of animals but the QBA scoring was more positive when the students had been told that the conditions of the animals were good than when they had been told that the conditions were less good. Although good training before assessment might reduce the likelihood of such a result, the value of QBA must be questioned in studies where there is a potential for bias. All scientific measurements are subject to being altered by observer bias during the process. Tuyttens *et al.* (2014) also found some effect of the observers' perceptions on other welfare measures. However, most measurements used as animal welfare indicators are likely to be altered only slightly by any bias. QBA could be altered greatly but might not be altered at all. It may be that methods for QBA use can be developed to prevent the influence of any bias.

How well do QBA results correlate with other welfare indicators? An increasing number of publications, such as Stockman *et al.* (2011b), reported correlations with widely used measures. However, using observations of video clips of dairy cows by eight experienced and by ten inexperienced observers, Bokkers *et al.* (2012) found no correlation between QBA results and other welfare measures. Similarly, in a comparison of QBA and a range of measures of welfare recommended for dairy cattle by Welfare Quality (Keeling, 2009), there was no meaningful pattern of relationship between these (Andreasen *et al.*, 2013). However, it may be that the QBA evaluation gives information about the animals that is different from that obtained by other means. If a wide enough range of welfare measures is used, there should be correlations between some of these and the output of QBA.

Wemelsfelder (2007) and Wemelsfelder *et al.* (2012) advocate combining QBA with other welfare measures. The use of QBA as a stand-alone measure seems inadvisable and QBA results are not valid if there is a potential for bias or if inter-observer reliability has not been checked. It is my view that QBA is valuable in three circumstances: (i) in studies where non-scientists are asked to evaluate the welfare of animals: (ii) in preliminary studies of work where welfare assessment is proposed; and (iii) as a supplement to well-established welfare indicators where QBA produces information that they cannot produce and provided that bias by observers can be eliminated. Following preliminary studies (ii), the next step is to identify the observations made during QBA to develop new quantitative welfare indicators and to combine the use of these with that of established indicators. It would be better for animal welfare scientists to limit the use of QBA to the three situations described above.

Studies of Preferences and Their Strength

Since aspects of motivational systems have evolved and exist now because they are adaptive, most of the strong preferences of animals are for resources or actions that benefit them, that is, that help them to survive and breed successfully. During development, individuals will have acquired further information that helps them to take decisions that lead to benefits. One consequence of this, pointed out by Duncan (1978, 1992), M. Dawkins (1983, 1990), Broom and Johnson (1993, 2000) and others, is that the assessment of motivational strength during tests of preference is important in any attempt to ensure that poor welfare is avoided and good welfare is maximized in animals kept or affected by man.

Some studies of preferences have involved observation of what animals choose to do or to obtain when they have a wide variety of opportunities for action. Stolba and Wood-Gush (1989) recorded how sows allocated their time and energy to different behaviours when put in an area of grassland and woodland but were provided with the same concentrate food, in the same quantity, as that given to confined sows. The sows spent 31% of their time during daylight grazing, 21% rooting, 14% in locomotion and only 6% lying. Any preference action requires the animal to make a sacrifice of some sort when it gains access to some quantity of the resource or spends time consuming it. These sows paid a price for carrying out each activity in that they could have done something else instead. Some examples of the many types of preference tests (D. Fraser and Matthews, 1997) are described below.

The majority of indicators of good welfare we can use are obtained by studies demonstrating positive preferences by animals. An early study of this kind was that by Hughes and Black (1973), showing that hens given a choice of different kinds of floor to stand on would choose to stand on one of them. In this simple choice test, the cost of choosing involves only the expenditure of the small amount of energy required to move from one floor area to another, so the major sacrifice made when choosing one floor type is not to be standing or lying on another floor type.

The choice test is of some value to the animal welfare scientist when comparing resources that satisfy the same or a very similar need. Even in that case, more information is required because, of two resources compared, both may be of low value to the animal or both may be of value but only as luxury items. As M. Dawkins (1993) explained, a person may choose caviar over smoked salmon, but welfare would not be poor if only one of these were available. Choice tests are of little value where resources associated with different needs, whose motivational basis is quite different, are to be compared (Kirkden *et al.*, 2003). The needs may vary, not only in motivational strength but also in the rate at which they can be satiated or the quantity of the resource required for satiation. Hence more sophisticated preference tests are needed.

As techniques of preference tests developed, it became apparent that good measures of strength of preference were needed. Taking advantage of the fact that gilts preferred to lie in a pen adjacent to other gilts, van Rooijen (1980) offered them the choice of different kinds of floors that were either in pens next to another gilt or in pens further away. With the floor preference titrated against the social preference, he was able to gain better information about strength of preference. A further example of preference tests, in which operant conditioning with different fixed-ratios of reinforcement was used, is the work of Arey (1992). Pre-parturient sows would press a panel for access to a room containing straw, one that was empty or one containing food and water (see Fig. 6.6). If it was cheap (in terms of energy and time) to get to the straw, the sows opened the door to the straw much more often than the door to the empty compartment. With straw on one side and food on the other, both were chosen if the cost, in terms of plate presses, was low. Increasing the ratio of presses to each door opening increased the cost of access. Up to 2 days before parturition they pressed, at ratios of 50–300 per reinforcement, more often for access to food than for access to straw. At this time, food was more important to the sow than straw for manipulation or nest-building. However, on the day before parturition, at which time a nest would normally be built, sows pressed just as often, at a fixed ratio of 50–300, for straw as for food.

The sow pressing a plate to open a door is carrying out an operant. In operant tests, a cost is imposed upon access to a resource by requiring the subject to perform a task. This task requires time and effort that could otherwise have been spent on other actions. In some cases, the task is unpleasant to the subject. The operant

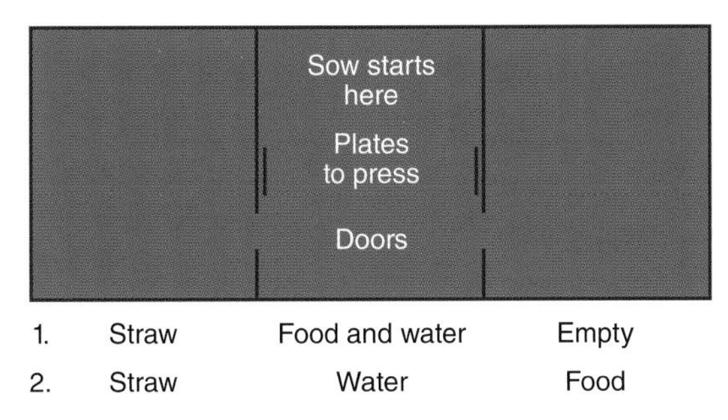

Fig. 6.6. Sows would press a plate in order to gain access to: 1. a pen with straw in it rather than an empty pen and 2. a pen with straw as often as a pen with food but only on the day when nest-building would normally occur (from Arey, 1992).

is varied, for example, by altering the reinforcement ratio as in Arey's study, in order to find out how great is the demand of the animal for the resource.

Other indicators of the effort that an individual is willing to use to obtain a resource are the distance a cow will walk in order to get to pasture (Charlton *et al.*, 2012) or the weight of a door that is lifted. Manser *et al.* (1996), studying floor preferences of laboratory rats, found that rats would rather lift a heavier door to reach a solid floor on which they could rest than lift a lighter one to reach a grid floor. This work started with an investigation of the choice of rats for a solid or a wire grid floor. Manser *et al.* (1995) found that rats in connected cages (see Fig. 6.7) would walk on either kind of floor but always rested on a solid floor if given the opportunity. The maximum weight of door that the rats would lift (see Fig. 6.8) in order to reach any resource was ascertained, and the weight lifted in order reach a solid floor for resting was found to be very close to that. In similar experiments, rats were found to choose bedding and a dark nest-box and to be willing to lift 150 g in order to reach a novel empty cage, 290 g to reach a cage with bedding material in it, 330 g to reach a cage with a nest box and 430 g to reach a cage with both bedding and nest box (Manser *et al.*, 1998a,b).

The terminology used when motivational strength is being estimated is that of micro-economics (Kirkden *et al.*, 2003):

> A resource is a commodity that the animal can use or an activity that it can carry out. The demand of the animal is the amount of action, for example, operant responses, which the animal shows in order to obtain a resource. The price that the animal pays is the amount of action required to obtain a unit of resource. The income of the animal is the amount of time, energy or other variable limiting the action that the animal has available to it.

If an animal has to indicate – for example, by pressing a plate – what its demand for a resource is, it is possible to find out the animal's demand when one varies the price of that resource by making the animal press one, five, ten, 20, 100 or 500 times in order to gain access to it. For example, the pig in Arey's study had to press the plate to gain a period of access to a pen with straw in it, and the rat in the studies by Manser *et al.* had to lift various weights to reach a cage with a particular floor or nest-site. If various prices are used, the animal's behaviour can be described using an inverse demand curve in which price is plotted against demand (see Fig. 6.9). The demand

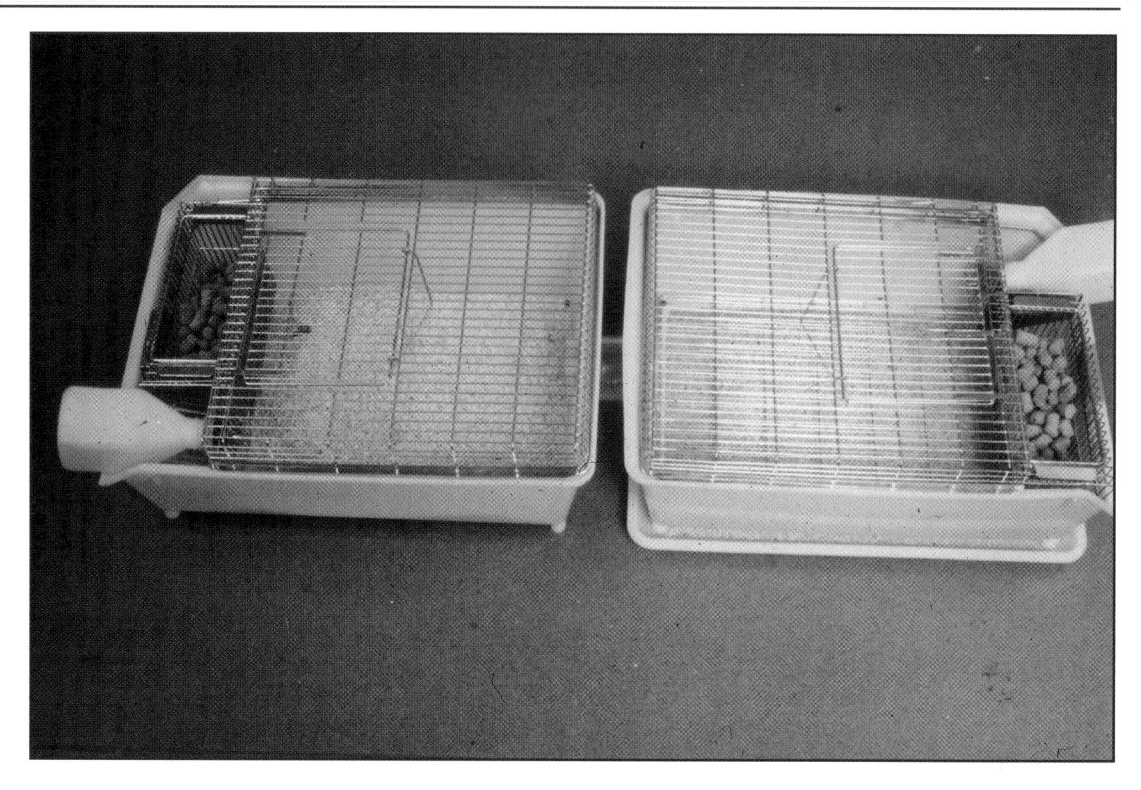

Fig. 6.7. A rat cage with a wire-grid floor is connected to one with a solid floor for a choice test. The rats preferred to rest on the solid floor (from Manser *et al.*, 1995; photograph C.M. Manser).

Fig. 6.8 When two rat cages were connected to this box, the rats had to lift the weighted barrier (an operant) in order to go from one cage to another with a resource in it (as used by Manser *et al.*, 1995; photograph C.M. Manser).

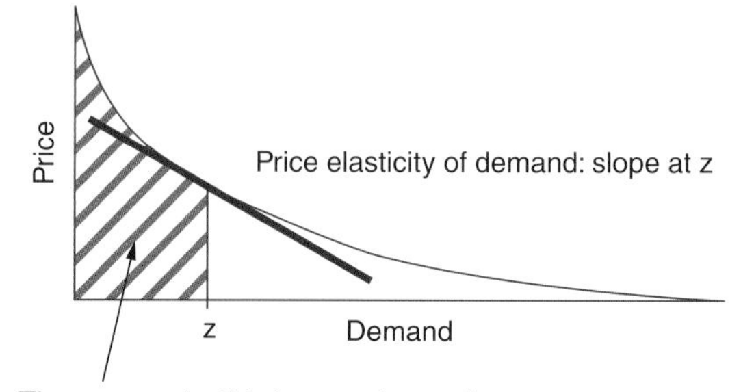

Fig. 6.9 An example of an inverse demand curve that shows how an animal's demand for a resource is related to the price that it has to pay for that resource. The price elasticity of demand and the consumer surplus are indicated for a given demand.

value is how often the animal carried out the work required to obtain the resource.

M. Dawkins (1983, 1988, 1990) proposed the use of elasticity of demand as an index of motivational strength. If the demand for a resource does not decline much when the price increases, the demand is inelastic, whereas if demand drops as price increases it is elastic. For example, if the price of coffee increases by

100% but most people still buy the same amount of coffee, those people's demand for coffee is inelastic. On the other hand, if the price of beef increases by 100% and most people stop buying beef or buy very much less of it, those people's demand for beef is elastic. *The price elasticity of demand is the proportional rate at which a subject's consumption of the resource changes with the price of the resource*, as indicated in Fig. 6.9 at a point on the curve.

Following Dawkins' suggestion of the use of elasticity of demand, Kirkden *et al.* (2003) explained how researchers used the rate at which demand changes with price, the rate at which expenditure changes with price, the rate at which expenditure or expenditure share changes with income and the slope of a bilogarithmic plot of expenditure against income (e.g. Matthews and Ladewig, 1994; Bubier, 1996; Cooper and Mason, 1997, 2000; Warburton and Nicol, 2001).

Kirkden *et al.* (2003) also describe in detail four shortcomings of elasticity of demand indices. First, since price elasticity of demand varies with price and successive units of a commodity are not worth the same as one another to an animal, it is not reasonable to assign a single value to the price elasticity of demand for a resource. Second, since individuals have a tendency to defend a preferred consumption level and to become satiated, the price elasticity of demand index will tend to overestimate the value of resources for which satiation occurs rapidly. Third, the price elasticity of demand tends to underestimate the relative value of resources whose initial consumption levels are high. For example, bread is consumed by humans at higher levels than salt, so demand for salt would be more inelastic than demand for bread, not because salt is more important to people than bread but because it is consumed in smaller quantities when cheap. Fourth, the income available to an animal will vary, but the price elasticity of demand does not take account of this. Since demand increases with income, the price elasticity of demand will underestimate the value of resources when a decrease in income causes an increase in demand, and overestimate it when an increase in income causes a reduction in demand.

Another index of strength of preference shown in Fig. 6.9, consumer surplus, is the area under the demand curve and can be readily measured whenever a demand curve can be generated. The first three shortcomings of elasticity of demand indices described above do not apply to the consumer surplus index. The income of the animal will affect the consumer surplus index so, in studies of motivational strength, income should be set at the level that might occur in real life. If this can be done, the consumer surplus is the best indicator of motivational strength and should be used instead of elasticity of demand indices. In some cases, elasticity of demand indices have provided a good indication of motivational strength, but the consumer surplus will always be more reliable. In the experiments of Duncan and Kite (1987), Manser *et al.* (1996), Mason *et al.* (2001), Olsson *et al.* (2002) and others, the reservation price, which is the maximum price paid for a resource (e.g. the maximum weight lifted by a rat), was measured. This value can be a useful shortcut to the consumer surplus in the circumstances where a demand curve cannot be generated. In a study of how important various resources are to rabbits, Seaman *et al.* (2008) used the consumer surplus to show that food and access to social companions were equally important, both were more important than access to a platform and this in turn was more important than access to an empty cage. All of the arguments presented here are explained in detail by Kirkden *et al.* (2003), and a useful example of comparisons of the different methods of assessing motivational strength is that of Mason *et al.* (2001). The general conclusion of this review of measures of motivational strength is that preference tests, including those requiring the animal to use quantifiable operants, are of particular value in trying to find out what is important to animals. The consumer surplus is the best index of motivational strength when a demand curve can be produced, and the reservation price is a useful indicator when this is not possible. Once such information has been obtained, better housing and management conditions can be designed, and these can be compared with existing conditions using direct indicators of welfare.

In an attempt to integrate preference measures and direct measures of welfare, Nicol *et al.* (2009) compared hens provided with what they chose and hens that were not. They found that measures of welfare associated with positive choice included lower body temperature, blood glucose, heterophil-lymphocyte ratio and response to novelty. Some other welfare indicators were not associated with positive preference, for example plasma corticosterone concentration. However, this is not surprising as elevation of corticosterone is a response to a short-term problem and not a response to prolonged positive or negative situations.

Risk Assessment and the Use of Welfare Outcome Indicators

Scientific studies of animal welfare generally have, as one of their aims, the investigation of what poses a risk of poor welfare and what benefits the individual and results in good welfare. However, formal risk and benefit assessment have been utilized only recently. In fact, the risk of the negative has been the main thrust of the studies, e.g. by the European Food Safety Authority (Müller-Graf *et al.*, 2007; Berthe *et al.*, 2012). This approach has been constructive and useful in drawing attention to what is known and unknown and to the quality of the information.

While research on animal welfare can use sophisticated equipment and prolonged investigations, an inspection of animal housing, transport, etc., is necessarily brief and can use only those measures that can be evaluated in the time available for the inspection. As a consequence, the aspects of the animals that indicate previous welfare and can be recorded by an inspector are welfare outcomes. Welfare outcome indicators include behaviour, such as whether an animal can walk normally, injury scores, signs of disease and mortality rate (Broom, 2014b). Like welfare indicators used in research, these measures are animal-based.

Further Reading: Fraser, D. (2008) *Understanding Animal Welfare: the Science in its Cultural Context.* Wiley Blackwell, Chichester, UK.

Section 3

Organization of Behaviour

7 Behaviour Towards Predators and Social Attackers; Anti-predator Strategies

All wild animals are subject to attack by predators and parasites, and there are substantial selection pressures promoting efficient means of countering these attacks. Domestic animals still have these anti-predator and anti-parasite mechanisms, and they affect animal welfare and animal management, so everyone involved with companion animals or farm animals needs to know about them. Some behaviour and coping mechanisms for dealing with or avoiding parasites are also relevant to pathogen attack (Hart, 2011). The everyday physiological responses are described in Chapter 6, while the behaviour of mammalian predators of domestic animals is described by Kruuk (2002). Anti-predator behaviour is summarized here and discussed at greater length by Broom (1981). Those who see sheep in the countryside will be familiar with the alert posture that they adopt when a human or dog is detected (Fig. 7.1). Exploration is also important in anti-predator behaviour and is discussed in Chapter 11. The first reaction when predator attack is estimated to be probable is fear. *Fear is a feeling that occurs when there is perceived to be actual danger or a high risk of danger.* Physiological and behavioural changes are characteristic of fear. *Anxiety is* a related concept, also *a feeling*, but one *resulting from a perceived risk of a specific or general danger or aversive event.* Fear, anxiety and tests used to evaluate their extent in domestic animals are reviewed by Forkman *et al.* (2007). Since there are many causes of fear and anxiety, and the optimal responses to these will differ, no single test will be universally relevant. Some widely used tests, such as the 'open field' or 'novel arena' test and the 'tonic immobility' test are particularly difficult to interpret and should never be used as the only test. In fact, they are of little value for many species. A combination of tests, selected to exemplify the responses of the species under consideration, should be used. The widespread reliance on certain anxiolytic drugs in anxiety research is also questionable unless the combination of test and drug has been validated for the species.

The two basic types of defensive mechanisms are primary defence mechanisms (Box 7.1), which operate regardless of whether or not there is a predator in the vicinity, and secondary defence mechanisms (Box 7.2), which operate during an encounter with a predator (Edmunds, 1974). Secondary defence mechanisms are used when a predator is detected, when the prey animal perceives that it has been detected, or when an actual attack occurs.

The most obvious avoidance reaction is flight. The anti-predator response will generally be stronger to a more dangerous stimulus. Hansen *et al.* (2001) found that sheep took longer to recover and showed a greater flight distance, i.e. the distance at which they showed escape behaviour, when exposed to a stuffed wolverine, lynx or bear on a moving trolley than to a man or an inanimate object. A man with a dog elicited a quicker and more prolonged response than a man alone. Flight may be socially controlled or uncontrolled. When herd flight is controlled, the animals flee in their normal travelling order (Sato, 1982), in which a high-ranking individual female is usually the leader. When panic occurs there is uncontrolled flight without commitment to any order. The promptness of flight in horses in their original habitat on the plains served as a vital survival tactic. Horses can be frightened by sound, but stimuli that suddenly become visible are more likely to cause an alarm reaction in the horse than are disturbing sounds (see Fig. 7.2). Fear may be shown by all animals, including predatory species. The fear behaviour of dogs when they hear loud noises such as fireworks or thunder includes hiding, seeking contact with familiar humans and prolonged barking (Blackwell *et al.*, 2013).

Avoidance reactions of cattle in response to threatening approach can be passive or active. Social submission, for example when attack by a conspecific is perceived to be likely, may vary from slight head depression with deviation away from the stimulus to the animal assuming recumbency and refusing to rise. This latter

Fig. 7.1. Alert sheep on a Welsh hillside (photograph D.M. Broom).

Box 7.1. Primary defence mechanisms (after Broom, 1981) include:

- Hiding in holes, e.g. rabbits resting in holes.
- The use of crypsis, e.g. moths that are difficult to see on a tree trunk or mammalian prey species that minimize their body odours.
- Mimicry of inedible objects, e.g. caterpillars that look like bird droppings.
- Exhibition of a warning of danger to predators, e.g. skunk coloration.
- Mimicry of individuals that warn of danger to predators, e.g. flies that look like bees.
- Timing activities to minimize the chance of detection by a predator, e.g. being active at night.
- Remaining in a situation where any predator attack is likely to be unsuccessful because of possibilities for secondary defence, e.g. rabbits feeding near their holes or small antelope grazing near thorn bushes.
- Maintaining vigilance to maximize the chance of detecting the advent of a predator, e.g. sheep spending time looking, sniffing and listening for predators.

Box 7.2. Some types of active secondary defence mechanisms (after Broom, 1981):

- Exaggerating primary defence, e.g. a camouflaged caterpillar remaining entirely motionless when a predator approaches.
- Withdrawal to a safe retreat, e.g. a rabbit running to its burrow or an armadillo curling up.
- Flight and evasion, e.g. a hare running and jinking when pursued by a dog.
- Use of a display that deters attack, e.g. moths moving wings to reveal large eyespots.
- Feigning death, e.g. American opossums and many birds, including chickens.
- Behaviour that deflects attack, e.g. some butterfly eye markings that deflect bird pecks away from the body, and broken wing displays by plovers that deflect predators away from the nest.
- Retaliation, e.g. biting, butting or chemical secretions.

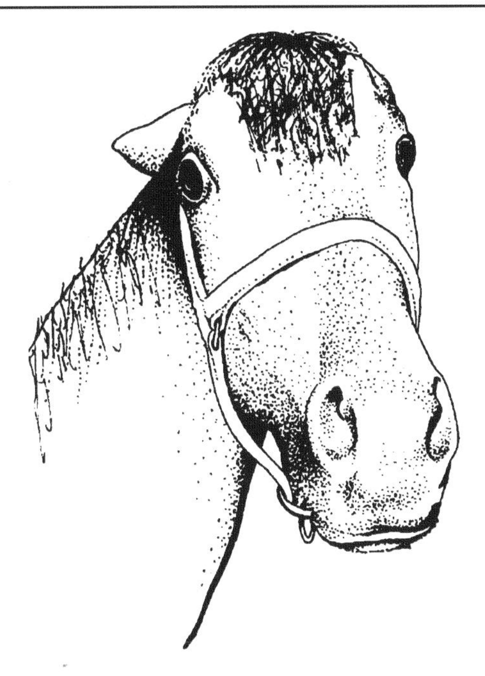

Fig. 7.2. This is a horse showing alarm, including putting its ears back, opening the eyes wide and opening the nostrils (drawing A.F. Fraser).

behaviour is a confusing condition in the presence of a concurrent illness contributing to recumbency. The characteristic feature of such submission in cattle is vigorous, low extension of the head and neck and an abnormally low level of reactivity when aversive stimulation is received. The recognition of submissive reactions is essential in handling sick or 'fallen' livestock of all species, to ensure that their condition is given appropriate consideration.

Agonistic Reactivity

Aggressive behaviour is mostly seen when groups of social animals are first formed. Although sheep seldom show fighting behaviour, rams compete at the start of each breeding season and sheep may show aggressive butting if intensive husbandry conditions increase competition over food or bedded areas. Butting in cattle, goats and sheep, biting of the mane or withers in the horse, pushing and biting in pigs and growling, hissing and biting in dogs and cats are common forms of agonistic behaviour (see Fig. 7.3). Agonistic behaviour embodies many of the behavioural activities of

Fig. 7.3. Goats butting one another (photograph R. Harnum).

fight-or-flight and those of aggressive and passive behaviour. Agonistic behaviour includes all forms of behaviour by an animal associated with conflict with another animal. Aggressive acts are most evident when the aggression of the initiating animal is countered with equivalent aggression. This is common in exchanges between individuals closely matched in status.

An interesting case of agonistic reactivity is found in livestock-guarding dogs (Coppinger and Schneider, 1995). These dogs socialize with sheep or other livestock and limit agonistic motor patterns towards sheep to play only, but show vigorous defence of their sheep flock-mates if the latter are menaced by wolves, coyotes or other predators.

The form of fighting varies from species to species. Horses are often unpredictable in the way in which they react aggressively. Their response to alarm or threat may be flight or attempted flight on the one hand, or attack on the other, depending largely on temperament. Sparring matches and real fights occur between young stallions. In sparring there is skirmishing in which they circle, sniff each other and stamp with a forefoot. In the fight each tries to force his opponent to the ground. In reaction to this, the defending animal uses neck movements to ward off the attack. Attempts are made to bite the opponent's muzzle, forelegs, neck, shoulder and ribs while whirling round, rearing and kneeling. Serious fights begin without skirmishing and start by attempting to bite, rear up and strike out with the forefeet. When an animal is losing a fight he takes flight, defending himself as he does so by kicking out with the hind feet. In stallion fights, loud bellowing vocalizations are made.

To minimize aggressive events, horses maintain large individual distances. The individual distance is that at which some response to the advancing animal is made.

In the case of the 'critical distance', the animal will be more likely to attack than take flight. These distances vary according to the typical reactivity of the animal resulting from its inherent temperament, its experience, domesticated training, competition, housing, feeding and so on.

Fighting between pigs is most severe among adults. Mixing adult pigs together, therefore, is an operation that should be avoided or carried out with care. If a strange sow is introduced to an established group of sows, the collective aggressive behaviour of the group directed at the stranger can be so severe that physical injuries may result in death. When two strange boars are first put together, they circle around and sniff each other and in some cases may paw the ground. Deep-throated barking grunts may be made and jaw-snapping engaged in as fighting starts. The opponents adopt a shoulder-to-shoulder position, applying side pressure against each other. Boars tend to use the side of the face permitting the upwardly directed, lower tusks to be brought into use as weapons. Boars attack the sides of their opponents' bodies in this fashion. Fighting may continue for an hour before submission in one animal. The loser then disengages from the conflict and runs away squealing loudly. With the other's dominance thus established, the encounter ends.

When fighting begins in cattle, the animals fight with their heads and horns. They try to butt each other's flanks. If one animal manoeuvres itself into a position where it can butt the flank of the other, the second animal turns around to defend itself and attempts a similar attack. Fighting sessions do not normally last longer than a few minutes, but in cases where the animals are equally matched the 'clinch' move may be repeatedly employed. This is a move where the animal being attacked from the side turns itself parallel to the other and pushes its head and horns into the region of the other's lower flank. This often arrests fighting for several minutes before action is resumed. When one animal submits, it turns and runs from the other, which then may assert its dominance by chasing after the former for a short distance. If neither animal submits, fighting may continue until both tire.

When a fight between two tomcats is imminent, they may circle one another vocalizing loudly. The aggressive contacts that cause injury are biting and raking the opponent's body with the claws of the forelimbs. The bite that is most dangerous is on the head and neck, and each individual endeavours to move so as to avoid being bitten in these vulnerable areas. As a consequence, bites on the shoulders and flanks are easier to make. A form of defence by a cat that is pushed over or grasped is to throw itself onto its back and strike at the opponent with the claws of the hind feet as well as the forefeet. Scratch wounds are a much commoner consequence of involvement in a fight than bite wounds. Fights between dogs are usually considerably quieter than catfights and, as the claws are not used as weapons, more pushing is involved. The large canines of a dog, similar in size and shape to its wolf ancestor, may be used for biting or for slashing. The fighting strategy of a dog will depend upon whether it is fast-moving, or heavy and powerful, in comparison with its opponent. Slashing bites tear with the canine teeth and are usually preceded and followed by rapid body movement. More prolonged bites involve the substantial muscles that hold the mouth shut.

As a feature of social reactivity, mock fighting is seen as a variant of play. The form of mock fighting is somewhat ritualized. The initial activity is one of solicitation: this usually takes the form of the approaching animal bounding towards the associate animal with jerky head movements. In cattle there is head-lowering and tail-raising. In horses, pigs, dogs and cats there is biting of the neck of the associate. The following phase in mock fighting is usually in the form of a contest, in which one animal pushes or applies weight to the other. During this it is common for the animals to circle. Such circling motions are a feature of mock fight behaviour in calves and piglets reacting to associates. Head butting also takes place. The termination of mock fights involves neither harm nor chasing. The development of motor patterns, such as those involved in fighting, can occur in the absence of opportunities for play during development, but the refinement of skills is likely when play-fighting is carried out (Pellis and Pellis, 1998).

Defensive Reactions to Man by Farm Animals

Man was a dangerous predator to the ancestors of our farm animals and, in reality, still is today. Hence people are often treated as a source of potential danger by present-day farm animals. This fact is often ignored by those who work on farms but competent stockmen learn to recognize such responses and to treat the animals in a way that minimizes their occurrence. If a person enters an animal house rapidly and noisily, the animals may show a violent escape response. This can occur in a calf, pig or poultry house, but the response of poultry can be the most damaging. A wave of escape behaviour, often called hysteria, can pass down the house and result in a pile-up of birds against the wall of cages or of the house. Many birds may die in this situation. If entering is preceded by a knock and avoidance of unexpected movements and noises, such problems can be minimized.

The qualities of a good stockman include the avoidance of causing panic, as described above, and an ability to act calmly and predictably when coming close to animals. Early studies on stockman behaviour and effects referred especially to dairy cows. Baryshnikov and Kokorina (1959) showed that milk let-down occurred faster if the cows saw the familiar milker, and Seabrook (1977) showed that milk yield was higher if the stockman moved in a deliberate, calm way, talked quietly to the cows and followed a regular routine. Management efficiency and milk production have been improved by studies of defensive behaviour of cows to stockmen and by amending aspects of milking parlour design (Albright and Arave, 1997). Recent studies of stockmanship emphasize the advantages of the good stockman, both for animal welfare and for production efficiency (Hemsworth and Coleman, 2010).

Further Reading: Hemsworth, P.H. and Coleman, G.J. (2010) *Human–Livestock Interaction: the Stockperson and the Productivity and Welfare of Intensively Farmed Animals.* CAB International, Wallingford, UK.

8 Feeding

Introduction

Feeding by predators or herbivores involves a complex series of decisions and often depends upon an elaborate array of mental, motor and digestive abilities. Wild animals, or free-ranging domestic animals, need to find the right sort of habitat and then to find concentrations, or patches, of food before they can start looking for particular food items. Finding the food source is also important for young animals, which have to find their mother's teats or another food source that they have not seen before. *Foraging is the behaviour of animals when they are moving around in such a way that they are likely to encounter and acquire food for themselves or their offspring* (Broom, 1981).

A hunger system comprises: (i) perceptual mechanisms for recognizing food; (ii) a central hunger mechanism for integrating causal factors for eating and coordinating necessary movements; and (iii) motor mechanisms for locating and ingesting food (Hogan, 2005). The initiation of feeding behaviour can be affected by diurnal rhythms and social factors, but inputs from monitors of body state are of particular importance. Inputs to the brain reported to be of importance in several species include visual input, input from taste receptors, input resulting from stomach contractions, insulin effects, plasma glucose detector input and fat store monitor inputs (Mogenson and Calaresu, 1978; Rowland *et al.*, 1996). Stimulation of the lateral hypothalamus can lead to eating by rats or cattle, but the lateral hypothalamus is not essential for eating to occur. Glucose levels are clearly of little importance to feeding in ruminants (Baile and Forbes, 1974).

Once food is found, the rate of ingestion will limit intake. This will depend upon: (i) oral mechanics and other abilities of the animal; (ii) the physical and mechanical properties of the food; (iii) the availability of water; (iv) the nutrient qualities of the food; and (v) the effects of disturbances such as those due to danger of predation, attacks by insects or competition from other members of the species. Both the efficiency of finding food and the various effects on the rate of ingestion will be modified according to the previous experience of the individual. The point at which ingestion of a meal ceases will depend on gut size and input to the brain from sensory receptors, such as those which signal that the gut is full. Booth (1978) proposed that eating occurs when the flow of energy from absorption becomes too small and disappears again when absorptive flow becomes adequate. Such explanations, whether or not referring to satiety centres in the hypothalamus, cannot explain the many different situations in which feeding is initiated or terminated. As Rowland *et al.* (1996) explain, the control of feeding depends upon several interacting systems. The roles of the lateral hypothalamus, lateral preoptic area, amygdala, temporal cortex, orbito-frontal cortex and striatum in the control of feeding in rodents and primates are reviewed by Rolls (1994).

During the delay before the next meal, the food is processed. The rate of processing, which depends on gut cross-sectional area, enzyme activity in the gut and food quality, will often be a major factor limiting food intake. The next meal may be delayed by more than the digestion time if there is input to the brain which indicates that the general metabolic state as indicated by, for example, the fat store level, is such that no further meal is needed. If extra food is needed due to metabolic needs, e.g. external temperature is low, then the onset of the next meal is accelerated. Another important factor that can affect the delay before the next meal is the quality of the ingested food. If this is insufficient then the animal may move to a better place before recommencing eating. The efficiency of digestion can be impaired by illness, parasites or by adverse conditions that lead to adrenal activity, as well as by the quality of food, so any of these factors can have an effect on intake. Physiological changes alone do not explain all of the variation in feeding behaviour. Feeding behaviour is strongly influenced by reinforcement, both positive and negative, from food palatability and by the environmental and social associations of feeding. It is necessary for concepts of motivation and reinforcement to be incorporated into any comprehensive view of food-intake control. It has been suggested that, as the animal develops, drinking and feeding may occur as natural complements of each other and they may occur frequently and in modest amounts, not because the animal is compelled to restore accumulated deficits but because it anticipates the pleasures of ingestion and thereby avoids the deficits entirely (Epstein, 1983). Berridge (1996) explains how different regions of the brain control two separable components of pleasure associated with eating, the pleasantness of the food itself and satisfaction of the desire to eat certain foods. Studies of various animals with food continuously available, including rats (Le Magnen, 1971), cattle (Metz, 1975), and 300,000 meals by broilers, ducks and turkeys (Tolkamp *et al.*, 2012) show that meal size is more often correlated with the interval before the following meal (post-prandial correlations) than with the interval since the last meal (pre-prandial correlations). These animals are not compensating for an accumulated deficit but are using a feedforward control system. They must have learned that food has a certain effect, so they consume enough for a certain future period duration.

The following sections will deal with: (i) grazing behaviour as an example of how feeding is organized; (ii) finding food; (iii) the ability to obtain food; (iv) meal size and food selection; (v) the effects of disturbance on feeding; (vi) social facilitation; (vii) competition and feeding behaviour; and, finally, (viii) some specific details about feeding in cattle, sheep, horses, dogs, cats, pigs and poultry.

Grazing and browsing behaviour

Any animal that wishes to feed must take a series of decisions about how to behave in order to find, ingest and digest food (Broom, 1981). A goat will often choose to browse on shrubby herbage (Fig. 8.1), while a sheep or cow will graze on pasture plants (Rutter *et al.*, 2002). A grazing cow must first find a suitable patch of herbage on which to graze. In doing this it will usually remember where such a patch may be found and it will do best if it returns to a patch that has been allowed to regrow since it was last harvested. There is evidence from the work of Favre (1975) on sheep grazing on mountain pastures that the flocks return to small areas of pasture at intervals that would allow effective regrowth. When a grazer is standing in utilizable pasture, it does not eat all green material at random but must still take a series of decisions that will allow it to harvest the food effectively (Broom, 1981; Penning *et al.*, 1995). The grazer (see Fig. 8.2) must assess the herbage and decide: to lower the head and take a bite, how large a bite to take, at what rate to bite, whether to stop biting and chew or otherwise manipulate the grass in the mouth, whether to swing the head to one side, whether to take one or more steps forwards, and whether to raise the head and carry out some other behaviour. Then it must decide when to start grazing again.

Patterns of eating, in which there are variations from nil to maximum rate, are characteristic of grazing behaviour in horses, cattle and sheep (see Fig. 8.3). The actual duration of active eating is influenced by food

Fig. 8.1. Goat browsing. Most animals that graze will also eat shrub and tree leaves and so can exploit the larger volume of edible material available from these than from pasture plants alone (photograph R. Harnum).

Fig. 8.2. Merino sheep grazing. Such behaviour often involves complex decision-making, including decisions about food selection (photograph D.M. Broom).

quality and availability (Rutter *et al.*, 2002; Penning, 2004). Grazing activity is sometimes mainly in the daytime, except when day temperatures are so high as to be aversive, and the onset of active grazing is closely correlated with the time of sunrise. Most of the daylight hours are occupied with grazing periods. These periods usually add up to more than half of total daylight time, but some night grazing is also practised. The most active grazing season coincides with spring in most temperate regions.

During very hot weather in summer, more night grazing occurs. Cold and wet spells of weather in winter can reduce grazing, but they do not have a very significant effect upon the ratio of day to night grazing. In winter, horses spend most of their time grazing while less time is spent grazing during very warm weather. Summer grazing behaviour in both cattle and horses is modified and reduced by fly attacks.

On arid ranges, sheep and horses have been observed to travel long distances each day to water. It is likely that usable range is determined by the furthest distance from available water that livestock are able to travel on a daily basis. Grazing animals can ingest snow as an alternative to water if the latter is difficult to reach, or is frozen, in winter. Range grazing animals, in the presence of snow, can afford to forage outside the usual watered territory. Cattle drink twice daily, on average, in warm weather, but once-daily drinking is more common in winter.

Grazing uses up energy for travel in addition to time. Dairy cows have been found to walk five times per day from a building to pasture 240 m away (Charlton *et al.*, 2012). The nature of the grazing territory or home range, and its quality, influence grazing travel. Horses may travel 3–10 km/day and spend about 2–3 h in grazing travel. Cattle move from 2 to 8 km/day in grazing travel distance and spend about 2 h in grazing travel time. Sheep on range travel about 6 km/day while grazing and spend 2 h on this travel. On good pasture, sheep may travel only 1 km/day. Range livestock also travel considerable distances regularly to salt-lick locations which should, therefore, be sited to allow sufficient visits. The grazing activities of milking cows are synchronized and arranged around the milking times. Active grazing bouts are followed by rumination, usually in sternal recumbency. Synchronized grazing may occur in response to environmental cues such as dawn, dusk, rain, management methods and social factors.

The investigation of grazing behaviour is greatly facilitated by the use of automatic recording methods. The rate at which grazers bite at the herbage and the duration of grazing and ruminating can be determined using a recorder with an elastic tube placed around the jaw of the animal; its stretching can be monitored and records produced such as those in Fig. 8.4. Biting, rumination and swallowing are distinguishable and can be recorded on a recorder carried by the animal, or telemetrically (Penning *et al.*, 1984).

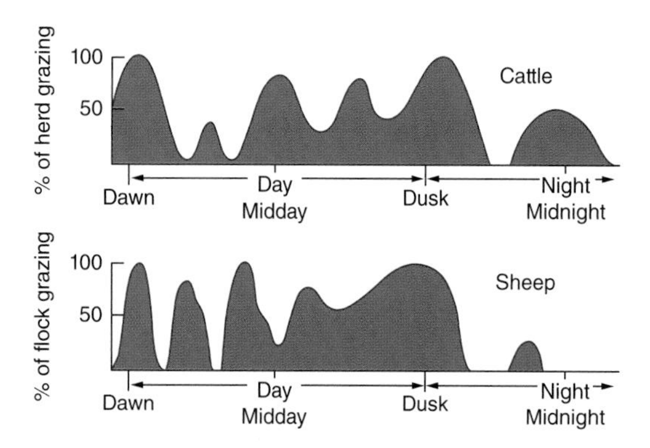

Fig. 8.3. Typical diurnal distribution patterns of grazing in cattle and sheep during spring, summer and autumn. The daily dips in activity are more noticeable in summer.

Finding Food

The methods of finding food in domestic animals are very diverse. For example, the feeding method in mallard or domestic duck is to up-end in the water and search for vegetable or animal material.

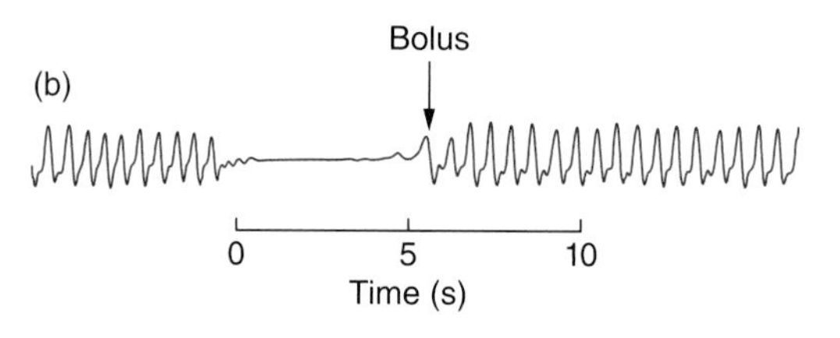

Fig. 8.4. Traces produced by the jaw movements of a sheep with a thin elastic tube, containing carbon fibre, around its jaws. The stretching or contracting of the tube changes electrical resistance and this is converted to a trace: (a) jaw movements during grazing; (b) the characteristic trace during ruminating and swallowing (after Rutter *et al.*, 1997).

Sheep, goats or cattle on sparse pasture often have to use much energy searching for plant material that is worth harvesting. They may have to travel long distances and remember where suitable patches of pasture are to be found. Since less suitable plants may be more readily available, foraging in such conditions may involve selection and is considered under that heading. On some occasions, however, suitable food is present but at very low density. Sheep in dry conditions in Australia may have so little green material available that it is worthwhile for them to dig in the ground for the shoots and seeds of subterranean clover *Trifolium subterraneum*. Given a choice, sheep in temperate Europe will select 70% of clover in their diet (Parsons *et al.*, 1994), but always take some grass, probably because of its fibre content (Rutter *et al.*, 2000). Cattle monitored using a GPS recording system (Orr *et al.*, 2012) preferred to graze on grass, rather than on rush, and on short grass rather than on longer grass that was flowering. They ate coarser grass in the evening when they would have longer to ruminate before grazing again. When food is in short supply, grazers may also turn to browsing on plants that are normally avoided.

When a predator such as a cat or dog has to look for its own food, it may use one of several strategies. Dogs are very flexible in their food finding and selection behaviour as they can utilize a wide range of nutrients (Manteca, 2002). Cats may move through an area where prey species are likely to be present, ready to give chase if one is detected. The chase will often be preceded by a period during which the cat attempts to get close to the prey individual by stalking it quietly. An alternative strategy is to wait at a place where prey individuals may appear, for example by a frequently used path or at a waterhole, and then to attack. The cat needs to know how best to overcome the prey species. A dog would also use either of the two methods described above. However, packs of wolves and dogs are generally much more successful at catching the larger prey by concerted action. Pack hunting may allow the effective capture of prey that would not be caught by individuals.

When a calf is born and left with its mother, it is important for its disease resistance that it obtains the colostrum, or first milk, from the mother. Dairy cows that have had several calves have large, pendulous udders and fat teats and this causes difficulty for the young calf when it makes searching movements that would result in teat-finding if the teats were higher and smaller (Selman *et al.*, 1970b; Edwards and Broom, 1979). In a study of 161 calvings, 80% of heifers' calves were successful in finding a teat and ingesting colostrum within 6 h, but 50% of calves of cows of three or more years of age failed to find a teat in this time (Edwards, 1982). Hence, it is important for farm staff to put the calves of older dairy cows on to a teat, preferably within 3 h of birth, as colostrum production and the ability of calves to absorb immunoglobulins from colostrum both decline rapidly after birth (Broom, 1983a).

There is seldom any teat-finding problem for calves of beef cows or for other farm animals. Lambs may, however, be deserted by their mothers and hence fail to obtain colostrum or milk. This problem is worse if twin lambs are born and is much greater in some breeds, such as Merinos, than in other breeds. Piglets are usually able to find the udder unless they are very weak at birth, but there may not be enough functional teats if the litter is very large. Due to the very brief milk ejection period and the competition between piglets for teats, piglets that are weak at birth, overlain by the sow or which become separated from the sow may fail to suckle. The problem of finding the food source may be even greater for young animals fed artificially. A young calf, lamb or kid will often not drink milk from a bucket unless actively trained to do so. Calves fed from artificial teats sometimes do not suck from these. In a study of dairy calves reared in groups from 24 h of age, an artificial teat was put in the mouth of each calf, but some calves were not readily stimulated to go to the teat and drink. Such problems are discussed further in the section on social facilitation.

Ability to Obtain Food

It is easy for a chicken to ingest a food grain, once it has found it, but the harvesting of pasture plants is more difficult because of the structure of the plants. Vincent (1982, 1990) has shown that grass leaves cannot easily be broken by the propagation of a crack across the leaf, following local damage, but require a considerable amount of force, as many fibres have to be broken. As a consequence of the relatively large amount of energy and time needed on each occasion that the grass, or other pasture plant, must be broken, long pasture is more worthwhile energetically to the grazer than shorter pasture. Even if the animal is offered cut fodder, larger particles may be more profitable energetically than smaller particles of the same digestibility, and cattle prefer unchopped silage to chopped silage. Presumably, the larger amount of material per mouthful that can be obtained is the reason for this. The ease of harvesting food is clearly a factor that the grazer takes into account when deciding how and what to eat. The mechanical difficulties associated with breaking growing plant material, the movements involved in gathering the food and the manipulations necessary before swallowing set limits on the rates at which grazers and browsers can eat. Such rates are of importance where animals are required to eat quickly, for example a cow in a milking parlour. A cow requires 2–4 min to eat 1 kg of grain or manufactured concentrates. Hence, high-producing cows fed mostly or all concentrates at the time of milking may have difficulty in consuming their ration before they are required to leave.

Feeding

The efficiency of feeding by animals is altered by experience. In studies of developing domestic chicks, Cruze (1935) showed that accuracy of pecking at food grains improved in all chicks as they matured but that practice had a considerable effect on pecking accuracy (see Fig. 8.5). The efficiency of grazing in young sheep also improves with experience. Arnold and Maller (1977) reared sheep without grazing experience for 3 years and then found that their intakes were considerably lower than those of experienced grazers on the same pastures (see Fig. 3.6).

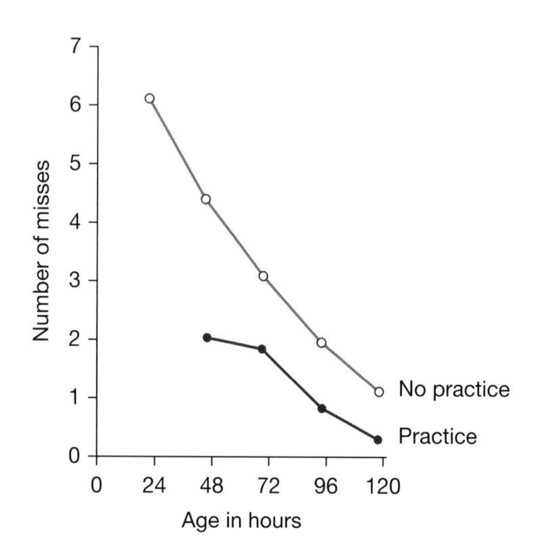

Fig. 8.5. The number of pecks that missed the grain declined with maturation of young chicks given no pecking practice, but at each age chicks given 12 h of practice were more accurate (from Cruze, 1935).

Domestic animals usually learn very fast when food is provided in a new place or when a new procedure for obtaining food is required of them. Modern methods for individual feeding are examples of such situations. There are systems where cows, calves or pigs wear around their necks a transponder that is recognized electronically and causes a door to open or food to be provided. The Callan–Broadbent gate system for dairy cows involves individual mangers being accessible through a movable, hinged gate that opens and allows the cow to lower its head into the manger, but only if the transponder on the cow's neck is recognized. Hence cows have to learn which gate is theirs and most do this very quickly. Systems where rations are provided on a daily basis at a single feeder, when a cow, calf or pig approaches and its transponder is recognized, are also readily learned. Farm animals are also remarkably adept at learning how to gain food when the stockman does not intend that they should do so!

Meal Size and Food Selection

Many studies of animal feeding show that animals can recognize the energetic value of foods, and can take account of the energetic cost of obtaining food when organizing their feeding behaviour, so it is clear that they have an appetite for energy. This does not mean that energy intake is always paramount in determining how feeding will occur, however. Nutrient quality, other functional systems including water balance, predator avoidance and social factors also influence feeding. The termination of feeding can occur because ingested food releases hormones, such as the neuropeptide cholecystokinin, from the gastrointestinal tract. Brain opioids also have some involvement because their inhibition (e.g. by naloxone, which is an opioid receptor blocker) can suppress ingestion. Plasma glucagon, produced in the pancreas, increases during feeding and injection of glucagon can reduce meal size, so this is another mechanism whereby meals can be terminated (Geary, 1994).

The amount of food eaten over a period of several days by a fully grown individual with free access to food is generally just sufficient to maintain body weight. The energy value of the intake, which we would measure in joules, is that which keeps the body fat stores at a set point (Baile and Forbes, 1974). If a pig, for example, is given food diluted with low-energy material, it eats more so that the energy intake remains constant (Owen and Ridgman, 1967). The energy intake is reduced, however, if energy intake has been higher than normal. The establishment of the set point occurs during rearing, so animals that are starved early in life may remain thin when adult even if abundant food is present; and young animals that are overfed may become obese adults. One consequence of the operation of the system controlling intake is that changing nutritional demands, such as those due to pregnancy or lactation, can be allowed for if adequate food is available, as can extra energy requirements due to adverse climatic conditions. In an analysis of a large number of studies of meal size and distribution in various species, Tolkamp *et al.* (2011) reported that in more satiated animals, the probability of ending a meal increases with amount consumed while in less satiated animals, the probability of starting a meal increases with time since the last meal.

Domestic animals require a variety of nutrients, often obtained from different sorts of food. Taste has an effect on food selection. Dogs have many taste receptors that respond to sugars and also show a rather weaker response to amino acids (Thorne, 1992). The cat, on the other hand, shows little response to sugars and is clearly an obligate carnivore highly adapted to meat eating and remaining on a high-protein diet (Manteca, 2002). In addition to choosing food, which gives the best net energy return, do homeostatic mechanisms exist to stimulate consumption of specific and essential nutrients, in proportion to their need by the body? The answer appears to be in two parts. First, regulatory systems exist for water and sodium, creating thirst and salt appetites (Fitzsimons, 1979; Booth *et al.*, 1994). There is also evidence for a specific calcium appetite in birds. Second, nutritional deficiencies in general do not have specific homeostatic methods of self-correction. Centres mediating thirst are located in the hypothalamus. Pigs, sheep and cattle rendered sodium deficient will consume appropriate amounts of solutions containing sodium ions and can be trained to show an operant response for a sodium reward (Sly and Bell, 1979). An ability to recognize calcium deficit within the body is present in domestic fowl, for calcium-deprived hens will choose a calcium-enriched diet even if that same diet without the calcium would be rejected as unpalatable (Hughes and Wood-Gush, 1971).

Salt deficiency results in freely available salt-licks for animals being put to full use. They provide one way of supplying trace elements that might not be ingested even if made equally freely available in another mixture. Sometimes, a salt appetite in animals can be very acute and can lead individuals or groups into long searches for salt. Grazing animals with access to the seashore can be seen foraging on the shore below the high tide line, where they will frequently ingest seaweeds and will lick and chew other salted material. Occasionally, in various species, voluntary salt ingestion can be excessive enough to create salt poisoning.

There are no regulatory ingestive systems for specific deficiencies of minerals or essential organic substances, but animals can learn that certain foods reduce illness. Garcia *et al.* (1967) found that thiamine-deficient rats learn to eat more of thiamine-rich foods. Animals can compensate for deficiencies and provide for special needs, e.g. during pregnancy, by trying a variety of foods not normally eaten and continuing to eat foods that have beneficial effects. A variety of species of animals learn much about what foods are acceptable or valuable for remedying dietary inadequacies from parents, siblings or social group members (Galef, 1996). The absence of most nutrients in the diet of farm animals is not recognized directly, and the animals must use trial and error learning to attempt to compensate for dietary deficiencies. The consequence of this is that the food selected may fulfil nutrient requirements but, especially when the food is manufactured, the wrong amounts of various nutrients may be taken. Tribe (1950) found that sheep offered linseed cake meal consumed twice as much protein as necessary but, if the protein was in the form of fishmeal, they took less than the necessary amount. Growing pigs may select an optimal amount of protein for weight gain in some circumstances or a non-protein diet in others.

Mineral deficiencies can lead to *pica, the seeking out and eating of objects and materials that are not normally food*; for example, the chewing of wood, bones, soil, etc. Pica is a notable feature of phosphorus (P) deficiency in cattle. Even when given free access to bonemeal, deficient animals still do not change the pica to selective ingestion of the appropriate foodstuff. When P-deficient cattle can eat such bonemeal they seldom eat enough to correct completely a deficiency great enough to have caused the pica. Horses have been found incapable of correcting mineral deficiency when given free access to a digestible mixture rich in the necessary mineral.

In obtaining food, animals not only have to obtain sufficient energy and nutrients but they have to contend with the defences of the food animals and plants. Natural selection has acted on both plants and animals so as to minimize the chances that they will be eaten. Physical defences include weapons and mechanical defence. The weapons of a group of buffalo menaced by a lion are obvious, but the thorns of an acacia or bramble and the irritant chemicals of a nettle or poison ivy are just as effective. Some animals have tough hides or bony plates to protect them, and plants can also have tough outer layers. As mentioned above, grasses and other pasture plants have developed rows of parallel fibres within their tissue, which makes the leaf and stem very difficult to break. Plants often have chemical defences (Arnold and Hill, 1972; Harborne, 1982) and may change their growth form so as to make grazing on them more difficult (Broom and Arnold, 1986).

Animals deal with poisons by recognizing that a poison has been ingested and getting rid of it from the gut, developing enzyme detoxification mechanisms or by learning to avoid consuming an amount that poisons them (Freeland and Janzen, 1974). Cats show such food neophobia (Manteca, 2002). If new foods are eaten it is important to the majority of animal species to be able to deal with poisons. The simplest method for avoiding poisons is to avoid eating any new food; the desirable behavioural characteristics are: (i) consume only small quantities of new food; (ii) have a good memory for different food characteristics; (iii) be able to seek out special foods; (iv) sample foods while eating staple foods; (v) prefer familiar foods; (vi) prefer foods with small amounts of toxic compounds; and (vii) have a searching strategy that compromises between maximizing variety and maximizing intake. Galef (1996) explains how aversion for some foods that are, or are perceived to be, toxic, can be learned as a result of social influences by various rodents and primates.

Food preferences can also serve a useful function in that they allow animals to avoid foods containing toxins. Foods may be avoided when first encountered because of their taste or other characteristics. It is also possible, however, that domestic animals could learn that certain foods lead to later illness. Laboratory experiments on rats show that foods containing poisons, which took up to 12 h to cause effects, were subsequently avoided by rats (Garcia *et al.*, 1966; Rozin, 1968, 1976). Studies on cattle, sheep, goats and horses by Zahorik and Houpt (1981) demonstrated that novel food whose consumption was followed by sickness and discomfort within 15 min was avoided subsequently but, if the delay before the discomfort was 30 min, the animals did not seem to associate that discomfort with the novel food and that food was not avoided subsequently. When animals are grazing they do very often show clear rejection of plants with toxins in them and, in many cases, this must be a consequence of learning from the effects, perhaps immediate, of eating the plants.

Chemicals whose presence led to rejection include tannins, coumarins, isoflavones and alkaloids (Arnold and Dudzinski, 1978). Grazers also avoid pasture contaminated by their own dung, a behaviour that reduces the likelihood of parasite or disease transmission. Since dung, especially slurry from cowsheds, is important as a fertilizer for pasture, this avoidance behaviour is important. Cattle offered clean pasture or pastures treated 7 weeks earlier with slurry preferred to eat the clean pasture (Broom *et al.*, 1975). If the only pasture available was slurry treated, the cows ate only the tops of the grass. They stopped grazing and walked more often than on clean pasture and they were involved in more competitive encounters (Pain *et al.*, 1974; Pain and Broom, 1978). Given the opportunity, all grazers are selective in their diet, but this selection depends upon net energy return from each plant species as well as on any toxic substances that might be present. Preferences for particular plants over others occur both when food is plentiful and when herbage availability is low, for example in sheep grazing annual pastures in Western Australia (Broom and Arnold, 1986). The senses used in selection of plants from pasture are sight, touch on the lips, taste and smell (Arnold and Dudzinski, 1978; Rutter *et al.*, 2002), but sight seems to be the least important sense, in this respect, for Merino sheep. Many man-made diets include several different components and animals will take some of these preferentially, and hence will not consume the supposedly balanced diet that is provided for them. For a review of food preference studies on various domestic animals, see Houpt and Wolski (1982).

Early experience can have a considerable effect on the food preferences of domestic animals. Arnold and Maller (1977) found that sheep reared on range-land areas of Australia had different preferences from those reared on sown pastures. Similar factors can result in animals refusing to accept foods that would be beneficial to them. Lynch (1980) found that up to 82% of cattle totally rejected food supplements. Arnold and

Maller (1974) showed that the reluctance of sheep to eat grain early in life was reduced if they were fed it as lambs. Keogh and Lynch (1982) found that a similar improvement in adult grain consumption could be obtained if lambs observed their mothers eating grain, even if they themselves did not eat it. Similarly, Oostindjer *et al.* (2011) showed that piglets that had seen their mother eating particular foods ate more of the same food 8–10 days later, whether or not they had themselves eaten it before.

The Effects of Disturbance

The times of starting and stopping feeding and the rate of feeding can be considerably affected by climatic conditions, predators, insects and competitors. Animals may refrain from eating during the hottest part of the day because they must seek shade at this time (Johnson, 1987), or may cease eating during heavy rain or high wind because the normal feeding movements are difficult in these conditions. If an animal detects the presence of a predator it will stop feeding, and all domestic animals maintain some vigilance for potential predators. A chicken or a sheep that spends much of its time looking out for possible predator attack may be unable to consume an adequate amount of food and may feed in a different way when it does feed. Man is often treated as a predator by domestic animals, and disturbance by people may have considerable effects on feeding. Animals may not feed normally when they have to come close to people to obtain food, and precise records of normal feeding behaviour are often not obtained by experimenters for this reason.

Insect attack may have very large effects on feeding behaviour. Cattle attacked by warble flies (*Hypoderma*) and sheep attacked by the sheep-bot fly *Oestrus* may show panic reactions that certainly affect feeding behaviour (Edwards *et al.*, 1939). Biting flies may also have considerable effects on where and when animals feed as well as on the number of interruptions during feeding. The stable fly (*Stomoxys calcitrans*) and other flies that bite or cause annoyance to cattle can impair growth rates or milk production (Bruce and Decker, 1958). This may be due to decreased intake, or increased energy demands caused by fly attack. These flies and others, such as the headfly (*Hydrotaea irritans*) that transmit disease, attack specific parts of the body of the animal (Hillerton *et al.*, 1984). Their attacks can be reduced considerably by the use of insecticidal ear tags (Hillerton *et al.*, 1986).

Social Facilitation

Many domestic animals are species that live in social groups, and members of a pack of dogs or a flock of sheep are often observed to eat at the same time as other group members. If cattle, sheep or pigs are taken from their group and housed individually they eat less (Cole *et al.*, 1976). This could be a response to lack of companions in general or to lack of companions at feeding time. Even when food is continually available, social animals usually synchronize their feeding (see Fig. 8.6). Hughes (1971) found that chickens in cages synchronized their feeding much more often than would be expected by chance. As a consequence of such effects, the duration of grazing is much more constant when animals graze in a herd than when they graze individually. Pigs also prefer to eat when other pigs do (Hsia and Wood-Gush, 1982) and piglets synchronize suckling. This is part of the general phenomenon of social facilitation. *Social facilitation is behaviour by an individual that is initiated or increased in rate or frequency by the presence of another individual carrying out that behaviour.* This is a more specific definition than the mere social enhancement of a behaviour in the presence of other individuals (Zentall, 1996). The rate of feeding is also affected by the presence of one or more companions. Chicks pecked more frequently and ingested more when a companion was present (Tolman and Wilson, 1965), and an apparently satiated hen took more food if a hungry hen was introduced to its cage (Katz and Revesz, 1921). A similar effect among calves has been demonstrated. A calf fed on a milk replacer was fed alone early in the morning, but its companion calf was not fed. If the companion was then reintroduced to the adjacent pen and given milk, which it drank, so that the first calf could see the second drinking, the first calf consumed more milk replacer. In a subsequent experiment, the first calf was again fed alone then the second calf was introduced to the same pen but was muzzled. When milk replacer was made available, the muzzled calf tried to drink and stimulated the first calf to take even more milk. The results are shown in Table 8.1. These results, together with observations of calves feeding when housed in groups of ten, showed that food intake by young calves can be increased when others

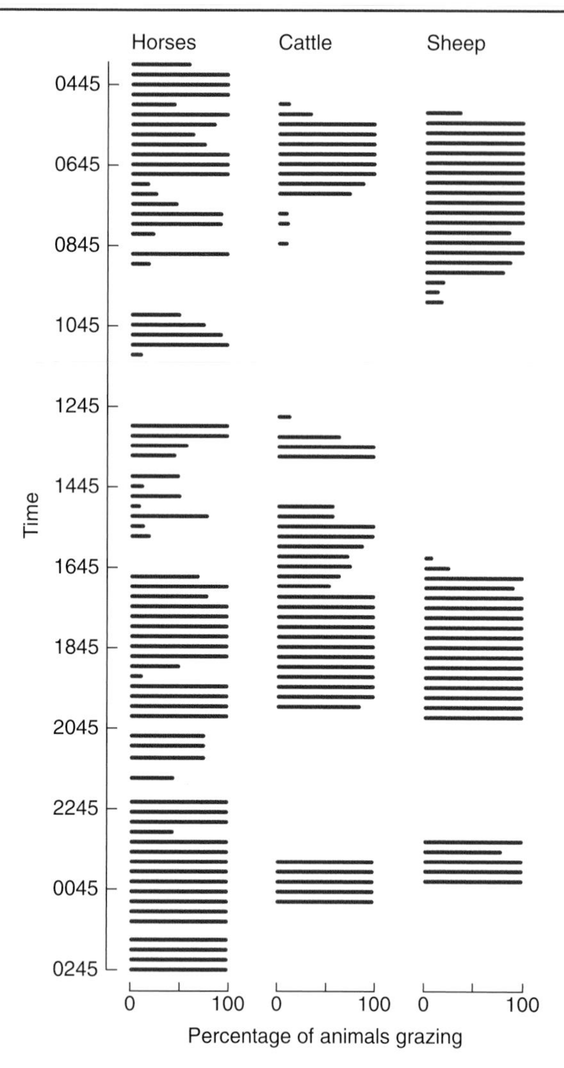

Fig. 8.6. The percentage of horses, Merino sheep and cattle grazing is shown at 15 min intervals during 1 day. Sheep, like cattle, graze as a group for concentrated periods but horses graze individually and for more of the day (after Arnold and Dudzinski, 1978).

can be seen and heard feeding. Hence, since competition is relatively unimportant in these calves, the chances that any calf will receive too little milk replacer are reduced if several teats are provided close together (Barton and Broom, 1985). One teat for two calves works well.

In dogs, the presence of other members of a group can lead to individuals consuming 50% more food than would occur if they were alone and, when previously satiated dogs were joined by hungry individuals, they often ate more food (Manteca, 2002).

Competition and Feeding Behaviour

Individuals competing for food may be successful because of their fighting ability, or because of threats that provide some information about their fighting ability. This is not the only ability that might lead to success, however, for competition for a food item is often resolved by the faster-moving individual acquiring it. The hen that runs and pecks fastest is most likely to obtain a grain thrown to a group of hens. If a limited amount of food is available, the faster eater often obtains more than slower eaters. This applies especially to pigs or cattle feeding from a trough. For many generations, trough-feeding of a limited quantity of food has resulted in selection for animals that can ingest food at a high rate. Even at pasture, animals graze at faster rates when they know that the herbage available is limited. Benham (1982) found that a herd of cows, moved to a new strip of pasture every day, did most of their grazing immediately after introduction to the new strip. The biting rates and the incidence of aggression were highest at this time.

In wild ungulates, high levels of competition for a particular food resource often lead to individuals moving to another source in the same area or to a new area

Table 8.1. Social facilitation of milk feeding by calves (from Barton and Broom, 1985).

	Mean milk intake (l)	Mean sucking rate (l/min)	Mean sucking duration (s/day)
Calf alone	5.5	0.68	487
Calf with hungry calf in next pen	7.5	0.62	724
Calf with muzzled calf in same pen	9.2	0.64	864

(Owen-Smith, 2002), but farm animals often cannot do this. The effects of competition on food intake have been apparent from many farm animal studies. Wagnon (1965) found that heifers kept with older cows lost weight, while those with older cows fed separately gained weight. The heifers were unable to feed adequately because the space available at the feed bunker or trough was inadequate for all of the animals, and the older cows actively prevented the heifers from feeding on many occasions.

All situations where animals in a group are unable to feed at one time should be avoided whenever possible. Feeding troughs should also be designed to minimize any fighting or threats at the time of communal feeding. Bouissou (1970) found that cattle would feed from a trough alongside other animals much higher in the competitive order if a barrier extending over the trough was present, but they would not do so if there was no barrier or an inadequate barrier. Where clearly demarcated feeding places with a place for each animal to put its head are provided, there should be enough of these places for each animal in the group. More space per animal is required if no barriers of any kind are provided. In a choice test following training of cows to discriminate low from high palatability food, Rioja-Lang *et al.* (2012) showed that a trough space of 0.6 m or more is needed to prevent weaker cows from being excluded (see Chapter 29).

Observation of fights and other competitive encounters often makes it clear that individuals recognize one another and consistently defer to some and take precedence over others. The competitive order that can be described need not be linear and need not be the same in all situations or when assessed using different measures (Broom, 1981), but the behaviour of animals at the top and bottom of the order are often very different when the animals are feeding. The frequency of disturbance while feeding and the consequent duration of feeding are both apparent when the paths of cows high and low in the competitive order are compared (Albright, 1969). Calves that did badly in competition were found to obtain less concentrate food and to gain weight less well when the trough length was insufficient for all animals to feed at once (Broom, 1982). The strategies used by farm animals during feeding will often vary according to the social situation. Held *et al.* (2010) found that pigs that had learned there was a larger and a smaller amount of food hidden in an area would normally go to the larger amount first. However, if there was another pig with them, their behaviour depended on the previous behaviour of that other pig. If the other pig had previously robbed the first pig of food, that pig would go to the smaller amount of food first and then have a chance of getting more of the larger amount of food. If the other pig had not competed with it for food, the larger amount of food was chosen as usual.

Hunger, Starvation and Inability to Obtain Food

When an animal is not able to obtain any food, at a time when feeding would normally occur, the motivation to search for food increases. If food is not found, further attempts to obtain food will gradually be replaced by the changed behaviour associated with reduced energy availability. In a previously well-nourished individual, the first metabolic change during a period of food deprivation will be utilization of food reserves. This will be associated with the increased presence of the metabolites of the food reserves in the blood. Once the readily available food reserves have been used up by the food-deprived animal, other body tissues such as muscle will be broken down in order to provide the energy required for survival or for other urgent processes. An urgent process for a pregnant mammal is to provide nutrients for the developing fetus, for a lactating mammal it is to provide milk for the young and for a bird about to lay a clutch of eggs it is to produce enough eggs for the clutch to be complete. When animals use up more food than they can obtain, their body condition becomes poorer. In sheep, poor body condition in mid-pregnancy was found by Morgan-Davies *et al.* (2008) to be the best predictor of reduced probability of survival.

Hunger is a term referring to motivation (see Chapter 4). If the levels of the set of causal factors relating to obtaining nutrients are high, actions that tend to reduce the causal factor levels will be promoted. An animal making great efforts to conserve energy because of food deficit or to obtain food is said to be hungry. The more the individual tries to do either of these, the hungrier it is considered to be. An animal may be hungry because food is not available at a time that its biological clock indicates that food is expected. It will be more hungry if the interval since the last meal is two or three times the normal inter-feeding interval, and very hungry if the deprivation is sufficient to initiate utilization of food reserves or other body tissues. Hunger may change behaviour because of the changes in motivational state, so that greater risks are taken in attempts to obtain access to a feeding place, or prey in the case of predators.

During a period of food deprivation, when does starvation start? A useful threshold point is that starvation starts when the animal starts to metabolize functional tissues. If the animal starts to metabolize muscle that is needed for effective body functioning, the welfare is poor because there is less ability to cope with the environment. There will also be changes in body condition so that the individual becomes thin. If functional body tissues, other than specially adapted reserve tissues such as glycogen in the liver, or body fat, are being metabolized because energy levels from food are absent or are too low, the animal is starving. Hence *starvation is the state of an individual with a shortage of nutrients or energy such that it starts to metabolize functional tissues rather than food reserves*. Starvation refers to a metabolic change associated with overall energy availability deficit. There may also be metabolic changes associated with the lack of a specific nutrient such as a mineral, vitamin or essential amino acid.

Starvation is most often a consequence of lack of food, but it can also occur because the energetic expenditure of the animal is more than the energetic input from that individual's food. Hence a cow that is producing so much milk that the energy input from food is not sufficient to provide the production may become thin as it is using up muscle, liver and other body tissues to allow the high level of milk production to continue. Similarly, a hen could be starving because egg production costs are greater than the energy income from food.

Some farm management practices lead to hunger frequently or even normally (Lawrence *et al.*, 1993; Lawrence, 2008), and some have sufficient impact to cause starvation. Most animals that have been selected for meat production, with rapid growth and a high level of feed conversion efficiency, can be fed *ad libitum* during the growing period of young animals. However, the animals used for breeding also have the genes promoting high food intake and fast growth. Hence, they have a large appetite for food and a potential to grow too fast, for good body functioning and health are poor: broiler breeders and breeding sows are examples of this. In order to reduce both food costs and the likelihood of pathological conditions associated with very fast growth, people managing these animals restrict their food to less than one-third of their voluntary food intake. As a consequence, all of these animals are hungry for much of their lives. Dairy cows may be hungry or starving because, as mentioned before, the metabolic output is greater than their input from food (Webster, 1993).

Neglect in providing food for domestic animals, or deliberate provision of no food or of too little food, can also lead to hunger and subsequently, perhaps, starvation. The most widely used method of detecting hunger is to assess the level of motivation to obtain and ingest food. One method of recognizing starvation is to assess body condition. If the individual has no areas of fat, muscles are reduced in size and bones such as the ribcage very readily seen, it is clear that starvation has occurred. Early recognition of starvation involves detecting the metabolites of muscle and other tissue break-down. Work on starved cattle by Agenäs *et al.* (2006) and subsequent studies (Heath, pers. comm.) have indicated that combinations of metabolites are needed to show that significant starvation has occurred.

Cattle

Cattle have to rely for food intake on the high mobility of the tongue, which is used to encircle a patch of grass and then to draw it into the mouth, where the lower teeth and the tongue are used to hold the bound grass while it is broken by a head movement (Fig. 8.7). The nature of a cow's eating process is such that it is virtually impossible for the animal to take pasture plants closer than 1 cm from the ground. After taking a series of bites, the cow manipulates the plant material, chewing only two or three times before swallowing. The head is swung and steps are taken so that the next bites can be

Fig. 8.7. Welsh Black cow grazing (photograph D.M. Broom).

taken from a new area. The bite size, rate of biting, number of head swings and rate of stepping are affected by the pasture height. Some data from a study carried out on a rotationally grazed pasture in Berkshire, UK, are shown in Table 8.2.

When similar measurements were made by Broom and Penning on cows on two set-stocked pastures, each with two average pasture heights, bite rates were not proportional to average pasture height. The cows avoided longer, coarser grass and grazed at a constant rate on shorter grass with a higher proportion of leaf on both pastures. Chacon and Stobbs (1976) found that as *Setaria* pasture in Australia was eaten down, bite size declined markedly and bite rate increased to a maximum. This and other work by Stobbs emphasized that cows show clear selection for leaf. On day 1 of the study, 32% of the herbage dry matter available was leaf, the rest being stem and dead material, but 98% of the intake was leaf. The intake was assessed using fistula samples. By day 13, only 5% of available herbage was leaf but 50% of intake was still leaf.

When cattle on free range have grazed down an area they will move to another area. Decisions about whether or not to move will depend upon the average return from that area as compared with what the animals know to be the average return from the habitat as a whole. Cattle in fields also use their previous experience in deciding how much energy to invest in attempting to graze after pasture has been grazed down. If they know that they will be moved to a new paddock or strip each day they graze very fast at the beginning of the day, thus competing with one another for the available herbage, and do not graze much in the latter part of the period on the day's strip. When animals are moved on after a few days when the pasture has been grazed down, they learn how to train farmers to move them at the appropriate time. The cue with which they train the farmer is the sight and sound of a row of cows standing by the fence and bellowing. For cattle, as for other ruminant grazers, the amount of energy obtained from the food is often limited by food processing time. The maximum rate of processing is limited by the cross-sectional area of the gut. A consequence of this is that it is important for the grazer not to waste digestion time on poor-quality food if good-quality food is available (Westoby, 1974). This is the reason behind the active selection of leaf over stem. It is also a reason for choosing to eat some pasture plant species rather than others. Selectivity results in certain plants being eaten down in the pasture while others are left untouched. In addition to those of poor nutritional quality, plants may be avoided because they are hairy, spiny or poisonous. Cattle in loose housing spend about 5 h/day eating and their rumination time is also reduced. Although cattle in feedlots are in a very unnatural environment, they still show diurnal rhythms similar to those evident in natural grazing, but their total eating time is much reduced. In place of natural grazing bouts, feedlot cattle have about 10–14 feeding periods, with approximately 75% of these occurring during daylight hours. If hay or silage is fed, 5 h/day may be spent on active eating, as in the loose-housing system. Eating time becomes reduced as roughage is reduced and the proportion of concentrate feed is increased.

Eating space is important in determining the number of cattle that can eat at one time, and this establishes the maximum amount of time during which a pen of animals may eat over a 24 h period. When eating space is restricted, feed intake of the group shows a compensatory increase in rate of consumption. Groups of feedlot beef cattle were fed experimentally in single stalls, with only one eating space provided for each pen of animals. The eating behaviour of the stall-fed groups was compared with that of trough-fed groups. The stall-fed cattle ate faster and differed in their diurnal eating

Table 8.2. Grazing by Friesian cows on ryegrass pasture of two different heights (in part from Broom, 1981).

Grazing characteristic	Short grass	Long grass
Mean length of longest shoot (cm)	13.0	30.0[a]
Time grazing in 24 h (h)	7.9	6.9[a]
Time walking in 24 h (min)	56.0	30.0[a]
Mean bite rate (bites/min)	51.0	47.0[b]
Mean chews per 100 bites	30.0	38.0[a]
Distance walked, head down (m/min)	2.5	1.9[a]

[a] $P < 0.01$; [b] $P < 0.02$.

pattern when compared with the cattle fed from troughs. The diurnal pattern of cattle waiting to eat from the single stalls did not differ from the diurnal eating pattern of trough-fed cattle. The ability of cattle to eat successfully from a single feeding space was related to the protection offered by the stall. Dominant cattle did not prevent subordinates from gaining access to the stall. Low-ranking cattle replaced higher-ranking cattle as frequently as they themselves were replaced by higher-ranking cattle (Gonyou and Stricklin, 1981).

Cattle feeding indoors also modify their feeding behaviour according to the food supplied, and show clear preferences. The time taken to consume food varies according to its volume, the concentrates that may be in it, whether it is wet or dry and the way it has been processed before being given to the animals. Alfalfa requires more chewing before ingestion than ground maize which, in its turn, requires more chewing than shelled maize. Given a choice between silage and hay, milking cows will spend more time at the silage, often two-thirds of the total eating time, while spending the remaining one-third at hay. In a study of food choice it was found that green fodders and roots were preferred to protein, while cereal chaff was preferred to straw. The cows in Fig. 8.8 are fed a nutritionally balanced total mixed ration but they select some components before others so the first to feed get better food than the last, even if there is food present throughout the day. Many studies of cattle in field situations show that they graze mostly during the hours of daylight and cover, on average, about 4 km/day. The distance travelled increases if the weather is hot or wet or if there is an abundance of flies around. During the season of hot weather, more grazing may be done at night than during the day.

The time cattle spend grazing during the 24 h period is 4–14 h. The number of drinks taken per day is between one and four, and the time spent lying down is usually in the 9–12 h range. These figures may vary between beef cattle and dairy cattle, tropical and temperate cattle, and restricted or free-ranging herds (Stricklin *et al.*, 1976).

European breeds of cattle drink more than tropical breeds. European cattle (*Bos taurus*) drink 30% more water per unit dry matter ingested at 28°C than Zebu cattle (*B. indicus*) and 100% more at 38°C (Winchester and Morris, 1956). This is because Zebu cattle conserve water better. Cattle fed on foodstuffs with a high level of protein drink much more than those on a lower-protein supplement. The amount of water consumed by pregnant heifers has been calculated to be 28–32 l/day, while the average daily intake of water by non-pregnant adults is about 14 l.

Grazing behaviour has been studied in the Chillingham herd of wild cattle in Northumberland, UK, which for centuries has been free from interference except during part of the winter when hay is given (Hall, 1983; Hall and Clutton-Brock, 1995). The general pattern of grazing during summer has major grazing periods around dawn and dusk, during mid-morning and early afternoon. During winter, little night-time grazing takes place. Grazing bouts, defined as periods of uninterrupted

Fig. 8.8. Dairy cows selecting from total mixed ration (photograph D.M. Broom).

grazing, can be up to 3 h in length. Grazing-bout length is longer for females than for males in summer, but this difference is not evident in winter. Grazing-bout length is correlated with time since the last grazing bout and with time to the next.

Following ingestion comes rumination, which allows cattle to regurgitate, masticate and then swallow food that they have previously ingested into the rumen. Thus, animals can continue their digestive activities at leisure, when away from a preferred grazing area or sheltering during bad weather. Cattle prefer to lie down during rumination except in bad weather. Rumination starts in young calves of 4–6 weeks of age after they have eaten solid, fibrous food such as pasture plants. The growth of normal gut papillae and gut enzyme development, typical of a herbivore, does not occur unless such food is eaten. A calf of this age deprived of solid food still tries to show a form of rumination behaviour.

The duration of rumination increases with the amount of solid food eaten. Fibrous food plays the major part in this control mechanism. During the 24 h cycle rumination takes place about 15–20 times, but the duration of each period may be only a few minutes or it may continue up to 1 h or more. The time spent ruminating amounts, on average, to three-quarters of the time spent grazing. The peak period for rumination is shortly after nightfall; thereafter, it declines steadily until shortly before dawn, when grazing begins. The factors that may disturb or cause the cessation of rumination are various. During oestrus, ruminating nearly always declines. Any incident giving rise to pain, fear, maternal anxiety or illness affects ruminating activities.

Sheep

Grazing activity by sheep is largely confined to the daytime, and the onset of grazing is closely correlated with sunrise. Grazing is punctuated by ruminating, resting and other activities. The number of grazing periods over each 24 h cycle averages four to seven and the total grazing time usually amounts to about 10 h.

The number of rumination periods may amount to 15 during the 24 h cycle. Although the total time of rumination may be from 8 to 10 h, the length of each period varies from 1 min up to 2 h. Where 3–6 l of water are consumed, the number of urinations and defecations total approximately 9–13 and 6–8, respectively. It is widely recognized that sheep prefer certain pasture plants (Arnold, 1964; Broom and Arnold, 1986). Sheep do not normally consume plants or grass that have been contaminated by faeces. On average, sheep consume food equivalent to 2–5%/day of their body weight.

In a study of grazing by unmanaged Soay sheep on a Scottish island with a diversity of ecotypes available to the sheep, the best predictor of spatial variation in sheep density was the total biomass of green grass of all palatable species (Crawley *et al.*, 2004). The sheep tended to avoid tussocks, which include dead grass stems and may include other plant species, and to graze the areas between them. Bite size was greater on long pasture including tussocks. Before April the pasture available to the sheep was lower than the amount that they could eat, so the sheep were less selective and had to eat plant material poor in energy quality. After April, plant production exceeded consumption. In years when the sheep population was too high for the food availability, sheep died and in some years there was a great drop in the sheep population to <50%. After a sheep population crash, the pasture plant population was significantly changed, emphasizing the effects on pasture plants of selective grazing by the sheep.

Sheep generally frequent a particular watering place. They often use specific paths to water sources and follow a recognized route rather than a direct one. As with feeding, the amount of water consumed varies according to breed, quality of pasture and weather conditions.

Horses

Horses graze by cropping the pasture close to the roots with their incisors. While grazing, they cover large areas and seldom take more than two mouthfuls before moving at least one step further, avoiding grass patches covered in dung. They maintain some distance between one another when grazing in groups. The young foal does not graze very efficiently until it is several weeks old. By about the end of the first week of life, however, the foal has begun to nibble the herbage in association with its dam.

Horses do not drink very frequently in a 24 h period and many may only drink once a day. When they do drink they typically consume very large quantities of water, taking up to 15–20 swallows. Horses were found to prefer grazing white clover patches to other species present, such as perennial ryegrass, timothy and cocksfoot (Archer, 1971). Less palatable pasture plants were red clover, brown top and red fescue. Observations on eliminative behaviour and grazing indicate that olfactory stimuli are important in directing feeding behaviour towards certain plants and away from dunging regions (Odberg and Francis-Smith, 1977). Most horses select the short, young growth of plants, and often show a preference for the more fibrous grasses. They also graze the higher-carbohydrate grasses in a mixed pasture.

Studies have been made on browsing in the ingestive behaviour of native Scottish ponies (Shetland and Highland) in various environments and geographical regions (D. Fraser and Brownlee, 1974) and of other horses (D. Fraser, 2003). The ponies showed preferential selection of rough grazing including privet hedges, dead nettles, burdock and fallen leaves, in autumn and winter particularly. Ash leaves seemed to be particularly favoured. The bark of trees was often eaten; in restricted areas of grazing, a number of trees could be debarked to a height of 2 m, the bark-eating commencing at the lower levels of the tree trunk. The bark of some trees was apparently sought in preference to others; while poplar was a principal choice, ash, oak and rowan were also favoured. Ponies were observed to paw out thistles and eat them together with their roots. Thistle and nettle eating were practised more noticeably in winter than at other seasons and were observed even in pastures where no scarcity of other herbage was notable.

Social facilitation strongly influences grazing in horses. There is transmission of feeding habits from the mare to young foals. Group size and leadership may also influence grazing by dictating the timing of group activity. Extremes of weather such as strong heat, wind or rain reduce the time that horses spend grazing. Season affects the grazing animal through seasonal changes in the weather and the state of the sward. Grazing at night is more common during the summer than the winter in ponies.

Pigs

Rooting is a salient feature of ingestive behaviour in pigs. Even when pigs are fed with finely ground foodstuffs, they continue to show rooting activities. The snout of the pig is a highly developed sense organ and olfaction plays a large part in its behaviour, not least in feeding activities. Pigs are omnivorous and, at free range, eat a variety of vegetable materials. They may also eat some animals such as earthworms. Under modern systems of husbandry, however, it is usual for pigs to be fed on compounded feedstuffs. In as little as 15 min of each day pigs consume a sufficient quantity of food of this type for 24 h.

When food is continuously available, whether or not searching is required to obtain it, pigs space their eating and drinking periods throughout the day. Pigs quickly learn to drink from mechanical devices that supply water when some plate or button is pressed. Water drinking is influenced by both animal size and environmental conditions. Under normal conditions of management, fully grown pigs consume approximately 8 l/day. Pregnant sows may drink in excess of 10 l/day and lactating sows up to 30 l.

The quantity of food that pigs consume is affected by the palatability of the feedstuff. Preference is shown for constituents such as sugar, fishmeal, yeast, wheat and soybean. Substances that reduce the intake of food include salt, fat, meat-meal and cellulose. As a general rule, pigs eat wet foodstuffs more readily than dry, though much depends upon palatability. Under management conditions where pigs are hand-fed they typically show hunger when feeding time approaches, and it is evident that the temporal arrangement of their feeding activities is very well defined. The speed of eating in the pig is found to increase as body weight increases.

Breeding sows must regain body weight after the end of lactation, and this is one reason why they are very competitive at feeding time. If the head and shoulders of sows feeding alongside one another are separated by a barrier, aggression is reduced. Hence, group-housing in pens with access to individual feeding stalls is an effective management system. Electronic sow feeders can also work well, probably with one meal of concentrates per day and access to low-energy-density roughage such as straw. The feeders should have a front exit so that sows do not have to back out – when they could be bitten by a queueing sow. The pig is a highly social animal and social facilitation is a common feature in its behaviour, including its ingestive habits. When a group of pigs unknown to each other are mixed, aggression initially is the dominant behaviour but this soon becomes increasingly inhibited and, as it does so, social order results. Social facilitation of feeding behaviour leads to intakes by sows in groups greater than those peculiar to any individual pig isolated from the group. In grouped growing pigs where food is provided for all at once, there should be enough room for all pigs to feed simultaneously. Even when the hoppers are all filled, if these are too few the pigs will not be able to avoid competition in obtaining their full daily quota of food.

Fighting among growing pigs in groups is significantly greater with single trough-space feeding than with long trough feeders. Competition can be minimized by giving groups of self-fed pigs access to several feeders at a time, and the incidence of behaviour such as tail-biting and ear-biting can be reduced considerably when these are provided. Growing pigs have a daily water turnover of about 250 ml/kg when fed dry pellets at the rate of 4–5%/day of body weight. Water intake is little changed when food intake increases. Both reduction of food supply to half its usual amount, and fasting, significantly increase drinking and water turnover rate. Pigs, therefore, consume more water when food is restricted, a behaviour attributable to hunger.

Some pig husbandry systems require food to be delivered at regular intervals at less than the *ad libitum* intake. Such a feeding routine can cause the water consumption of some pigs to increase by up to five to six times its normal level. Polydipsia also occurs as an aberration when pigs are kept in close confinement.

Poultry

Pecking and swallowing is the main ingestive behaviour of the fowl. Free-range poultry, when they grasp a large food object in the bill, may run with it while calling. Again on free range, the domestic hen typically makes two or three backwards scratching movements with alternate feet before stepping back one pace to peck at the area of ground that has just been disturbed. Poultry typically peck at their food with jerky head movements directed like small hammer blows. Sometimes, poultry will sweep their bills in hoppers of soft feed, displacing it from side to side. In the typical food-pecking action, the bird's eyes are closed at the time of the strike; the pecked item, such as grain, is then grasped between the mandibles and the head is jerked upwards and backwards as the food is swallowed.

Fowls usually eat most at the start or at the end of the day. Laying birds tend to eat more at the end of the day than non-layers, and non-layers more in the morning. Reproductive state appears to be the most important single factor causing variation in feeding patterns. The type of feeding pattern shown depends mainly on how much is stored in the crop at the end of the day and how hungry birds are in the morning. With non-laying birds, an increase in feeding at the end of the day depends on an ability to predict the onset of darkness, but with laying birds it can also be a direct consequence of the timing of oviposition or egg formation.

Poultry drink frequently each day, and some studies have shown that fowl will visit a drinking fountain in their pen 30–40 times per day. As birds get older and, usually, crowding increases, fewer, larger-volume drinks are taken.

Dogs

Dogs are largely carnivorous, but will also eat a variety of foods of plant origin. Like humans, they cannot digest cellulose, which forms the cell wall in plants, but their gut structure and enzymes deal with bones and indigestible fibre better than the human gut. Associated with adaptation for a high-protein diet, dogs select meat and other foods rich in protein. Their ability to find, chase and catch active animal prey is well known. The hunting behaviour of wolves, the ancestor of the domestic dog, involves individual searching for small prey such as mice or pack hunting for large animals such as deer. Much hunting is carried out at night and at dawn and dusk. Cooperative hunts are often led by an older female and several older females may take turns in leading.

Domestic dogs may be deprived of aspects of chasing and other behaviour that their ancestors would have carried out during hunting. Dogs need to carry out the behaviours associated with feeding, as well as the ingestion of nutrients. A balanced diet is sometimes obtained by eating small quantities of grass or other materials that are not parts of the dog's everyday diet.

Cats

Cats are almost entirely carnivorous so they cannot thrive on a vegetarian diet. They hunt for active prey, by moving through their environment and chasing what prey animals they encounter, or by lying in wait in an appropriate place and pouncing on prey that arrive there. When animal prey are ingested, bones and inedible fibre are either processed by the digestive system or ejected from the mouth.

As with dogs, hunting behaviour is a necessity for the cat as well as the actual nutrients obtained from the animal food. Cats may be inactive for long periods after taking a large meal.

Further Reading: Penning, P.D. (ed.) (2004) *Ingestive Behaviour.* Herbage Intake Handbook. British Grassland Society, Reading, UK, pp. 151–176.

9 Body Care

Care of the body, through skin hygiene, evacuation and actions that regulate body temperature, is important for survival. Indeed, the provision of appropriate environmental conditions so that animals can maintain themselves is very important for good welfare. Acts of body care, such as scratching, shaking and licking, are usually brief but these acts are numerous and their total occurrences, per day, constitute a significant proportion of total activity. As well as grooming and preening, body care includes defecation and urination, sheltering from wind, shading from sunshine, bathing and wetting the body in heat. Grooming, fly-avoidance, other body care and the various other disease-minimizing behaviours in non-human animals are of great value for survival and have parallels with the substantial efforts that humans make to minimize disease (Hart, 2011).

Animals take care of their body surfaces in an organized, deliberate way. Body care contributes to the attainment of comfort and the avoidance of discomfort. Animals usually seek and secure comfort as a matter of high priority. Cats spend long periods grooming themselves, provided that they receive adequate food. Dogs groom and scratch themselves frequently. Members of both species select resting places and respond to sensory information about their environment in ways that facilitate body care.

Cattle may stand in the cool water of a pond or stream in high temperatures, turn their faces away from strong chill winds or turn their backs into driving snow. The entire herd often adopts a similar orientation in the course of thermoregulatory tactics. Open, sunlit places are used freely and preferentially by cattle in mild temperatures, e.g. about 23°C, but shade from direct sunlight is sought in temperatures over 28°C by most European breeds of cattle.

In high temperatures, pigs will bathe in water to head height and wallow in mud, giving their bodies muddy and moist surfaces suitable for heat loss and protection from the sun. Sheep and goats will shelter in buildings (Fig. 9.1), under hillside ledges and alongside hedges when cold. Also, animals in cold conditions will huddle to conserve body heat. Shorn and unshorn sheep in early summer will seek shade from direct sunshine and shelter from wind according to fleece cover and ambient temperatures (Alexander *et al.*, 1979; Johnson, 1987). Horses often generate herd gallops in snow, thus raising body temperature.

Shelter from flies is often difficult for grazing animals to achieve unless they can find an area receiving air currents, for example high ground. Fly-avoidance behaviour is of very high priority for many animals because of the high risk of disease transmission (Hart, 2011). Cattle respond to flies that irritate or bite by agitating the head and ears, shaking folds of neck skin, other skin movements, tail-switching, kicking and stamping. Hillerton *et al.* (1986) and Harris *et al.* (1987) found that the frequency of ear flicks by cows was proportional to the numbers of the fly *Musca autumnalis* on the faeces of the cows and that the numbers of kicks and stamps were proportional to the numbers of the biting fly *Stomoxys calcitrans* on the legs. If the cows had fenvalerate-impregnated ear-tags, fly numbers were reduced and fly-dislodging behaviour was reduced. Horses show similar behaviour when troubled by flies, often shaking manes and forelocks as an additional means of dislodging flies and creating air turbulence to deter settling. There may also be social responses to a fly problem; cattle and horses may gather closely together and with tail-switching they can set up a fairly proficient fly screen for themselves. In the presence of very dense fly populations, cattle will be more likely to stand in close groups with their heads together. Occasionally they lie with their underparts – belly, brisket, neck and throat – on the ground. This 'grounding' action seals off many sensitive skin areas from exposure to fly irritation. Such behavioural defences against flies may be very important since flies

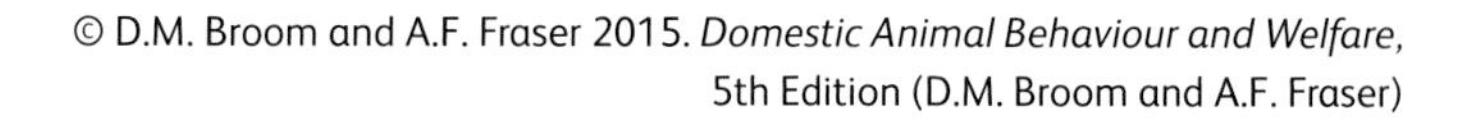

Fig. 9.1. Pygmy goats sheltering during snow-fall (photograph R. Harnum).

may carry disease. For example, the fly *Hydrotaea irritans* carries the pathogens that cause summer mastitis in cattle (Hillerton *et al.*, 1983; Hillerton, 2004).

The rectum and bladder are normally emptied when they become distended, and postures of defecation and micturition or urination are adopted that limit soiling of the hind limbs and tail. Grazing or penned animals avoid soiling the sites that they have chosen for eating and sleeping by evacuating their urine and faeces at other selected locations. The ruminating species do not generally localize defecation, but horses and pigs show very clear forms of eliminative behaviour which minimize the soiling of chosen eating and sleeping sites (Odberg and Francis-Smith, 1977; Petherick, 1983a). They avoid lying down where they have eliminated, when afforded space to do so. Horses often urinate at the edge of restricted grazing areas and will return regularly to defecate at a specific site chosen for this until, in some cases, large faecal piles are formed. The controlled eliminative behaviour of the pig is remarkable and is well organized, even in very young pigs, in that they have clearly demarcated dunging areas. These are normally located in the coolest, dampest part of their pen. It is only in very crowded or spatially restricted housing that pigs cannot form and use a dunging area. Hence, contrary to the popular view, pigs actively attempt to keep clean, in that they avoid their own faeces. Their need to wallow when overheated may result in earth on the skin.

The behaviour of grooming has certain characteristics common to many species. Scratching the head with a hind foot is one; licking certain accessible parts is another. There are few body areas that are not scratched or groomed in this way by dogs, cats, cattle, horses and pigs. Other grooming activities are peculiar to the species: cats use their tongues for most of their fur-care behaviour; horned cattle frequently rub their horns and horn bases against accessible solid structures; rolling in horses is a form of skin attention; chickens will dust-bathe if sand or other suitable material is available but will dust-bathe in dry food or feathers if there is no alternative. When prevented from dust-bathing, they will show the behaviour with a lower latency and a higher frequency than if they had not previously been thus prevented (Vestergaard, 1980).

The eye, face, nose and nostrils of ungulate animals receive hygienic attention by the animal rubbing its face up and down the side of the appropriate foreleg, which may be held out in front of the other to be more accessible. In nasal cleaning, horses do not use their tongues to clean out their nostrils as do cattle, but they snort to do so. Nasal secretions in livestock can be considerable; in severely cold weather these secretions can freeze and become obvious on the muzzle. Abnormal nasal discharge in all livestock may accumulate in some illnesses. This is partly due to their excessive production and partly due to the fact that body care behaviour is suppressed in most illnesses, so nasal cleaning may cease.

Organization of Body Care

Some of the tactics employed in general body care are practised with remarkable precision with regard to location, position, posture group density and duration. The timing of grooming behaviour is affected by various causal factors including hormone levels; for example, prolactin induces grooming. Comfort behaviour may be opiate-related (Cools *et al.*, 1974). Prolactin may stimulate dopaminergic turnover in some brain areas, including the nigrostriatal system inducing grooming (Drago *et al.*, 1980).

When farm animals are sick or depressed, the body care activities may become reduced or arrested. In many illnesses the coats of affected animals lose their normal clean and orderly appearance. Such coats, lacking the effects of friction from rubbing, moisturizing from licking and removal of debris from scratching and brisk shaking, become 'staring' or 'harsh' in appearance. Sick animals, with reduced body care motivation, may not discriminate in choice of resting place or time and place of evacuation. Since they are probably also lying for longer periods than usual, their coats may become heavily soiled, particularly about the hindquarters. Similar deterioration in coat condition is also the result of prolonged stall-housing or enclosure with inadequate bedding, as these are circumstances which prevent or impair the operation of body care behaviour. For example, veal calves kept in crates that do not allow turning round cannot groom normally, often being unable to reach the hindquarters at all and having difficulty in carrying out other grooming movements. Since grooming generally follows a pattern, starting at the head and moving gradually to the hindquarters, if some grooming movements are difficult or impossible the result may be frustration and repetition of the part of grooming that can occur. This leads to excessive grooming that may be a stereotypy and may result in hair ingestion and hair-ball formation in the gut (Broom, 2004; Chapters 25 and 29).

Comfort-shifts occur periodically in resting phases, including recumbency. Many of them are minor positional changes such as partial rotation of the trunk; most are small postural adjustments of the limbs, and tail movements. Such comfort-shifts will be familiar to those who have watched sleeping dogs. Some dogs adjust their body position frequently during sleep. Other movements during sleep are identifiable as components of sniffing, agonistic or other social interaction and have been found to be associated with the rapid-eye-movement (REM) sleep that is associated with temporary loss of slow-wave, e.g. delta, sleep. Such REM sleep is generally associated with dreaming in humans who are awoken after showing it. Grooming is occasionally shown during REM sleep in dogs.

The frequency of comfort-shifts in domestic animals is commonly several times per hour, but this is reduced during some illnesses. Their absence or reduction can lead to the development of oedema, necrosis, ulceration or abscesses over points of pressure, e.g. the tarsal joint. In other instances the occurrence of comfort-shifts in behaviour increases significantly, and this is indicative of discomfort. Such discomfort may be associated with pain, as in colic, or the first stage of parturition, for example. Frequent comfort-shifts can relate to less specific forms of discomfort associated with poor quality of accommodation. Comfort-shifts can become so frequent and intense as to constitute a state of agitation in conditions of extreme pain or discomfort.

Special and Species Features of Body Care

Grooming

Cattle and goats lick and thereby clean every part of their bodies that they can reach (see Fig. 9.2). To groom inaccessible parts they often rub parts of their bodies against trees and fences and use their tails to keep off flies and brush their skins. The value of grooming is

Fig. 9.2. Self-grooming activity in Golden Guernsey goat. Most parts of the body surface can be reached by these acts. Inaccessible areas are groomed by friction against objects (photograph D.M. Broom).

seen in that it helps to remove mud, faeces, urine and parasites and thus greatly reduces the risk of disease. It has been estimated that calves groom themselves on 152 occasions and scratch 28 times per day for a total time of almost 1 h daily. The importance of grooming and fly-avoidance behaviour to cattle welfare is discussed further in Chapter 29.

Grooming behaviour includes auto-grooming (self-grooming) or allogrooming (of others). Autogrooming can be contact between two parts of the body, as in licking, scratching with a horn or foot, or rubbing one part on another, such as head on leg. Alternatively it may be environment-based contact, for example, when an animal rubs against a post, a tree, a stone, a wall, a fence or a cooperating association animal. Allogrooming includes mutual licking, nipping with the teeth, massaging, nose-rubbing and body-rubbing. When dogs groom one another the groomer is often, but not always, subordinate to the individual being groomed. Allogrooming in cats is usually between preferred associates; one cat may solicit grooming from another (Wolfe, 2001) and it may occur when feral cats reunite after hunting (Barry and Crowell-Davis, 1999). Dogs are able to groom themselves using the mouth and all four paws to reach all parts of the body but also rub their body on objects (Manteca, 2002). Cattle in stable, long-lasting groups often form grooming partnerships.

Horses show self-grooming, mutual grooming and 'environment-based' grooming when they rub and roll their bodies on the ground (A.F. Fraser, 2003). When the horse is about to roll it sets itself down, with some care, on a selected spot of ground. It rolls on to its side and rubs its body on to the ground surface. The rubbing lasts a few moments and the horse rotates towards sternal recumbency, from which position it rolls once more on to its side to rub again. This process is usually repeated and in some of these rolls the horse rotates on to its back, holding this position long enough to twist its back once or twice, working the skin of the whole of its back on to the ground. From this position the horse usually rolls back to the starting position, but occasionally the entire rolling episode may terminate with the horse rolling from the supine position on to its other side, thereby going through 180 degrees of rotation. At the conclusion of rolling, the horse stands and carries out very vigorous shaking of the whole body. Each shake begins at the anterior end and passes down the body to the hind limbs. During this shaking the animal's entire hide ripples and dislodges debris, including dust picked up in rolling.

This skin-rippling effect is revealed by slow-motion replay. Another form of equine grooming is nipping of accessible body areas with sharp, repetitive biting actions. Mutual grooming between horses in pairs is common. The horses face towards each other and nip areas of the other's back not accessible to themselves, usually behind the withers. This behaviour can be sustained for many minutes (Fig. 9.3). Horses groom their crests by rubbing them to and fro beneath a manger, tree branch, etc. Manes can become damaged in this way under some circumstances.

Yet another form of equine grooming is scrubbing of the buttocks. A swaying action of the rump is used against a convenient structure such as a post, tree, building, gate or fence that may become adopted and be broken down by continual use. While this behaviour can be a sign of parasitism in the horse, it is also normal grooming with no clinical relevance.

Preening and dust-bathing are forms of body care in poultry. In preening, the bird cleans its plumage with its beak, using brisk and repeated head actions as feather vanes are combed and separated. Ducks oil their feathers by such action, which has a waterproofing effect because, first, the feather with fully knitted barbules does not allow water through and, second, the oil from the preen gland further prevents wetting of the feather surfaces. A duck that is unable to carry out preening and oiling behaviour can become wet and hence cold in cooler conditions. Ducks need water on the head and body for proper preening to occur. If water is supplied from a nipple drinker only, proper preening does not occur. For many water birds, it seems that entering water to bathe is needed for fully effective preening and oiling. As explained further in Chapter 32, if preening is inadequate, the ducks are at greater risk of being cold and wet and of succumbing to disease (Jones and Dawkins, 2010; O'Driscoll and Broom, 2011; Liste *et al.*, 2012b).

Through dust-bathing, chickens and other land birds are able to improve the condition of plumage inaccessible to the beak by friction against a suitable substratum. On a loose substratum the bird can excavate a shallow depression wide enough to contain its body. Within this the bird can use vigorous body movement and work the loose material into its feather cover, even scooping such material over itself while lying on its

Fig. 9.3. Mutual grooming between the two horses on the left is stopped by the intruder on the right (photograph A.F. Fraser).

side. After such dust-bathing activity, the bird will stand erect and forcefully shake out free debris from the general plumage to complete this body care. Deprivation of dust-bathing, for example in birds on a wire-mesh floor with little or no particulate matter available, leads to frustration behaviour and plumage damage. Hens on wire-mesh floors may attempt the early parts of the dust-bathing behaviour sequence, even when the only particles present are pieces of food or feathers. Vestergaard and Lisborg (1993) saw hens remove feathers from other birds and use them in the typical first movements of dust-bathing, so inability to dust-bathe could sometimes lead to increased likelihood of feather-pecking.

Thermoregulatory behaviour

This form of body care is employed when the environmental temperature, wind speed or precipitation presents the animal with a challenge to its comfort. Dogs that are, or are likely to be, hot or cold beyond the tolerable level will endeavour to find a cooler or warmer place in which to position themselves. When too cold they increase the volume of insulating air trapped in their pelage by erecting their hair. When too hot, since they cannot cool themselves by sweating, they pant vigorously. Panting initially involves drawing in air over the convoluted membranes around the turbinate bones in the nose so as to cool them by conduction and evaporation and then expelling the warmed air rapidly from the mouth. Dogs, cats, ruminants and birds pant (Sjaastad *et al.*, 2003). Cooling also occurs in the moist surfaces of the lungs. The cooled areas are generally well supplied with blood, so the circulating blood is cooled. When more overheated, air is taken in through the mouth as well as through the nose, in order to increase the flow rate.

Cats also select appropriate conditions to facilitate thermoregulation and pant when overheated. They may also anoint themselves with saliva, which cools adjacent surfaces by evaporation. Cattle often stand broadside to the sun's rays on a cool day. They are more likely to seek the shade of trees, avoiding direct solar radiation on a hot day (Gonyou *et al.*, 1979). Poultry seek shade from hot, direct solar radiation and may crowd dangerously in such flocking if the shade area is restricted. Poultry take up roosting positions with their wings held out from the body when heated. In experimental and farm situations, farm animals can learn to operate switches that switch heaters on or off and hence control their environmental temperature (Baldwin, 1972; Curtis, 1983).

The temperature at which animals must produce heat within the body in order that the cooling effect of the environment is slowed down or stopped is called the lower critical temperature. This temperature is generally lower in large animals than in small; lower in animals that are metabolizing rapidly (because of production demands or other impacts on their physiological state); and lower in well-insulated animals (Sjaastad *et al.*, 2003; Table 9.1). The temperature at the upper limit of the thermoneutral zone is the upper critical temperature.

Sheltering behaviour and orientation to cold wind, rain or snow are shown in specific ways by domestic animal species. Sheep shelter below ledges in the terrains of highlands and moorlands in high wind or driving snow. Cattle turn their backs into driving rain or snow, closing their inguinal region by adducting their hind limbs closely below them. Horses also turn their hindquarters towards strong winds. Out-wintering livestock make strategic use of tree clumps, woods, walls, buildings, etc., affording leeward protection from chilling winds. In circumstances of communal sheltering, the individual space normally maintained by the group members, as a fairly constant feature of their social organization, is usually surrendered. This facilitates huddling, a behavioural arrangement which, in itself, is a proficient thermoregulatory tactic. It is important to the animal and of high priority in times of critical metabolic conservation.

Table 9.1. Lower critical temperatures in selected mammals.

Species	Lower critical temperature (°C)
Lamb (newborn, wet)	38.0
Piglet (newborn)	27.0
Lamb (newborn, dry)	25.0
Sheep (10 mm-thick wool)	22.0
Piglet (25 kg)	20.0
Dog (short-haired)	15.0
Pig (100 kg)	15.0
Calf	9.0
Dairy cow (maintenance condition)	0
Sheep (50 mm-thick wool)	−5
Beef cattle (growing)	−7
Dog (Husky)	−10
Sheep (70 mm-thick wool)	−18
Dairy cow (30 l milk/day)	−30
Arctic fox	−40
Reindeer (winter condition)	−40

Heat stroke, the development of a body temperature at which vital bodily functions are endangered, should be distinguished from heat exhaustion. The former is due to a breakdown in heat-regulating mechanisms, whereas the latter is not the result of failure of heat regulation but rather of the inability to meet the price of heat regulation. Heat exhaustion is a state of collapse due to hypotension brought on by depletion of plasma volume following sweating and by extreme dilation of skin blood vessels, i.e. by decreases in both cardiac output and peripheral resistance.

The reactions of cattle to the threat of overheating depend on whether they are *Bos indicus* humped cattle such as Brahman, or *B. taurus* such as Aberdeen Angus. Though the behaviour of both Aberdeen Angus and Brahman cattle in cloudy, overcast conditions is much the same, differences become apparent when the two breeds are in conditions of direct solar radiation with little or no air movement. Aberdeen Angus seek out shade and consequently spend less time grazing than do Brahman cattle. On the other hand, adaptive behaviour of the Aberdeen Angus is much less marked if there is good air movement, regardless of the change in temperature. Cattle living in tropical rainforest and equatorial areas of the world show a greater need for shade than those living in sparse or semi-arid areas where rainfall is low and shade limited. It has been found that Dwarf Shorthorn cattle in Nigeria may spend as long as 4–5 h resting in the shade during the day and, to compensate, as long as 3–5 h grazing during the night. In this way, their behaviour may be nearer that of the temperate Aberdeen Angus, Hereford and Holstein than that of the Brahman.

The ability of newborn piglets to adapt to their environmental temperature is very limited, as they are prone to lose body heat rapidly (Sjaastad *et al.*, 2003). They can increase body heat production two- to fourfold by shivering, but this uses up energy. The behavioural mechanism for dealing with this problem is huddling. From the time of birth, young piglets display huddling behaviour as an organized attitude for most of the day (Signoret *et al.*, 1975). During huddling they lie parallel to each other, often with head and tail ends alternating to some degree along the row. While lying closely together side by side, they usually have their limbs tucked underneath them. When a group is large, some of the piglets in the middle may overlie others. Interruptions in huddling result from occasional 'comfort-shifts' in individuals. The result of this huddling behaviour is that the quantity of heat lost by the piglets is much less than would otherwise occur.

Although huddling behaviour is characteristically shown in the litter early in life, it is nevertheless a behavioural pattern that is retained by the pig into adult life as a means of conserving body heat and deriving tactile comfort. Older pigs have a thick layer of subcutaneous fat and also, partly as a consequence of that feature, an absence of loose skin. Sweat glands are very few and their distribution is almost entirely confined to the snout. These factors tend to cause a progressive accumulation of body heat in warm surroundings. The relatively poor body cover of hair makes the animal particularly vulnerable to the effects of direct solar radiation.

Pigs with limited experience of direct sunshine are highly prone to heatstroke when suddenly exposed to it. In heatstroke the animal is hyperthermic, prostrate and breathing at its maximum rate while its associated circulatory responses fail increasingly to cope with the physiological emergency. The duration of unrelieved heat stroke may be only 1–2 h before terminating in death. The circumstances associated with heat stroke are predictable and are those that expose the animal and force it into physical exercise in a high ambient temperature. Clinical cases are therefore most commonly found in association with transport, inadequately shaded pens and with the effort of farrowing, mating or being driven on foot.

The pig's heat-regulating systems, apart from respiratory responses, operate through cooling behaviour such as wallowing, moisture and shade-seeking (see Fig. 9.4). Wallowing in water is highly effective in alleviating hyperthermia. A pig wallowing in mud quickly acquires a thick coating of mud over its lateral and ventral surfaces and its limbs. After the pig has left the wallow this coating adheres and dries out to form, over much of the animal's body, a protective insulation against the sun's rays. By absorbing heat from the pig, this superficial layer of caked mud can also assist in relieving hyperthermia. These factors permit pigs such as breeding sows to graze and forage actively in an environment that otherwise could not be economically utilized for extensive pig husbandry.

Among free-ranging sheep, sheltered areas are constantly being identified and confirmed as a result of exploration. Unshorn and recently shorn sheep choose different resting locations. During calm weather, in daylight, sheep largely remain as one flock but in windy weather, and at night, most shorn sheep congregate in a shelter while unshorn sheep remain away from shelter (Alexander *et al.*, 1979). The inclusion of recently shorn sheep in a flock of unshorn animals does not lead to an increase in the proportion of unshorn sheep likely to lamb in shelter. Unshorn sheep in the presence of shorn sheep appear to avoid sheltered areas, and the colder or higher areas of grazing are preferred spots for them (Mottershead *et al.*, 1982).

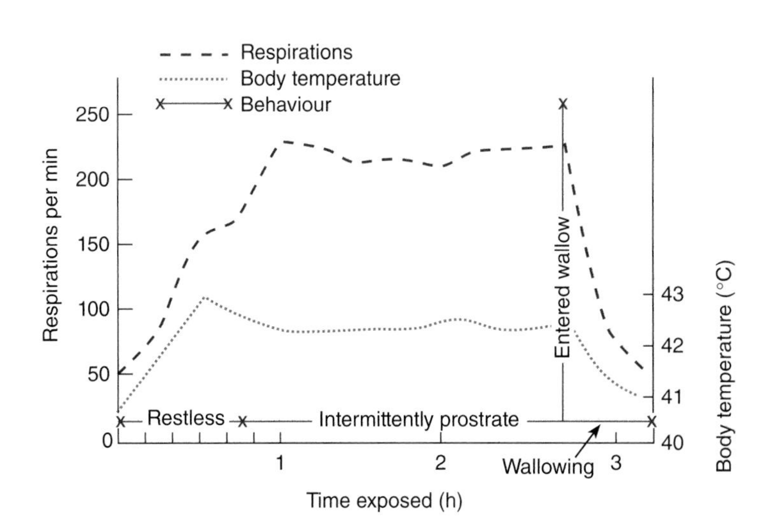

Fig. 9.4. Behaviour in relation to incipient heat stress in a pig. Wallowing is seen to control the syndrome after physiological responses have reached high levels without relieving the condition.

In hot, sunny conditions sheep will often use the shade of a tree or other shelter. Sherwin and Johnson (1987) recorded a significant positive correlation between the proportion of the day spent shading and the daily maximum air temperature. Such behaviour is an important means of regulating body temperature, since for newly shorn animals the heat load due to solar radiation can be similar in magnitude to the metabolic heat production of the animal (Stafford-Smith *et al.*, 1985). There is much individual variation in the strategies used by sheep being heated by direct sunlight; however, Johnson (1987) found that while some sheep in a group he studied spent much time in shade during the hotter part of the day, others stayed in the sun and allowed their body temperatures to rise, but had lower respiratory rates and consumed no more oxygen than those staying in the shade.

Defecation

With few exceptions, such as rabbits, animals avoid the ingestion of their excrement and avoid grazing where there is faecal contamination. Horses and pigs feed in a different area from that in which they defecate, and cattle refrain from grazing close to dung pats. Where slurry from a cowshed was spread on a field and left for 7 weeks, cows preferentially ate clean herbage if it was available but grazed the tops of the herbage over the slurry if there was no alternative (Broom *et al.*, 1975).

Cats

In the wild and when kept as pets, cats will bury urine and faeces, and Beaver (2003) suggests that this behaviour serves the function of reducing the odour and limiting the risk of disease and parasite spread. They also bury faeces in litter trays if they have been trained to do so. If domestic cats accustomed to defecating outside their owners' house are unable to go outside, they may not immediately recognize a litter tray as a suitable site for defecation. In multi-cat households, cats are more likely to use litter trays if there are separate trays for each cat, and preferably an extra tray also (Hetts and Estep, 1994), especially if there are social tensions between the cats (Overall, 1997). House-soiling problems can also result from changes in litter type (Crowell-Davis, 2001; Beaver, 2003), insufficiently frequent emptying and leaving of the litter tray (Landsberg *et al.*, 2012), or disease in the cats (Horwitz, 2002).

Dogs

Dogs need to urinate and defecate for the elimination of unwanted waste, but also use these actions as a means of marking. Urination behaviour is sexually dimorphic, with males usually lifting the hind leg laterally and females usually squatting. Manteca (2002) reports that, in addition to any abnormalities of urination posture caused by pathology, 3% of male dogs and 2% of female dogs use the normal posture of the other sex when urinating. The defecation posture is similar in the two sexes.

Horses

While horses are defecating or urinating they usually cease other body activities. Stallions show careful and deliberate selection of the spot where defecation is to occur. Following defecation and urination, a stallion usually turns and sniffs the spot where this has taken place. After defecation, in the case of both the stallion and the mare, the muscles of the perineum contract and the tail is lashed downwards several times.

While urinating, the stallion and the gelding adopt a characteristic stance, the hind legs being abducted and extended so that the back becomes hollowed. The mare, when urinating, does not show the same marked straddling posture as is shown by the stallion; nevertheless, the posture is similar in that the hind legs are abducted from each other. Following urination by the mare, the vulvar muscles contract. More elaborate patterns of urination are shown by mares in oestrus.

As already mentioned, horses typically show care in selecting areas for defecation (Fraser, 2003). They return again and again to the same patch, and these patches can accumulate large quantities of faeces during a grazing season. Adult animals defecate 6–12 times per day, depending on the nature of the feed eaten. Normally, urination occurs less often during the day and horses have been noted to urinate as few as three times per day. Urine is passed commonly when the animal is freshly bedded and during rest periods in the hours of darkness.

Bovines

Although the eliminative behaviour of cattle seems not to be directed at a certain area, large amounts of faeces are often placed closely together. At night and during harsh weather cattle tend to bunch, and this appears to be the main reason for the close deposition of faeces. The animals pay little attention to their faeces, often walking and lying among excreta but avoid grazing close to faeces.

There is evidence that, in some dairy cows, allelomimetic behaviour occurs: when one alarmed animal defecates or urinates, others may commence to do likewise.

The normal defecation stance for both male and female animals is one in which the tail is extended away from the posterior region, the back arched and the hind legs placed forward and apart. The posture assumed is such that there is the least likelihood of pelage contamination. Calves appear to take more care than adults to expel the faeces well away from the body. Unlike the female, the male bovine animal is able to walk while urinating and displays only a slight parting of the legs while doing so. The posture assumed by the female while urinating is very much the same as employed while defecating, and the urine is expelled more forcefully by the female than the male.

During the 24 h daily cycle, cattle normally urinate about nine times and defecate 12–18 times. The number of eliminative behaviours and the volume that is expelled, however, vary with the nature and quantity of food ingested, the ambient temperature and the individual animal itself. High consumers such as Holstein cattle may expel 40 kg of faeces in the 24 h cycle, while lower consumers such as Jerseys are found to defecate much smaller amounts under the same husbandry conditions.

Pigs

Evacuation in piglets occurs on the day of birth, using postures characteristic of older animals. As the piglets' neuromuscular coordination improves so they are able to hold these postures without falling over, trembling or sitting down. The posture for defecation is the same for males and females: the animal squats, curls its tail up over its back, flattens its ears and half-closes or fully closes its eyes; this same posture is used by females when urinating. Males stand with the front legs slightly advanced, causing the back to depress; they urinate in a series of squirts unlike the female, which ejects a continuous stream. The characteristics of ear fattening and eye closure are less pronounced when urination is occurring. Piglets take some time to develop normal dunging behavour. Elimination does not take place at random in the pen, for specific sites are chosen by pigs for defecation and urination (Whatson, 1978; Amon *et al.*, 2001). In spite of a reputation to the contrary, pigs are extremely clean in their habits if the system of husbandry imposed upon them gives them the opportunity

Fig. 9.5. Duroc pigs feeding but able to use the wet area as a wallow to cool themselves (photograph D.M. Broom).

to express their normal defecatory behaviour. Pig premises that are appropriately designed to create dunging areas are usually properly used by pigs. Even in the most limited quarters, pigs reserve an area for sleeping accommodation and an area for defecation. This sleeping area is kept as clean and dry as possible. Under conditions of crowding, it is sometimes difficult for groups of pigs to maintain organized eliminative behaviour as, for example, when growing pigs are allocated less than 1 m^2 of floor area each.

Pigs kept in extensive conditions with access to a wallowing area (Fig. 9.5) use the wallow to regulate body temperature. When penned pigs are exposed to high ambient temperatures and their normal behavioural methods of controlling hyperthermia cannot operate, it is commonly found that they defecate close to the water supply and use that area to cool themselves if necessary. In farrowing pens, the piglets usually defecate near to a wall and particularly in the pen corners (Petherick, 1982). They avoid defecating in the area used for resting. Fattening pigs also tend to defecate near to a pen wall and orientate parallel to, or with their hindquarters towards, the wall. The size of the dunging area is increased at higher temperatures (Aarnink *et al.*, 2001).

Further Reading: Hart, B.L. (2011) Behavioural defences in animals against pathogens and parasites: parallels with the pillars of medicine in humans. *Philosophical Transactions of the Royal Society, B* 366, 3406–3417.

10 Locomotion and Space Occupied

Introduction

The essence of behaviour is bodily movement. Action patterns, locomotion, displays and other movements give the animal many of its qualities. All the functional systems such as body temperature regulation, feeding and reproduction incorporate movements. Descriptions of bodily motility refer to the bases of physiological activities such as breathing, anatomically localized actions such as blinking, specific goal-related acts such as drinking and gross activity of the whole animal such as locomotion, positional alteration or stretching.

The original ecological niche of each species demanded different behavioural strategies and, consequently, species differ in the amount and type of movement required to maximize effective use of their natural habitat. Animals move to sustain life such as in the search for food and the avoidance of danger. Most of this chapter concerns land animals but aquatic animals also have complex locomotor mechanisms. In most species of fish, locomotion depends on the body musculature causing fins and body trunk to exert force on water, while other fins allow changes in direction (Domenici and Kapoor, 2010). If fins are damaged, locomotion is impaired. Eels are farmed and, while their locomotion is usually swimming, because of their ability for oxygen exchange through their skin, they are also capable of locomotion with similar movements on land. Most mammals can swim.

The topics discussed in the rest of this chapter are postures and movements at rest, gaits during locomotion, distances travelled and the need for exercise in land animals. It seems that freedom to move is a pleasure for animals. Observing calves in spring turned out to grass or racehorses anticipating their morning gallop gives the impression that freedom to run is a pleasure in itself to the animal.

Postures and Movements at Rest

Postural adjustments are needed during rest, grooming, feeding and display and there is a minimum amount of space that is needed in order that such adjustments are possible. Stretching enables an animal to keep its joints and muscles in a state such that they can be used effectively when required. Stretching is often performed after a period of rest or after a period when the limbs are folded. Most stretching occurs in a series of actions as follows: flexion at the throat, arching of the neck, straightening of the back, elevation and movement of the tail and full extension of one and then the other hind limb (A.F. Fraser, 1989). Extension of the forelimbs or wings, singly or together, is a related exercise. Animals that are prevented from stretching, and subsequently provided with an opportunity to do so, will spend a much longer than normal period showing the behaviour, i.e. a rebound behaviour. For example, hens confined in a battery cage so that they cannot stretch their wings will spend longer stretching their wings after a long period of confinement than after a short period (Nicol, 1987b). Similarly, calves that have been closely confined show more vigorous activity when given space than calves reared with more space to move around (Friend and Dellmeier, 1988). Horses can rest for long periods in the standing position because they have a 'stay apparatus', a system of muscles and ligaments that temporarily locks the main joints into position without expending much energy (A.F. Fraser, 1992; Pilliner and Davies, 2004).

Other important movements not involving locomotion are those involved in standing up and lying down. Characteristic sequences of movements are involved (see Fig. 10.1), and the animal needs a certain amount of space for these. Petherick (1983a) calculated how much room a pig needs for lying on its sternum and for lying with the legs stretched out. This space is shown for a range of body weights in Fig. 10.2. Baxter

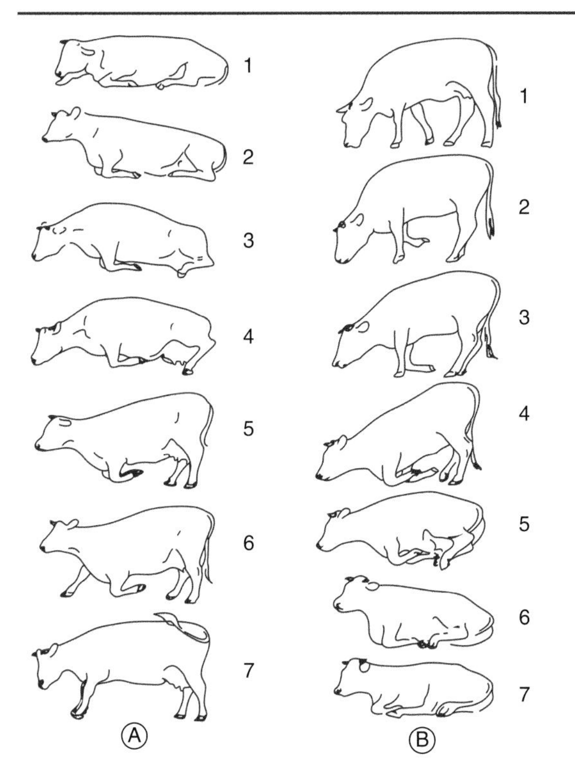

Fig. 10.1. The typical sequence of movements occurring when a cow stands up (A) and lies down (B). In A, a recumbent cow stands up by lifting its body on hind limbs first; in B, a cow investigates the ground first and then lets its weight fall on to its flexed forelimbs.

and Schwaller (1983) filmed the movement of pigs when standing up and lying down (see Figs 10.3 and 10.4) and were able to calculate the minimum space requirements of pigs from this. A calculation of the space requirements of pigs in order to stand, lie and show normal interactive behaviour was made by EFSA (2006b).

Locomotion and Gaits

In land animal locomotion, the limbs act synchronously in any one of a variety of patterns, each of which is termed a gait. Two forms of gait pattern exist: symmetrical and asymmetrical. In symmetrical gaits the movements of limbs on one side repeat those of the other side, but half a stride later. In asymmetrical gaits the limbs from one side do not repeat those of the other. Symmetrical gaits include the walk, the pace and the trot. Asymmetrical gaits include the various forms of the canter and gallop, including the lope and the rotary gallop.

The full cycle of leg movement during the phases of support, propulsion and movement of the body through the air is termed a stride. A stride is a cycle of movement of a limb, while stride length is the distance covered between successive imprints of the same foot. Within each stride each limb for a time acts in a support phase and in a non-supportive or swing phase. In the walk the support phase is longer in duration than the swing phase and determines the stability of this gait. As the walk speed is increased the duration of the support phase decreases while that of the swing phase increases.

The sound produced when a foot strikes the ground is the beat. If each foot strikes the ground separately, the gait will be a four-beat gait. If diagonal pairs are placed down simultaneously, as in the trot (Fig. 10.5), the gait is two-beat since only two beats will be heard for each stride. The canter is a three-beat gait. A lead leg is the leg that leaves the ground last during the canter or gallop. The gallop resembles the canter, but since it has more propulsion it has an extra 'floating phase' at the end of each stride. The walk, the trot and the canter are the forms of quadruped motion involved in locomotor behaviour. These have been described in definitive terms for the horse by Waring (1983) and Barry (2001). The animal changes from walking to trotting at a speed below which walking requires less energy than trotting and above which trotting requires less energy than walking. The same principle is true for changes between the other gaits, and thus the natural gait at any speed incurs the least expenditure of energy. Within each gait there is a speed at which energy expenditure is minimal (A.F. Fraser, 1992; Back and Clayton, 2001).

Walk

The walk is defined as a slow, regular symmetrical gait in which the left legs perform the same movements as the right, but half a stride later and in which either two, three or sometimes four legs support the animal at any one time. The support role is more important in the forelegs, which are nearer the centre of gravity, and the propulsive role is more important in the hind limbs. Approximately 60% of the static weight is supported by the forelegs. The sequence of leg movement in quadrupedal walking is: left front (LF), right hind (RH), right front (RF), left hind (LH), which always results in a triangle of support that includes the centre of gravity. Animal statues and models often show impossible leg positions.

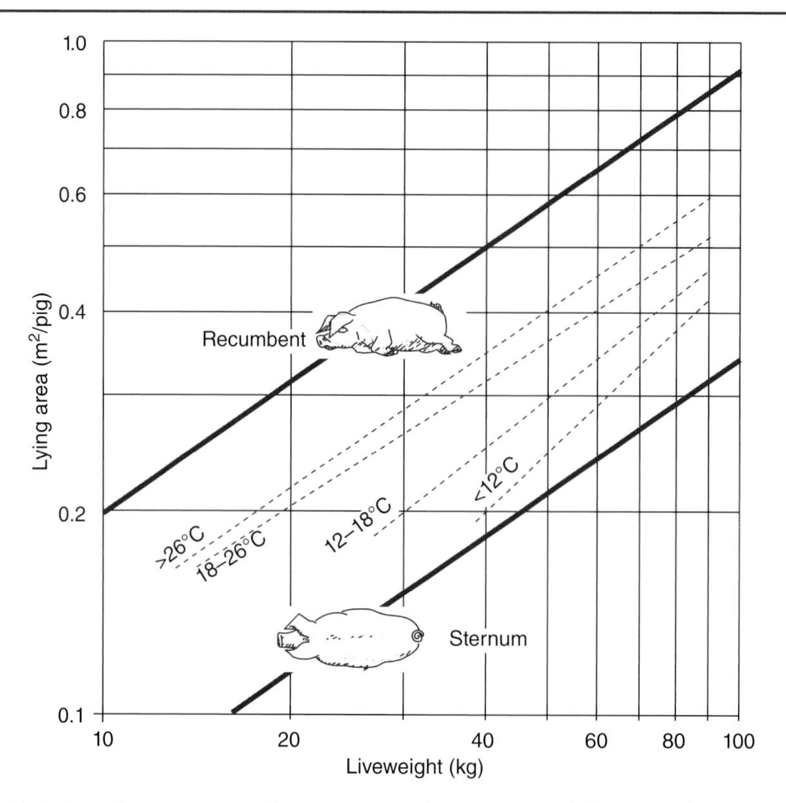

Fig. 10.2. Relationship between floor area occupied by resting pigs and air temperature (after Petherick, 1983b).

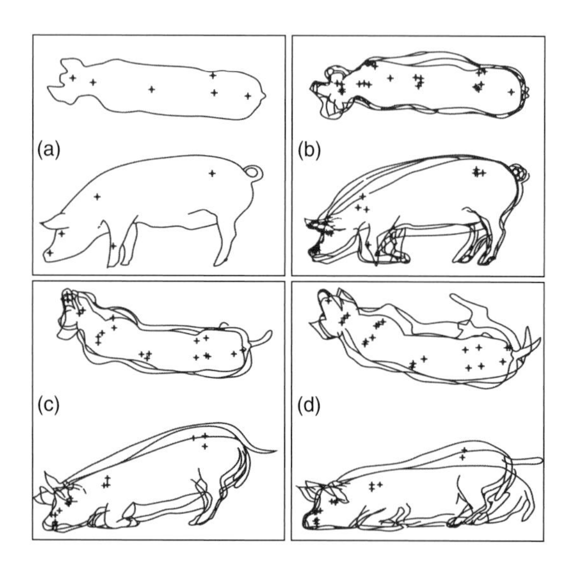

Fig. 10.3. Stages of movement in lying down by sows (from Baxter and Schwaller, 1983).

Trot

The trot is a symmetrical gait of medium speed in which the animal is supported by alternating diagonal pairs of limbs, so the sequence is LF and RH, then RF and LH (Fig. 10.5). The forelimbs are free of the ground longer than the hind limbs to allow the front feet to clear the ground in advance of the placement of the hind feet, on the same side. If there is a period of suspension between the support phases, the gait is referred to as a flying trot. A slow, easy, relaxed trot is called a jog or dogtrot. Sometimes, the term dogtrot is used to mean that the animal travels in a straight line with the hindquarters shifted to the left or right. The trot is occasionally also classified as ordinary trot, extended trot and collected trot. Standardbred horses use the extended trot in racing when the limbs reach out to increase stride length and speed. The hackney uses the collected trot, which is characterized by flexion and high carriage of the knees and hocks, a gait suited for snow or marshland running.

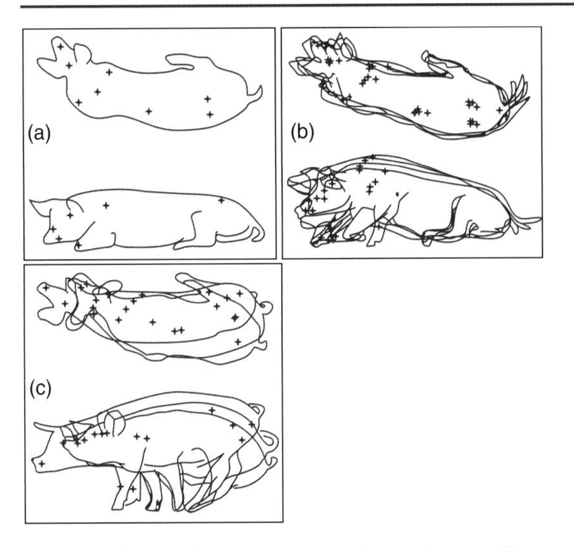

Fig. 10.4. Stages of movement in standing up by sows (from Baxter and Schwaller, 1983).

Canter and gallop

The canter is essentially a slow gallop. It is a three-beat gait with one diagonal pair of limbs hitting the ground simultaneously. The hoof falls are typically as follows: (i) one hind foot; (ii) the other hind foot and the forefoot diagonal to it simultaneously; and (iii) the remaining forefoot. Hence, the sequence written in comparable form to those above is: RH, LH and RF, LF. The canter is a gait in which the horse can use its neck muscles to advantage by the accentuated upward swing of the head, which helps to raise the forequarters and to advance the leading forelimb. As the horse tires in the canter it will bob its head more to utilize this minor auxiliary system.

The gallop is an asymmetrical gait of high speed in which the animal, during parts of the stride, is supported by one or more limbs or is clear of the ground. If the limbs are placed down in a circular order with two close together in time, then a gap, then the other two, e.g. RF LF, LH RH, the gait is referred to as a rotary gallop. If the hind limb leads through to a forelimb of the opposite side, e.g. RF LF, RH LH, the gait is termed a transverse or diagonal gallop. Horses seem to prefer the transverse gallop while dogs use the rotary gallop. In the transverse gallop the placement of the limbs, and therefore the support pattern, is transferred from the initiating hind limb diagonally to the forelimb of the opposite side. As in the rotary gallop, two forms of this gallop exist, dependent on the order of footfalls, e.g. LH RH, LF RF (a right lead transverse gallop) or RH LH, RF LF (a left lead transverse gallop).

Fig. 10.5. Horse in diagonal trot showing the soles of the diagonally opposite hooves (photograph R. Harnum).

The jump

Horses have large, bulky bodies and relatively rigid spines so they are not well adapted for jumping (Pilliner and Davies, 2004). Horses avoid jumping over ditches and obstacles only 60 cm high unless they have been encouraged to do so or have had experience in doing so. However, horses are capable of jumps that are impressive in longitudinal distance and height. As they prepare to jump, all horses strut with the forelimbs, thereby changing the momentum of forward movement into vertical impulse. In horizontal locomotion, the hind limbs hit the ground separately so as to maintain the rhythm and fluidity of stride. In jumping, however, these limbs hit the ground at a similar distance from the jump (approximately

2.00 m) and are spatially separated by only 0.09 m. This action of the hind limbs allows a more symmetrical and balanced upward propulsion. This synchronous and symmetrical action of the hind limbs at take-off allows these limbs to coordinate their movement in the air phase when the hind limbs are raised in unison following take-off (Leach and Ormrod, 1984).

In the approach to a jump, if the forelimbs landed in a synchronous manner while the horse was moving at this speed, it is likely that a smooth transition from horizontal to vertical movement would not be possible. Disruptive changes would result in reduced overall speed and efficiency of the jump performance. The lightened weight of the forelimbs, and the recoil resulting from the elastic storage of energy in these limbs from strutting, probably help the forelimbs to clear the jump.

Distance travelled

Animal routes are often seen, most clearly from the air, as pathways criss-crossing grazing ranges. They are even found in most paddocks. Animal walks occur frequently, even regularly, in normal domestic conditions.

The daily distances that grazing herbivores travel are largely affected by the location of food and water. In paddocks where water is close to good-quality food, the distance walked by animals is small and occurs mainly as movement while grazing, but if water is some distance from food then walking to and from water contributes a greater part of daily distance travelled. The location of water and the quality of forage have a significantly greater influence on the distance travelled by cows than forage quantity, stage of lactation of the cow, body condition of the cow or time of year. Horses at range, for example, may travel up to 65–80 km daily to water. Sheep will also travel great distances regularly to water where vegetation and water are both scarce. In normal conditions, sheep travelled 9–14 km/day in range conditions (Squires *et al.*, 1972). Daily distances travelled by cattle range from 0.9 km on a 0.1 ha paddock to 24 km under drought conditions on rangeland.

The Need for Exercise

Domestic animals have an apparent need for movement and try to exercise frequently (A.F. Fraser, 2010). Maintaining a minimum level of activity keeps animals physically prepared for any necessary movement. The high incidence of lameness in intensively kept farm animals is a consequence of high body weight in relation to leg strength in broilers (Knowles *et al.*, 2008) and genetic selection for high milk yield in dairy cows (Oltenacu and Broom, 2010). Lameness may sometimes be a result of lack of exercise preventing proper activation of limb joints, muscles and pedal tissues. The productive farm animals require daily opportunities for exercise. Horses certainly need exercise on a daily basis. If horses do not have exercise with the consequent focus on bone, tendon, ligament and muscle, they can develop in maladaptive ways if young, or lose normal function if older. Too much exercise can also cause problems (Goodship and Birch, 2001). It is now clear that excessive restrictions result in anomalous behaviour. The occurrence of and importance of exercise movement *in utero* is apparent in the studies on fetal behaviour.

Movement as a form of exercise is detectable in the mammalian fetus and in the chick embryo. Each normal fetus makes many thousands of movements during gestation. These movements have been likened to isometric exercises that promote good development of muscles and joints and prepare the fetus for post-natal life. Husbandry constraints, which inhibit locomotor activity, impose two main deficiencies of perception; reduced stimulus input and diminished sensory receipt of body movement feedback. The various sense organs in the animal's moving parts, such as tendons, joints and muscles, respond to mechanical action, movement, position, touch and pressure and constitute a major part of the sensory input of animals.

Episodes of wing-flapping in the red jungle fowl and the domestic chicken demonstrate that the frequency of wing-flapping in the chicken has not been altered by domestication. Vigorous wing-flapping still occurs in heavy, flightless strains of chickens. Evidently, neuro-behavioural processes can be stable throughout domestication even though the motor performances of motion can become diminished or eliminated.

If humans fail to take enough exercise during several weeks or months because they are in zero gravity in a spaceship, or because they are elderly and have too little energy or confidence to exercise, the consequence is osteopenia, reduction in bone mass. Lack of exercise in laying hens that are confined in a cage, where they cannot flap their wings and can take no more than a few steps, results in weaker leg bones and wing bones that have only 52% of the strength of the bones of hens that can flap their wings (Knowles and Broom, 1990). The bones of these hens are more likely to be broken when the hens are handled (Knowles *et al.*, 1993). Similarly, sows that are confined in stalls and have very few opportunities to exercise have bones 30% weaker than those of sows that live in more space in group-housing (Marchant and Broom, 1996). Sows do not often break their bones, but are more vulnerable to bone breaking if their bones are weaker. Hens or sows that have weak bones are less well able to cope with their environment, so their welfare is poorer than that of animals with normal bone strength. The welfare of those hens or other animals whose bones are broken is substantially poorer than that of animals with intact bones.

In many species, certain kinds of exercise are necessary for the efficient development of movements that promote survival. For example, a cat that has the opportunity to climb will have better muscle development for both prey-catching and movements that help in predator avoidance.

Further Reading: Domenici, P. and Kapoor, B.G. (2010) *Fish Locomotion: An Eco-ethological Perspective*. Science Publishers, Enfield, New Hampshire.

11 Exploration

Introduction

All domestic animals are strongly motivated to explore and investigate when they encounter a new environment. Only when the environment has become very familiar to them does the exploratory behaviour subside, but it reappears in an animal's behaviour after any change in its environment. Hence an animal appears to maintain a potential for generating activities that focus the senses upon additions, changes, salient features and novelties in its close environment (Fig. 11.1). Habituation to familiar stimuli occurs, although familiar areas may be revisited, presumably to check for changes. Exploratory behaviour equips the animal with a system of behavioural adjustability that can be brought readily into operation (Syme and Syme, 1979).

On some occasions, the animal is in a novel situation and this causes some degree of fear, so exploration may be motivated in part by this fear. Inglis (2000) explains the general importance of uncertainty reduction when the individual responds to feeling fear, perhaps by exploration behaviour. Because novelty or unfamiliarity can result in both fear and exploration, Hogan (2005) proposes that fear and exploration are a unitary system involving approach at low levels, withdrawal at moderate levels and immobility at high levels. This may be true for the example Hogan describes in which a young domestic chick is confronted by a novel mealworm, but the exploratory system is wider than this and is not always associated with fear, as explained below. In a study by Håkansson *et al.* (2007), two genetic strains of junglefowl reared in the same conditions differed in fear behaviour but not in exploratory behaviour.

Exploratory System

The exploratory system in behaviour is evident through many animal activities. This system can be outlined most simply as causal factors and consequent activities as follows:

- Need within the animal for perception of environmental characteristics.
- The activation of the exploratory behaviour.
- The receipt of sensory feedback from the environment and hence information that can be used.
- Reduction in the causal factor level as a result of sensory input.
- The return of the cycle to a basal level of readiness with the lodgement of the information in the memory.

Hence, *exploration is any activity that has the potential for the individual to acquire new information about its environment or itself.*

The role of investigative behaviour in the facilitation of learning is considerable (Toates, 1982, 2002). When a new stimulus is perceived, a state of heightened awareness is induced, more information is gathered and an addition to the animal's experience occurs. Some activities involve environmental testing and investigations, including sniffing or touching landmarks such as trees, bushes and turf; scraping earth and snow; rooting into soil and bedding; head-pressing on fences or upright structures; licking hard surfaces; and looking from vantage points.

The functions of exploratory behaviour

Each of the functional systems of animals requires that some exploration and investigation occur before it can operate effectively. The efficient exploitation of food sources must be preceded by location of these and estimation of efficient routes to use when acquiring food. Water sources and places where water loss is minimal must be found. Physical hazards must be located if they

Fig. 11.1. Young goats investigating new circumstances (photograph D.M. Broom).

are to be avoided successfully. Exploration of other individuals and of own ability must occur if adequate sexual and social behaviour is to occur. Most of all, exploration is necessary if effective anti-predator behaviour is to be shown. Animals that are cryptically coloured or that need to hide must find a good resting place. Those that flee or show other active responses to predators must investigate the characteristics of their surroundings and learn about them so that they can escape in the right direction and take advantage of opportunities for impeding pursuers. If predator attack is imminent, animals must scan their surroundings and be ready to respond to environmental changes.

Given the obvious selective advantage of showing exploratory behaviour, it is not surprising that animals give a very high priority to such behaviour. Barnett (1963) reported that rats explore a novel arena even if they have been deprived of food and are offered both food and the opportunity to explore. Cattle also show such investigation (Kilgour, 1975). The kind of exploration shown varies considerably from species to species. All animals must explore their own abilities. They need to determine their own characteristics, abilities, limitations, social status, etc., since they expend much activity investigating these factors. Animals constantly test their powers of reach and prehension in feeding. They can learn of their strength through pushing. They can also learn, through play, their personal powers of flight, attack and defence (Hall, 1998). Depriving animals of some sorts of exploratory behaviour is extremely aversive.

Active movement in a new area and responses to novel objects depend on sensory usage. A rat or dog put in a novel area walks around sniffing, but a bird or ungulate may explore by looking around. Objects are manipulated by primates but may not be touched at all by animals that cannot gain information in this way. Wild animals are much more cautious in their responses to novel objects than most domestic animals.

Factors affecting exploratory behaviour

Most animals put into a new open area are likely to follow the boundary before exploring the interior of their enclosure. In these initial activities groups they may be bunched in a closer intra-group spatial organization than usual.

A factor often associated with close confinement, is the absence of possibilities to explore, investigate and interact with social companions. Livestock held in social isolation show a variety of abnormalities of behavioural development. Isolation-reared lambs tend to avoid the situation of a novel environment by withdrawing and avoiding interaction with it. This demonstrates that socially deprived lambs have deficits in exploratory capacity. A deficit of this nature would not provide the animal with good survival prospects in a natural environment. Similar observations have been made on foals. Naturally mothered foals in novel situations are more active and vocal than orphan foals (Houpt and Hintz, 1983). Incomplete exploration is a poor basis for adaptation, and an animal with such a behavioural deficit must be at a biological disadvantage in many situations. Exploratory deficits will result in cognitive deficits, and associated fearfulness will probably create stress predisposition in various demanding husbandry situations.

Exploration and awareness

The key function of exploration is preparation for what might happen in the future. Will a predator attack and what shall I do? Will there be a shortage of food or competition for food and how shall I act so as to maximize the chance of obtaining sufficient food? If it becomes cold, windy or wet, where shall I hide? When such questions are put in the first person, it becomes clearer that the actions of the individual that explores are preparations for possible, perhaps predicted, future events. A substantial degree of awareness is needed to be able to evaluate and prepare for what might happen in the future.

Further Reading: Inglis, I.R. (2000) The central role of uncertainty reduction in determining behaviour. *Behaviour* 137, 1567–1600.

12 Spacing Behaviour

Introduction

Spacing behaviour is of considerable importance for social species that live in closely associated social groups. A pack of wolves or dogs, which can be considered as one species, *Canis lupus*, are clustered in an extended family group in relatively small areas for much of their lives. Dogs that have been living as companions to people rapidly form such groups when allowed to do so. The great fortune of humans is that they are sometimes allowed to be a significant part of such groups. In social species, cross-specific relationships of a mutually beneficial kind are frequent. Many dogs identify humans as important social partners and many humans identify dogs as important social partners. Occasionally, the individual regarded as a partner views the member of the other species as largely a provider rather than a partner. However, in this chapter on spacing behaviour, the social grouping perceived by the members of the group is relevant, so it is meaningful to describe the human–dog distance and interactions between the two. The dependence of many humans on canine company and the dependence of many dogs on human company is important to the lives of both, and studies of spacing of the dog and human partner are likely to demonstrate that proximity is monitored by both partners in many instances. Each exploits and benefits the other. Occasionally, humans force reluctant dogs to associate with them. There are parallels with humans doing the same to other humans.

Spacing of animals falls into two general types: (i) individual space that is defined in terms of the individual and hence moves with it; and (ii) home range and territory, which refer to a static area used by the animal. In order to appreciate why spacing behaviour occurs as it does, the ecology of the animal must be considered (Hediger, 1963). Spacing of social animals gives much information about social organization and is subject to dynamic change as individuals adjust their relationships continually. The spacing of the members of a social group at any moment depends upon the activities of the group members, as is clearly demonstrated by the work of Keeling and Duncan (1988) on domestic fowl.

Dogs, horses, rabbits, cattle, sheep, goats, pigs and domestic fowl allow fairly close physical proximity between one another, except in special circumstances related to sexual, maternal and aggressive behaviour. The distance they maintain between themselves and other animals, especially potential predators, is much greater. This latter flight distance is the radius of space within which the animal will not voluntarily permit the intrusion of man or other animals that might be dangerous without escaping. In domesticated animals, the flight distance to man shrinks with appropriate husbandry and human socialization (Hutson, 1984). Reactions to intrusions include startle, alarm, fight-or-flight display and vocalization.

When animals limit their movements to a home range this will include resources such as food. These resources will often be defended so that an area may become basically a substitute goal, representing the food, etc., it could provide. When an area is defended, it is called a territory. The term agonistic behaviour is often used for behaviour involving threat, attack or defence. However, groups of farm animals – especially those kept for long periods in extensive areas such as common grazings, ranges and ranches – devote little energy to fighting or threat. Through systems of social organization, group harmony is a prominent feature of collective behaviour (Broom, 2003). Conspicuous features of group behaviour are social facilitation and synchrony of action, so that the members of a group are often involved in the same activities.

Home range

The home range is the area that the animal learns thoroughly and that it habitually uses. In some cases the

home range may be the animal's total range. Within a home range, such as an extensive area of pasture, there may be a core area. This core is the area of heaviest regular use within the home range. The core area generally includes resting areas.

Territory

A territory is an area that is defended by fighting or by demarcation, which other individuals detect so that the mark or other signal is a deterrent to entry. It need not be permanent, but would often provide for requirements of nutrition, shelter, resting, watering, exercise, evacuation, periodic movement and defensive shifting. In many species, territories are used to attract a mate. A further advantage of defending an area in which the individual and its social group live is to reduce the risk that diseased animals will enter the area and transmit infection (Hart, 2011). Humans are not the only species that utilize quarantine.

Individual space

Most animals actively preserve a minimum distance from themselves and attempt to prevent others from entering this space. The minimum distance within which approach elicits attack or avoidance was called the individual distance by Hediger (1941, 1955). It includes the physical space which the animal requires to occupy for its basic movements of lying, rising, standing, stretching and scratching. An example is seen whenever birds such as gulls or swallows are standing or sitting in a line along a rail or wire. This space is somewhat expanded in the head region to accommodate the greater amount of movement of the head in the course of ingestion, grooming and gesturing. Such spacing is also evident in shoaling species of fish. Each fish maintains sufficient distance from its neighbours to allow swimming movements, except in shallow water (see Fig. 12.1). This is a distance both vertical and horizontal.

The advantages of maintaining individual distance to produce an individual space (Broom, 1981) can include reductions in: (i) damage to the body due to contact; (ii) interference and competition while feeding; (iii) impedance when starting to flee; (iv) disease or parasite transmission; and (v) the possibility of rape. Individual space can vary according to activity; for example, Walther (1977) reported that male Thomson's gazelles walk 2 m apart, rest 3 m apart but graze 9 m apart. Walther considered that the high figure when grazing could be due to the similarity of the grazing posture to threat display, but problems of interference with grazing behaviour may also be a cause. The term social space is sometimes used for the individual space maintained by an active animal.

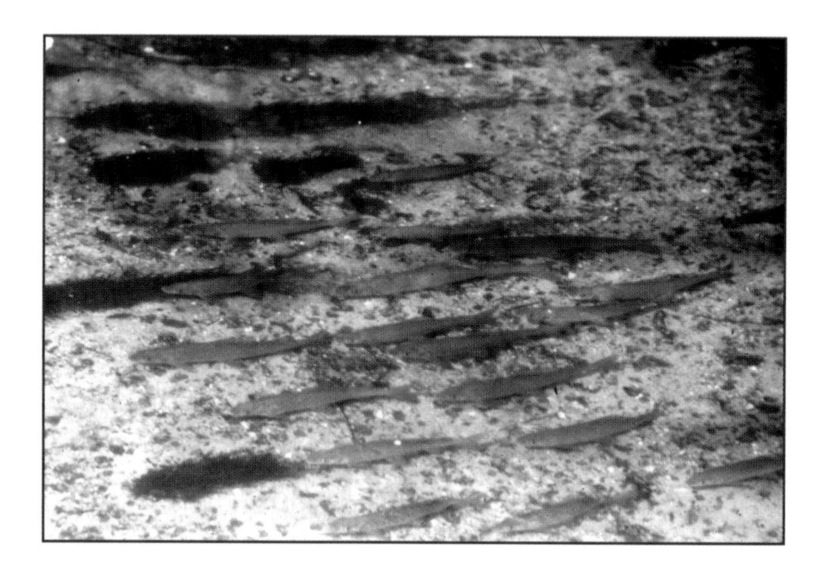

Fig. 12.1. Trout resting in shallow water and maintaining an individual distance from their neighbours (photograph D.M. Broom).

Spatial Features

Spatial features may be defined by local geography in significant ways for animals. The edge of a river, lake, wood or cliff could be important to an individual. A series of caves, an open space where no predator could approach, a marshy area or a refuge tree could be selected by an animal as a place to spend time. In man-made environments, too, particular types of space may be selected or avoided. For example, dairy cows milked in a milking parlour may show a preference for one milking place or one side of the milking parlour (Hopster *et al.*, 1998; Paranhos da Costa and Broom, 2001).

Association versus avoidance

Although domestic animals maintain individual space and sometimes defend territories, they also actively remain close to certain other individuals. Some of such association is between mother and offspring (see Chapter 19). Other association is between animals reared together or between animals that form an attachment later in life. Calves reared together are much more likely to be recorded as associating than are those from different rearing groups (Broom and Leaver, 1978). Pairs of cows that spend much time close together and often also mutually groom one another have been described by Benham (1984).

A special type of spacing behaviour results when individuals with close bonds arrange themselves so that they establish a particular distance to the nearest neighbour. Domestic herbivores, when grazing, maintain close contact with one and possibly more individuals, but the distance to that nearest neighbour varies much less than distances to other individuals in the group. In 60–70% of animals, an individual is the nearest neighbour of its nearest neighbour. This gives spatial structure of pairs of individuals within a group. The distance to second-nearest neighbour can vary considerably. Animals that are associated may move closer together when there is danger. The trout fry in Fig. 12.2 schooled closely when they detected a human observer.

Spatial needs

Spatial needs in domestic animals are both quantitative and qualitative (Box, 1973; Petherick, 1983a). Quantitative needs relate to space occupation, social distance, flight distance and actual territory. Qualitative needs relate to space-dependent activities such as eating, body care, exploration, kinetics and social behaviour. It is often necessary for animals to remove themselves from visual contact with others. Hence, the quality of space includes the presence of barriers, places where the head can be put so that others are not seen, or dark places where concealment from an aggressor can occur.

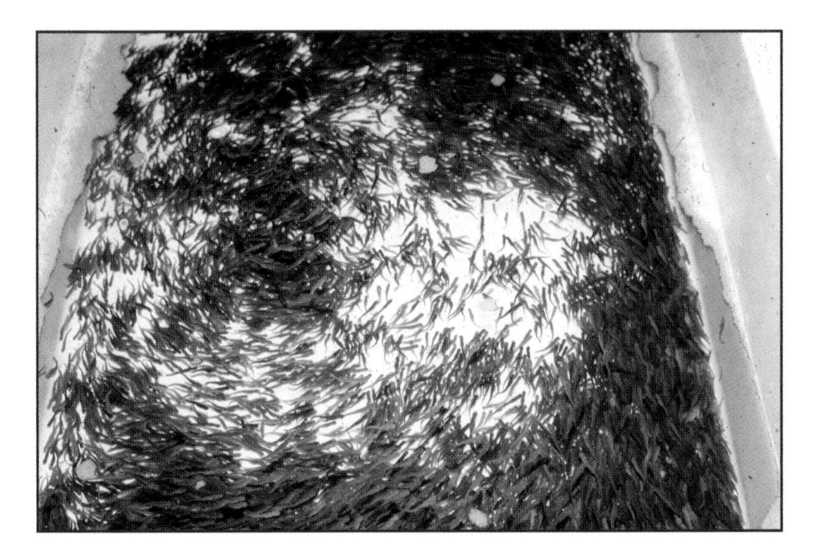

Fig. 12.2. Trout fry on a trout farm keeping close together when danger threatens (photograph D.M. Broom).

The minimum spatial need is for that amount of room that physical size and basic movements require. Each animal needs distances of length, breadth and height in which to stand, lie and move its major parts, including head, neck and limbs. In the acts of raising and lowering the body, the animal is required to make forwards and backwards movements (Baxter and Schwaller, 1983). During these the weight, or the centre of gravity, is shifted to or from the forequarters or hindquarters. The animal uses the weight either as a counterbalance in rising or as a direct pull in lying. Another need for length is in stretching the head and neck forwards and one or the other hindlimb backwards. Lateral articulation of the head and neck is yet another frequent and necessary form of movement. In lying, the animal's need for space is increased through partial rolling of the body and the extension of the hindlimbs.

As well as physical space, an animal needs space for social reasons. This space is used to keep some separation between itself and its conspecifics and to carry out avoidance behaviour (Mendl and Newberry, 1997). It is preserved in many instances by gestures of threat or intention. As with other spatial needs, individual space – including space to move the head – can be surrendered briefly without aggression occurring in a variety of circumstances such as short-term crowding and resting. The space occupied by an animal is a function of its weight and of its need for various activities. During a short period of confinement – for example, for weighing – domestic animals may be able to tolerate confinement in a space scarcely greater than that occupied by the body. However, a wild animal or domestic animal that has never experienced close confinement may be so disturbed by that confinement that its welfare is very poor and there is a substantial risk that it will die.

The relationship between space occupied and weight is a constant multiplied by the body weight to the power 0.67. This constant changes according to: (i) the extent to which the body shape of the animal deviates from a sphere; and (ii) the activities that are important for the animal to carry out. In calculations of the space required for pigs, Hans Spoolder (in EFSA, 2006b) modified equations for space occupation and took account of evidence for the space required by various activities in order to calculate what space a growing pig needed.

For all animals kept by people, the space provided must allow for normal locomotion if the conditions are to be called good. A general, arbitrary rule is that the animal should be able to carry out its normal form of locomotion for a period of about 30 s. For a cow, sheep, or pig in a field, the animal would run quite slowly about 130 m in that time so a field with sides of about 35 m and an area of 1200 m^2 would suffice. A parrot in an aviary might fly at least 250 m in the time but could turn several times so an area of 1200 m^2 might also suffice to meet its needs for exercise. A mouse might be able to turn twice as often as a parrot in 30 s so the distance travelled might be 3 m, and the area needed approximately 4 m^2. All of these figures are approximations but the area needed could be calculated using measurements of the speed of locomotion and turning ability for the species.

Crowding

Groups of individuals whose movements are restricted by the physical presence of others are said to be crowded. A high density means more likelihood that one animal will come closer to another than its individual distance. As a consequence, the intrusion into individual space may result in an aggressive response or an avoidance reaction that, in turn, results in a further such intrusion. Crowding does not necessarily result in increased agonistic behaviour but it often does so.

If a high social density causes adverse effects on the fitness of individuals then the term overcrowding is used. The distribution of resources is of great importance in determining whether or not individuals are adversely affected by high density. Crowding in the presence of many food sources, shelter sites, etc., can lead to no adverse effects, while local crowding around a single food source can be harmful to some of the animals present. In all situations it is essential to consider the quality of the living space (Box, 1973) as well as the social density.

Crowding has an effect on the extent to which animals move about. As broiler chickens get older they fill the space in which they are kept. Newberry and Hall (1988) monitored the movements of chickens from 4 to 9 weeks old at the low stocking density of 0.134 m^2 per bird. As the area occupied per week declined from 0.134 to 0.049 m^2, the distance moved per hour declined from 4.5 to 2.3 m. If the decreased space allowance decreased from to 0.134 to 0.067 or 0.041 m^2 per bird, there was a rapid decline in movement.

Very high densities of rodents can lead to adrenal hypertrophy, high blood pressure, kidney failure, impaired immune responses and reproductive failure

(Christian, 1955, 1961). The effects of crowding on production and welfare of farm animals are discussed in Chapters 29–33.

High social density is not the only factor in grouped livestock that can increase competitive behaviour and lead to adverse effects on individuals. A large number of individuals in the group can also have such effects. Al-Rawi and Craig (1975), working with domestic hens in battery cages, reported that for a constant cage floor space allocation of 0.4 m^2 per bird, the frequency of aggressive pecks was three times higher if group size was 28 than if it was 8. Individual hens interacted with each other bird and, at this density, such interactions involved much pecking. When group size was increased further in each cage, however, birds could not move around to interact with other individuals. Craig and Guhl (1969) found that the frequency of competitive interactions did not increase when group size was increased from 100 to 400 at constant stocking density. In a review of the effects of increasing group size on the amount of aggression in farm animals, Estevez *et al.* (2007) found that the social structure had an effect but, in general, increased group size is associated with reduced aggression. Although it is more demanding in terms of the cognitive ability of animals to live in large groups, our domestic animal species have sufficient ability to cope with this (Croney and Newberry, 2007).

Spacing Behaviour of Domestic Animals

Dogs

The behaviour of dogs when other dogs or humans enter the space that the individual dog regards as constituting a serious risk to its person or its territory is well known to the other dogs and to dog-owners. A growling dog may use its weapons, so continuation of approach increases the risk that it will bite. The attack may be presaged by louder and more frightening signs of intent. The speed of movement associated with any bites is often a surprise to a human, although probably not to another dog. Domestic dogs living in houses may defend a very small territory, such as a sleeping box or just the square metre around a food bowl. However, other dogs may be just as vigorous in defence of a house, a garden or commercial premises.

It is this latter propensity that has led to the widespread use of dogs to guard property. Dogs encouraged or trained to defend an area may do so when the intruder is a human whom the trainer does not intend to be excluded or attacked; for example, an employee, a small child or a public employee on legitimate business. Many postal service employees know of the risky dogs on their delivery round. One might wait by the letterbox to grasp and maim the hand that puts letters through the box. Another might attack and bite any person who enters the front gate, ripping with the teeth at the legs or even the face if able to do so. A few dogs have been deliberately trained to do so. Others have been accidentally trained to do so or have not been prevented from doing so because of poor owner control. The biting and shaking with the canine teeth shown in territorial fights between dogs is very occasionally addressed to humans.

Male dogs mark with their urine the space that they defend, as well as space that they visit but do not defend. The urine mark conveys information about identity, sexual condition and sometimes quality and size. A large male dog can urinate higher than a small dog and higher levels of testosterone may be evident from the height. The responses to these marks will have an effect on the spacing of the animals. An individual dog, or a pack of dogs, may use such marks to help to defend a territory.

Cats

Cats mark certain individuals and objects by means of their facial and tail-base glands and also, in the case of males, by urine-spraying. The gland on the side of the face, between the eye and the side of the mouth, is evident as a darker region in tabby cats. The glandular area at the base of the tail may also be darker in tabbies and some other breeds. The mark deposited indicates that this particular cat has been present and in control of an interaction at that point. Many cat owners do not realize that the cat rubbing against them may just be indicating ownership. Cat owners are usually more disturbed by the urine spray-marking of tomcats.

Active, aggressive and noisy nocturnal defence of territory by cats is a feature of human urban environments. The resolution of disputes can lead to injuries and may result in some individual cats showing very great reluctance to venture outside the human habitation in which they are relatively safe. Cat owners may fail to understand that for there to be six cats in 20 adjacent, small human houses causes much violation of feline territory. The behaviour during attacks is not often observed, but includes some clawing and much biting to the head area.

Cattle

Cattle show territorial aggressive acts by butting or threatening to butt. When another individual's head is close, they 'hook' with sharp, oblique head-swings. In small paddocks, an older bull becomes strongly territorial: in this territorial display the bull digs the soil with its forefeet, scooping loose soil over its back. It horns ruts into the soil, rubbing its head along the ground. At the conclusion of the display it will stand and bellow repeatedly. It is common for a specially selected site, or stand – which may be a prominence of land – to be used for this display.

Beef cattle under high crowding conditions in pens show clear preferences for certain locations within their pens. They tend to position themselves around the outer edges and corners of pens, the central areas often being less occupied. The fact that the animals use the outside suggests that the ratio of perimeter to area of enclosure is important. The more crowded the animals, the greater the importance this relationship is likely to be in alleviating adverse effects of crowding.

Distances between subgroups in cattle increase as forage conditions deteriorate but individual space is maintained, even on rangeland and on intensive pasture. Bulls are often located on the margins of the herd. Bulls require a great social distance at a relatively younger age than steers, and the presence of heifers increases the distance between bulls, at the same time reducing herd scatter. Cows, when lying, mostly keep within 2–3 m of one another, compared with 4–10 m when grazing. In unlimited grazing space, bulls maintain an individual space averaging 25 m radius.

Horses

In territorial aggression, horses fight with their own typical offensive and defensive weapons (Arnold and Grassia, 1982.) They may bite, kick out with their hind feet and strike with their forefeet when the flight distance is violated. Both hind feet may be kicked out, for example, after backing up to the intruder. The double kick is delivered directly to the rear without aim. Kicking out defensively with one hindlimb is sometimes called the mule kick, although it is used by equids generally. The mule kick is a precisely directed kick with one hind foot. Horses usually share a common grazing area with other grazers without aggression, but may show aggression to cattle and sheep grazing with them by biting, kicking or chasing them. Free-ranging horses spend much of their time grazing, about 12 h or more being common. In relatively arid areas, horses often travel very long distances daily to water. Their grazing range, under these conditions, is probably limited by the availability of water. When snow is on the ground, horses can obtain fluid by eating it and are then independent of running water and can utilize different ranges. In free-ranging horses, trips to water and salt each day are a requirement. Horses will also show preference for certain areas of shade if there is a choice.

Equine behaviour includes territorial rituals. Stallions pass faeces in specific sites where their dung may become heaped. All horses use certain territorial spots for defecation. These areas are not grazed. A restricted grazing territory for horses soon becomes divided up into 'lawns' and 'roughs'. The lawns are the areas closely cropped and the roughs are the dung areas, which remain ungrazed except in starvation conditions.

Sheep

Sheep threaten one another by means of head movements. If no submissive response occurs they may push, butt or tug at wool. Among animals that know one another well, however, a look suffices as a threat and leads to avoidance. In sheep, spacing among individuals varies considerably with breed and location. In extensive moorland and mountain country, sheep keep a greater distance apart than sheep on lowland ground. This may be a matter of adaptation or the result of the wider dispersion of suitable food. In situations of dispersion, e.g. mountainous areas, the modal distance to the second-nearest neighbour is three times that of the nearest neighbour on lowland grazing. This is the result of the persistence of 'pairing', which is a feature of permanent hill flocks. If a flock is dispersed for any reason, pairs and small groups still go around together, so social cohesiveness is preserved. Within each flock home range there is a series of overlapping home ranges, each with its group of sheep.

The associations formed between individual sheep in natural flocks of Dorset Horn, Merino and Southdown sheep were studied in Australia (Arnold *et al.*, 1981). When the Dorset Horns were grazing, the associations between individuals were within 'feeding' home ranges. In the Southdown, the individuals associating together used widely dispersed areas of the paddock rather than one general area. Merinos usually remained in one group, dispersed into subgroups only under extreme food shortage, and then the sex and age groups segregated as in the other two breeds. Scottish Blackface form subgroups under nearly all conditions irrespective of breed, flocks of sheep drawn from different sources not rapidly integrating into a socially homogeneous group. Even when sheep of the same breed, but from different flocks, are mixed they may take a long time to integrate (McBride *et al.*, 1967). This effect is shown in Fig. 12.3.

Modal distances between nearest neighbours in sheep of different breeds show a range from 4 to 8.6 m on mountain and moorland and from 3.4 to 4.4 m on lowland grazing. These figures show that pairings are readily formed by sheep and that they can be managed better in pairs. For example, the design of the transition zone from a pen to a race can be improved by allowing two sheep to enter the race in parallel.

Pigs

Pigs threaten to obtain or defend space by means of head movements, which may be followed by barking, pushing with the nose and shoulder barges if no submissive response is given (Jensen, 1982, 1984). The stocking density in groups of pigs is known to have various effects upon their behaviour. Social encounters in penned pigs take place most often at or near the food source. These social encounters lead to non-injurious results when a hierarchical system has previously been well established. When growing pigs are allocated inadequate space, there is a rise in severity of social encounters. When the stocking density is too high it is found that individual pigs that are low in the social hierarchy are unable to avoid the consequences of aggressive encounters. The productivity of the unit is thus adversely affected.

Pigs, more than other farm animals, practise intense contact behaviour and show comparatively little territorialism except under feral or semi-feral conditions. The space in the region around the head, however, tends to be preserved by most animals. In addition, members of a group often observe a 'social limit', which is the maximum distance any animal will move away from the group. They do require space, however, to employ avoidance tactics that mitigate aggression. The recognition, when managing pigs, of an 'avoidance system' in the social behaviour is very important.

Avoidance is an inverse response in agonistic situations and it is the positive factor affecting agonistic control. The idea that aggression establishes and operates the social system is inadequate. The avoidance behaviour, which often occurs in the absence of any aggressive act, is the vital component of the behavioural mechanism that generates social stabilization. Of course, avoidance is an overall strategy, which calls for specific behavioural tactics, and these tactics need space for their practical operation.

Poultry

Territorial behaviour changes with season in feral domestic fowl. In the breeding phase, well-defined territories are held by dominant males where they mate with females. These raise a brood nearby in a home range. For the rest of the year each male has a harem, and the harems utilize overlapping home ranges (McBride *et al.*, 1969; Duncan *et al.*, 1978). Chickens also display home range behaviour under intensive conditions.

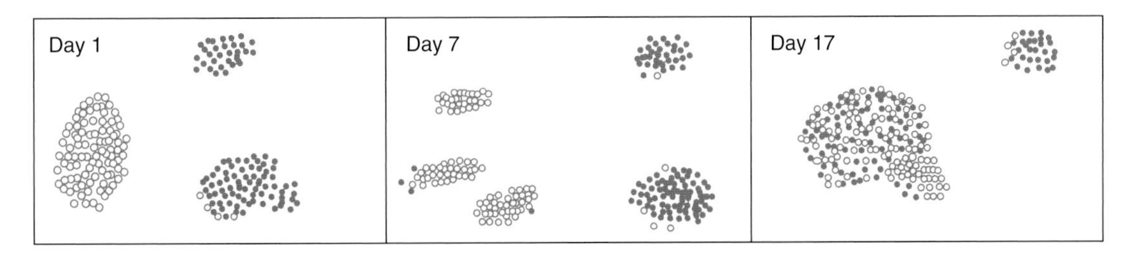

Fig. 12.3. When two groups of 100 Merino ewes (• and ∘) were put together at a density of 15/ha, their resting distribution was as shown on the 1st, 7th and 17th days. Complete flock integration took 20 days (from Broom, 1981, modified after McBride *et al.*, 1967).

Cock-crowing occurs more in establishing territory. With adequate territory, cock-crowing is usually limited in time mainly to early morning. Where cockerels are crowded together and actual territory is nil, as in cage systems for artificial insemination purposes, excessive crowing can occur.

Threats by males depend on the distance between them and the orientation of the subordinate towards the dominant bird. High-ranking birds are frequently in front of and beside the food dispenser, but perches may be used heavily by the low-ranked animals. The perch seems to serve as a sort of refuge rather than as a controlling position.

Further Reading: Croney, C.C. and Newberry, R.C. (2007) Group size and cognitive processes. *Applied Animal Behaviour Science* 103, 215–228.

Rest and Sleep

Introduction

Passive punctuation of behavioural sequences with less active periods interspersed among the most energy-consuming phases produces rhythms of living (Ruckebusch, 1974). In their simplest forms, self-conserving tactics include ad hoc resting and short-term inactivity. A significant proportion of life is spent at rest (Meddis, 1975; Valros and Hänninen, 2009). Resting and sleeping are governed by timing controls more obviously than some other cyclic activities:

1. Standing and lying. These are terms describing body position in relation to the ground. At some times, animals assuming these positions may carry out no other recognizable behaviour except for occasional limb- and body-shifting.
2. Drowsing. A state of wakefulness alternating with light sleep with head movement and eye closure. The animal may have a fully upright stance, such as is usual in the horse. Sitting on the sternum is a common postural form of drowsing in several other species, including cattle. In the latter case, no rumination would occur. Dogs will lie on the sternum with the forelegs flexed under or extended with the head positioned between them. In poultry the neck is withdrawn and the tail held down.
3. Resting. Typically, rest is taken in a recumbent posture with evident wakefulness. The posture adopted results in reduced energy utilization, but sleep does not occur and other activities such as grooming and rumination can be carried out during resting.
4. Sleeping. Sleep is defined by brain changes and by loss of behavioural responses to many stimuli. True sleep occurs in the form of both 'brain sleep' (with electroencephalogram delta waves) and 'paradoxical sleep'. In the latter, rapid eye movements (REM) can be seen below the closed lids. Minor leg movements also occur, especially of the distal limb parts including the digits.

Detailed monitoring of sleeping individuals has led to a better understanding of physiological events during sleep (Shepherd, 1994; Smolensky, 2001). The function of rest and sleep may originally have been to minimize the danger of predation when active behaviour was not necessary. An immobile individual in an inconspicuous position is less likely to be detected. Sleep is shown by predators as well as by prey, although this does not prove that minimizing danger is not a function of sleep, as almost all predators are also subject to predation, especially when young. A second function must be energy conservation. For some kinds of animals and in some circumstances a function may be restorative, allowing metabolic recovery. A third function may be the replaying of events during waking life, now known to involve the hippocampus (Bendor and Wilson, 2012). Whatever the evolutionary origin, this system, requiring periods of passive behaviour, is of importance in the maintenance of the animal and receives high priority in environments to which the animal is adjusted.

A biological clock measures time and issues temporal signals to the rest of the body in each domestic animal. Rhythms of brain activity also occur (see Table 13.1). Rhythmic activity is also a feature of the endocrine system. A system in the body exists that adapts the rhythms of organs and cells to the environmental temporal cues such as photoperiod stimuli (see Chapter 2). Cells in the lateral hypothalamus show a diurnal rhythm of responsiveness to stimulation. The reticular formation may programme sleep and wakefulness, but the suprachiasmatic nucleus in mammals and the pineal in birds are clearly of major importance in the control of rhythms. Two forms of sleep are slow-wave sleep (or quiet sleep, SWS) and paradoxical sleep (or rapid eye movement, REM). In SWS, the EEG is characterized by synchonous electrical waves of high voltage and slow activity; in paradoxical sleep, the EEG shows low voltage and fast activity similar to that seen in the wakeful state, but there is very little muscular activity

Table 13.1. Types of EEG waves in the mammalian brain.

Type	Rate	Site	Behaviour
Alpha rhythm	Medium	Neocortex	Relaxed wakefulness
Beta activity	Medium/fast	Neocortex	Alert wakefulness
Delta waves	Very slow	Neocortex	REM sleep; deep sleep
Theta activity	Slow	Hippocampus, etc.	Deliberate activity; orientation
Sleep spindles	Medium	Neocortex	Drowsing; moderately deep sleep
Olfactory rhythm	Fast	Olfactory bulb; pyriform cortex	Alert

and the animal is more difficult to arouse than when it is in slow-wave sleep.

The following sleep stages have been defined:

- Awake: normal responsiveness to stimuli. EEG either beta activity or alpha rhythm.
- Drowsing: a transitional state in which the animal is sluggish; the EEG is irregular in rhythm and of variable low to medium voltage in amplitude.
- Light sleep: EEG shows sleep spindles; animal is readily awakened.
- Moderately deep sleep: sleep spindles interrupted by delta waves and there is REM.
- Deep sleep: the EEG tracing consists largely of delta waves. This stage can include REM.

The depth of sleep is usually determined in terms of the intensity of stimulation – usually of a sound stimulus – required to awaken the sleeping animal, and increases in the order given above. REM sleep occurs in all mammals, but not all the manifestations of REM sleep are exactly alike in all species. Animals are more difficult to arouse from REM than from other sleep states. Heartbeat and breathing tend to be more irregular and, on average, faster during REM than during non-REM sleep. Even though phasic movements are common, postural muscle tone diminishes, especially in the muscles of the trunk, neck and shoulder. While the rate of cerebral metabolism is slightly lower in non-REM sleep than in the awakened state, in REM sleep it is slightly higher. The duration of sleep varies among species. While a fox may sleep for 14 h/day, sheep, rabbits and horses typically sleep for 8, 7 and 5 h/day, respectively (Sjaastad *et al.*, 2003).

While it may be perfectly valid to count as secondary benefits of sleep the recovery of tired body systems, or the behavioural advantage of staying out of sight of enemies, there is evidence that processes occurring during sleep contribute to the health of the brain. There are two possible ways in which sleep can serve cerebral functions: (i) it could be that during sleep, neural materials that were consumed during waking are recovered or re-synthesized; or (ii) it may be that, during sleep, neural waste products that have accumulated are eliminated. However, the sleeping brain is not inactive, but glucose metabolism is reduced in both SWS and REM states. Young animals sleep more than older ones and adults, and also spend a large fraction of their time in REM sleep. The time spent in REM and total sleeping time varies between adults, but it is relatively stable for any one individual. Lost REM time disturbs animals. The problem with the brain recuperation theory is that animals of some species sleep much less than others and some individual people are able to go for many months without sleeping. Perhaps the recuperative processes can occur during non-sleep states, even if they are normally best carried out during sleep.

Postures during Sleep and Sleep Deprivation

A range of postures are used by animals about to sleep. The cat in Fig. 13.1 is lying with the back of the head on the ground, a common posture in cats. Cattle and sheep take up stationary resting attitudes, sometimes standing while ruminating. Hoofed livestock can adopt the upright inactive stance during a form of rest. No other type of foot is as suitable as the hoof in stationary positions on surfaces of various types and conditions, ranging from mud to ice.

The single hoof of the horse is better than the cloven hoof for high-speed locomotion, but the equine limb is also well designed for standing. In the horse, support in standing is provided by arrangements of

Fig. 13.1. Cat resting with top of head on the ground (photograph R. Harnum).

ligaments and tendons, which form strong, flexible, elastic supports. This helps the animal maintain a standing position with very little muscular effort. Both fore- and hindlimbs have the 'stay apparatus', which is mainly ligamentous, while the 'check' and 'reciprocal' apparatuses in the fore and hind, respectively, are composed of structures that are either completely tendinous or are tendinous projections of certain leg muscles. Both stay and check apparatuses function mainly in supporting the fetlock joints, and bind and brace the sesamoid bones. The suspensory ligaments of both fore- and hindlimbs assist in this. The reciprocal apparatus is both muscular and tendinous, being made up of an extensor and a flexor muscle with the long tendinous insertion of each. These muscles, peronous tertius in front and superficial digital flexor behind, are in combination with the distal ends of both the femur and the tarsus to form a parallelogram, which causes the stifle and hock joints to articulate in unison. Thus, if the stifle is maintained in extension, the distal leg will bear the weight of the horse with little additional muscular effort. Horses are thus able to drowse, and even engage in SWS, while standing (see Fig. 13.2).

Cattle, unlike horses, are unable to rest satisfactorily in an upright stance for extended periods. They therefore become fatigued more severely when movements or husbandry disturbances prohibit recumbency. Sternal or lateral recumbency positions allow recovery. The state of wakefulness occupies 85% of the 24 h period in the herbivorous species but only 67% in pigs. Ruckebusch (1972a, 1972b) found that horses slept only during the night, and cows and sheep mostly at night. With regard to position, horses spent 80% of this time standing; sheep 60%; while cows and pigs assumed a recumbent attitude 87% and 89% of the time, respectively. In addition, confined pigs spend a very high proportion of their daylight hours in recumbency. Ruckebusch and Bell (1970) observed that sleep was generally distributed in two or three periods during the night, and during each of these periods transition from SWS to REM sleep was usually repeated three or four times. REM occurred in horses, cows and sheep only during the night-time, but in pigs it was not restricted to this period and these animals had a higher incidence of sleep than the other species, except when preoccupied with food.

Pigs exhibited the greatest number of periods of REM sleep both during the 24 h period (33) and during the night-time (25), those of shortest mean duration being 3.2 and 3.0 min, respectively. The smallest number was recorded from sheep (seven), and horses showed REM sleep periods of longest duration (5.2 min). The ratio of REM sleep to total sleep was highest in horses and lowest in sheep. Pigs also exhibited a high ratio of REM sleep to total sleep.

Patterns of sleep, for example cyclic and breathing disorders (Vandevelde *et al.*, 2004) can be used as an indication of stress. Disinclination to lie at rest is seen

Fig. 13.2. Horses resting in the standing, sternal recumbency and lateral recumbency positions (drawings A.F. Fraser).

in horses with orthopaedic conditions. The normal sleep and resting characteristics of animals should be appreciated for purposes of assessing welfare so that abnormalities, which may have symptomatic significance, can be detected. A horse lying down at night in a normal posture is probably asleep. An adult horse that lies down during the day (unless in the company of its foal) or rests in sunshine may be abnormal and should be carefully observed for other evidence of illness. Significant clinical signs include sleeping fitfully and rising from resting postures frequently.

Management practices should interfere as little as possible with normal circadian patterns of maintenance behaviour. Interruption of activity cycles and loss of sleep may play an important role in the aetiology of the stress-related diseases associated with newborn management, livestock transport, mixing of strange animals and introduction of new animals into established groups.

Species-typical Features

Dogs

Dogs spend about 50% of their time sleeping and about 20% of total sleep in REM sleep (Manteca, 2002). The characteristic posture of the sleeping dog is lying with the head turned back on one side. Dogs may make leg movements and vocalize during REM sleep. In general, dogs are more likely to sleep during the hours of darkness than during the day, but may be more active at night if hunting for food or a mate require it.

Cats

Cats spend 47–65% of each 24 h asleep and, for more than 20% of this time, the sleep is REM sleep (Manteca, 2002). They may sleep on the sternum with the legs partly folded or laterally with the legs stretched out. Sometimes, if a perfect sleeping structure is found, the legs may point upwards (Fig. 13.3). Young cats of <17 days of age have a higher proportion of REM sleep than older animals. The proportion of time during day and night that cats are active depends upon the extent to which they have to, or want to, hunt, as most serious hunting is nocturnal.

Cattle

Rest and sleep are important for cattle, as for other farm animals, and a free opportunity for such behaviour is a necessity. At pasture or in a cubicle-house system where one lying place is allocated per animal, individual cows have ample opportunity for lying and resting. In a cubicle-house system when there are more cows than there are cubicles, this condition changes (Wierenga, 1983). Under these conditions the lying time of cows is reduced in proportion to the amount of over-occupation. The effect is reduced lying time among the low-ranked cattle due to a shortage of lying places during the night, when cows prefer to rest. However, they are capable of indulging in non-REM, but not REM, sleep while standing. By keeping them standing for 24 h, selective REM sleep deprivation can occur (Webb, 1969). Preventing recumbency at night results in partial sleep deprivation, since food intake during the day reduces available sleep time. In non-REM sleeping episodes, cattle progressively

Fig. 13.3. Dog and cat sleeping (photograph A.F. Fraser).

exhibit irritability towards the presence of personnel and the time spent in drowsiness is halved.

During the day, cattle usually rest in sternal recumbency while ruminating. A total time of less than 1 h is likely to be spent in lateral recumbency, and episodes of rest in this position are normally brief. These may be associated with periods of sleep. The forelimbs are curled under the body and one hind leg is tucked forwards underneath the body, taking the bulk of weight on an area enclosed by the pelvis above, and the stifle and hock joints below. The other hindlimb is stretched out to the side of the body with the stifle and hock joints partially flexed. They will occasionally lie with one or other foreleg stretched out in full extension for a short period. Cattle will also occasionally lie fully on their sides, but do so only for very short periods while holding their heads forwards, presumably to facilitate regurgitation and expulsion of gases from the rumen. Adult cattle take up the sleeping position seen in calves with their heads extended inwards to their flanks. This is certainly a normal resting and sleeping position taken by adults periodically, although it is also a posture typically adopted in milk fever.

Horses

Horses are polyphasic animals as regards sleep or rest periods: 95% of horses have two or more such periods per day. The total length of time spent recumbent per day is approximately 2.5 h, with slight variations associated with age and management. Twice as much time is spent in sternal recumbency as in lateral recumbency, and normal adult horses rarely spend more than 30 min continuously in lateral recumbency: the mean time spent continuously in this position is 23 min.

Horses lie down and rise in a specific manner. In descent, the forelegs are flexed first, then all four limbs; the head and neck are used for balance as the animal lowers itself. In sternal recumbency, horses do not lie symmetrically: their hindquarters are rotated with the lateral surface of the bottom limb on the ground. In lateral recumbency, the upper limb is invariably anterior to the lower forelimb, which is usually flexed. The hindlimbs are usually extended with the upper limb slightly posterior to the lower limb. To rise, the horse extends the upper foreleg. The forequarters are raised so that the lower foreleg can be extended. At the same time both hindlimbs begin to extend, but the main thrust of rising comes from the hindlimbs.

A sound horse seldom remains lying when it is approached, probably because a standing horse is better able to flee or to defend itself. Horses may be able to drowse and even engage in SWS while standing by means of the unique stay apparatus of the equine limbs. REM sleep, however, occurs almost always while lying down. Adult horses do not lie for very long periods.

Mares with young foals tend to lie longer than usual when the foal is nearby and sleeping in full lateral recumbency. Mature horses are unable to lie in this flat-out posture for long periods of time as their respiratory functions become impaired. The full weight on the thorax of the horse, when laid flat, appears to be such that circulation to the lungs becomes inefficient after about 30 min. This is not the case among foals and young horses, however, and these subjects can be seen spending many hours in the day sleeping on their sides at full stretch.

During the day, the horse is awake and alert over 80% of the time. At night, the horse is awake 60% of the time, but drowses for 20% of the night in several separate periods. Stabled horses are recumbent for 2 h/day in four or five periods. Ponies are recumbent for 5 h/day and REM sleep occurs in about nine periods of 5 min average duration.

Management practices can affect equine sleep patterns. When moved from stable to pasture, they do not usually lie down during the first night and total sleep time remains low for 1 month. If horses are tied too short in a stall, so that they cannot lie, they do not have REM sleep. Horses stabled only at night may limit sleep until they are free during the day. Sleep deprivation may occur when they are transported long distances, especially if tied. Diet also affects sleep time in horses, as it does in ruminants. If oats are substituted for hay, total recumbency time increases; fasting has the same effect.

Pigs

Pigs may spend much time resting and sleeping if their movement is restricted but are more active in good environments. Pigs confined in stalls may be recumbent in rest or sleep for as much as 19 h/day. However, pigs free to move around sleep much less. Of the total sleep time, SWS occupies 6 h/day on average and REM sleep averages 1.75 h in about 33 periods.

Pigs are characterized by extreme muscle relaxation during sleep. While this is difficult to appraise in a mature sow, young piglets exhibit profound relaxation.

Sheep

Sheep are awake for about 16 h/day; SWS occupies 3.5 h/day and REM sleep occurs in seven periods of an average total of 43 min. Sheep can exercise for 5 h/day without showing signs of substantial fatigue (Cockram *et al.*, 2012).

Poultry

The natural resting characteristics of poultry have a number of definite characteristics. Drowsing, whether sitting or a standing position, consists of two stages: (i) the neck is more or less withdrawn, the head is moved regularly and the eyes are open; the tail is slightly down; and (ii) the neck is withdrawn, the head is motionless and sometimes drooped; the eyes are closed or are slowly opened and closed; feathers are slightly fluffed; the wing may droop; the tail is down; and, while standing, a slight crouching posture is adopted. During sleep, whether sitting or standing, the head is tucked into the feathers above the wingbase or behind the wing; feathers are slightly fluffed; sometimes the wings droop; while standing, a slight crouching posture is shown and the tail is down.

When perching for the night, domestic fowl may regularly choose a particular place. This can result in competitive behaviour during the phase of initial perching. When the hen assumes the sleeping posture, it moves its head slightly forwards then turns it backwards. It puts the beak at the proximal end of the wing and pushes the head between the wing and the body while making a vibrating movement with the head. This movement also occurs in ducks. While sleeping, jerky movements of the head and neck occasionally occur and sometimes there is a soft, peeping noise. External disturbances during drowsing cause an alert posture; arousal from sleeping requires stronger stimulation than from drowsing.

Resting postures are very uniform among birds. In the sleeping posture, the head is above the centre of gravity, and this may give stability to the bird in situations of muscular relaxation during physiological sleep. The sleeping posture could be a behavioural adaptation, reducing heat loss via comb and wattles and closing the vent. Although all poultry rest in a sitting as well as in a standing position, sitting is most common. Sitting is a very stable position, as poultry possess a mechanism by which the feet close around the perch when the animal sits. Sitting is advantageous from the point of view of energy conservation, as metabolism is 40% less and heat loss decreases by 20% as compared with standing.

While most maintenance behaviours are concentrated in the light period, preening occurs during the night, probably because this is the only behaviour that can be performed adequately and efficiently on a perch in the dark. During the night, all poultry normally rest on perches. Perching during the day is associated with a short resting or preening bout, and the lower perches are preferred for this. About 1 h before sunset, birds start perching for the night. Usually, it takes 30–60 min for the whole flock to take their places. Much flying on and off the perches is seen during this period. Fowl prefer to use the higher perches, but some individuals also use the lower ones. Poultry resting and sleeping on perches draw the head and neck close to the body and grasp the perch firmly with their feet, maintaining this position for several hours. In cages with sloping floors, there is a preference to sleep on the highest available place on the slope, though this is apparently a poor substitute for a perch, which is preferred whenever present.

In poultry, the genetic strain is an important factor affecting perching behaviour. Birds of some strains seldom use high perches. These strain differences are due neither to weight disparity nor to social pressure. Some birds perch whereas others are non-perchers. Appleby (1985) has shown that early experiences affect later perch use. The type of material from which the perches are constructed, such as wood or wire, has little effect on behaviour, but perch shape and size do have an effect.

Further Reading: Valros, A. and Hänninen, L. (2009) Behaviour and physiology. In: Jensen, P. (ed.) *The Ethology of Domestic Animals*, 2nd edn. CAB International, Wallingford, UK, pp. 25–37.

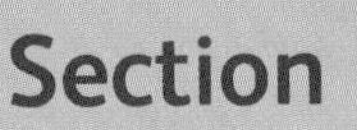

Section 4

Reproductive and Social Behaviour

14 General Social Behaviour

Introduction

Most domestic animal species live socially when wild and the individuals actively associate with one another when they can do so in domesticated conditions. Animals living in long-lasting social groups tend to show social facilitation of behaviour and to be involved in synchronized activities. The social living and social behaviour associated with this are generally beneficial to individuals and promote their survival. Where a species has complex social organization, the individuals have some dependence on the others in their group and may find it difficult to adapt to isolated living conditions.

Animal relationships with man, when successful, range from the near familial to the habituated, but in all events the domestication of various species has been helped by a capacity for social affinity. Domesticability ensures its own type of survival, different from natural survival perhaps, but no less biologically real.

Some terms used in the description of social behaviour are listed below (after Broom, 1981). These concepts are discussed further in this chapter.

- *Physical structure: the size of the group and its composition in respect of age, sex and degrees of relatedness of group members.*
- *Social structure: all of the relationships among individuals in the group and their consequences for spatial distribution and behavioural interactions.*
- *Group cohesion: the duration of association of the members of the group and the frequency of fission in which one or more members leave the group.*

Some terms used in the description of social structure concern the roles of individuals in the group:

- *Leader: the individual that is in front during an orderly group progression.* Other colloquial uses of this word are best replaced by different terms in order that the descriptions used in behaviour research should be as precise as possible.
- *Initiator: the individual that is the first to react in a way that elicits a new group activity.* This new activity may be similar to that of the initiator but need not be. For example, a movement to a new feeding place may elicit the same movement by others but an alarm call may elicit freezing behaviour.
- *Controller: the individual that determines whether or not a new group activity occurs, when it happens and which activity it is.* The controller can also reduce the likelihood that individuals show certain activities. The control may sometime be exerted by force or threat, but the more common situation is that in which members of a group look to the controller before changing activity or moving as a group. Examples include the old mare in a group of New Forest ponies, the hind in a group of red deer or the male in a group of gorillas or baboons that determines when the group will move to a new area (Schaller, 1963; Kummer, 1968; Tyler, 1972). Initiators may try to start such group movements but be unsuccessful unless the controller decides to move. In small groups of ducklings, the initiator of a movement towards a bathing area was the individual most likely to be followed by the others (Liste *et al.*, 2014).
- *Competition: the situation where individuals seek to obtain the same resource.* This need not involve any physical confrontation between rivals for the fastest mover may often succeed in obtaining a food item. In other circumstances, the cleverest rather than the strongest or the fastest may be successful.
- *Hierarchy: an order of individuals or groups of individuals in a social group, based upon some ability or characteristic.* The term is most frequently used where the ability assessed is that of winning

fights or displacing other individuals. The hierarchy might involve just two levels, as in the case of a despot who wins against a set of equal individuals. Usually, hierarchy refers to a series of levels like a linear peck order (Schjelderup-Ebbe, 1922) or to a linear series with some triangular relationships in it. Such an order might reflect an ability to dominate other individuals – the subordinates – restricting their movements and access to resources. However, the term dominance order is seldom appropriate as orders are usually based on specific measurements such as winning fights or displacement from a feeder rather than an assessment of real dominance. Numerous behavioural phenomena are seen in domesticated animals in the course of social interactions. Mated pairs often show prolonged association and interactions that help to cement the pair bond. Other pairs of individuals may also associate and interact for their mutual benefit. Social relationships between and within sexes and age groups are prominent in each species and are aided by communication. Most interactions in groups of animals are cooperative, friendly or at least tolerant and very few are aggressive or competitive. Interactions such as those in Fig. 14.1 are frequent in stable groups. Unfortunately for many farm animals, humans often disrupt the formation of stable groups.

The first social relationship is often with the mother. Newly hatched chicks are attracted to the hen by warmth, contact, clucking and body movements. This attraction is greatest on the day of hatching. They learn to eat, roost, drink and avoid enemies in the company of their mother (Fölsch and Vestergaard, 1981). Chicks rapidly learn the characteristics of their environment on the first day of life (Bateson, 1964; Broom, 1969a). Thereafter they are more likely to approach the familiar and avoid the unfamiliar, but this rapid learning period can be extended if no appropriate stimuli, such as mother or siblings, are detected. The set of brain processes involved in this rapid learning, sometimes called imprinting, develop in the course of interaction with stimuli received from the environment of the young chick (Bateson and Horn, 1994; Bateson, 2003). The attachment to the mother is further strengthened as her voice and detailed appearance are recognized. The period spent with the mother is terminated as she rejects the chicks by pecking at them when the down starts to disappear from their heads. Soon after this the brood is dispersed.

Among confined animals, modified social behaviours are seen. The domesticated animal's social interaction with people varies considerably from species to species, depending on the system of keeping and whether the animal has experienced unpleasant human actions. There can be a special relationship involving the positive association of the animal with man. Interspecies affiliations can occur in various forms.

The ancestors of most domestic animals lived socially, and social living continues to be an important

Fig. 14.1. Affectionate head-overlapping in two colts (photograph A.F. Fraser).

part of life for such animals used today on farms, as companions or for work and entertainment. Exceptions include cats, golden hamsters and mature males of some species. Cats will sometimes live in social groups, for example, some feral cats in cities, but most associate with offspring or mates only. Golden hamsters live separately and may fight if forced to associate. Sexually active bulls may isolate themselves except when consorting with females, when there is a potential for mating.

When living in a herd or pack, dogs, sheep, cattle and horses maintain visual contact. Pigs show more body contact and keep in auditory communication. When disturbed suddenly, most sheep and horses first bunch together and then run in a group from the source of disturbance. Pigs, dogs and cattle move in looser groups. During the bunching of animal groups in natural or high-density situations, individuals may be forced to violate the personal space of others. Social interactions at such close quarters may be suppressed, may depend upon friendly or aggressive interactions or may be a result of physical ability and aggressiveness. Social relationships, especially those that are stable, have important effects on social interactions in all social groups.

The same responses tend to be shown very consistently in all encounters between conspecifics. Consistency of social interaction requires that individuals are able to recognize one another and that their social positions have not been altered as a result of confusion by large groups, illness or temporary removal. This stability of social relationship requires:

- Recognition between individual animals.
- Established social positions.
- Memory of social encounters that establish social status.
- Memory of observations of the behaviour of social group members.

Commonly given estimations of the total number of group members that can be recognized or remembered by each individual are 50–70 in cattle and 20–30 in pigs. A great variety of learning is possible in social groups, as individuals may copy the actions of others, establish cooperation or alliances because of social experiences and be able to learn more readily because of the stability that social relationships and support can bring. Sophisticated learning is much more likely to occur when responses are encouraged by social bonds. For example, learning by a parrot to use words occurred in such situations (Pepperberg, 1997).

Communication

As explained in Chapter 3, communication starts very early in life: domestic chicks and some other birds start communicating while still in the egg. Mammals *in utero* can be affected by the mother's activities and events in her world. Young fish that still have their yolk-sac can interact with one another. Communication implies the transmission of information from one individual to another and is discussed in most chapters of this book. It depends on the sensory world of the animal, as discussed in Chapter 1, and on discrimination and learning abilities, for example those considered in Chapter 3. Animals that live in social groups generally have complex communication systems.

While some information transmission results from involuntary movements, most of it is controlled by the sender and can involve deceit. An individual in pain may avoid informing a potential predator of the pain and an individual involved in a courtship display may attempt to exaggerate its strength or attractiveness. An example of a well-known means of communication is tail-wagging in dogs. This may be used by a dog in a situation in which it wishes to indicate submission and to minimize the risk of being attacked. It may also indicate a desire for friendly interaction, a request for a specific action by another individual or general happiness. Quaranta *et al.* (2007) found that the tail-wagging of dogs had a bias to the left side when in the presence of an unfamiliar dog behaving as if to dominate, or when in situations where there may have been some uncertainty about what would happen. The tail-wagging had a bias to the right side when the dog was in the presence of its owner, an unfamiliar person or a cat. These results may be associated with brain lateralization of processing. The visible differences are likely to be evident to a watching dog, in which case the information communicated would be different.

As explained in Chapter 1, much communication by mammals, fish, insects and some birds is olfactory. The worlds of many animals are dominated by smell and what is seen or heard is of secondary importance. All who are responsible for animals should be aware of this. The sense of smell is used by cows to recognize other cows, even in pair-competitive situations; the vomeronasal organ (VNO) may be responsible for this. It has been found that social hierarchy, determined by conventional inter-individual aggression contests, is distinctly changed by experimental inactivation of the VNO. Steers that were higher ranking pretreatment generally lost rank, and lower-ranking steers gained rank over controls. This has led to the postulate that the VNO has a role in social aggressive behaviour that contributes to social hierarchy in species where it is present.

Auditory communication is important to many domestic animals and those who care for them usually have some ability to evaluate the state of the individual from the sounds that it makes. As a consequence of research on the causes of mortality in young piglets, their vocalizations have been much studied. It has been possible to analyse the sounds spectrographically for many years but the greater computing power now available allows more sophisticated statistical analysis of the acoustic properties of piglet calls. This has been done by Tallet *et al.* (2013) who found that calls could be put into either two or five categories. The classification of call types by human observers listening to pigs have resulted in the same two or five categories. Observation of the behaviour of sows and of other piglets suggests that they respond differentially to the call types so probably use the same categories. Illmann *et al.* (2013) describe how the calls made by a piglet that is isolated or being crushed, as can happen if the heavy mother lies on it, are identifiable and functional.

Associations and Social Networks

In groups of social animals it is common for animals to associate; that is, they spend time closer to one another than the mean group inter-individual distance. Within herds, it is often found that discrete pairing through mutual selection of each other's company is a common social strategy which operates to the advantage of both, particularly in agonistic situations involving other, dominant animals. The associative characteristics of animals are now recognized as a clear manifestation of their choice for company that must represent a basic need. In a group of donkeys, mules and ponies, the most frequent associations were with individuals of the same species (Proops *et al.*, 2012). Association measures, such as recording the nearest neighbour, are most appropriate when animals can choose whether or not they are close together. When 15 hens were in 3 m^2 pens, they did not have particular individuals as nearest neighbours (Abeyesinghe *et al.*, 2013). However, as the animals were all in close proximity, they could see and interact with all in the group so may have considered all 15 as neighbours.

Social network analysis addresses complex biological questions as it describes the direct and indirect relationships occurring among individuals within a group (Wasserman and Faust, 1994) and quantifies the social ties and influences among connected individuals. Importantly, social network analysis has been used to identify the roles of key individuals in the group (Lusseau and Newman, 2004) who possess higher numbers of contacts and interactions and thus the power to influence social relationships within and between groups. Cañon Jones *et al.* (2010, 2011) studied social networks in groups of farmed salmon. Social network analysis of aggressive interactions revealed that feed-restricted groups had denser and less distant networks indicating that aggression was a social interaction rapidly transmitted within the members of the network. Most importantly, the out-degree and in-degree centrality differences revealed that feed-restriction divided fish according to their levels of aggression, resulting in fish becoming either initiators or receivers of aggression. Initiators of the aggression had higher out-degree centrality and therefore were extensively involved in interactions within the network, having more ties with other fish within the group (Fig. 2.5). These fish are more influential and are more likely to gain resources. On the other hand, receivers of aggression had fewer interactions and their spatial positions within the group were dependent on initiators. These high in-degree and low out-degree values of receivers indicated that these individuals seldom initiated aggressive interactions and did not retaliate or counterattack their aggressor. There was more aggression, more fin-biting and more fin damage at higher stocking densities.

In free-living sheep and goats, the females and the juveniles use certain fixed areas, which may change with season, with the males associating in bachelor groups in other areas but joining the females during the breeding season. The basic social unit in these groups is the female with her most recent offspring. Associated with this unit are likely to be related animals, for example a yearling and the female's mother. The stable unit is made up of numbers of subgroups of related individuals. In large bachelor male groups, the associations between individuals are looser and the composition of the groups may change frequently, but in small bachelor groups there is much cohesion.

Dogs and cats

As explained by Braastad and Bakken (2002), the larger canids normally live in social groups based primarily on kinship and a flexible social organization. Pack size depends upon the size of the main prey species, since larger prey can only be caught by large groups of dogs or wolves. Individual dogs may spend much of their lives associating with certain other dogs in their pack. Although fewer cats are associated with other particular individuals, some choose to live in groups, and hence a cat may often be found associated with one other or a small group of other cats.

Cattle

Domestic cattle in a free-range situation move from place to place in groups in which individuals maintain close proximity to one another (Phillips, 2002). They are closer together when driven or escorted. Dairy and beef cattle often lie in groups and grazing animals often stand within a few metres of one another, seldom moving out of view of the rest of the herd. Associations during movement, other active periods and resting periods are not random. A study of young dairy heifers by Broom and Leaver (1978) showed that associations were much more likely among some heifers than among others. In this study, animals that were reared together as calves were more likely to associate when adult, and Bouissou and Hovels (1976) obtained the same result. However, other associations also formed and the occurrence of such associations, in a suckler herd of mixed ages, is described in detail by Benham (1982b, 1984). The animals in the suckler herd often allogroomed, licking one another, and spent most of their waking time together. It is likely that social licking has effects on the psychological stability of the animals concerned, as well as on simply cleaning the skin and hair.

Sheep

The development of social organization in sheep has been described by Arnold (1977) and Lynch *et al.* (1992). The first social bond a sheep develops is with its dam. Once the bond is established it remains in females unless broken by separation. Ewes and lambs during the first 4 weeks of the lamb's life are found to stay within 10 m of each other for over 50% of the time. The formation of 'weaner' flocks breaks this social bond and a new social organization has to be developed with the formation of small groups, in which inter-animal distances are low. Gradually, these groups become larger until a flock is eventually formed. The size of subgroups increases with age from weaning up to 4 months old; this is unrelated to the size of the paddock or space available to the animals. Even as late as 11 months of age, subgroups may be formed. Normal adult flocking behaviour appears to be established by 15 months of age. The flock identity is strong for adult sheep, and members within contact distance immediately run together when disturbed.

Three characteristic flock structures have been described as follows:

- A tightly knit flock.
- A flock widely dispersed but with uniform spacing between individuals.
- A flock split into subgroups but that remains a social entity with membership of subgroups continually changing.

When resting, social distance is greatly reduced, and an analysis of 72 flocks showed that, when resting, sheep occupied an area of 10 m^2 per sheep. Distance to nearest neighbour is one attribute of social arrangement; the cohesion of all members of a flock is another. This cohesion varies with environmental factors. The average distance between neighbouring sheep when grazing varies from 4 m to more than 19 m; the greatest distances are for hill breeds of sheep and the smallest for Merinos. The average distance of the nearest neighbours among sheep in all breeds is within 5 m, but breeds differ on this basis and fall into four classes of dispersion: Merinos are the closest, lowland breeds are next, hill breeds are further apart, with the mountain breeds being the furthest apart. Most data on nearest neighbour distances could be affected by the presence of human observers – the sheep would tend to stay closer together when aware of humans.

Horses

While being a typical herd species, horses also show a marked preference for certain individuals of their own species. Two horses encountering each other for the first time show much mutual exploratory behaviour. Exploratory behaviour at introduction involves an investigation of the other's head, body and hindquarters using the olfactory sense.

Horses show a form of social order when they live in groups, and a social hierarchy can become established within these groups. The older and larger animals have been reported to be high in the dominance order. Stallions do not necessarily dominate geldings or mares but have a significant role in defence of the group (McDonnell, 2002). A dominant individual may influence the movement of the herd through the grazing area and will sometimes break up exchanges between other horses. Socially dominant horses are sometimes found to have more aggressive temperaments than the others. Horses running at pasture show special features of behaviour if the group contains a stallion and breeding mares. Stallions usually drive younger male animals to the perimeter of their groups, but will not show any aggressive attitudes towards them if they remain there. The stallion attempts to herd a group of brood mares together. The normal size of a 'harem' among horses is about seven to eight mares. The colts tend to form a bachelor group after splitting off from the herd at the age of about 1–2 years. Fillies may or may not join this group. However, most horse groups spend very little time interacting in a competitive way.

Mares that have been kept together will continue to associate closely and consistently when put with other mares. In mares from different studs, such close associations are not formed, although most individual mares are found to associate closely with certain individuals. In herds of both sexes, colts and fillies tend to separate from the mares and stallions. The stallion usually attacks members of the younger group if they approach too closely. The stallion will round up the mares on the periphery of his herd or 'harem', but will ignore or repel fillies.

In free-living ponies close groups are formed (see Fig. 14.2). Most groups are family groups, with fillies remaining with their mothers for 2 or more years. When they leave their mothers young mares frequently change groups, often joining older mares with foals. Stallions in winter form bachelor groups. Groups of young males have a loose social organization, with members leaving to form other groups for a period and then rejoining the original. Stallions without mares often live as solitary individuals. The formation of close social bonds in horses is essential for group stability and it is important to plan for these in the management of domestic horses.

Pigs

The social organization of groups of pigs is known to include the establishment of various friendly relationships and a social hierarchy (Jensen and Wood-Gush, 1984). For the social hierarchy to function properly, the size of a group and the space allocated to it are important. It is also necessary for the members of the group to be capable of prompt recognition of each other. Sensory

Fig. 14.2. A sub-group of ponies in a free-ranging herd (photograph A.F. Fraser).

clues such as olfactory stimuli are involved in the maintenance of the social structure. It is also evident that pigs in an established group are quickly able to recognize an alien in the group, but pigs are not territorial (Horrell and Hodgson, 1985, 1992a,b). Visual and olfactory cues seem to be the principal differentiating features of pigs for each other.

Poultry

The domestic chick shows early social responses, while still in the shell, to adult calls and to chicks from other embryos (Vince, 1973), and it may give low-pitched distress calls if cooled or rapid twitterings of contentment if warmed. Chicks that are hatched at slightly subnormal temperatures give distress calls as their moist down dries and they lose contact with the eggshell. Contact with a broody hen or other warm object prevents these calls. Newly hatched chicks are attracted to the hen by warmth, contact, clucking and body movements, and this attraction is greatest on the day of hatching. They develop the behaviours of maintenance, in particular to eat, roost, drink and avoid enemies, in the company of their mother.

In chicks, the most sensitive period for learning the characteristics of the mother is normally between 9 h and 20 h after hatching. The attachment to the mother is further strengthened as her voice and appearance are recognized. The clutch is the basis of flock organization and, even after it has dispersed, chickens need company. A chick reared in isolation tends to stay apart from the flock. Flock birds eat more than single birds due to social facilitation.

Adult flock formation depends on tolerant association. Strange birds are initially attacked and are only gradually integrated into the flock. Newcomers are relegated to positions near the bottom of the peck order and only active fighting will change this. Hens and cocks have separate peck orders, as males in the breeding season do not peck hens. However, most associations, even at feeding places, involve little fighting or aggressive pecking (see Fig. 14.3).

Fig. 14.3. Mutual acknowledgement between hens (photograph R. Harnum).

Leadership

In their affiliative movements, animals often respond to the initiative of a lead animal by following (Squires and Daws, 1975; Sato, 1982). All herding and flocking animals show 'follow reactions' in various social circumstances involving movements. Leadership among sheep is often provided by an old ewe: older animals are more likely to lead, and the status of the animal in the social hierarchy may not be a determining factor. In the movements of cattle during their husbandry routines, the order of following often remains very consistent.

Several observers have reported that the voluntary order of cows going for milking at each milking session is fairly consistent (Reinhardt, 1973). Dairy cows organize themselves into a specific order for entering the milking shed according to their milk yield, with high-yielding cows taking an advanced position in the order. The movement order to milking is quite consistent, though the rear animals have more fixed positions than the 'leaders'. Evidently, the follow reaction in an order system can be influenced by the 'reward' of milking. Types of leadership in cattle have been subdivided into three categories as follows:

- Leadership during movement to and from locations of eating, drinking and sleeping. This establishes movement order.
- Leadership in the initiation of grazing and resting.
- Leadership in direction during grazing activity.

In the suckler herd studied by Benham (1982b), some individual cows always moved at the front of the herd and ran around to the front if the herd changed direction: it was clear that they were leaders, but not controllers, of movement. Summer temperatures may reduce the development of grazing, while association will be increased to avoid the attacks of biting flies.

Sato (1982) observed that the summertime lowering of grazing and the development of association frequently caused heifers to behave less as followers and to show dispersive grazing formations. Since grazing essentially involves dispersive movement, 'follow reactivity' is readily obscured. Aggregative movement, on the other hand, may amplify following behaviour.

Changes from grazing to resting or from resting to grazing take place as follows. When an individual cow initiates activities different from the remainder, it returns to the activities of the rest of the herd if the remainder do not follow. Drifting occurs when a neighbour begins to follow, until the behavioural policy of the whole group is changed. Each animal is dependent on the herd influence, and leadership ranking is not simply a measure of individual preference in effecting reactivity. The milking order may not be the same as the order when travelling; the travel order may differ from the order going through gates, doors or out of yards, into races or other linear controls.

Some use is made in animal management of the 'Judas' animal to lead groups in to slaughter premises. Using the natural movement pattern of the species concerned, sheep, cattle and horses can all be trained to lead. Under free-range conditions, the older grazing stock can transfer to their offspring information about seasonal pathways, areas of good pasture and watering places if this familial bond is not disrupted before weaning. In this way, home range areas can be established efficiently. Sheep in extensive pastures may establish separate home range areas on the basis of familial leadership and its social cohesion. Such cohesion contributes to social facilitation.

Social Facilitation

Within groups, the activity of certain individuals – usually rapidly followed by the majority – seems to direct behavioural policy for all. This group effect serves as a basis for the holistic strategies of group behaviour. Social facilitation in flocks and herds is involved in daily movements and in stampedes, marches and migrations that persist as outstanding behavioural phenomena in animals. Social facilitation is more likely where there is adequate association, ability to communicate and react, a potential for mimicking activities, similarity of motivational state and suppression of intra-species aggression.

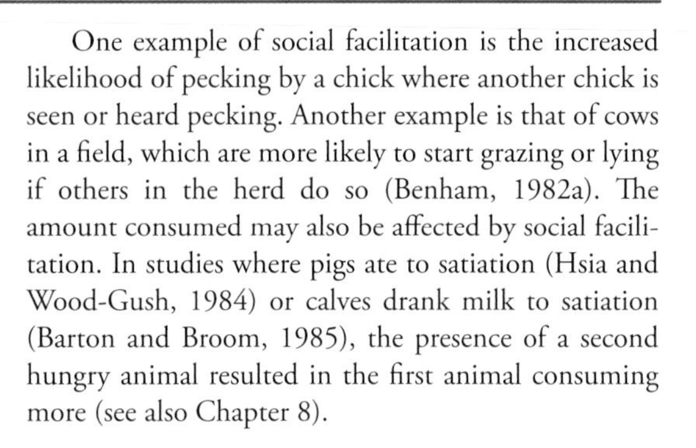

One example of social facilitation is the increased likelihood of pecking by a chick where another chick is seen or heard pecking. Another example is that of cows in a field, which are more likely to start grazing or lying if others in the herd do so (Benham, 1982a). The amount consumed may also be affected by social facilitation. In studies where pigs ate to satiation (Hsia and Wood-Gush, 1984) or calves drank milk to satiation (Barton and Broom, 1985), the presence of a second hungry animal resulted in the first animal consuming more (see also Chapter 8).

Social order

In groups of animals that have been together for some time, there may be a hierarchy. This can result in maximal group-bonding and minimal aggression, creating the social stability that is a vital requirement in good animal husbandry (Schein and Fohrman, 1955). Social interactions in the development of a 'peck order' usually involve aggressive acts such as biting, butting or pushing. The introduction of a new individual is disturbing, perhaps hazardous, due to the number of separate aggressive encounters that the newly introduced animal will receive from most members of the group in turn, or sometimes in concert. A social hierarchy is not an inviolable structure; it is merely the state of settled-out relationships between individuals. Linear social orders have been described in groups of rabbit bucks and does (von Holst *et al.*, 1999). However, efforts to describe social structure in domestic animals in terms of hierarchies are usually misleading over-simplifications, just as they are in groups of humans.

Of greater importance are other relationships, which need not involve any aspect of competition. Friendship pairs are formed and these involve mutual tolerance. In a study of donkeys, Murray *et al.* (2013) found that most donkeys frequently associated with another individual and preferred that individual, rather than another group member or a stranger, if given a choice in a Y-maze. The function and durability of the social structure is dependent on a set of relationships and avoidance tactics. 'Dominance hierarchy' is not an appropriate description of a social relationship based in a mixture of social arrangements. For example, in groups of dogs, the complex social relationships are seldom usefully explained by considering the group to have a single social hierarchy (Manteca, 2002). It seems that animals use organized tactics of stable social living,

which may prove to be systems of pacts. Affiliative behaviour is generally more important than aggressive behaviour. Such relationships may lead to stable bachelor grouping and association in which there is matriarchal guidance to group members.

In domestic fowl, the peck order is most clearly seen in competition for food or mates, and subordinate hens may obtain so little food that they lay fewer eggs. Dominant hens mate less frequently than subordinate hens, but dominant males mate more often than subordinate males. However, most relationships involve recognition without competition (Fig. 14.3). The birds acknowledge one another but there is no sign that either bird has been involved in a fight. Birds in a flock kept in a state of social disorganization by the removal and replacement of birds eat less, may lose weight or grow poorly and tend to lay fewer eggs than do birds in stable flocks. Additional feed and water troughs distributed about the pen enable subordinate hens to feed unmolested, and an adequate number of nesting boxes gives these birds the continuous opportunity to lay. If space permits, flocks of over 80 birds tend to separate into distinct groups, and different social orders may then be established. Large, dense flocks may be subject to hysteria.

When a turkey identifies a potential competitor, the most common social interaction is a simple threat. One bird may submit to the other, otherwise both birds warily circle each other with wing feathers spread, tails fanned and each emits a high-pitched trill. One or both turkeys may then leap into the air and attempt to claw the other. The one that can push, pull or press down the head of the other will usually win the encounter. Such bouts usually last a few minutes. Haemorrhages may occur during a tugging battle, since richly vascularized skin areas may be torn, but actual physical damage is usually slight and birds do not fight to the death. An injured lower-ranking bird must be separated from the group, however, until its wounds heal, as other birds will tend to peck and aggravate the wound, with fatal results.

Formerly, it was thought that social order was the result of social dominance through aggression, and for that reason it was commonly termed the 'aggression order'. The use of dominance should be restricted to the phenomenon that can occur in every pair of animals in which one member can inhibit the behaviour of the other (Sato, 1984). The order of the group is therefore the sum of all such inhibitory relationships. Dominant animals have probably been aggressive in the past to obtain their dominant positions, but a dominant animal need not be aggressive subsequently (Reinhardt and Reinhardt, 1982). Measures of the dominance position of animals in a group should be based on observations in the particular herd, contain sufficient observations to be reliable and reflect the actual magnitude of differences among animals. Relationships are the result of learning, with many different factors being involved in the formation of a relationship. Once learned, dominance relationships can persist for a long time (Syme and Syme, 1979).

Social dominance is not usually exerted when social animals are grazing or are resting. In horses, however, subordinate animals deliberately avoid moving too close to dominant animals and dominant animals frequently threaten subordinates while eating. Social dominance is exhibited in competition for supplementary feed given in a restrictive place or at water troughs in cattle, sheep and horses. Much behavioural observation on 'aggression orders' in domestic animals, and in pigs particularly, is now suspect since the subtle behaviours of avoidance-submission – such as 'head-tilting' in pigs – went unrecognized. These behaviours provided all the social exchange involved in maintaining social stability. In a permanent herd of free-ranging sows, an 'avoidance order', seemed to be maintained mainly through the behaviour of the subordinates. They performed the most social acts, mainly submissive and flight behaviour, and these were mainly received by the dominant animals. The view is expressed that these facts support the use of the term 'avoidance order' as a measurement of social dominance, instead of the 'aggression order'. Jensen (1984) observed that confinement and semi-confinement decreases the social activity, measured as number of observed interactions per time unit, and leads to unsettled dominance relationships combined with high aggression levels. Semi-confinement systems do not provide sufficient space for a stable social system, and the frequencies of aggressive behaviours were actually highest in semi-confinement. Systems in which animals are in small groups having individual feeding stalls and an area of secondary space provide enough area for the sow to settle dominance relationships and to keep the aggression level fairly low. Within systems that provide security of position at feeding and secondary space for resting, it is possible for a constantly settled avoidance order to exist.

Social dominance effects can be very important in cases of high stocking densities or poor farm design.

Inadequate trough space, narrow races, inadequate space in indoor housing or lack of feeders can mean that dominant animals command resources at the expense of subordinate animals. The latter will suffer and health and general production can be affected. Documented examples include the higher internal parasite load of subordinate grazing stock during droughts when scarce food is commandeered by dominant stock. However, most associations are amicable. The cattle in Fig. 1.1 feed together regularly and carefully so as to avoid injuring one another with their large, sharp horns.

Since aggressive behaviour is most seen when groups of pigs, cattle or horses are first formed, the frequent changing of group members should be avoided. The adverse effects of mixing together individuals from different groups are of particular importance in pigs (Arey and Edwards, 1998; Turner *et al.*, 2001; Turner and Edwards, 2004). The extent to which aggressive or other injurious behaviour occurs in groups of domestic animals is of great importance in the management of some species. A dog shelter or holding centre can be a place where serious injuries occur as a result of fights, while damaging injuries also occur on poultry and pig farms. The impact of aggression on welfare, and how to manage aggression, are considered further in Chapters 29–38.

One principle relevant to the causation of aggression problems is the effect of mixing individuals unfamiliar with one another, as mentioned above. Other principles concern the effects of group size and space allowance on the likelihood of social problems. Taking pigs as an example, aggression does not increase with increasing group size unless resources are limiting (Turner and Edwards, 2004). Pigs in large groups may form subgroups or restrict their space use to certain areas. If group size is larger, at a given space allowance per individual, larger groups mean greater likelihood of opportunity to move over greater distances and hence exercise better, so aggressive interactions may be reduced.

When space allowance for pigs was 0.55, 1.65 or 2.0 m^2 per pig, much more time was spent lying inactive or biting the tails of other pigs at the smaller space allowances (Guy *et al.*, 2002). This occurred in part because, with more space, a more complex environment was possible, but Beattie *et al.* (1996) found that pigs showed more locomotor behaviour such as running and jumping when more floor space was provided in enriched pens. Such possibilities for behaviour reduce the likelihood that injurious behaviours will occur. Production of milk and other physiological responses can be affected for several days while aggressive social interactions are taking place. Sometimes tranquillizers have been used to aid social tolerance when strange pigs have to be penned together or when wild horses have to be collected.

Further Reading: Lusseau, D. and Newman, M.E.J. (2004) Identifying the role that animals play in their social networks. *Proceedings of the Royal Society of London B (Suppl.)* 271, S477–S481.

---

15 Human–Domestic Animal Interactions

Human attitudes to animals are said by Serpell (2004) to have two kinds of motivational determinant: affect and utility. The interactions that occur between humans and other species, including companion and farm animals, will depend on the relative importance of these for the human involved. Attitudes of people may change so a person who is initially solely an animal user may develop empathy for animals in the course of interactions with them. Inter-specific interactions between non-human animals may also change with the experience of the individuals involved. Some domestic animals may attack people, especially mothers defending their young (Fig. 15.1).

Many wild animals spend time in multi-species groups. Flocks of forest birds, shoals of coral reef fish and herds of grazing mammals may include members of several species that are able to respond to, and take advantage of, the alarm signals and food-finding indicators of the various species present. However, close inter-specific relationships are not normally reported to occur. Animals of one species may be able to gain much information about individuals of another species without forming any relationship with them. For example, mice are known to be more disturbed by the odour of male humans than by that of female humans (Sorge *et al.*, 2014) and they have the capability to distinguish individual humans but do not normally show individualistic reactions to humans. Animals of different species are more likely to form such relationships if the associations are prolonged, as they are when humans have companion animals in their homes. Dogs respond to pack members and their relationship with familiar humans is similar in some ways to that with other dogs in their pack. A dog is more likely to respond to a signal by a human if that human is a familiar individual and hence a known quantity to the dog. A human who is gazing or pointing at some object may have information, useful to the dog, about the object. If the dog can appreciate this, the object may be investigated (Ittyerah and Gaunet, 2009).

Animals kept by humans are more likely to show inter-species relationships because of the conditions imposed upon them during development and during adult life (Waiblinger, 2009). As a consequence, inter-specific friendships or companion bonds may develop. Interactions and relationships between domestic animals and humans are of particular interest because of their relevance to the welfare of the domestic animal and of the human involved, and because animal management may be helped or hindered by them. There have been many studies of the effects of companion animals on their owners. After early studies indicating that stroking a familiar dog or spending time looking at a tank of fish had calming effects on heart-rate and blood pressure of people, further studies of effects on human health have been carried out. These are reviewed by Friedmann *et al.* (2010) who point out some of the difficulties associated with such studies, as well as the evidence for positive effects of pet ownership. If the affiliative behaviour, and consequent relationship, functions from pet to human as well as from human to pet, there should be effects of the human on pet physiology as well as effects of the pet on human physiology. Odendaal and Meintjes (2003) investigated effects in both directions. A group of dog owners and dogs were blood-sampled before and after a period of friendly interaction between dog and owner with the owner's attention focused on the dog. The concentrations of β-endorphin, oxytocin, prolactin, β-phenylalanine and dopamine increased in both species after this interaction while that of cortisol decreased in the humans only. The lack of decline in cortisol in the dogs may have been a consequence of the unfamiliar environment disturbing the dogs a little but not disturbing the humans because they were prepared for it.

When bonds exist, for example between people and dogs or cats, communication between the human and the dog or cat may be sophisticated (Miklósi *et al.*, 2005). For example, dogs looked at humans whom they had discovered knew how to do task A in the task A

Fig. 15.1. This sow has piglets and is about to attack the person who is approaching too close to her (photograph D.M. Broom).

situation but looked more at another person, who had been observed to solve other problems, in a different task situation (Horn *et al.*, 2012). The dogs could adjust their behaviour towards individual people according to their previously acquired information about those people.

The benefits to humans of interactions with companion animals is a rapidly expanding subject (e.g. Podberscek *et al.*, 2000; Serpell, 2002). One benefit for people is the companionship – many would say friendship – that can be the result of the interactions between a person and a dog, cat, horse or other animal with which the person spends time. This benefit is derived by very many people as approximately half of households in the developed world have pets (Westgarth *et al.*, 2010) and while some people interact little with the animals in their household, others have close bonds with their pets. A further benefit is increased exercise. Dog owners go out for walks 23–120% more than non-owners, and pregnant women who have dogs are 50% more likely to do the 3 h per week of exercise that is recommended for them (Westgarth *et al.*, 2012). These benefits would seem to outweigh any small disadvantage associated with increased risk of disease transmission from pet animals as the risks of the major killer diseases are reduced by exercise and increased by obesity.

The removal of the young animal from its own dam to be raised by a human leads to social attachment to humans. The optimum time for such relationships and developmental processes to be formed varies with the species. With precocial species such as cattle and sheep, which are behaviourally competent at birth, the optimum time is from birth to 4–6 days. A leader–follower relationship may occur in which the animal follows the human for food or companionship. If the attachment to humans has been very exclusive, such hand-raised animals may later relate sexually to humans. In species that develop a dominance–subordination type of social structure, it is important that the human caretaker be dominant, particularly when the animals may be dangerous as adults. The dairy bull asserts increased dominance with maturation and growth. For such an animal, dominance is best established at the appropriate time for the species, usually early in life when no punishment may be needed. Since the social dominance interactions are fairly specific for individuals, the fact that one particular person dominates an animal is no guarantee that another will be able to do so with the same animal.

The hand-rearing of animals can be very beneficial to animal and owner in that individuals survive and thrive that would not otherwise have done so. However, the problems that can arise as a consequence of bonding socially with man at the sensitive period in development must be considered in later management of the animals. Animals that are handled more by man may change their social responses to members of their own species as well as to man. Waterhouse (1979) handled regularly some young dairy calves in a row of hurdle pens and found that these calves showed less social interaction with their neighbours. There were no long-term adverse effects on social interactions in this study, but there are reports of regularly petted animals and of entirely hand-reared animals not adjusting well to social conditions.

During the rearing of puppies, contact with humans during the period from 3 to 12 weeks of age is necessary if the dog is to develop friendly relationships with humans (Manteca, 2002). Increased contact between 2–9 week-old kittens and humans results in owners reporting less fear of humans and more emotional support from the cat when adult (Casey and Bradshaw, 2008).

It would seem inadvisable to handle some young animals in a group much more than others (Broom, 1982). The problems that can arise are of: (i) misdirected sexual behaviour; (ii) playful behaviour towards a person by an animal that has grown large enough to be dangerous; or (iii) aggression towards people by an animal that has little fear of man because of its early social experience with man. This problem is soluble by taking care to socialize the animal with its own species at an appropriate time, rather than keeping it with only human company.

Having pointed out that domestic animals can form social attachments to people, it should be emphasized

Table 15.1. The effects of handling treatments on the level of fear of humans and performance of pigs in four experiments (from Hemsworth and Barnett, 1987).

Experiment and measures	Mean for handling treatment		
	Pleasant	Minimal[a]	Aversive
1. Time to interact with experimenter (s)[b]	119	–	157
Growth rate from 11–22 weeks (g/day)	709	–	669
Free corticosteroid concentrations (ng/ml)[c]	2.1	–	3.1
2. Time to interact with experimenter (s)[b]	73	81	147
Growth rate from 8–18 weeks (g/day)	897	888	837
3. Time to interact with experimenter (s)[b]	10	92	160
Growth rate from 7–13 weeks (g/day)	455	458	404
Free corticosteroid concentrations (ng/ml)[c]	1.6	1.7	2.5
4. Time to interact with experimenter (s)[b]	48	96	120
Pregnancy rate of gilts (%)	88	57	33
Age of a fully coordinated mating response by boars (days)	161	176	193
Free corticosteroid concentrations (ng/ml)[c]	1.7	1.8	2.4

[a]A treatment involving minimal human contact.
[b]Standard test to assess level of fear of humans by pigs.
[c]Blood samples remotely collected at hourly intervals from 08.00 to 17.00 h.

that lack of contact with man is a much more important problem on farms. The most common relationship between a farm animal and the people whom it sees is that the animal is afraid of the people. This fear is extreme in poultry and may be extreme in sheep, pigs and other animals. If the stockman treats animals roughly or inconsistently then there are effects on welfare and there may be effects on production. Milk production by dairy cows is much affected by the behaviour of the stockman towards the animals (Seabrook, 1984, 1987).

The work of Hemsworth, Barnett and collaborators has shown that early handling of all the members of a group of pigs can have effects on the later responses of the pigs to man, the ease with which the pigs can be handled when older and the reproductive performance of the animals (Hemsworth *et al.*, 1981a,b, 1986a,b; Gonyou *et al.*, 1986; Hemsworth and Barnett, 1987). Using a standard test of latency of and amount of approach to a person, it was found that sows on different farms varied greatly in their responses to man. The average level of fear of human beings was negatively correlated with the reproductive performance of the pigs on the farm. In several experimental studies, pigs were either: (i) minimally handled or stroked and patted when they approached (pleasant handling); or (ii) slapped or prodded with an electric goad when they approached (adverse handling). The pleasant handling resulted in the smallest behavioural and adrenal cortex response to man, highest growth ratio, highest pregnancy rate in gilts and earliest mating responses in boars (see Table 15.1). It is clear from these results that the welfare and production of pigs is substantially affected by the extent of controlled human contact with the animals. The same must be true of all farm animals.

A study of the effects of handling by farm staff on the later responses of dairy heifers was carried out by Bouissou and Boissy (1988). Animals that had been handled on 3 days per week during months 0–3 or 6–9 showed some improvement in ease of handling at 15 months when compared with unhandled controls. A substantial improvement in ease of handling and reduced heart-rate and plasma cortisol responses to novel situations were shown if the heifers had been handled for 3 days per week each month during months 0–9.

The training of stockmen and other people who look after animals can have great effects on animal welfare. This effect is almost always beneficial to the animals as a greater knowledge about animals and methods of husbandry results in better care. Training methods are reviewed by Hemsworth and Coleman (2010) and Coleman and Hemsworth (2014).

Further Reading: Waiblinger, S. (2009) Human–animal relations. In: Jensen, P. (ed.) *The Ethology of Domestic Animals*, 2nd edn. CAB International, Wallingford, UK, pp. 102–117.

Seasonal and Reproductive Behaviour

Introduction

Reproductive activities are not ever-present features of behaviour. Their induction requires processes of maturation and stimulation, which enable the animal to produce efficient reproductive activities and responses. *Reproductive effort is all of the resources expended by an individual on reproduction in a season.* This includes all preparation for reproduction and parental activity as well as parental care (Box, 1973). The evolutionary basis for parental behaviour is discussed by Clutton-Brock (1991). Most reproductive behaviour depends upon hormonal state and also upon sensory stimuli. The process of acquiring reproductive capability is dependent on a wide range of factors including neural mechanisms, hormones, pheromones and the sensory reception of a variety of stimuli. The state of reproductive capability often overshadows other classes of behaviour by its motivational high priority. This capability often exists during only one season of the year and its initiation depends on a stimulus or a combination of stimuli.

Sensory Factors

Reproductive responses of mammals are much affected by the olfactory sense. Awareness of odour can result in preparation of animals for reproduction, or other activities, as well as directing individuals to a particular immediate response (Sommerville and Broom, 1998). As explained in Chapter 1, the dog has a threshold for olfactory concentrations about one million times more sensitive than humans (Stoddart, 1980; Manteca, 2002). Odour can have a stimulatory value in initiating male sexual behaviour. The importance of the role of pheromones in the breeding behaviour of animals has been described in Chapter 1. For example, the secretions of the submaxillary salivary glands and preputial fluids in the boar (Patterson, 1968; Signoret, 1970; Perry *et al.*, 1980). Boar odour and the odour of oestrous sows can advance and synchronize puberty in gilts (Pearce *et al.*, 1988; Pearce and Pearce, 1992). A variety of male animal odours are detectable even to man, for example the smell of the billy goat or boar, which has its principal effect on female members of the same species (Dorries *et al.*, 1995). Odour plays a large part in the establishment of the strong bonds between a mother and the newborn mammal. These bonds are dependent first on mutual recognition through odour (Alexander *et al.*, 1974). Although odour is the principal means by which early recognition occurs between mother and young mammals, visual recognition soon takes over as the secondary means of mutual identification, and auditory recognition is also important to sheep (Alexander and Shillitoe, 1977).

It has been known for some time that the relative length of the light period of each day is a factor in determining breeding behaviour in some domestic animals. Seasonal breeding, for instance, is largely determined by the changes in the daily photoperiod. Photoperiodism operates in two principal ways:

1. Some animals exhibit their reproductive activities during that portion of the year during which the daily light period is long. It is widely known that for horses the normal breeding season commences in the spring – that period of the year when light is becoming stronger and the number of daylight hours greater – and continues through summer.
2. Some animal species confine their breeding behaviour to that portion of the year characterized by the minimum amount of daily light: sheep and goats are examples of this. Most breeds of sheep and goats commence their breeding seasons in the autumn, when the daily photoperiod is less than the dark period and the light period is diminishing further day after day (A.F. Fraser, 1968).

Clearly, the natural light stimulus for those domestic animals that show seasonal breeding is a complex one involving the absolute quantities of light and dark as well as relative quantities of light each day. Although

gradual, day-to-day changes in the photoperiod are important in the initiation of breeding, it is also clear that the fixed nature of the photoperiod is important; that is, seasonal breeding animals maintain their breeding activities as long as an adequate quantity of light, or of dark, is delivered. When the photoperiod fails to provide adequate stimulation for the animal, a refractory period develops during which the breeding performance is arrested.

Male dogs and cats are sexually active throughout the year. Female dogs are more likely to come on heat during spring than winter. Male dogs patrol a home range and put urine marks consistently in certain places, or over the top of other dog's marks, often using the raised leg display that is not shown by females (Pal, 2003). Female cats often come on heat when the day length increases around the spring equinox. Female mink and foxes are strictly seasonal and come on heat in spring (Sjaastad *et al.*, 2003). Visual stimuli are often combined with olfactory stimuli in eliciting reproductive behaviour. In several species, the form of the stationary oestrous female is the key stimulus for the male to initiate copulatory behaviour.

In male cattle, mounting behaviour is apparently released by a form presenting a dorsal surface and supports. Bulls will mount dummy cows which are of very simple form, having only a metal frame and a covered body. Most bulls tested with a dummy cow will mount it. Similarly, boars will normally readily mount a dummy sow consisting of a tubular frame with a padded covering. The simple form of the mounting releaser helps to explain why some bulls and other male animals with inadequate mating experience will attempt to mount from the side or even the front of the female subject.

In newborn ungulates, attempts at suckling are released and directed to the mammary region and the teat after being orientated first to the darkened underside. The young calf, lamb or foal may find the angular area of this underside by nosing up a vertical limb, then moving to the prominent mammary region and finally to the lactiferous sinus and the protruding teat. The young animal soon learns the direct route to the mammary gland.

Hormonal and Pheromonal Facilitation

Reproductive behaviour is based upon the sexual differentiation in the brain, which occurs at an early age. Processes of sexual behaviour depend on stimulation by oestrogen in the female and testosterone in the male. Reproductive behaviour, which varies among species, is associated with courtship, mounting and coital action in males; and courtship, soliciting and coital acceptance in females. During development, sensitive periods exist for the determination of sexual behaviour. Sexual differentiation in sheep, for example, is determined in the fetal phase of life. During the development of young mammals, the action of gonadotrophin releasing hormone (GnRH) initiates some sexual changes. Male lambs injected monthly with GnRH showed increased activity but such injections did not have this effect in female lambs (Evans *et al.*, 2012).

The neuromuscular mechanisms for most of the components of sexual behaviour are present in both sexes. Given appropriate stimulation, typical heterosexual behaviour can develop. The type of sexual behaviour displayed is the result of the degree of hormonal or other stimulation applied to elicit it (Hurnick *et al.*, 1975). In companion animals, the variation in the magnitude of the effects of castration is remarkable. If animals are castrated so that the production of testosterone is terminated, 80% of cats cease to be able to ejaculate but 84% of dogs can still do so (Hart, 1985; Manteca, 2002). In female sheep and goats, male-like behaviour induced by testosterone includes most of the distinguishable male components including mounting and pelvic thrusting. In the intact ewe the difference between male and female behaviour appears to be one of length of hormonal stimulation (Lindsay and Robinson, 1964). A dose of either oestrogen or testosterone can induce female oestrous behaviour in ewes within 24 h. Continuous stimulation by either of these hormones can result in a progressive change from female to male behaviour. Ewes treated in this way are effective as males in inducing the 'male effect' in anoestrous ewes (Signoret *et al.*, 1982).

Sex pheromones act through the olfactory system, which includes the vomeronasal organ and the olfactory bulbs. Cats can respond to conspecific urine scent marks by becoming reproductively active, using the vomeronasal organ for detection of the pheromone (Meredith and Fernandez-Fewell, 1994). The pheromones may be produced in secretions of the genital organs or skin glands, or occur in the urine, faeces or saliva. The steroid compound produced in the boar and transmitted by saliva foam to the female to produce the immobilization reflex is a notable example of a pheromone facilitating behaviour. The steroid is released in the saliva when the boar courts the sow and results in the characteristic immobile stance. A sow in oestrus will give a standing response when this steroid in an aerosol is directed towards her snout. With domestic sheep also, it has been suggested that rams (Fig. 16.1) may stimulate oestrous activity in non-cycling ewes through olfactory receptors in the ewe. Extracts from the fleece of rams that can mate with many ewes have a bigger stimulatory effect on oestrous ewes in respect of their later lamb production than extracts from rams that mate with few ewes (Al-Merestani and Bruckner, 1992).

Seasonal and Climatic Breeding Responses

Sheep characteristically limit their breeding activities to specific seasons. Whether 'breeding' is taken to mean mating or parturition it is plain that, when seasonal breeding occurs, the newborn is provided with environmental circumstances favouring its survival to puberty. In cases where the young are born at an inclement period it is found, as a rule, that the rate of maturation is relatively slower in such species and considerable time is needed for the maturation of the young before they are subjected to the stresses of their first full winter. The periodicity of breeding is governed by the necessities of the young. For the most part, however, seasonal breeding is a consequence of the timing in mating behaviour.

Fig. 16.1. The presence of a ram like this Orkney, detectable by odour, sound and vision, is often enough to initiate cycling in ewes (photograph D.M. Broom).

In a seasonally breeding species, for example the horse and goat, the reproductive activity by both sexes is intensive in the breeding season and subdued, reduced or absent during the remainder of the seasons. The duration and intensity of oestrus have been observed to alter with seasons. In the mare, oestrus is normally longer in duration during the season of full breeding than at other times. In some breeds of cattle it has been variously noted that significant differences in the duration of oestrus occur with season. In breeds of sheep that breed throughout most of the year, the intensity and duration of oestrus have significant seasonal variations. When the breeding season is a very limited one, the intensification of motivation for mating is evident in the male animals.

Temperature effects

Even with non-seasonal breeders, temperature changes associated with climate may affect reproductive behaviour. It is a frequent observation in cattle-breeding organizations that a sudden spell of cold weather is associated with a drop in the numbers of cows in oestrus. Some other species, such as goats, indigenous to areas having equable climates but with marked periods of rainfall, show degrees of seasonal breeding in relation to rainy seasons. It is often observed in indigenous cattle that breeding becomes intensified with precipitation and the associated rapid growth of herbage. Some native breeds of cattle in West Africa are reported to show increased reproductive activity in relation to rainy seasons. Cold daily temperatures are considered to have some slightly beneficial effect on reproductive behaviour in sheep.

The effect of low environmental temperatures on the initiation of the breeding season in sheep is real, though slight. Cold days apparently hasten slightly the onset of reproductive activity. It is widely observed in the practice of artificial insemination of cattle that a sudden spell of cold weather is associated for a short while with a drop in the number of reported animals in oestrus. In a study of 46,000 inseminations in cattle in central Europe, weather conditions and occurrence of oestrus were correlated. Good weather led to more cows coming on heat but poor, deteriorating weather led to a reduction in the incidence of oestrus.

In other cases, daily peaks of mating activity are apparently associated with fluctuations in male sex drive. Bulls and boars exposed to high temperatures during the summer in hot climates show marked reductions in libido. Libido is inhibited in bulls in air temperatures of 40–50°C. These effects are transient, however; if affected animals are cooled by wetting, libido quickly returns to normal. It is common in South Africa to find boars extremely inactive during the hot hours of the day, sometimes totally ignoring the presence of oestrous sows. Liberal wetting with cold water is usually effective in improving the sexual activity of such boars. The conclusion can be drawn that excess body heat specifically inhibits libido but is only temporary in effect. The heat of summer days in central and southern Europe can adversely affect reproductive functions in dairy cattle. This may also be true of many other animals.

Seasonal breeding rhythm

As stated earlier, breeding rhythm is not just a response to variation in the environment: environmental factors only create the capacity for a rhythm, which is endogenous, the environment acting as a zeitgeber or time-giver; this applies particularly to yearly periodicity. The better the endogenous factor – or inherent rhythm – is developed, the more will the zeitgeber function merely as a synchronizer of events. Breeding periodicity results from the interaction of the two agencies of environmental and internal rhythm together, although the goat and the sheep both tend to maintain internal rhythm for about 1 year when transferred across the equator. Among ewes sent from the UK to South Africa, some immediately change to the southern hemisphere breeding season while others change slowly. Within 2 years, all such ewes switch to breeding during the autumn months of the southern hemisphere. An animal does not possess just one rhythm but has a multitude of rhythms in its physiological organization, with each behavioural element having its own relationship with the environment.

Daily patterns

Sexual behaviour in several species tends to occur at particular periods of the 24 h day. Sheep, for example, are observed to mate mostly around the hours of sunset and sunrise, particularly the latter. In Welsh Mountain sheep, most mountings by rams occur in the early morning, about the time of sunrise. This is mainly due to the fact that activity occurs commonly about the hour of sunrise and also at sunset, and this tendency is most evident in the early part of the breeding season. The onset of mating activity appears to become more uniformly distributed in time as the breeding season progresses.

Although first signs of oestrus in ewes are at dawn, evening twilight exists as a secondary peak period of initial oestrous behaviour. The rams are active throughout most of the 24 h period, going around the paddock investigating ewes every 20–40 min. It seems unlikely that the significant diurnal incidences of mating in sheep reflect fluctuations in libido. Rams are usually found to have satisfactory libido when hand matings are carried out in the hours of the day when matings are seldom seen to occur under natural conditions. This points further to the role of the female rather than the male in the crepuscular mating character of the sheep.

In the Red Sindhi breed of cattle in Pakistan, the onset of oestrus in 60% of cases occurs most often at night. A preference for mating during the night is evident in Brahman cattle and in various breeds of *Bos indicus*, but this by no means precludes daytime breeding. However, nocturnal mating is the rule in the swamp race of the Asiatic buffalo (*Bos bubalis bubalis*). By contrast, the river race of Asiatic buffalo practises daytime mating.

Some domestic animal species are long-day breeders, for example, the horse and donkey, while short-day breeders include sheep and goats. The domesticated horse breeds in the spring and early summer but the Przewalski horse, the closest remaining link with the prehistoric horse, breeds in the summer. Mares do not have an absolute anoestrous season in winter, and it has been estimated that only about 50% of mares go into true anoestrus at this period. Many breeds of sheep at tropical and near tropical latitudes breed at all seasons. In general, among those species that manifest seasonal reproduction, long-day breeders are those having long gestation periods (9–11 months), and short-day breeders have short gestation periods (5–7 months).

In most species, the breeding season appears to be controlled not by a single factor but by a combination of external stimuli, including behavioural ones. These vary in different species but, nevertheless, act through sense organs and mediate the internal rhythms of the individual through the hypothalamus and pituitary.

Illumination

Among the external factors in seasonal reproduction, photoperiod is the principal agent. In female sheep, the natural stimulus for seasonal breeding is shortening day length. Experimental attention to this phenomenon has been principally directed at bringing sheep into heat outside their normal breeding season by exposing them to schemes of diminishing periods of light per day. Oestrus can be induced in all ewes treated with 16 or 17 h of darkness for 1 month (A.F. Fraser, 1968). Extra-seasonal breeding can also be induced in goats treated with an artificially controlled light:dark ratio. Gradual light reduction can induce heat periods in goats. Conversely, increasing light results in the cessation of oestrous cycling in the goat. Goats seem to need daily light and dark, in a ratio of approximately 1:1, or 1:more than 1, for full breeding function. This approximate ratio is available virtually all the year round in the tropics and between the autumnal and vernal equinoxes in the high latitudes. In the tropical locations, no season is encountered in which reproductive reduction amounts to impotence. However, it is noteworthy that in wet weather in the tropics the general behaviour of goats is of an inactive character, and precipitation can create a demarcated breeding season.

The breeding seasons of the sheep and goat are determined in most instances by the times when the majority of females come into oestrus. Rams have been found to be lacking in libido during the anoestrous period, but rams penned and subjected to controlled light breed out of season. The sexual behaviour in male goats also demonstrates seasonal variation. In the autumn sex drive is strong and, in the period from spring to autumn, sex drive is relatively weaker.

Responses to artificial photoperiodism have been observed in the horse, but since this animal is a long-day breeder it is found, as expected, that increased light is the positive stimulus. Mares of the Shetland breed, a breed with a very restricted breeding season, can be made to undergo sexual rhythm by irradiation with strong artificial light. It is found in the northern latitudes of Japan that an additional 5 h of artificial light after sunset during the month of November improves sex drive in the stallion.

Further Reading: Sjaastad, Ø.V., Hove, K. and Sand, O. (2003) *Physiology of Domestic Animals.* Scandinavian Veterinary Press, Oslo.

Sexual Behaviour

Introduction

Three characteristics of female mammals are of particular importance in their effect on the likelihood of successful mating behaviour: attractiveness, proceptivity and receptivity (Beach, 1976). *Attractiveness, or attractivity, is measured by the extent to which a female evokes sexual responses from males.* This will depend upon odours that she produces, visual qualities, etc., and on her *proceptivity, the extent of invitation or soliciting behaviour. Receptivity is the willingness of the female to accept courtship and copulatory attempts by the male.* In normal situations, the female mammal is very selective but domestic animals often have only rejection of acceptance as options and sometimes acceptance is forced by human action. Variations occur in the degree of attractiveness, proceptivity and receptivity of female mammals in oestrus.

In poultry, the female's selectivity is often obvious. The male displays to the female (Fig. 17.1) who may or may not respond.

The Female

Oestrus is the state during which the female seeks and accepts the male. The behavioural features are synchronized with various physiological changes of the entire genital system essential for mating and fertilization. The signs of oestrus are characterized for each species (see Table 17.1), but variations occur between individuals. The usual routines of behaviour are disturbed during overt oestrus and, typically, there is a reduction in ingestive and resting behaviour, while locomotor, investigative and vocal behaviour are increased.

The term oestrus applies principally to behaviour, but it must be acknowledged that it also describes some internal physiological processes. Although the two facets of oestrus can occur separately, this is rare, and it is normal for them to exist simultaneously.

Features of oestrous intensity

The dichotomy of the discrete physiological and the overt behavioural characteristics of oestrus is evident in the condition known as silent oestrus. Oestrus can sometimes be so subdued as to be virtually undetectable. This constitutes a problem in pig, cattle and horse breeding (Hemsworth *et al.*, 1978a). Sub-oestrus indicates a very low intensity of oestrus. The term 'silent heat' is unsatisfactory for the condition in which oestrous behaviour is apparently absent in an animal that is, nevertheless, undergoing the ovarian changes typical of oestrus, since 'heat' implies intense oestrous behaviour.

In cows and mares, silent oestrus and sub-oestrus can be detected by repeated manual palpation of ovaries per rectum so as to ascertain progressive cyclical changes in the ovary of the anoestrous subject. It is important to differentiate between sub-oestrus and anoestrus, the latter having quite different causes and implications. Paradoxically, the animal in sub-oestrus is potentially fertile although mating capacity is absent.

Anoestrus is the condition in which a female animal fails to show cyclic recurrence of oestrus. Anoestrus normally occurs, of course, when the animal is pregnant or when it is in its non-breeding season. Some animals in the non-breeding season and some pregnant animals will occasionally show oestrus. Sub-oestrus and anoestrus can occur in high-producing, metabolically challenged animals. For example, Esslement and Kossaibati (1997) in the UK and Plaizier *et al.* (1998) in Canada have reported that a high proportion of dairy cows failed to conceive as a consequence of failing to show oestrus. In a comparison of cows showing more or less obvious oestrous behaviour, those that showed more obvious oestrus were much more likely to become pregnant after artificial insemination (Garcia *et al.*, 2011).

Fig. 17.1. Cockerel displaying to hen (photograph R. Harnum).

Table 17.1. Behavioural characteristics of oestrus in farm animals.

Animal	Typical oestrous behaviour
Horse	Urinating stance repeatedly assumed; tail frequently erected; urine spilled in small amounts; clitoris exposed by prolonged rhythmical contractions of vulva; relaxation of lips of vulva. Company of other horses sought; turns hindquarters to stallion and stands stationary
Cow	Restless behaviour; raises and twitches tail; arches back and stretches; roams bellowing; mounts or stands to be mounted; vulva sniffed by other cows
Pig	Some restlessness may occur, particularly at night from pro-oestrus into oestrus. Sow stands for 'riding test' (the animal assumes an immobile stance in response to haunch pressure). Sow may be ridden by others. Some breeds show 'pricking' of ears
Sheep	May be a short early period of restlessness and courting ram. In oestrus proper ewe seeks out ram and associates closely with him; may withdraw from flock. Remains with ram when flock 'driven'
Goat	Restless in pro-oestrus. In heat most striking behaviour includes repeated bleating and vigorous, rapid tail-waving; poor appetite for 1 day

Cattle and goats generally employ vocalizations during oestrus, presumably to summon and maintain the attendance of the male. In addition, the vocalizations of many males appear to have considerable effect on the manifestation of oestrus in the female.

Cattle and goats bellow and bleat while in oestrus and the ewe also can occasionally be heard to bleat during heat. The sow is reported to make grunting sounds in heat and it is in this species particularly that the effect of male vocalization on the manifestation of oestrus in the female has been most closely studied. Although the influence of boar odour on heat manifestation appears to be considerable, the chant de coeur of the boar is a more effective stimulus than odour in the induction of sexual receptivity in the sow.

Duration of oestrus

Few animals have a normal oestrous cycle that is so notoriously irregular in duration as the horse, and this irregularity is found in the horse at all latitudes. Although the mare usually remains in heat for fully 1 week, it can last much longer. A common range in the duration of oestrus in the mare is 4–10 days. Shorter heats more often occur as the breeding season progresses (see Table 17.2).

Although it is considered that the true oestrous period of cattle lasts 18–24 h, some variation is reported in them also. A significant seasonal difference occurs in the average duration of bovine oestrus. Oestrus has an average duration of 15 h in the spring as compared with 20 h in the autumn. The oestrous period is considered to be shortened by mating. Heifers, when mated to vasectomized bulls, tend to ovulate earlier than those not mated. The period of receptivity is shortened in many cows after natural service, by as much as 8 h when natural circumstances are provided and repeated matings take place. It is also observed that, when some female animals are 'teased' with vasectomized males, the duration of oestrus is slightly shorter than otherwise. This substantiates the recognition that oestrus is not under endogenous control alone and that its manifestation is subject, in part, to environmental factors including biostimulation (Esslemont *et al.*, 1980).

The heat duration in the sheep appears to be under influences very similar to those in the cow. No seasonal difference in duration of oestrus apparently occurs in Merino ewes, but Merinos have shorter oestrous periods than some other breeds. Mating appears to reduce the heat period in the sheep (see Table 17.3). The determination of the duration of oestrus in the ewe has been based on the period throughout which the male will mate with the ewe. Several experiments have shown, however, that a decline in the ram's libido, specifically for the individual stimulus ewe, occurs after a number of matings and, when a fresh stimulus ewe is presented, there is restoration of libido. This suggests the possibility that duration of oestrus in the sheep, as estimated by the course of mating with a given ram, may be frequently underestimated. Heat periods of 3 days' duration have been noted in Suffolk and Cheviot ewes.

The length of the period of oestrus in the goat is sometimes longer than the usual estimates in sheep. The limits of the goat's heat period are more easily determined because of the clearer signs of heat in this species. The period of oestrus is 1–3 days (average 34 h) in most goats.

Post-partum oestrus

Since complete involution of the uterus requires several weeks following parturition in all animals, there are no physical grounds for expecting oestrus to occur soon after parturition. However, several species show normal heat very soon after parturition.

The mare is the best-known example of a species showing early oestrus after parturition. This early heat is called the 'foal heat' and occurs on average about 9 days after the birth of the foal. The foal heat is often short and is of low fertility, but is otherwise apparently normal. It has been observed that 65–69% of mares of most breeds show oestrus not later than the 20th day after parturition.

Table 17.2. Temporal features of oestrus in the mare.

Feature	Occurrence		Remarks
	Average	Range	
Age at first oestrus (months)	18	10–24	Breed variations occur
Length of oestrous cycle (days)	21	19–26	Length of cycle is largely dependent on length of oestrus, e.g. 5-day oestrus has 21 days' average cycle; 10-day oestrus has a proportionally longer cycle (26 days)
Duration of oestrus (days)	6	2–10	Heat periods early in breeding season are usually long (10 days) and tend to get shorter as breeding season advances
First oestrus post-partum (days)	4–9	4–13	Ninth day after foaling is common time
Breeding lifespan (years)	18	16–22	Breed variations occur
Breeding cycle	Seasonally polyoestrous; natural breeding season is spring and summer season (i.e. seasons of increasing light) in either northern or southern hemispheres; extended breeding time in tropics; nearer the Poles the breeding seasons are very restricted, also a feature characteristic of northern breeds, e.g. Shetland pony		

Table 17.3. Temporal features of oestrus in sheep and goats.

Feature	Occurrence		Remarks
	Average	Range	
Sheep			
Age at first oestrus (months)	9	7–12	Usually occurs in first autumn when well grown
Length of oestrous cycle (days)	16.5	14–20	Very long intervals usually indicate intervening silent heat
Duration of oestrus (h)	26	24–48	
First oestrus post-partum	Spring or autumn		Some ewes show oestrus while lactating
Cycle type	Seasonally polyoestrous		7–13 heats per season according to breed, 'silent heat' commonly precedes overt oestrus
Breeding lifespan (years)	6	5–8	Short breeding life for hill ewes
Breeding season	Precedes shortest day of year but varying in extent according to breed		Northern breeds (e.g. Blackface) have shorter season than southern breeds (e.g. Suffolk, Merino); latter can breed biannually
Goat			
Age at first oestrus (months)	5	4–8	Kids born in spring show oestrus in autumn of same year
Length of oestrous cycle (days)	19	18–21	Short, infertile cycles (e.g. 4 days not uncommon); short cycles in tropics
Duration of oestrus (h)	28	24–72	Seldom less than 24 h
First oestrus post-partum	Autumn		Tropical breeds can sometimes be bred while lactating
Cycle type	Seasonally polyoestrous		8–10 heat periods
Breeding lifespan (years)	7	6–10	Shortest in tropical breeds
Breeding season	Commences around autumnal equinox		September–January in northern hemisphere; extensive season in tropics

Heat is not rare in sows about 3 days after farrowing. Cattle show a much more delayed heat after calving, and this heat is further delayed if the subject is nursing a calf or is being milked frequently each day. In Red Sindhi cows after calving, cows whose calves were weaned at birth returned to oestrus in 110 days as compared with 157 days for cows whose calves were not weaned. In most European breeds of cattle, the first oestrus occurs at an average of 31.7 days after normal calving. The wide variation in the heat interval in dairy cattle is commonly taken to be associated with the rate of milk production, with high-producing cows taking longer to show the first post-partum heat. Cows running with a bull after calving show heat about 27 days earlier than similar cows without exposure to a bull. The heat interval is commonly about 6 months after calving among dairy cattle that nurse their calves.

A number of sheep of various breeds that have seasonal breeding rhythms show heat quite soon after lambing, before the onset of their anoestrous season.

Oestrous behaviour

Dogs

The bitch becomes more excited during oestrus but does not normally show new patterns of behaviour. In the presence of a potential mate (and not all male dogs elicit such behaviour) she presents her hindquarters with some lateral movements. In general, the pre-copulation behaviour is passive (Manteca, 2002).

Cats

The female cat increases the level of activity, is more vocal and shows more marking behaviour. She is more tolerant of male presence and shows lordosis. This involves keeping the ventral surface in contact with the ground and raising the hindquarters (Manteca, 2002).

Sheep

Relatively quiet heat is the rule in sheep rather than the exception. Oestrus in the ewe is extremely difficult to detect if there is no ram in the flock. But, with a ram in the immediate environment, it is usual for the ewe in oestrus to associate closely with it for most of a 24 h period. Ewes frequently initiate the first sexual contact by seeking out the ram and thereafter following the ram while heat persists. Sometimes the ewe will rub herself against the ram. Further behavioural evidence of oestrus sometimes appears as tail-shaking, but this occurs only at mating. Different grades of intensity occur in the sex drive of ewes, although the general mating pattern is similar in all. Many ewes in oestrus actively seek out the male and, when a choice of ram is possible, it is the most active ram that is generally chosen. Since ewes in oestrus visit a penned ram, the use of funnel gates can allow their identification because they become trapped in a pen close to the ram.

Goats

In sharp contrast to the sheep, the manifestation of oestrus in the goat has very marked behavioural signs. Goats of any breed probably show more conspicuous oestrous behaviour than any other farm animal. For the 1–2 days of oestrus the female demonstrates a rapid tail-waving: the upright tail quivers vigorously from side to side in frequent bursts of flagging. There is repeated bleating throughout oestrus, the animal eats less than usual and has a tendency to roam.

Cattle

It is uncommon for very intense manifestations of oestrus (for which the term 'heat' should perhaps be reserved) to be observed in cattle. Signs of oestrus, in addition to vulval mucous discharge, include general restlessness, raising and switching of the tail, arching or stretching of the back, roaming and bellowing (Garcia *et al.*, 2011). Where bulls and cows are together, the bull follows the cow in oestrus (Fig 17.2). The most noticeable element of behaviour, however, is the mutual riding that takes place between the oestrous subject and conspecifics. These are often the closest social associates of the subject. Ultimately, it is the cow in oestrus that stands to be ridden, but this does not wholly preclude the subject, in its turn, riding others, male or female.

Fig. 17.2. Highland cattle: the bull is following a cow in oestrus (photograph D.M. Broom).

The behavioural signs of oestrus in cattle include the following points:

- There may be an increase in general activity in ways that could be generally termed restlessness.
- The oestrous cow bellows more than usual.
- Grooming activities, in the form of licking other animals, are increased.

Typically, the oestrous cow frequently makes mounting attempts on cattle, sometimes preceded by butting and chin-resting. When several cattle in a group have been prompted to mount each other, through the initial activity of the oestrous cow, it may become difficult for an observer to identify the cow in the group that is in oestrus, but when one animal in particular is standing to be mounted by others it is usually the animal in oestrus.

- Oestrous cows often show jerky movements of the vulval region; arching of the back and stretching of the back are also frequently shown, sometimes with licking or sniffing of the perineal region.
- Some degree of appetite reduction occurs during the most intense manifestations of oestrus.
- The flehmen response may be shown.

Oestrus lasts for a period of 12–24 h and is commonly observed to be of shortest duration in young cattle. Considerable variation occurs in both the intensity of oestrus shown and the degree of notice taken of an oestrous subject by other cattle in the group. Only a small proportion of the group participate in mounting the oestrous animal. When two or more heifers are in heat, one is usually the more attractive to other heifers. For cows, on average, no great similarity in mounting attraction or expression of oestrus at successive heats is shown. The stage within the oestrous period does not appear to influence the mounting behaviour of others. A few cows are consistent in the intensity of their heats.

Ovarian dysfunction alters the nature of reproductive behaviour. Typically, it manifests itself either by anoestrus or by abnormal oestrus. Follicular cysts are the most common cause of both types of syndrome. Most other forms of ovarian dysfunction suppress oestrus. With oestrus suppression, two types of syndrome occur, namely sub-oestrus and anoestrus. The two syndromes are sometimes confused but have very different significance.

In sub-oestrus, the ovary of the animal exhibits the normal cyclical changes without overt oestrus occurring. Some cases with histories of anoestrus can reveal evidence of full ovarian function on rectal palpation by the presence of either an active corpus luteum or a graafian follicle: these cases are properly diagnosed as sub-oestrus. Sub-oestrus is of common occurrence in the early post-partum period. Poor or inadequate observation by attendants, for example, may result in erroneous use of sub-oestrus because oestrus is short or occurs at night. 'Silent heat' is a stockman's term used synonymously with sub-oestrus.

Anoestrus is only a symptom and is not in itself an independent disorder. It must always be remembered that pregnancy is a principal cause of anoestrus. Anoestrus under other circumstances requires special consideration as a symptom of several types of ovarian dysfunction.

Arrested ovarian function without a responsible lesion is not infrequently associated with anoestrus. In such cases no gross physical abnormality may be present in the ovaries, but these are found to be smooth-surfaced, being free of corpora lutea and graafian follicles. The condition may be a consequence of high production pressure or other environmental adversity. It often has the status of a herd problem in very large dairy herds.

Horses

The mounting behaviour between females so typical of oestrus in cattle is not observed in mares. Oestrous behaviour in mares shows a range of characteristics peculiar to this species (A.F. Fraser, 1970, 2010). The intensity of oestrous behaviour varies. A mare in oestrus frequently adopts a urinating stance. During these periods of straddling, mucoid urine is ejected in small quantities, which may splash at the animal's heels. Following this, the animal maintains the straddling stance for a time with the hind limbs abducted and extended. The tail is elevated so as to be arched away from the perineum. The heels of one or other hind hoof are commonly seen to be tilted up off the ground, so that only the toe of that hoof remains touching the ground. While this stance is maintained, the animal shows flashing of the clitoris by repeated rhythmic eversions of the ventral commissure of the vulva (see Fig. 1.3). The duration of equine oestrus is 4–6 days on average, but varies, some lasting only 1 day and others lasting up to

20 days. The company of other horses is sought and particular interest is shown towards the male. In the presence of the stallion, the mare in typical heat will direct her hindquarters towards him and adopt a stationary stance. It is found, however, that some mares of fractious temperament, though in heat, may kick forcefully on being mounted by the stallion.

There are several types of irregularity in equine oestrus. In the mare, as in the cow, a form of ovarian dysfunction can be clinically recognized in which there is follicular development in the ovary without an associated behavioural oestrus being detected. This suboestrus is commonly referred to as 'silent heat'. Normal ovulations usually occur in these cases.

Excessive displays of oestrus or 'nymphomania' are encountered in mares, but this condition is not associated with true cystic ovarian disease. Nymphomania seems to be a condition of 'transient persistence' of one or more follicles, which eventually regress or ovulate spontaneously. The term nymphomania may also be used by horse breeders to describe excessive oestrus manifestations within a normal oestrous period. A mare with normally recurring oestrous cycles could be in heat 22% of the time. There is great variation in the age at which young mares attain normal ovarian activity, as evidenced by polyoestrus.

Seasonal anoestrus is not uncommon in the winter in northern countries: European countries often report an incidence of about 50% of winter anoestrus in mares. The incidence in North America is at least as high as this and, in Canada, it is considered normal for most mares to be in seasonal anoestrus during the winter.

Oestrus in the mare can, in some cases, be shown without associated follicular development on either ovary. In some of these cases the behavioural signs of oestrus can be quite intense.

Pigs

One salient feature of oestrous behaviour in the sow is the adoption of an immobile stance in response to pressure on the back. In pig-breeding practice this is often supplied by the animal attendant pressing the lumbar region of the sow or sitting astride the animal. The onset and the termination are gradual and of low intensity, but the 'standing' period is well defined and lasts less than 1 day on the average. The oestrous sow is sometimes restless when enclosed, this being rather more noticeable during the hours of night. Some breeds, particularly those with erect ears, show a conspicuous pricking of the ears in full heat. The ears are laid close to the head, turned up and backwards and held stiffly. 'Ear pricking' is often shown when some movement is taking place behind the animal. Mutual riding is very much less common than in cattle, but the subject in heat is often ridden by other females. Occasionally, among groups of sows, one particular sow will perform most of the riding.

While the level of oestrous behaviour expressed by the gilt influences her fertility, factors exist in the environment that are capable of affecting the expression of oestrus. The level of oestrus expressed can be determined by the back-pressure test in the presence of a boar, based on the willingness of the gilt to stand and on the duration of this standing response.

Oestrous stimuli

It is now recognized that, for oestrous responses to be shown in complete form, in many domestic animals it is necessary for some form of stimulation to be provided. Male attendance, which supplies prompting behaviour, is now appreciated as being an important contributor to oestrous displays in females. This male influence on oestrous behaviour parallels the 'Whitten' effect, in which oestrus in mice can be synchronized and induced by the introduction of a male mouse into a colony of females. The majority of the females come into oestrus as a result of being exposed to a specific odour from the male animal. This phenomenon occurs in several domestic animal species.

The introduction of a ram to a flock of ewes can influence the commencement of the breeding season in that flock, even without the ram having physical contact with the female members of the flock (Fig. 17.3). Ram influence, being provided under practical conditions by 'teaser' rams, undoubtedly brings on seasonal breeding activities in ewes more rapidly than occurs when they are left in an all-female group. The masculine stimulus provided by the sight or by the sound or odour of the ram can influence the breeding behaviour of ewes some distance away from the ram.

Boars provide a range of stimulating factors in the form of elaborate nudging and highly specific vocalization, together with the production of pheromones, in order to induce maximum oestrous responses in sows.

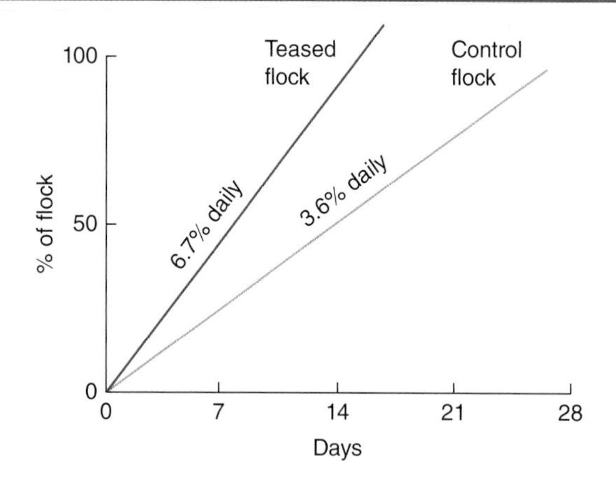

Fig. 17.3. Daily rate of lambing in teased and control (non-teased) ewes. Following stimulation, both flocks were joined and one ram was introduced; this achieved close synchronization of pregnancy.

In several experiments in which teaser bulls are run with groups of newly calved cows, the so-called teased animals show signs of oestrus much earlier than similar cows in control groups. Breeding behaviour may be shown 4 weeks earlier in the teased group than among controls.

Oestrus can be prompted or induced to some degree by genital stimulation in ways resembling phenomena of the 'Whitten' type. Genital stimulation is indicated by nuzzling, nudging and licking about the perineal region in the pre-coital behaviour of cattle, sheep and goats. Stimulation afforded by the presence of the male in a group of females, in addition to inducing oestrus, has the added result of affecting most of the female population simultaneously. In consequence oestrus, which has been prompted by biostimulation, shows varying degrees of synchronization (see Fig. 17.3).

When the ram is turned out with a flock there is a high incidence of oestrus 18–20 days later. The introduction of a ram serves as a stimulus to terminate the anoestrous season and results in a degree of synchronization of heats. Vasectomized teaser males in goats and sheep prompt the manifestation of oestrus in 91% and 97% of the females of the two species, respectively, after a latent period of 1 month. Stimulation by the ram is especially effective during the transition from the non-breeding season to the breeding season of the given breed, and also at the end of the breeding season. This synchronization of oestrus tends to synchronize parturition so has the result that many lambs are born at the same time. Where predation of lambs is heavy, lamb survival is more likely if many are born in a short period. If rams stayed with ewes, synchronization would be much less likely to occur, so a gene that promoted the separate flocking of males and females prior to the breeding period would spread in the population. It is clear that this has happened in the past.

The Male

Libido

Libido is an internal state that is measured by the likelihood of showing sexual behaviour given appropriate opportunity. This develops at puberty and is primarily dependent on sensory input and the production of hormones such as testosterone.

Measurements of sex drives in male animals have included: (i) number of ejaculations in an exhaustion test; (ii) the reaction time (delay before ejaculation); (iii) proportion of failures to mount; and (iv) proportion of failures to ejaculate.

In dogs, sexual activity occurs at relatively low concentrations of plasma testosterone. Fifty per cent of male dogs show sexual behaviour when their plasma testosterone concentration is only about 12% of that typical for intact male dogs (Nelson, 2000). In any attempt at measuring sex drive, a constant stimulus is needed. However, bulls, in the presence of a constant stimulus animal, show a gradual decrease in the number of ejaculations per unit of time until no further responses occur. The loss of response to a given stimulus animal may not interfere with the degree of response to other animals, but the recovery of sexual response to the same stimulus animal ranges from poor to complete after an interval of 1 week.

Although sexual reaction times in male animals can be influenced by a variety of somatic and psychological factors, estimation of the reaction time provides a simple and, ordinarily, reliable measurement of libido. Reaction times of bulls, rams and stallions remain almost constant over intervals ranging from 1 to 20 days. After the first ejaculation, reaction time increases with successive copulations as repeated breeding progresses. It has been found that the mean reaction time of four random observations spaced well apart gives an extremely reliable indication of the long-term sexual responses of the animal. Some variations in libido can have quite considerable consequences in farm economics. In the bull, libido varies in degree between age groups and between breed types. In general, a lower level of libido is found in beef than in dairy breeds. Comparing the species, the highest levels of libido are generally noted among the seasonally breeding animals, for example rams (Banks, 1964). Those species that concentrate their breeding season into relatively short periods require high levels of libido for effective reproduction during that time.

The level of libido may change as a consequence of various factors. Animals of some strains have low libido; unwise selection can permit its propagation and there is evidence that this has occurred in some of the beef breeds of cattle. Young rams usually show low libido after introduction to a new group (Holmes, 1980). Also, there are physical changes that occur in ageing bulls that are known to reduce their sex drive. An animal experiencing discomfort or even pain during movement in mounting will, in time, have his breeding behaviour impaired. Obesity in stud animals often contributes to low libido. Some skeletal defects such as arthritis are also a common cause of poor motivation in breeding behaviour. Seasonal fluctuations are widely observed in the libido of the male goat and sheep. During each breeding season, a level of libido is reached. In tropical locations, seasonal waxing and waning of libido may not occur in some sheep. In Africa, the mean libido is consistent throughout the year in the breeding activities of rams and goats. Rams and male goats are found to show seasonal reductions in libido, occurring mainly in the summer season (Fig. 17.4).

When entire male animals experience complete loss of their sex drive for some reason or other, they may seek out the company of other male animals of the same species. The bachelor groupings that result when such male animals gather together are a phenomenon that has been observed in many species of free-living animals. Bachelor groupings can be noted when large numbers of bulls run together. They are also seen among rams during the long non-breeding season of the year.

Courtship and Mating

Courtship

Courtship behaviour in dogs is often relatively brief and involves the male drawing attention to himself, sometimes with apparently playful, bowing movements (Manteca, 2002). Pal (2011) reported that more prolonged courtship by dogs was associated with a higher frequency of mounting. Various types of mating systems were recorded in this study of free-ranging feral dogs, including monogamy, polygyny, polyandry, promiscuity, opportunistic accepted mating and rape.

Courtship in ungulates involves male and female behaviour, but males are often more active. Many pre-coital components of behaviour tend to be species-specific. These include nosing of the female's perineum, nudging, flehmen (see below), flicking out of the tongue, striking out with a forelimb and low-pitched bleating. Butting of the female's hindquarters is also occasionally seen in ungulates. Mounting intention

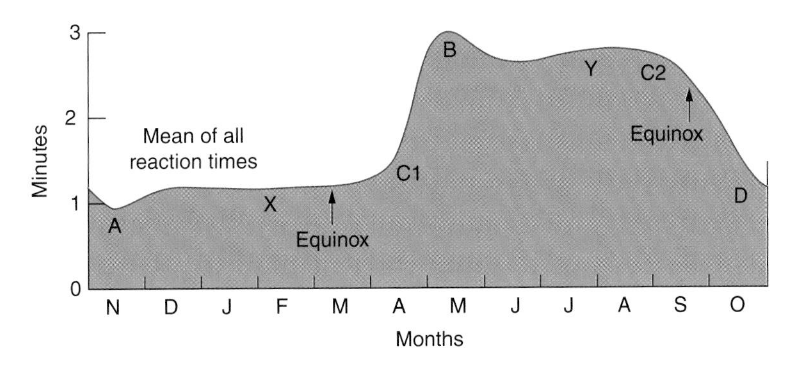

Fig. 17.4. Annual mean reaction (ejaculation) times in a group of 12 male goats in twice-weekly semen collections under controlled, experimental conditions. The breeding and non-breeding seasons are marked by plateaux (X and Y) in responsiveness. Changes (C1 and C2) in libido are closely linked to the two equinoxes, showing that a photoperiod with a greater quantity of dark than light per diem was stimulatory in these animals. Highest and lowest responsiveness (B and A) are brief phases commencing and terminating the breeding season in this species in northern latitudes.

movements are sometimes shown by males, as in all livestock. The behaviour of courtship includes three of its salient components, viz. female-seeking, nudging and tending. The seeking by males of females in oestrus goes on almost continuously under free-breeding conditions. Nosing the perineum and the hindquarters of females is a common male activity. Many male animals, such as rams and boars, will actively pursue females in pro-oestrus. Nudging behaviour prompts the female to move forwards. In oestrus, the female may respond with a stationary stance. This facilitates mating and provides reciprocal stimulation for the male.

In tending behaviour, the male maintains close bodily contact and association with the female. Both sexes contribute to this temporary alliance. In the tending–bonding of ungulates there are often phases when the male animal rests his chin over the hindquarters of the female. This chinning behaviour is notable in cattle and represents testing by the male of the female's receptivity by tactile sense.

Male sexual behaviour

During the sexual display of a male cat he emits characteristic vocalizations and shows much urine marking. He investigates the vaginal secretions of the female, using flehmen responses, and may grasp the neck area of the female (Manteca, 2002).

One of the behavioural components of male sexual display in all hoofed stock except the pig is flehmen. In this, the animal fully extends the head and neck, contracts the nares and raises the upper lip while taking shallow breaths. It occurs most usually subsequent to smelling urine and nosing the female perineum and is a form of odour-testing.

The sheep and the goat have components in their courtship activities that are extremely similar (Lindsay, 1965; Price *et al.*, 1984b). They include the following acts: nosing the perineum, nudging, olfactory reflex, flicking of the tongue, striking out with a forelimb and low-pitched bleats. The male goat has a large repertoire of such acts, including some others such as: (i) urine-spilling on to the forelegs, resembling urine-spilling on to the hind feet in reindeer; (ii) butting the female's hindquarters, resembling a component of bull behaviour; and (iii) false mounting attempts, such as occur in some other species, particularly the horse. Bulls often pump their tailheads and pass small amounts of faeces in pre-coitus. Male activities include: (i) threatening and displaying; (ii) challenging and contesting; (iii) sign-posting and marking; (iv) searching and driving; and (v) nudging and tending.

The threat display of the bull occurs as a physiological state of fight-or-flight. In this state, the animal arches his neck, shows protrusion of the eyeballs and erection of hair along the back. During the threat display the bull turns his shoulder to the threatened subject.

The threat display of the stallion involves rearing on his hind legs and laying back the ears. Threat displays are rarely shown by rams towards humans but, nevertheless, forms of threat are exhibited in the presence of other potential aggressors and in these circumstances the threat display usually involves vigorous stamping of a forefoot. The readiness with which threat in any form is displayed varies and reveals something of the male's temperament.

A challenging bull may show the following behaviours: (i) the bull paws vigorously at the ground with a forefoot, with the head lowered; this pawing or scraping breaks up soil, which is scooped over the animal's withers, where it may gather as an earthy mantle; (ii) the animal rubs the side of its face and its horns into the area bared by pawing; this is done with the animal kneeling and with some vigour; and (iii) the bull stands stationary, repeatedly bellowing with a broken voice.

Male animals may engage in behaviour that defines a defended area. Pawing and horning behaviour by bulls creates bare patches of earth, and these patches located throughout his territory are clearly a claim to possession of a given area ('stamping ground'). Stallions also claim territory: they do this at pasture where they urinate and defecate at selected spots. Given suitable territory, they mark it in this fashion, returning from time to time to defecate and urinate again in the same places. These activities apparently serve the dual purpose of marking male territory and priming female sexual responses by pheromones.

Searching for breeding females, and then driving them before them, is an activity of some male animals. Male seeking females in oestrus may do so actively. Nosing the perineum and the hindquarters of females is a frequent male activity in various mammalian species.

The key stimulus required to elicit sexual behaviour in the male is simple in the case of the bull. Bulls will mount and attempt copulation with a dummy consisting of a covered frame. The basic releaser for mounting in the bull is in the form of an arch or bridge of appropriate dimensions. A simple dummy or 'phantom' cow is used in some programmes of bull testing. Of 2500 bulls, 90% mounted the dummy. When given a choice, however, bulls preferred a live cow. The visual stimulus is considered to be of paramount importance in the bull. Bulls that are blindfolded will still serve, however, and clearly other stimuli operate. Olfactory cues are of great importance in the bull and the stimulus provided by the sight of a sexual object may be enhanced by the olfactory stimuli released by an oestrous female in her urine. The internal and external stimuli, involving conditioning and all the senses, are temporarily cumulative and the threshold may be exceeded by one, some or many stimuli according to the responsiveness of the subject.

In practice it is found that changing the stimulus animal or the setting results in a greater number of ejaculations per unit time in the bull and sheep. Sexual responsiveness in the male sheep can be readily restored, after exhaustion with one ewe, by presenting a fresh, unmated ewe. Changes of species can have an adverse effect; for example, the responses of jackasses are poorer to mares than to jennies. Young boars with experience of natural mating take longer to respond to the dummy sow than do similar young boars without experience of natural service. In certain circumstances the male animal can be abnormally stimulated. Restriction and homosexual groupings of intensively confined animals are liable to produce abnormal intraspecific sexual behaviour.

Some breeds of domestic sheep and goats have such an intense breeding period that there are resemblances to rutting. The fighting that occurs between Blackface rams at the outset of their breeding season closely resembles a rut. Head-on charges between rams can be so forceful that ruptures sometimes result from the impact. Fighting between breeding males can be merely mock battles from which no serious injuries result; but, on the other hand, such contests can be very real, vigorous and potentially harmful to the participants. Among domestic stock it would be extremely unwise to allow adult breeding males to fight on the assumption that the contest will be a mock, ritual event. Fatal trauma too often results in such circumstances.

Mating behaviour

Coitus in most domestic mammals has its timing so arranged that most spermatozoa are introduced into the female genital tract before the ovum is liberated from the ovary. Initial copulations are centred mainly in the earlier part of oestrus, with repetitions of mating occurring in the later part.

Mating results from the emergence of oestrous behaviour in the female and the activation of male sexual behaviour. Copulation in cats is relatively brief but is repeated. In dogs, however, copulation is prolonged by a 'genital block' lasting 10–30 min, during which the penis cannot be withdrawn (Manteca, 2002). Both sexes of ungulates make equivalent contributions; the male contribution to mating is summarized in Table 17.4. Pre-coitus, the basic response of the male, is orientation towards the female. The male ungulate typically aligns himself behind the female, in the same long axis, until mounting occurs. Deviated stances occur in close association with impotence characterized by diminished drive. In the deviated pre-coital stance the male animal, while keeping his head at the hindquarters of the female, turns his body away from her alignment. Such disalignment sometimes precedes impotence, thereby allowing the latter to be forecast under conditions of controlled mating. Deviated position is held characteristically for lengthy periods of several minutes, and the angle of disalignment appears to be greater with more complete forms of impotence.

The salient component in mating behaviour is mounting. The neuro-ethological mechanism for the activity of mounting differs from that mediating intromission. Mounting by both sexes is seen in some species; it is particularly common in cattle. Female cattle will mount each other even when there is a male present in the group. The pig is another animal that exhibits mounting by females when one of their number is in oestrus. Mounting by females on other females in oestrus is a very rare event in the horse. As a rule, it does not occur in sheep.

'False mounting' attempts by the male animal are commonly seen in courtship. In these instances, dismounting subsequently follows quickly without any forelimb clasping or pelvic thrusting movements. False mounts show that the mechanics of mounting and of intromission are separately controlled. False mountings are to be seen in the mating patterns of stallions, sheep and the goat. In the stallion, it is believed that some two or three false mounts are normal before effective mating is achieved.

In a detailed study of mounting behaviour in the stallion, it was found that older animals mounted more quickly than younger ones, that blindfolded subjects mounted more quickly than others and that about two

Table 17.4. Coitus in farm livestock.

Male reaction time (min)	Pre-coital behaviour of male	Manner of intromission
Stallion		
Averages about 5	Noses genital region; genital olfactory reflex; bites croup region; penis erects fully	One to four mounts; several pelvic oscillations; terminal inactive phase
Bull		
Mode 2 Mean 12 Mean of beef breeds 20	Noses vulva; genital olfactory reflex; alignment; licks hindquarters	Single pelvic thrust coordinated with clasp reflex
Boar		
1–10	Approaches sow giving series of grunts; noses vulva vigorously; champs jaw and froths at mouth	Short protrusions of spiral penis repeated till intromission occurs; pelvic oscillations followed by somnolent phase
Ram		
0.5–5.0	Noses vulva; genital olfactory reflex; paws with forefoot	Very quick single pelvic thrust with forelimb clasping
Goat		
0.2–1.0	Bleating; stamping with forefoot; rapid licking; genital olfactory reflex	Very quick pelvic thrust

false mounts seemed to be usual before intromission and ejaculation were effected. In the stallions, which were blindfolded, it was observed that mounting was undertaken after the male had 'shouldered' the female. The mounting behaviour of the stallion when presented with a dummy mare was also studied and it was noted that young, inexperienced stallions mounted the dummy more readily when they were blindfolded than when they could see, but still more mounted the dummy when it was visible and sprinkled over with urine from an oestrous mare. All of this points to the positive stimulus of odour and the weak effect that visible features may have on mounting behaviour in this animal.

Bulls, which have a significantly protracted reaction time and which appear to have their mounting responses inhibited by environmental factors when presented with oestrous cows, sometimes respond to blindfolding by mounting promptly. In the ruminant species, intromission consists of only a single pelvic thrust, which is followed by dismounting. In the stallion, intromission is maintained for a period of 1 min or more during which there is repeated pelvic thrusting and subsequently the adoption of a fairly static posture, after which dismounting occurs. Clasping by the male during intromission and mounting is an important component of coital behaviour (see Fig. 17.5).

The stallion and the bull effect tight clasping of the respective female with their forelegs adducted into her flanks. In the case of the bull, this clasping increases in intensity at the moment of penetration and ejaculation. Vigorous clasping also takes place in mating between sheep but, when the male and female are heavily covered by wool, clasping is inevitably impaired to some degree.

The manner of intromission in the pig is unique. In this species the male mounts and makes thrusting actions with the penis, which repeatedly makes semi-rotatory actions. Only when the spiral glans penis of the boar becomes lodged tightly in the firm folds of the cervix does this action stop and ejaculation commence. It is clear, in fact, that the locking of the penis in the cervix acts as the essential stimulus to ejaculation in the boar. Although clasping is not easily shown by the boar, this animal employs a 'treading' action with his forefeet on the back of the sow while mating. Large boars on small sows or gilts will clasp, but in these cases it is common for the female to collapse and effective intromission or mating does not occur.

Treading is an important feature of mating in poultry. In the mating behaviour of poultry 'waltzing'

Fig. 17.5. Stallion mating with mare (photograph A.F. Fraser).

by the cock is the main pre-copulatory behaviour. In waltzing, the cocks adopt a stilted walk around the female bird, tilting the body to one side. In freely mating poultry, a high frequency of treading occurs in social relationships where the male chases the female frequently, or where the male waltzes to the female frequently. Treadings most often follow crouchings by the females, chasing the female and previous treadings by other males. In ducks, mating may take place on water or land, and sometimes the mating drake will be assisted by another male pressing the female's neck on to the ground when the mating drake is mounting.

The erect penis on intromission has increased sensitivity to tactile stimuli and these, in their turn, effect ejaculation. Spinal reflexes are also involved in ejaculation. The release of oxytocin, on intromission, contributes to the ejaculatory process. Significant amounts of oxytocin are present in the bloodstream immediately before and after mating.

Repeat mating behaviour

Following ejaculation, male animals show a refractory period, which is a state of sexual exhaustion. The state of sexual exhaustion is not principally a physical one, however, and refers mainly to the loss of stimulus quality by the female. A quick return to mating behaviour is shown by male animals when they are given an opportunity to mate a new oestrous subject.

It is normal among some domestic animal species for repeated matings to occur with any given female. Cats mate again 5–15 min after the initial mating (Manteca, 2002). Stallions probably re-serve oestrous mares five to ten times in each heat period and most rams are noted to re-mate with ewes three or four times. Bulls are seen to re-mate with oestrous cows repeatedly perhaps on five or six occasions. Boars normally serve sows several times over a period of 24–48 h. Among poultry repeated matings are frequent. The boar is capable of a great number of services before exhaustion occurs. Boars can serve oestrous sows up to 11 times in one heat period. It was observed in a test that each of three boars ejaculated eight times during a 2–2.5 h test period in which nine oestrous females were available. In a test of mating ability in the bull, however, it was reported that 75 ejaculates were collected from one particular bull in a 5 h period of testing (Almquist and Hale, 1956). The average ram mates 45 times per week in a large flock.

Variations occur in the degree of receptivity in oestrous female animals. Some studies have shown that many ewes permit about six matings during each heat period. When competition between ewes exists for a limited number of rams, older ewes are usually more successful than maiden ewes in obtaining repeated matings.

Natural matings have the effect of shortening the duration of oestrus in cattle. The period of receptivity in cattle may be shortened by as much as 8 h when natural circumstances are provided and repeated matings take place. When some female animals are 'teased' with vasectomized males, the duration of oestrus is slightly shorter than otherwise. Oestrus is not under endogenous control alone and its manifestation is subject, in part, to environmental factors.

Further Reading: Fraser, A.F. (2010) *The Behaviour and Welfare of the Horse*, 2nd edn, Ch. 10. CAB International, Wallingford, UK.

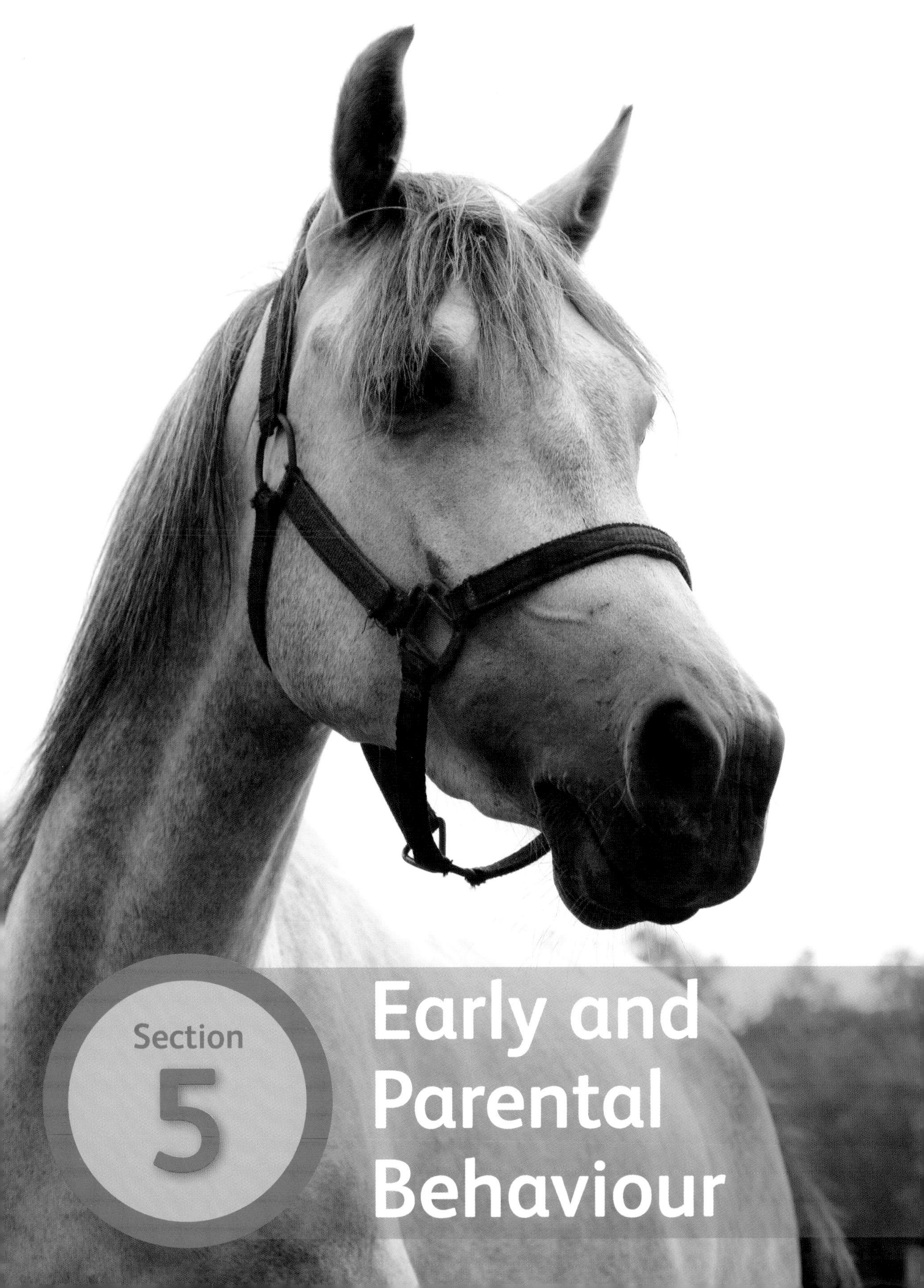

Section 5

Early and Parental Behaviour

Fetal and Parturient Behaviour

Introduction

Features of behaviour emerge and change during the development of the animal. Gene action in the mother mammal may be switched on by the effects of stimuli during pregnancy and hence result in a different environment for the developing fetus. This can change the likelihood of survival for one or more fetuses, the hormonal environment of the fetus and hence its sex, or the nutrient availability, size and vigour of the fetus and neonate. Gene action in the fetus may be switched on by stimuli from its environment, including the presence of other fetuses and the nutrient and hormonal environment (Broom, 2003). This can result in a different set of developmental changes relevant to what is optimal for each fetus. This is not necessarily the same as that which is optimal for the mother. Hence, there can be conflict in the action of maternal and fetal genes in programming the development of the fetus. When a fetal or neonatal animal is developing, the characteristics are not just those that will eventually lead to an effectively functioning adult.

Behaviour in the developing animal reveals phenomena of the neural programme relevant to fetal and neonatal survival (Cowan, 1979). An appreciation of such phenomena is important to optimal conditions in the prenatal and perinatal periods. Mortality rates in newborn animals are great and cannot be controlled without progressive management, taking account of their needs, including those to show certain behaviours.

Our knowledge of the effect of the prenatal environment on postnatal behaviour has been increasing (Bateson *et al.*, 2004). Precocial neonates, such as foals, calves, lambs and piglets, have already been exposed to sensory stimulation of some kind before birth and it is therefore likely that sensations of position, gravity, touch, smell and taste occur within the uterus. It is known that sheep fetuses can hear some extra-uterine sounds (Vince and Armitage, 1980; Vince *et al.*, 1982). The fetal ear of ungulates is completely formed and functional before birth. The quantity of amniotic fluid in fetal nostrils at birth is usually considerable, indicating that the olfactory mucous membrane has been in close intimate contact with the material, which will later be expelled at the birth site.

In the embryonic domestic chick, movements of the head, body, wing and limb begin as jerky actions at 4 days of incubated age. These continue as uncoordinated movements in irregular sequences, reaching a peak at the middle of the incubation period and continuing until a few days before hatching, when smooth and coordinated movements develop. When the chick is ready to hatch, synchronized limb movements resembling walking and wing movements that resemble flapping are present (Broom, 1981; Provine, 1984; Chapter 3, this volume).

Features of behaviour in the human fetus have mostly been derived from studies of the living aborted fetus subjected to probe stimulation during a brief extra-uterine life while suspended in isotonic saline solution. Fetal behaviour in domestic animals has been studied by palpation, surgery, ultrasound, radiography and fluoroscopy.

Action Patterns and Movement Sequences in the Fetus

In addition to movement of one part of the body and patterned movement, episodes of great fetal activity occur. During the study of naturally occurring behaviour in a fetus, the concept of the 'action pattern' (Broom, 1981) is useful. When gross fetal activity occurs, this is the result of groups of complex fetal movements quickly following one another. It is therefore appropriate, in all types of fetal activities, to recognize the functional units of these. The action patterns of the fetus in rotation, extension and pronation seem to be non-signal in nature.

Periodic mouth opening and closing has been reported in the human fetus as a local reflex, a simple fetal movement. Continuous fluoroscopy on the sheep fetus has shown that jaw movements, which are of high frequency in nature, constitute a notable item of terminal fetal activity. Slow and rapid jaw movements can be distinguished, the former as early as 40 days pre-partum. Rapid jaw movement is seen in the mature fetus in the form of rapid rhythmic mouth opening and closure. Most of these exceed ten in number per brief episode of oral activity. This rapid, rhythmic jaw movement is carried out with the lower jaw moving in the vertical axis in relation to the skull, and represents vigorous sucking activity by the fetus (see Fig. 18.1). Such episodes occur frequently in very late gestation. Involved with these are minor actions of head extension with the mouth open, and ventral head flexion with the mouth closed. It has been determined that such actions represent swallowing. In addition, some rapid jaw movements are seen in the last week, where the lower jaw moves at right angles to the upper jaw, with the mouth evidently closed. This has been assumed to represent 'chewing' mouth movement. It is clear, therefore, that the mature fetus engages in a considerable amount of oral behaviour, including sucking and swallowing. These observations dispel the notion that the 'suck reflex' is a phenomenon that comes into existence for the first time post-natally. The first sucking activity, in the form of rapid jaw movements, is seen in the sheep fetus, on average, 10 days before full term and occasionally as early as 16 days before birth. Fetal sucking has been observed at some point during the week preceding birth in all of the fetal lambs recently studied for this phenomenon. Typically, complex fetal movements contain three to five individual component actions. In the terminal phase of gestation complex fetal movements in cattle, horses and sheep increase in frequency. Some episodes follow closely on each other, giving lengthy spells of fetal activity. Extremely intense fetal activity is frequently followed by an extended period of fetal quiescence. In the radiographic study using fluoroscopy, complex fetal movements are seen to have quality in addition to quantity, in that they contribute to processes of action.

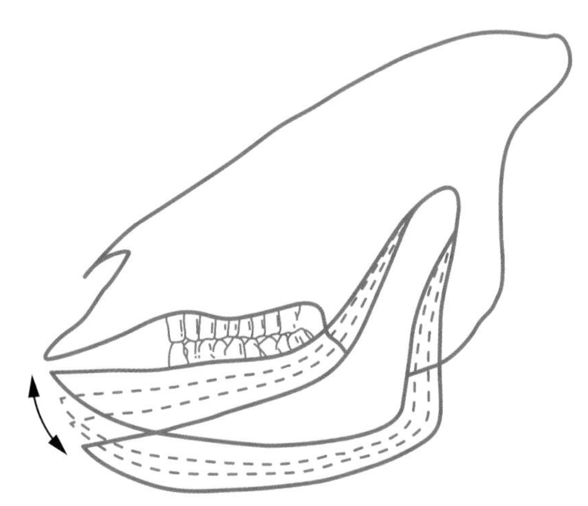

Fig. 18.1. Drawing of rapid jaw movement as seen by fluoroscopy in the sheep fetus during the 2 weeks before birth. This activity represents sucking, which is therefore not simply a postnatal development (drawing R. Bowen).

Righting behaviour can occur in the fetus which, in the case of sheep, is otherwise held in a supine position, the fetal spinal column resting on the maternal abdominal floor. The principal items in righting behaviour are:

- general activity;
- extension of the carpal joints and digits of the forefeet towards the maternal pelvis;
- elevation of the head and neck;
- rotation of the head and its extension towards the maternal pelvis;
- rotation of the anterior trunk from the supine position through 180 degrees; and
- attainment of a position of pronation.

Essentially, these righting reflexes dictate a conversion from general flexion to strategic extension of certain parts; they evidently occur during the 2 days before birth in the sheep, 1 day pre-partum in the bovine fetus and in first-stage labour in the equine. General activity, involving approximately 3500 movements, coincides with these fetal righting actions. Typically, this ceases before birth, when a phase of quiescent fetal behaviour prevails. If all of the essential righting reflexes are not accomplished, the fetal posture may be abnormal. The common abnormalities are the result of: (i) failure of one or both forelimbs to be extended into the maternal pelvis before birth commences; or (ii) deflection of the head to one side – a position that is exaggerated when the second, or expulsive, stage of birth forces the fetus into the birth canal. These circumstances can lead to the most common forms of fetal malposture encountered in dystocia in sheep and other farm livestock.

As a result of periodic rotations, the fetus terminally adopts the prone and squatting posture appropriate for birth. This occurs several times before the birth processes become imposed upon the fetus. Eventually, there is extreme extension of the head at the atlanto-occipital joint, and this ensures 'engagement' of the extended skull and forelimbs within the maternal pelvis. The primary righting act of carpal joint extension has the effect of directing the forefeet into the maternal pelvis. The elevated and extended head follows, having changed from a cranial presentation to a facial one. When the head and forefeet are fully engaged in the birth canal, ewes began to show the behavioural signs indicative of pain in the first stage of labour.

As birth of the anterior parts of the fetus progress, all the hind limb joints are flexed. These remain in this posture until the fetal abdomen is in the birth canal. At this time, the hind limbs convert to parallel posterior extension. Fetal expulsion is complete soon thereafter.

The postures of 1255 bovine fetuses were determined using a direct surgical examination in a caesarean survey. The fetuses had incidences of 87% with anterior presentation, 90% with head extension, 96% with extended forelimbs and 57% with extended hind limbs in posterior presentation. The application of these data relates mainly to the aetiology of fetal malposture and the recognition of dysfunction of posture as causes of bovine dystocia.

Parturient Behaviour

The process of birth passes through three very definite stages, which are clear in species in which single births occur. The first stage refers to the dilation of the cervix and the associated behaviour of the animal. This is the latter part of the pre-partum period during which behavioural changes occur. The second stage is the expulsion of the fetus itself. The third stage is the passage of the afterbirth or fetal membranes, which extends into the more general behavioural period postpartum (see Table 18.1).

Pre-partum behaviour

The pre-parturient period extends from late gestation to the beginning of the first stage of labour.

Apart from certain changes in attitude towards any previous offspring still being nursed, there is generally little of significance in the animal's behaviour until parturition itself is very close. Once parturition is imminent, many animals separate themselves from the main group and select a site for the birth. However, bitches that are about to give birth do not normally do this. They do find or make a nest in which the puppies will be produced. Cats will also find a suitable quiet place for the birth. After giving birth, the mother cat usually eats the afterbirth (Manteca, 2002).

Sows that have had the opportunity to do so will have already found a suitable site and built a nest during the 24 h period before parturition. In each case, the site is one where the birth may occur unhindered. The nature of these nest-building activities will depend on the material with which the sow is provided. The sow will attempt to clean and dry her selected birth site and will chew long grass or straw to provide bedding, carrying it a considerable distance if necessary. The sow uses her forelegs to move the bedding about. In general, sows adopt and maintain a lying posture at rest before birth. It has been observed that free-living sows choose a wooded area to build a den with dry vegetation. The dens are lined with chewed-up undergrowth and leaves and the site of birthplace is made dry and sheltered. In a concrete pen, the sow will still pursue her nesting tendencies using any material that is provided for her use as bedding. She will often resist human attempts to disturb or relocate her nest. The amount of time taken over the building of a nest varies from one sow to another but they nearly all make use of straw, hay and any other dry material that is provided.

The grazing ruminants often appear to withdraw from the grazing group when birth is only 1–2 h away but, in some cases, the parturient animal has simply failed to keep up with the grazing drift of the main herd or flock. For example, pony mares in heath and woodland areas may move to a quiet, sheltered area beyond the normal home range (Tyler, 1972) or they may stay close to the group. Goats are more likely to give birth in the overnight camping area of the flock and choose a site with overhead cover and shelter from the wind where possible. According to Lickliter (1985), 76% separate themselves from herd-mates prior to giving birth. Cows attempt to give birth in long vegetation where there will be cover for the calf during the period when it remains concealed. Deer will move into woodland or other sheltered areas to give birth, for their young also spend some time in concealment after birth.

Two phenomena widely reported to occur in lambing ewes are isolation-seeking and shelter-seeking. The fact

Table 18.1. Principal events in parturient behaviour.

Animal	Immediate pre-partum period	Parturition	Postpartum period
Mare	May be shy of interruption; anorexia only evident immediately before foaling; whisking of tail is shown as parturition commences; no 'bed' prepared	At first: restlessness and aimless walking, tail swishing, kicking and pawing at bedding; later: crouching and straddling, kneeling; finally: mare lies down to strain powerfully and regularly to expel fetus, latter must rotate from a supine position through 180° for delivery	Frequently remains lying on side for a period of about 20 min; does not eat fetal membranes; usually oestrus shown by ninth day ('foal heat')
Cow	May separate from herd; seek screened locality; anorexia develops (1 day); restlessness begins (several hours); no 'bed' prepared	Pain and discomfort expressed in very restless behaviour (alternately lying and rising); animal more usually lies during expulsion of fetus; expulsion of fetus slower than in mare	Will eat fetal membranes if fresh, occasionally before uterine separation is complete; maternal responses such as licking calf usually initiated promptly
Sow	Gathers bedding material and attempts to make a nest; becomes restless; lies on side in full extension to farrow	Shows signs of pain; strains when fetuses are in pelvis; tail actions signal each birth; the final expulsion of each piglet is usually done with one wave of abdominal straining	Eats fetal membranes and fetal cadavers or parts of them; lies extended on side for long spells accommodating frequent suckling by litter
Ewe	Interest in other lambs; may separate from flock (66%); most breeds seek some shelter; restless, paws ground	Repeated rising, lying; when straining, usually stands; lies during early expulsion	Grooms lamb attentively licking fluids, nibbling any membranes on lamb; remains at birth site usually
Doe goat	Sluggish walking; becomes restless, agitated	Shows clear signs of pain; repeated straining, occasional bleating	Variable attention to kid; grooms by licking and nibbling; if a 'leaver', may seem to have weak maternal drive

that a ewe, when first seen with a newborn lamb, is often separated from the flock does not necessarily indicate that the ewe actively sought isolation from the flock prior to parturition. While 60% or more of most ewes lamb in isolation from the flock, the remainder lamb in the area where the flock is grazing or camping. The majority of isolaters seek isolation actively while the rest are passively isolated, as they are left behind by the flock. Most ewes lamb on the site where the birth fluids spill and most do not move away from the site where they commence labour. Many of those come back to the site to complete the birth process. Fewer Merino sheep seek isolation and 90% lamb at the area where the flock is grazing or lying at the time labour starts. Fine-wool Merinos rejoin the flock as soon as they can after lambing, and show poorer maternal behaviour than other types of Merinos (Stevens *et al.*, 1982); most other breeds are better mothers than Merinos. Active isolation is strong in some other breeds. Evidently, sheep of different breeds behave quite differently in seeking isolation at lambing.

The biological advantages of parturition in isolation are twofold: the risk of interference by other ewes that are in the pre-parturient state is reduced and the best opportunity for developing a close bond with the young is provided. The newborn lamb orientates itself towards the nearest object, preferably a moving one, which is most likely to be the maternal ewe. Isolation-seeking is a behavioural tactic in sheep which, in the process of domestication, has apparently persisted in most breeds. Shelter-seeking by ewes, which may be associated with isolation-seeking, depends on weather conditions and the availability of shelter. Welsh mountain ewes, for example, seek shelter at lambing time in wind speeds of over 11 km/h. At higher wind speeds, where there is also a cooling chill factor, they choose progressively more sheltered sites for birth. In Soay sheep marked isolating and sheltering behaviour occurs shortly before parturition, with the ewe lying down in a sheltered position independent of what the other ewes are doing. Similar effects of wind speed and chill influence shelter-seeking by parturient ewes of other breeds, but the Merino is a notable exception, showing flock cohesion.

Under intensive housing conditions natural pre-parturient behaviour is not possible. For instance, the feral sow builds a nest of grass in a shallow pit (Jensen, 1986), while in farrowing crates sows show major changes in behaviour within 24 h of parturition. Sows will work hard to gain access to nest-building materials on this day (Arey, 1992). With sows in bedded farrowing pens of the traditional type it is observed in the pre-parturient phase that, while most of the time during the 3 days before the onset of labour is spent sleeping and feeding, an increasing amount of nest-building behaviour is also shown; this is usually evident in the form of collecting bedding. The first changes in pre-partum behaviour are restlessness, with the sow frequently altering her position, either from side to side when lying down, or from lying to standing. This activity gradually increases until the sow changes position every few minutes. Intermittent grunting, champing of the jaws and increased respiratory rate are also prominent features. During this period the sow may attempt to make a nest, using its forelegs vigorously in attempts to accumulate bedding into a heap.

In the immediate pre-partum phase of all domestic animals, which is the 24 h before parturition, indicative behavioural features begin to emerge. The animal becomes increasingly restless and frequently alters her position and possibly also her disposition. Phantom nest-building may occur, and sows provided with cloth strips manipulate them extensively at this time (Taylor *et al.*, 1986). For further details of nest-building and other behaviour see Marchant-Forde (2009). Gradually, still greater restlessness becomes evident until a stage is reached where the animal changes her position every few minutes. Recognition of this pre-parturient behaviour in the pregnant animal allows the time of birth to be predicted accurately in the great majority of cases. Forewarning such as this allows the corrective husbandry of parturient animals and their neonates.

Some similar behaviour to that in the parturient sow is seen in the cow. In beef cows, changes in behaviour appear from a few days to a few hours before parturition. Sometimes, restlessness is accompanied by the cow licking its flanks or swishing its tail. In dairy cows, some aspects of behaviour change as early as 6 weeks pre-parturition, when the cow avoids social exchanges of butting or being butted. Such changes may relate to the physical burden of pregnancy as a result of which the cow becomes less agile in advanced pregnancy and so is less able to sustain social rank through agonistic behaviour. Two weeks before calving, pushing at a crowded feed trough is no longer seen. The cow tends then to feed and drink when few other cows are at the trough. She also stays on the periphery of the herd when grazing and lying.

Shortly before parturition, some females show interest in recently born young of other females in the group. This behaviour indicates the early onset of the maternal behaviour under new hormonal influences. It is seen most frequently in ewes, since there is a high probability in large flocks that ewes with newborn lambs will be mixed with pre-partum ewes. Pre-lambing maternal interest has been observed as much as 2 weeks before lambing, but it usually occurs in the 12 h preceding birth. It can lead to a high incidence of mis-mothering and poaching of alien lambs, especially at high stocking densities. The incidence may be up to 20% with some breeds of ewe and is less in primiparous ewes. In Merino flocks pre-lambing maternal interest often occurs within 2 h of lambing. Interest varies from a brief inspection to cleaning, suckling and, on some occasions, to attempted adoption. Such ewes normally lose interest in other ewes' lambs at the commencement of parturition. 'Stealing' of other young in this way occurs in mares and cows and may pose serious problems in group-housing conditions (Edwards, 1983).

Increases in restlessness and withdrawal from group activity are characteristic of pre-partum behaviour in sheep, cattle and horses. Active separation from the group sometimes occurs but, in a study of dairy cows, many did not isolate themselves before parturition (Edwards and Broom, 1982; Edwards, 1983). In sheep and cattle, pawing and sniffing the ground may also occur, but pre-partum pawing is most notable in sheep. The length of time prior to parturition during which these changes in behaviour occur varies with the species; in sheep they occur quite close to parturition. Many ewes display uneasiness for the first time within 1 h of birth, while virtually all will lamb normally within 2.5 h of such signs.

The timing of parturition

Parturition does not occur randomly throughout the 24 h in most species that have been studied in detail. A higher incidence of birth at night has been reported for various mammals, including man, horses and pigs. Conflicting results have been obtained for sheep and cattle, but Edwards (1979) studied 522 Friesian cows and reported no peak of calving at night or during the daytime (see Fig. 18.2). It was of particular interest in Edwards' study that, while heifers calved evenly throughout the day, older cows were less likely to calve at milking times (see Fig. 18.2). This avoidance of the time when other cows in the herd were regularly milked could be due to extra disturbance at this time but heifers showed no such avoidance, despite being subject to the same disturbance. It is possible that some internal rhythm created during the lactation period is affecting behaviour. Whatever the cause, it is clear that the cows were able to avoid milking time, presumably by actively delaying the onset of the final stage of parturition.

Although ewes lamb at all times during the 24 h period, it has been observed over a period of time, covering several lambing seasons, that a disproportionately large number of ewes lamb during the 4 h period lasting from 19.00 until 23.00, and also during the early morning hours from 05.00 until 09.00 (see Fig. 18.3). The distribution of births over the 24 h appears to vary slightly with the breed of sheep. Merinos have a small peak between 17.00 and 21.00. Suffolk ewes show a peak between 08.00 and 12.00.

The foal attempts sucking within 1 h of birth and begins to trot after about 4 h. About 80% of mares foal at night and in the open, if possible (Rossdale and Short, 1967). Parturition is also usual at night in stabled horses, perhaps because this is when there is least human disturbance. The peak foaling hour in stabled mares is prior to midnight. In free-ranging mares birth is often in the dark hours of early morning. This gives the mare and foal all day to establish a bond and allows the myopic foal to travel in daylight.

Parturient synchronization points to a form of maternal protection, which may be either conscious or unconscious. For example, it has been found that many mares foal during the early summer when nutrition is at its height. Mares mated early in the breeding season will often foal at the same time as those that mate later in the season. Apart from this broad synchronizing effect on gestation, it has been found that seasonal factors may also affect the time of birth. Comparing two groups of mares, it was found that the average length of time of gestation was 8 days longer in those that foaled in the spring than in those that foaled in the autumn.

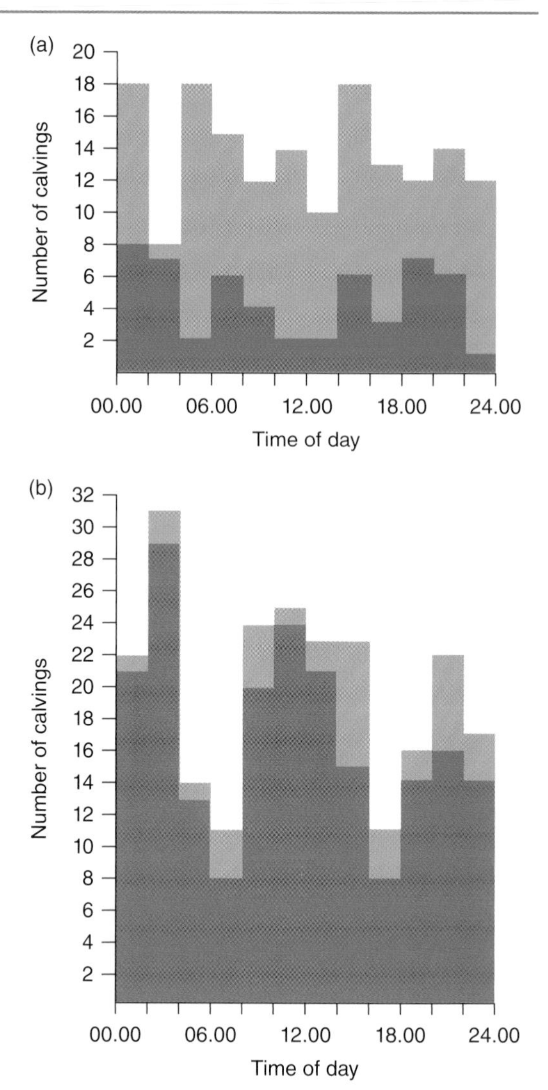

Fig. 18.2. The distribution of calving times during 24 h according to the parity of the cow: (a) first calvers; (b) third to eleventh calvers (median fifth). Unassisted calvings (light shading), assisted calvings (dark shading) (after Edwards, 1979).

Birth

During birth, pain is most evident during the period of fetus and fetal membrane expulsion. However, the extent of this pain is likely to vary considerably among species. In dogs, cats and pigs that have several young,

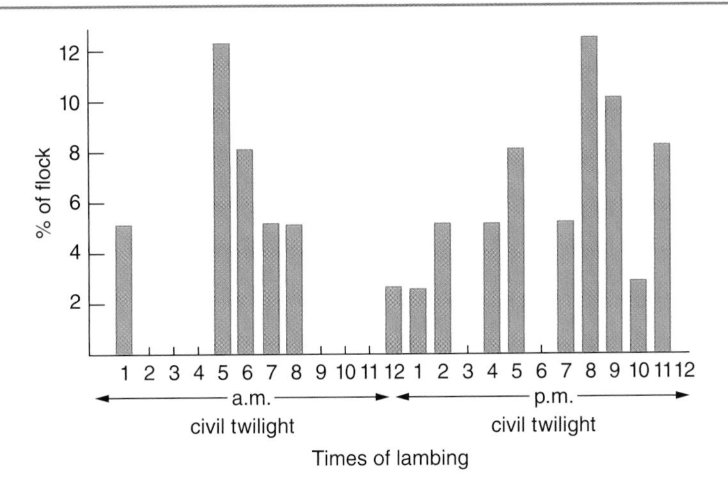

Fig. 18.3. Histogram of lambing times in flocks of Cheviot and Suffolk ewes observed experimentally over a 3-year period. The bimodal distribution points to two peaks before and after morning and evening twilight. Parturition in the ewe appears to be loosely related to the dark side of the crepuscular period.

the production of each individual involves less stretching of tissues and straining by the mother than that which occurs when a ewe has her one or two lambs or when a cow or mare has her single calf or foal. Difficulty in giving birth can result in very poor welfare and the use of analgesics is often needed to reduce pain at this time. Most of the details below refer to the species for which parturition is more traumatic.

The outer fetal membrane remains adherent to the uterine wall and, during the course of physical straining, becomes rent with pressure. This allows the amniotic bladder containing the fetus to bulge into the vagina, lubricated by secretions of cervical mucus and chorionic fluid, thus effecting further dilation so that the fetus enters the pelvic canal. At this stage of labour the contractions of the uterus are regular. Even at the end of this stage, they can be very strong and frequent. These events terminate the second stage with an acceleration in the expulsive efforts of the dam. Provided there is no impediment to its delivery, the fetus is then expelled by a combination of voluntary and involuntary muscular contractions in the abdomen and uterus. Repeated straining, particularly abdominal straining, is therefore the principal feature of maternal behaviour in birth. The straining efforts increase in number and recur more regularly when the second stage of labour has begun. At this time the strong reflex abdominal and diaphragmatic contractions are synchronized with those of the uterus.

The straining sessions are punctuated by resting intervals, each lasting a few minutes. Further extrusion of the fetus is not necessarily achieved with each straining bout. The course of extrusion is subject to arrest and even retraction of the fetus back into the dam. One of the main obstacles to single birth is the passage of the fetal forehead through the tight rim of the dam's vulva opening. Once the head is born, the rate of passage of the fetus is greatly accelerated. The shoulders follow the head within a few minutes and, immediately after this, the remainder of the neonate very quickly slips out of the birth passage. The mother's vigorous straining usually ceases when the fetal trunk has been born. Often, there is a short resting period at this point while the hind limbs of the neonate are still in the recumbent mother's pelvis, an event commonly occurring in unassisted horse births. During birth, the posture of the dam varies a great deal, depending on species and individual. Some remain recumbent throughout birth; in others there is alternate lying, standing and crouching.

Sheep

Most lambing ewes display signs of nervous and restless behaviour: lying down and getting up again, paddling with the hind feet and other classic signs of discomfort. In 17% of ewes no initial signs of parturition were evident in one study, although the sheep were kept under close supervision and almost constant observation at the appropriate times during two lambing seasons. Scraping the ground with a forefoot is common immediately before and after lambing. Although there may be frequent lying down and rising before birth, most ewes remain recumbent until the fetus is partially or completely expelled. In cases of twin and multiple births the neonates usually follow each other within a matter of minutes.

The mean joint duration of the first and second stages of labour is 80 min. The process of parturition in ewes would therefore normally seem to be a fairly swift one. The standard deviation in time of lamb delivery is about 50 min, and the lambing times would follow the normal distribution were it not for cases of parturition of >2 h due to difficulty in birth (dystocia). There is no apparent difference in duration between breeds or ages. Most lambs are born within 1 h of first showing at the vulva. Some ewes have prolonged labour lasting >2 h: this is due, usually, to dystocia. With twins, labour is much shorter for second-born lambs than for firstborn. In the ewe that is giving birth, the fetus has flexed elbows and shoulders. This was a constant feature in a study of 79 ewes by A.F. Fraser (see Fig. 18.4).

Goats

In a herd of Angora goats, during one season, two-thirds of the kids were born within a 4-day period (A.F. Fraser, pers. obs.). Immediately before birth a doe often appears fretful and nervous. The doe will paw bedding and will repeatedly lie down and stand up with signs of straining. Birth is usually complete within about 1 h of first behavioural signs of first-stage labour. Afterbirths are normally passed from about 20 min to 4 h after birth. Does often eat fetal membranes.

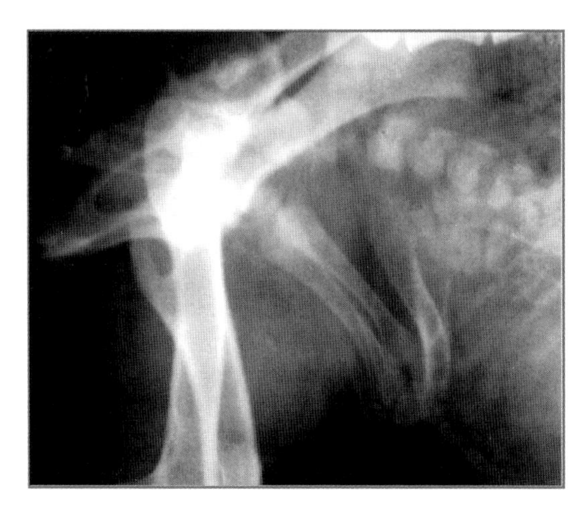

Fig. 18.4. Radiograph of sheep fetus at the start of birth (photograph A.F. Fraser).

Cattle

Until the first stage of labour has begun, physical changes rather than behavioural ones are apparent in the pre-parturient cow. Ingestive behaviour is, however, reduced at the time that labour is about to start. It is also at this time that the animal begins to show regular periods of restlessness; feeding sometimes recommencing between these periods. The cow in Fig. 18.5 is lying now but has been paying attention to a young calf. Eventually, the restlessness gives way to behaviour that is very similar to that encountered in conditions of colic. The cow appears apprehensive, looking all round her sides and turning her ears in various directions. At this period the cow will also walk excessively if she is at all able. She will also occasionally examine patches of ground and sometimes even paw loose litter or bedding as though gathering it into one spot.

The first stage of labour becomes apparent when the cow shows intensive comfort-shifting. She repeatedly goes through the motions of lying down and getting up again. She may also kick at her abdomen, repeatedly tread with her hind feet, look around to her flanks and shift her position frequently. About this time, also, the cow begins to pass small amounts of faeces and urine at intervals while arching her back and straining slightly. Cows tend to show these bouts of slight straining earlier in parturition than the other farm species. With time, the spasms of evident pain become better defined and more frequent. Finally, they begin to appear regularly about every 15 min and each spasm lasts about 20 s. The spasms are manifested by several straining actions in quick succession. After some bouts of straining, the allantochorion or 'first waterbag' is rent and a straw-coloured, urine-like fluid escapes. After this there is usually a short pause in the straining and muscular contractions. This pause terminates the first stage of labour, which may vary in duration from 3 h to 2 days, though a period of 4 h is a more common (modal) average time. About 1 h later the more powerful straining of the second stage of labour becomes evident and the amnion appears at the vulva.

At this time, straining occurs about once every 3 min and lasts for about 30 s; it grows more powerful and more frequent when portions of the calf, such as its forefeet, become extruded at the vulva. At this stage the cow either adopts the normal resting position or lies on her side. In a study of calving in 82 Friesian cows by Edwards and Broom (1982), all were recumbent when calving unless there was human assistance. Her upper

Fig. 18.5. This cow, in very late pregnancy, has been paying attention to the calf of another cow. Calves may fail to obtain enough colostrum if separated from, or stolen from, their mother (photograph D.M. Broom).

legs may even swing clear of the ground if she strains while lying flat on her side. Straining is virtually continuous until the head and trunk of the calf are extruded. Most cows that give birth easily and unaided remain recumbent until calving is completed. In a study by Metz and Metz (1987), 92% of such cows were still lying at the expulsion of the fetus but, when calving was difficult, 64% were standing by this time. The birth is completed with the breakage of the umbilical cord. Occasionally, a cow rising after the main period of extrusion will do so with the pelvis of the fetus still lodged inside her own and the retained fetus may swing from her for a period of time before dropping to the ground. Second-stage labour, i.e. birth, is usually completed within 1 h. Cows with some degree of calving difficulty show more posture changes and raise their tails for longer than do cows with little difficulty (Barrier *et al.*, 2012). Possible problems associated with parturition are reviewed by Sheldon *et al.* (2004).

Horses

The first sign of labour in the mare occurs when she becomes increasingly restless. She may perform circling movements, look around at her flanks, get up and lie down spasmodically and generally show signs of anxiety. At the onset of parturition feeding ceases abruptly. The mare rises and lies down again more frequently than before, rolls on the ground and slaps her tail against her perineum from time to time. Subsequently, she adopts a characteristic straddling position and crouching posture, frequently urinating at the same time. The mare may also show flehmen, especially after the allantoic fluid has escaped with the rupture of the allantochorion about the end of the first stage of labour when extremely vigorous straining – typical of the mare alone – occurs for the first time. Just before straining starts, an unusually high raising of the head is sometimes observed. But, when straining begins, the mare soon goes down flat on her side and the expulsive efforts become intensified. From the first signs of sweating it may be deduced that the first stage of labour, which may last up to 4 h, has begun, but false starts are not uncommon. After some straining the waterbag (amniotic sac) becomes extruded; within it one fetal foot usually precedes the other. The bouts of straining become more and more vigorous until the muzzle of the fetus appears above the fetlocks. Although the straining bouts at this period are very vigorous, the amniotic sac does not rupture. Most of the delivery time is normally taken up with the birth of the foal's head. Soon after this the remainder of the foal, except its hind feet, is expelled from the vagina. The reflex head movements of the almost wholly born foal finally burst the amniotic sac; the foal begins to breathe and its further reflex limb actions may extract the remainder of its hind legs from the dam. Although 98% of mares give birth in the lying position, final expulsion of the legs may be caused by the mare rising.

The duration of this second stage of labour is, on average, about 17 min, although in normal circumstances it may vary from 10 to 70 min (Rossdale, 1968). Following the completion of birth, mares often lie, in apparent exhaustion, for 20–30 min. As previously stated mares do not eat their afterbirths although they do groom their foals. The usual length of time for the extrusion of the fetal membranes is about 1 h.

Pigs

When farrowing is only a few hours away, the sow alternately utters soft grunting noises and shrill whining sounds. As parturition approaches she begins to grunt more intensely and emits loud squeals. During the process of delivery the sow is normally recumbent and lies on her side, although there are occasions when some sows adopt a position of ventral recumbency (lying on the sternum). Vigorous movements of the sow's tail herald each birth, and the piglets are expelled without a great deal of evident difficulty. Although polytocous species, such as pigs, exhibit only two stages of parturition, relatively little afterbirth is passed until all the piglets are born.

During farrowing, the recumbent sow may occasionally try to stretch out and kick with the upper hind leg or turn over on to her other side. These movements force out fluids and a fetus may be expelled at this time. Sometimes, during the birth of a piglet, the sow's body trembles and if she is of nervous temperament she may emit grunting and squealing sounds. Piglets are expelled at an average rate of one every 15 min. Nervous sows often stand up after the birth of each piglet; this may be associated with temporary reduction in pressure in the reproductive tract. The entire farrowing process normally lasts about 3 h. The sow pays little attention to her young until the last one is born and, when finally she rises, she sometimes urinates.

Postpartum behaviour

In the immediate postpartum period, the dam is engaged in the third stage of labour, which concerns, behaviourally, the expulsion of the placenta and the grooming of the neonate. Fetal membranes are usually passed effortlessly by the dam during the first few hours following the birth. Many cows occupy themselves by eating the afterbirth after its final expulsion (placentophagia). Not all animals are placentophagic: cows and sows are, while mares are not. It seems that the two groups do have certain general behavioural characteristics that differ. The species that are placentophagic usually keep their newborn close to the birth site for several days at least, while those that are not placentophagic lead their young away from the birth site very early on in the post-parturient period.

It has been noted that, in natural circumstances, mares often foal at night and in the open so that, by daybreak, the foals can trot and be led away by their mothers. Maternal behaviour is therefore related to the neonatal characteristics of staying in the nest area, following or hiding. While the newborn kitten or puppy is altricial and not able to move out of the birth site, the young of the grazing ungulates are precocial and can walk after birth. While deer calves or fawns and cattle calves are hiders, remaining for a day or more at or close to the birth site, lambs and foals are followers that walk or run with the mother and her companions very soon after birth.

In nest or hider species, parturition is followed by periods of isolation of the neonate from its mother. Follower species, such as the horse, maintain close and frequent contact between mother and offspring after parturition. These strategies of postpartum behaviour are different means of minimizing predation on neonates. Altricial young depend on being in a nesting place where they are hidden and may be defended by the mother. Follower species characteristically have rapidly developing, highly mobile neonates and sometimes can provide some form of efficient defence of their young. Hider species give birth where cover is available, for the vulnerable, non-mobile neonate is camouflaged and keeps still as a tactic against predation. Although the parturient animal of a hider species tends to choose a good birth site affording shelter and security, the mother may or may not remain close to the neonate (Hudson and Mullord, 1977).

Following parturition, individual female goats may either remain in the vicinity of their nesting neonate or leave the neonate alone while they forage with the herd. The individuals exhibiting these strategies are characterized as stayers and leavers. Stayer females tend to be older than leaver females. Female goats with twins are more likely to be stayers than are the mothers of singletons (O'Brien, 1984).

The cow often licks up her uterine discharges before the birth is completed. Once the calf is born, she rests for a variable length of time and then rises and licks the fetal membranes and fluids from the calf. She usually eats the placenta, and sometimes the bedding contaminated by fetal and placental fluids as well.

The sow expels fetal membranes in batches of two to four after all the piglets have been born, although some small portion is often passed during the process. Many sows will eat all or part of the expelled afterbirth unless it is removed promptly. Following the parturition they will often call their litters to suck by emitting repeated short grunts and may emit loud barking grunts if an intruder disturbs the nest. The sow rarely licks or grooms her young, but sometimes appears to try to position the piglets near her udder or draw them towards her teats using her forelegs in scooping actions.

It is important to provide close observation if a sow is particularly nervous, for it is in such an animal that cannibalism is most likely to occur. With such a sow also, the young are sometimes crushed by sudden and erratic movements. The piglets can be removed immediately after birth if the sow shows signs of dystocia. After the process is over and her piglets are returned, she will usually display normal responses towards them.

Further Reading: Barrier, A.C., Haskell, M.J., Macrae, A.I. and Dwyer, C.M. (2012) Parturition progress and behaviours in dairy cows with calving difficulty. *Applied Animal Behaviour Science* 139, 209–217.

19 Maternal and Neonatal Behaviour

Introduction

In many animal species parental care involves both parents. However, in most domestic animals the father plays little part in care (see also Chapter 5). There may be some defence of the group if the father is allowed by human action to participate in this. In this chapter the role of the mother is explained. The aspects of parental care are summarized by Clutton-Brock (1991), Paranhos da Costa and Cromberg (1998) and below in Box 19.1.

With the birth of the newborn or hatching of the egg, the vital relationship between the mother and neonate develops. This is established by the soliciting behaviour of the young animal and the acceptance of this by the mother. The maternal behaviour develops quickly perinatally and is, in general, characterized by collaborative behaviour unmatched in any other phase of the animal's existence. Commonly, the commencement of maternal behaviour is in the pre-partum period. Dogs and cats will already have found, or will find, a suitable place for the young pre-partum. In the sow, attempts at nest-making are usually shown at least 24 h before parturition and, in some cases, this nest-building is evident 3 days before farrowing.

Even among domestic animals there is some variation in the extent of parental behaviour shown. Animals vary in their life history strategies from the extreme (R) strategy of the mosquito, which lays many eggs with little investment of effort into each, to the extreme (K) strategy of the sheep, which invests much in each lamb. Trivers (1974) explained the concept of parental investment and how it helped to explain post-mating strategies and parent–offspring conflict: *parental investment is investment by a parent in an individual offspring which increases the offspring's chance of surviving and reproducing at the cost of the parent's ability to invest in other offspring.* A ewe shows high parental investment in its lamb, not only carrying it during pregnancy but giving it much care after birth. The benefit of parental care declines as the lamb develops and becomes better able to survive by itself.

The cost of parental care increases with the age of the young, for example because bigger lambs need more milk than newborn lambs. Mother birds that are feeding their young also have to work harder as the young grow. Hence the benefit:cost ratio decreases as the young get older (see Fig. 19.1). The point at which the parent should cease care is when the benefit:cost ratio is 1. The point at which the offspring should want the parent to cease care is half of this because, considering the degree of relatedness of parents, offspring and siblings, the offspring should be worth twice as much to themselves as the value of siblings. This topic is discussed in detail by Broom (1981). Parental behaviour can be manipulated by the offspring and is not just a matter of the parent deciding how much care to give.

Following parturition, the maternal mammal acquires a novel repertoire of behaviours orientated towards acceptance and maintenance of the newborn. The parturient animal has been behaviourally primed by an increased and special hormonal output. It appears that both the concentration of the reproductive hormones and their relative proportions to each other create the state of maternal behaviour. In the sheep, progesterone increases gradually during pregnancy then drops rapidly at birth; oestradiol and prolactin increase rapidly shortly before birth and drop around the time of birth (Poindron and Lévy, 1990; Broom, 1998). Mother birds also have to show complex responses to their young. If young domestic chicks show distress, the mother hen may react by warming them, protecting them or indicating where there is food. Edgar *et al.* (2011) experimentally induced minor distress in young chicks and found that the mother hen increased alertness, decreased preening behaviour and reduced eye temperature.

Most animals that are normally units of a flock or herd seek some degree of separation from their group when birth occurs. This arrangement allows the immediate

Box 19.1. Parental behaviour – objectives and abilities (modified after Tokumaru, 1998):

- Prepare the body for development of eggs, embryos.
- Prepare the body for care of the young after parturition or oviposition.
- Provide nutrients during egg and embryo development.
- Prepare an appropriate place for birth or egg-laying.
- Protect eggs or infants against predation and conspecifics.
- Help eggs and infants to thermoregulate.
- Provide food or help young to find and obtain food.
- Help young to learn about important aspects of their environment.
- Recognition of parent by young, or vice versa, or both.
- Defence of the young against predators and conspecifics.
- Maintenance of continual or intermittent contact between parent and young.
- Assessment of when to reduce and terminate parental care.

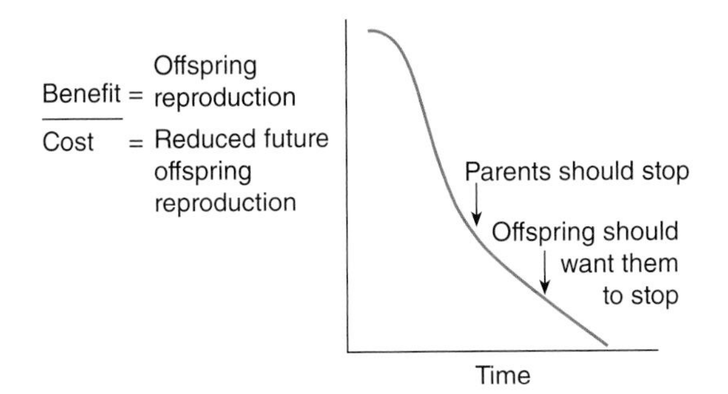

Fig. 19.1. How should parental investment change as the young get older? The ratio of benefit to cost of parental investment is shown. It is advantageous for the parents to cease care before the time that it is in the interests of the young for it to cease (modified after Trivers, 1974).

association between a mother and her newborn animal to develop during their own sensitive periods, which persist for several hours following parturition. The pair, in comparative isolation, bond together very quickly and efficiently. The phenomena of sensitive periods and bonding have already been discussed. With an established bond the mother acts largely in the interests of the neonate as it develops. Mother mammals often produce specific vocalizations postpartum. During grooming, both dam and offspring may vocalize and this is important for the development of the maternal–neonatal bond. Three distinct sounds are made by cows: (i) loud, open-mouthed bellows are made during licking; (ii) soft, throaty grunts are made later with the mouth closed; and (iii) a similar but louder sound is made when the calf wanders away. When nursing their offspring, ewes have a low-pitched gurgling call and goats give a series of short, low-pitched bleats. Sows give characteristic grunts in a rhythmic series while nursing their litters as a prelude to milk let-down.

Initiation of Maternal Behaviour

After giving birth, the majority of mother mammals lick the neonate. The new mother removes the amniotic fluids covering the neonate by thorough licking. Neonatal heat loss by conduction through thick amniotic fluid is reduced. Much grooming activity becomes progressively directed from the dorsum of the neonate and its head to ventral areas and limbs. These maternal attentions arouse the neonate and draw its primary interests towards its mother. In the course of grooming, the mother inevitably spreads considerable quantities of her saliva over the surface of her offspring. This soon dries out, but the dried saliva may impart a familiar pheromonal identity to the newborn. It is now clear that oral pheromones in the vehicle of saliva are important in the social exchanges of animals and, most notably, between young and mature individuals in recognizant situations. It is likely that the role of grooming is compound; it is under the influence of prolactin, the hormone that mediates much maternal behaviour, including grooming. The changes in calf-licking by the mother during the first 6 h are shown in Fig. 19.2 (Edwards and Broom, 1982).

This decline in licking over 5 h was shown when calves were licked by alien cows (Edwards, 1983), so the decline in licking is a consequence of the change in the stimulus characteristics of the calf rather than being due to a hormonal or other change in the cow.

The groups that do not show maternal licking of the neonate include the camel, pig and seal families. Most pigs and seals have less fur, which could be matted if no licking occurs, than other mammal families. Camelids do not normally require their fur to be unmatted in order that effective thermoregulation can occur.

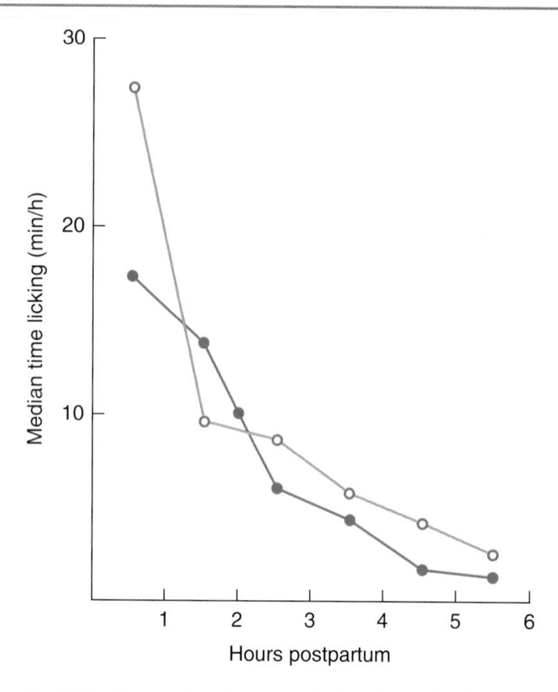

Fig. 19.2. The median time spent licking their calves by dairy cows during the first 6 h postpartum. Filled circles, heifers; open circles, second-plus calvers (after Edwards and Broom, 1982).

While grooming is maintained until the newborn has been well attended, it soon ceases. In the cases of animals with multiple births, maternal grooming behaviour may be directed more to those born early in the sequence, especially in Merino sheep. Maternal experience plays a part here since ewes inexperienced with twins are more likely to neglect grooming the second-born than would experienced ewes (Shillito-Walser *et al.*, 1983).

Once the young animal stands, the mother may act in a way that helps it to remain close to her and find the milk supply or she may show behaviour that is not helpful to the young. In studies of the responses of dairy cows to their calves, Broom and Leaver (1977) and Edwards and Broom (1982) described aggressive behaviour and inappropriate maternal behaviour as being shown by some heifers. In the first study, such behaviour was more frequent if the heifers had themselves been reared in isolation for 8 months while they were calves, and corresponded with inadequate social responses to animals of their own age. In the second study, heifers were found to be more likely than were older mothers to butt or kick the calf as it approached. Some heifers repeatedly turned to face the calf, thus making teat-seeking difficult. Heifers were also more likely to make movements that interrupted suckling. The likelihood of showing maternal behaviour is affected by hormone concentrations at the time. It has been suggested that concentration of oestradiol affects maternal behaviour in ewes. Blackface ewes have higher oestradiol concentrations than Suffolk ewes and show more lamb grooming, more low-pitched bleats and less lamb rejection. However, the greater amount of maternal behaviour in multiparous ewes than in primiparous ewes is not associated with differences in oestradiol concentration, or oestradiol to progesterone ratio, so this difference must be due to differences in experience (Dwyer and Smith, 2008).

Often, several hours elapse between the first and the last piglets in a litter, and the sow lies during the delivery of all of them. Her lateral recumbency is also the nursing position and, by this means, newborn piglets have access to the mammary region whenever they rise, usually only a matter of minutes after birth. All of this precludes grooming in its imperative form, characteristic of the other farm animals. Sows do nose their piglets, however, when the births are all over.

Following grooming, the mother has learned much of the identity of her young. Olfactory, gustatory, visual and auditory recognition have become established and will be reinforced progressively thereafter. From this time on, a mother showing normal maternal behaviour will care for the young animal and defend it with much intensity whenever she assesses that there is significant chance of succeeding in doing so. The young animal maintains an intimate association with its mother, by vocalizing for assistance or support, by numerous suckling attempts and by physical contact. Such reciprocity is rewarding for the mother.

Ewes separated from their lambs immediately after birth usually accept them on their return after absences of up to 8 h. Even when lamb interchanges are made, ewes can accept lambs earlier put to them after a lapse of some time without them. It appears that the ewe will lick the first recently born lamb or lambs presented to her within a period of several hours postpartum, and that licking for a period of 20–30 min establishes an attachment and a basis for recognition.

Removal of a kid or lamb at the instant of birth and then returning it to the mother after 2.0–4.5 h results in most of the young being totally rejected by their mothers. When goat dams are separated from their young for 1 h immediately after parturition they subsequently fail to show the normal maternal care. Sheep and goats will generally allow only their own young to suckle and will drive away alien young, suggesting that acceptance of young depends on the early sensitive period, which is restricted to the first few hours postpartum. Modern opinion dismisses the legend that handling young animals will lead to their rejection by their mothers. It is temporary removal from the dam and not handling, per se, which can lead to maternal rejection.

When the animal has been born, dried and has gained its feet, almost all of its activities are initially concerned with teat-seeking. While the newborn is exploring its immediate environment in the course of teat-seeking, the dam does not simply receive the soliciting approaches passively but shows positive orientation in accommodating them. The manner in which this is done is not always obvious at first, principally because the dam's behaviour is, superficially, one of inactivity. Typically, the dam will take a stationary position immediately adjacent to the newborn and will hold this position, permitting the progressive exploratory approaches of the young animal. Occasionally, the dam may make an alteration in position or stance to correct the young animal's orientation. Inexperienced mothers tend to overcorrect at first.

Some modifiability in the sensitive period in sheep and goats has been shown experimentally. Sheep and goat mothers, between 2 and 12 h postpartum, in close proximity to alien neonates, were restrained until they permitted the alien young to suckle freely (average 10 days). All adoptions were successful although some sheep were given kids and some goats given lambs. These results showed that enforced contact between dam and young, after the normal sensitive period of the first hours postpartum had elapsed, can prolong the period effectively.

Immediate separation of kids from their mothers at parturition for periods as short as 1 h, led to the later rejection of the kids by the mother goats. With only 5 min of contact, however, between doe goat and young immediately following the kidding, the rejection after separation could be prevented. A form of olfactory priming seems implicated in the process, as it affects the phenomenon of the maternal critical period in goats. Sheep- and goat-herders would agree that the odour of the young is the dam's prime criterion of acceptability.

A marked difference is observed in the responses of experienced and non-experienced goat mothers on occasions of kid separation. Previous maternal experience may serve to increase the likelihood of activation and maintenance of maternal responsiveness in domestic goats. The use of washing coupled with close confinement appears to be an effective aid to the fostering of lambs onto ewes, especially within the first day or two after birth (Alexander *et al.*, 1985a). Fifty-five per cent of ewes 1–3 days after lambing failed to accept their own unwashed lamb after being separated from it for only 40–48 h.

The mother hen also has to recognize her young and respond to them rapidly or they will not survive. After hatching, the chicks are close to where she has been incubating the eggs and hatching is generally synchronized. The hen responds to the cheeping call of the chicks by brooding them and hence keeping them warm (Fig. 19.3). Subsequent maternal behaviour includes defence of the chicks, leading them to safe places and indicating food by pecking at it.

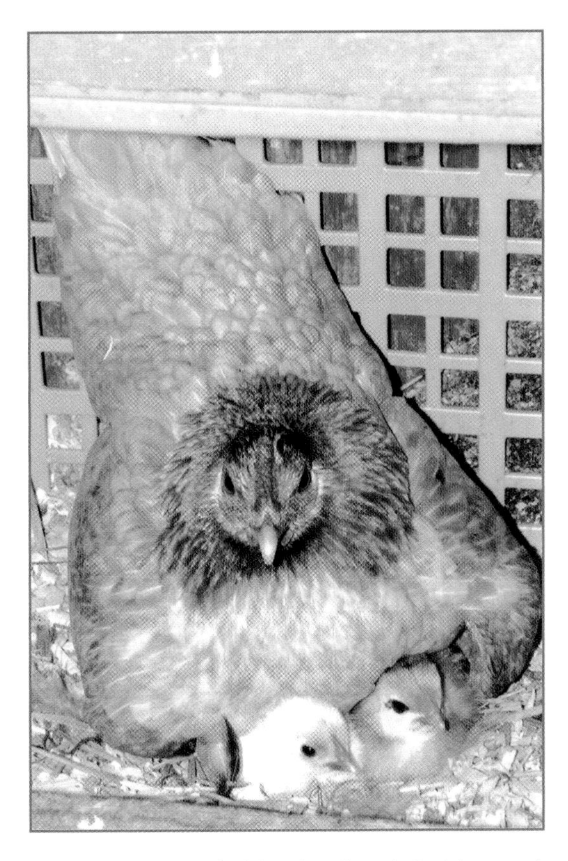

Fig. 19.3. Domestic fowl hen brooding chicks (photograph R. Harnum).

Maternal Motivation

If the mother does not become bonded with the young, maternal motivation does not fully develop but terminates fairly quickly. In dairy cattle, from which the newborn calves are sometimes removed without an opportunity for grooming, little or no evidence of a maternal motivation is shown 1–2 days postpartum. With some contact between cow and calf, however, the maternal motivation becomes strong and enduring. If a cow is left with her own calf for a 24 h period and the calf is then removed, the cow will still possess strong maternal motivation 5 days later, at which time fostering alien calves onto the cow is still a practical possibility. This, in fact, is the basis of the one-calf fostering method, which has been found to be a suitable one for multiple calf-rearing on cows under commercial circumstances. It has been found that this fostering procedure serves to strengthen the bond between the cow and her adopted calves. A relationship can thus be made between the development of maternal responsiveness and maternal bonding in dairy cattle. In cattle, mother–offspring bonds are weaker with twins than with singles (Price *et al.*, 1985).

In sheep, the success of fostering appears to be positively related to the persistence of sucking attempts by the lamb, and negatively related to the aggressiveness of the ewe. Attempts to foster different or extra lambs on to a ewe often involve the application of birth fluids, for example on a cloth jacket that is put on to the alien lamb (Price *et al.*, 1984a). Birth fluids still elicit a positive response from the ewe after they have been frozen and thawed. Other actions that promote alien lamb acceptance include: (i) stretching the ewe's cervix prior to introducing the young (Keverne *et al.*, 1983); (ii) using a masking odour; and (iii) restraining ewe and lamb together and tranquillizing the ewe. The effective fostering method used by shepherds for many years when the ewe's own lamb had died is to skin the dead lamb and put its fleece on the lamb to be fostered; the lamb's odour is detected by the ewe, who then allows the alien lamb to stand near her and suckle.

Maternal motivation can modify the normal reactivity of the mother, causing her to show changes in reaction towards man. After a period of successful association by the mother with her young, maternal motivation may persist for months. Towards its termination, reduction in maternal behaviour becomes apparent. The mother shows increasing indifference to care-seeking behaviour from her young. This is the onset of the period of parent–offspring conflict (Trivers, 1974), when the reproductive potential of the mother is maximized by reducing or terminating parental behaviour, but that of the offspring is maximized by encouraging the parent to continue parental care. This topic is discussed in detail by Broom (1981). Parental behaviour can be manipulated by the offspring and is not just a matter of the parent deciding how much care to give.

In domestication, some maternal behaviour appears to become pathologically altered by the duress of some forms of husbandry. Extreme aggression can be shown by newly farrowed sows towards attendants. This warrants a cautious approach to such animals at this time, particularly if their piglets are to be handled. In some instances pathological behaviour in sows takes the form of cannibalism of their own litters. Immediate sedation of such cases is required; many of these will show normal maternal behaviour on recovery from sedation. Abnormal maternal behaviour is shown by a small proportion of individuals in all the domestic species. Much of this leads to rejection of the newborn animal and depriving the newborn animal of intake of colostrum which, for optimum passive immunity, should be ingested promptly after birth under calm conditions (Broom, 1983a). It is believed that the aetiology of several neonatal diseases is complicated by such perinatal behavioural malfunction.

One feature of maternal behaviour observed in some ruminants is the concealment of the newborn (Lickliter, 1982, 1984). Free-ranging goats that have well-covered terrain in their habitat usually give birth to their young in some area with dense cover. After the first successful nursing, the female frequently leaves the young animal near the birth site and goes off at a considerable distance to graze. Periodically the dam will return to nurse the young animal, but it may be some days before the dam is prepared to allow the neonate to remain with her constantly. These young animals also require this hiding time to become developed sufficiently to become competent to associate actively with their mothers. Many of the species showing this characteristic of concealment of young also demonstrate synchronized parturition. They also tend to have specific territories reserved for parturition. As a result of large numbers of females giving birth about the same time to their young in specific sites, pools of newborn animals are formed.

The domestic cow, for example, calving out of doors, will often do so in some concealed position and, having nursed the calf once, she will move away some distance while the satiated newborn animal remains lying. Stockmen occasionally have difficulty in finding the newborn calves of these cows. Not infrequently the calf is found more than 100 m from the spot where the cow is observed. The link to the dam is initiated at the time of first nursing and improved on subsequent occasions when she returns to the calf in order to feed it. The mare shows quite the opposite relationship with the 'follower' newborn, refusing to leave it from the moment of birth.

With any mammalian species, after the first few days postpartum, the nursing pattern is determined by the young and by the mother, the dam often making herself available for nursing almost continuously. As a parallel development, the maternal behaviour becomes primed for the defence of the young. This defensive disposition causes some mothering animals to be singularly aggressive towards all other animals, including man. Maternal aggression against intruders, perhaps better called defence, was found to be positively correlated with the extent of oxytocin-receptor binding in the lateral septum area of the brain (Caughey *et al.*, 2011). This suggests that oxytocin action in this area may facilitate the expression of the behaviour. Studies of feral dogs in India have shown that both fathers and mothers will defend the pups against attack. Parental behaviour in dogs can also include feeding after lactation has finished. Mothers regurgitate food for the pups and fathers may also do so (Pal, 2005).

Milk 'Let-down'

All nursing and suckling behaviour is directed at the transference of milk from the mammary gland of the dam to the stomach of the young animal. The removal of milk from the mammary gland, however, is not by simply mechanical withdrawal alone: it requires also a positive milk-ejection process in the dam. Milk ejection occurs as a reflex and is an active, unconscious process. This milk-ejection reflex is commonly termed the 'milk letdown'. The reflex is manifested by a sudden rise in milk pressure after stimulation, which usually includes that of sensory nerve endings in the teat. The entire reflex path is a neuroendocrine arc, which usually commences with udder stimulation and passes via peripheral and central nervous systems to the hypothalamus and the posterior pituitary, where the arc is continued with the output of oxytocin. Oxytocin travels in the bloodstream to the udder, where a sudden rise in milk pressure then occurs. With the establishment of this pressure, passive withdrawal of milk by the sucking animal is facilitated. It seems that the elevation in plasma oxytocin is associated with a feeling of pleasure in the mother mammal.

The mechanism clearly requires activation, e.g. by vigorous nosing of the udder by the young (Fig. 19.4), and also requires time for oxytocin to circulate. The latter explains the slight delay in milk let-down that is commonly noted, e.g. in the suckled cow. Evidence of

Fig. 19.4. Belted Galloway cow with calf nudging her udder from posterior position (photograph D.M. Broom).

milk let-down is shown when the young remain stationary, rapidly gulping the milk released as a result of the rise in milk pressure. A circulation time after oxytocin is administered by intramuscular injection is required before let-down occurs. Latency periods from time of administration of oxytocin to time of milk let-down in certain species have been recorded as follows: goat, 14 s; sow, 2 s; cow, 50 s. Many cows will not let down milk unless the calf is present.

Although the principal stimulus to milk let-down is undoubtedly local physical stimulation of the mammary gland, there may be other factors – such as odour, sounds or visual stimuli – contributing to it. Conditioning occurs in the stimulation of milk let-down; for example, cows respond to many sounds and sights in the milking parlour by milk let-down. The let-down of milk can be prevented by factors leading to adrenaline (epinephrine) release. The complexity of the mechanisms involved in milk secretion makes it difficult to determine the details of the physiological derangement, but the inhibition of milk ejection, resulting from loud noises or other disturbing stimuli, is caused by the resulting circulation of adrenaline (epinephrine) acting directly on the mammary gland. Electric shocks given to cows experimentally inhibit milk ejection if the shocks are administered 1 min before milking commences, but they have little effect if given after milk flow has begun. Practical experience supports the finding that milk let-down can be easily prevented but not so easily arrested.

The pig has complex nursing and suckling behaviour, consisting of several distinct phases of suckling by the piglets and a characteristic pattern of grunting by the sow. Experimental findings suggest cause-and-effect relationships between the different elements of sow and piglet behaviour and the relationship of the behaviour of milk ejection. Synchronous features of pig nursing and suckling behaviour promote an even distribution of milk among all littermates (McBride, 1963; Hemsworth *et al.*, 1976).

Nursing and suckling

Nursing refers to the behaviour of the dam while the young are suckling. As a general rule, the postures are typically those of open, upright stance in monotocous species and of recumbent extension in polytocous species. The position adopted by the bitch or mother cat is that which allows comfortable access to the teats by the relatively immobile young puppies and kittens. The sow lies on her side and, in the first hours of life, colostrum is available continuously. The first-born piglets may go from teat to teat, taking colostrum (Hartsock and Graves, 1976). Later-born piglets are less likely to obtain sufficient colostrum for a variety of reasons (Broom, 1983a). After about 10 h, milk let-down becomes synchronized and periodic. The sow gives a characteristic grunting call (McBride, 1963; Whittemore and Fraser, 1974; Fig. 19.5) and milk is let down simultaneously from each teat. The piglets learn the call and the periodicity so they are ready on the teats. The milk let-down lasts only 10–25 s and occurs every 50–60 min. This system ensures that, provided there are not too many piglets for the teats, all can suckle satisfactorily. There are normally 14 teats.

The nursing position of the sow is full lateral recumbency. Some sows lie on either side while others favour one particular side. A small number of sows stand at most sucklings. There may be some adverse consequences of these postures. Some sows that habitually lie on one side to nurse have a reduced milk supply on that side. Sows that nurse while standing often have lower milk yields than those that lie. It may be that lactogenic udder massage by piglets is restricted in abnormal postures and that this results in reduced lactation. This seems a probable explanation, for milk flow can be quantitatively stimulated by udder massage (A.F. Fraser, 1980a; Algers, 1989). Mother sows may feed neonates hourly throughout the day and night and mothers of other species allow frequent suckling. Later, the

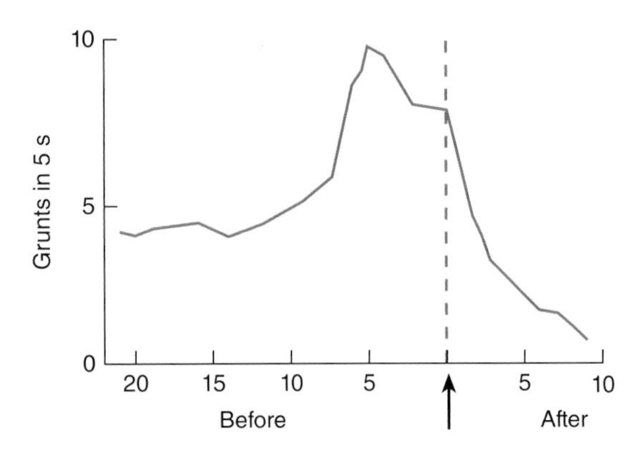

Fig. 19.5. Mean number of grunts per 5 s by six sows whilst nursing four or five times. The number of 5 s intervals before and after the onset of rapid sucking by piglets is indicated (redrawn after Whittemore and Fraser, 1974).

mother's nursings may become less frequent. At this time, although most mothers will have accommodated their young on demand until then, the maternal–offspring association may undergo some equalization in which the mother may not permit suckling at every instance. Suckling attempts are not only responses to a need for milk but may reflect instead a need for comfort. When lambs are alarmed, for example, they rush to suck their mothers, and the ewe tends to stand her ground and accommodates the lamb until her own flight distance is significantly encroached upon.

Maternal behaviour is a complex phenomenon. A high degree of variability exists both among mothers of different species and among mothers of the same species. Certain aspects of maternal care show improvement with successive parturitions. It was observed by Alexander *et al.* (1984) that maternal behaviour in the ewe can change in the same subject as a result of learning from one lactation to another.

The 'maternal care complex' is capable of progressive alteration as a consequence of the interaction of genetic, physiological and experiential factors (Wolski *et al.*, 1980). The status of the mother in a social hierarchy also influences her maternal abilities. Those females that occupy lower positions in the social hierarchy may have poorer access to the food supply and hence produce less milk. In sheep, the poorest mothers are usually those that are poorly self-maintained.

The Behaviour of the Neonate

As explained in Chapter 3, there are great differences among mammals and birds in the development at birth or hatching. The altricial young of the dog and cat can do much less at birth than the precocial young of ungulates and game birds. The neonatal kitten is able to respond to warmth and to find and suck a teat; the neonatal puppy is only slightly more developed. It is many days before the stage of development of the neonatal lamb or calf is reached. Many of the behaviour changes described here for precocial mammals are shown later by altricial mammals.

Parturition exposes the newborn animal to a greater quantity and range of stimuli than it has previously experienced. Exploratory behaviour involves response to fresh environmental features, including the feel of a solid substratum and sounds, sights and smells. Almost everything in its environment is unfamiliar except, perhaps, for some maternal features such as voice and the smell of the amniotic fluid.

Immediately after its birth, the neonatal mammal lies in extension. It soon raises the head and neck, flexes the forelegs, completes rotation of its sternum and flexes its hind limbs to rest on its sternum and one haunch. The head is shaken side-to-side, following which the limp ears may become mobile and later erect.

In the second stage of postnatal behavioural development, the newborn animal attempts to rise to a full upright stance in a series of movements typical of rising in that species. Such movements occur much later in the relatively altricial kitten or puppy than in the precocial ungulates. The newborn puppy shows the 'rooting reflex' that increases the chances of contact with the mother's teat. Tactile and olfactory stimuli and licking by the mother direct the teat-finding movements (Manteca, 2002). Lambs and calves raise their hindquarters before their forequarters. Foals begin to rise by extending the forelegs and raising the forequarters before the hindquarters (see Fig. 19.6).

The labyrinthine neck reflex initiates this rising process, and the anti-gravity function of the inner ear system is functional in this behaviour. More than one attempt is usually made to rise before the newborn establishes upright equilibrium. A few falls occur and it is common for a 'half-up' posture to be attained and held momentarily as a first measure of successful rising. Calves and lambs may have their hindquarters fully erected while they are still stabilized on their knees.

Fig. 19.6. First rising attempt by newborn foal (drawing A.F. Fraser).

Foals hold themselves partially up for a while on one fully extended foreleg. In the foal, extension and muscular tension of the forelimbs and neck are pronounced features of the early upright stance before competent upright mobility is established. When an upright posture is established, in these several species, the limbs are usually splayed to some degree before fine adjustment of stance is acquired.

When the upright stance has been secured, attempts at walking are quickly initiated. The first tentative walking steps occur as the typical four-step form of slow ambulation. In neonatal ambulation, unsteadiness is a prominent feature. In ungulates, this is apparently exaggerated by the presence of the eponychia, or collagenous pads over the sole of the fetal hoof. These pads have a protective role during fetal movements and are still present on the plantar hoof surfaces during the first walking activities. This soft eponychial tissue rapidly becomes shredded and removed from the soles by wear during early ambulation. The unsteady form of locomotion may further secure maternal attention. Neonatal motility in general appears to stimulate maternal concern and may promote the formation of the maternal bond.

Neonatal domestic mammals vary in visual ability, from the kitten whose eyes are not yet open to the ungulate that may be rather myopic, with a high degree of close visual competence at the expense of long-range vision and visual accommodation; or may be able to see well at all distances. Exploratory activities towards the dam typically take the form of close examination, by nosing with the head and neck fully extended on the same level as the trunk. Such actions facilitate the learning of maternal and hence species characteristics. There is variation among species in the details of their responsiveness, but these include close visual examination in the species that can see, the use of tactile hairs or vibrissae on the neonate muzzle and the olfactory function in this context. Undirected sucking towards the mammary gland may be facilitated by discrete maternal adjustments in posture. Once a teat-like protrusion is encountered, grasping and sucking occur. These neonatal activities are components of teat-seeking. Teat-seeking persists for two or more hours in the absence of reward. However, these attempts will wane in time and the orientation of the newborn towards the mother can then diminish and terminate. After successful suckling, teat-seeking becomes reinforced and learning occurs.

Foals

The average times taken by newborn foals to carry out various movements and activities are shown in Fig. 19.7 and Table 19.1. A great deal of exploratory behaviour is shown by newborn foals. Much attention is directed towards the pasture, the ground, the premises and their boundaries and other objects in the environment within touching distance. In the course of this exploratory activity, the foals nibble and mouth unfamiliar objects. Through such keen exploratory behaviour they acquire familiarity with allocated space, and this area is quickly adopted as the home range, for which they then have affinity.

During the first week, foals spend most of the day resting. In the next 2 or 3 weeks they rest about half of the time. They typically rest in a flat-out recumbent posture. Over 6 months, physical and physiological

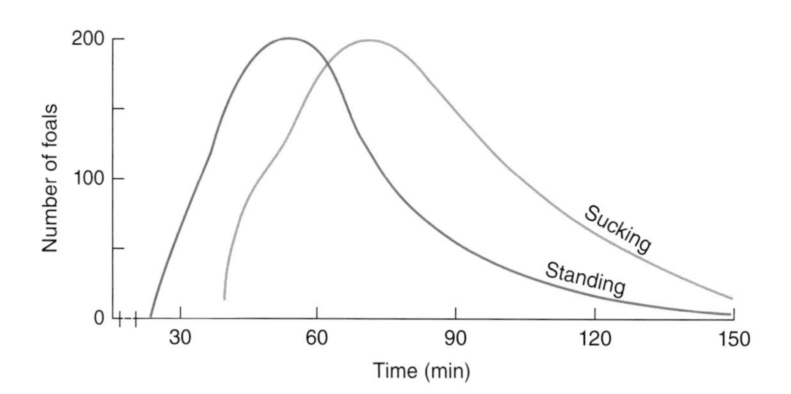

Fig. 19.7. Estimated time taken for newborn foals to stand and suck. Data taken from initial suckings observed in 435 foals, mostly Thoroughbreds.

maturity of the chest and lungs do not allow them to lie flat so often and they then lie in sternal recumbency more frequently. Groups of foals tend to lie down together, and this group effect of resting behaviour is very noticeable in larger breeding herds.

After standing, the foal suckles after an average interval of 21 min. In the following few days it seldom goes more than 1 h between sucklings (see Fig. 19.8). By 6 months of age the foal's suckling episodes are reduced to about ten per day. From 1 week onwards, foals gradually begin to eat grass. At first the young foal spreads and flexes its forelegs to reach grass and for a week or two, until its neck has grown, it is unable to walk and graze simultaneously. Foals graze about 15 min/h by about 3 months of age.

Table 19.1. Sequence of equine neonatal activities.

Activity	Time postpartum (min)
Head-lifting	1–5
Sternal recumbency	3–5
First rising attempt	10–20
First standing	25–55
First suckling attempt	30–55
First successful suckling	40–60
First defecation	60
First urination	90
First sleep	90–120

Calves

The newborn calf is licked vigorously by its mother, often being lifted partially from the ground by the mother's tongue, and may be encouraged to stand by this stimulation (Broom, 1981; Edwards and Broom, 1982). First standing usually occurs between 30 and 60 min after parturition, except where birth has been difficult (Edwards, 1982; Fig. 19.9). In studies of Zebu (*Bos indicus*) calves, the time to standing was 50 min in males and 40 min in females (Paranhos da Costa and Cromberg, 1998). The calf adopts a crouched stance with legs partly spread, shoulders lowered and the head and neck fully extended. It soon approaches a vertical surface like the leg of the mother or a wall and nuzzles against it. The nuzzling is concentrated at nose height and occurs more vigorously if a horizontal surface joins the vertical surface at about this height. This behaviour results in exploration of the underbelly of the mother or, if encountered by the calf, of the underside of a table 75 cm high. The point where the leg meets the horizontal surface elicits the most attention.

Selman *et al.* (1970a,b) emphasized that the shaded underbelly of the cow is attractive to the calf during its exploratory nuzzling. If the calf starts nuzzling at a foreleg it may spend some time exploring the axilla and, if the exploration is by the hind leg, the

Fig. 19.8. Foal suckling behaviour often includes tail movement at the time of milk let-down (photograph A.F. Fraser).

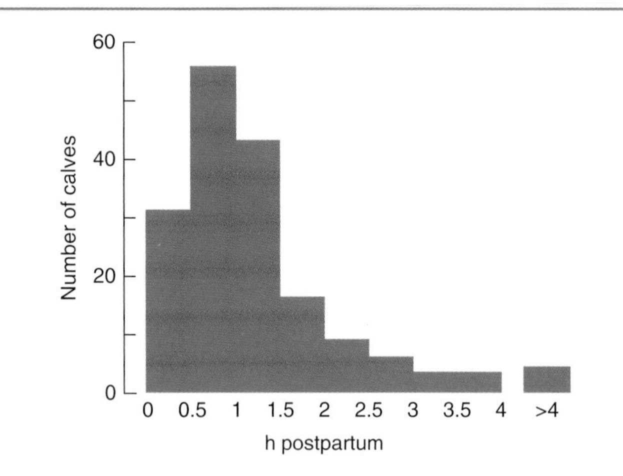

Fig. 19.9. The time to first standing for 171 dairy calves (after Edwards, 1982).

inguinal region may receive close attention. Calves lick and suck at any protuberance they encounter during their exploration. If a teat is found, the calf will take it in its mouth and suckling will begin. The exploratory nuzzling behaviour is usually referred to as teat-seeking, as it is terminated by finding a teat, but it may be prolonged if no teat is found. Many calves fail to find a teat during their first search and some fail after several attempts. Edwards and Broom (1979) and Edwards (1982) found that half of all calves of cows in their third or later parities failed to find a teat within 6 h of birth (see Fig. 19.10). This is because the pendulous udder and fat teats of older cows make teat-finding difficult. The calf's searching results in success in more than 80% of heifers' calves, since the udder is high and teats small, but often fails if the searching is directed at body underside level and the teats are much lower down. A variety of other factors affect the length of the delay before successful suckling and colostrum ingestion (see Fig. 19.11). For example, twin calves are often weaker than single calves and require two to three times as long to begin suckling.

The calf benefits from being licked by the dam, in order to stimulate physiological functions such as urination, defecation and general awareness. Newborn calves normally suckle five to ten times a day, with each nursing session lasting up to 10 min. The number of suckling bouts usually decreases with age, but this may vary depending on the rate of growth of the calf and the milk yield of the cow. Older calves graze as well as suckling from the cow. Calves at 6 months of age suckle about three to six times per day. Dawn is a common

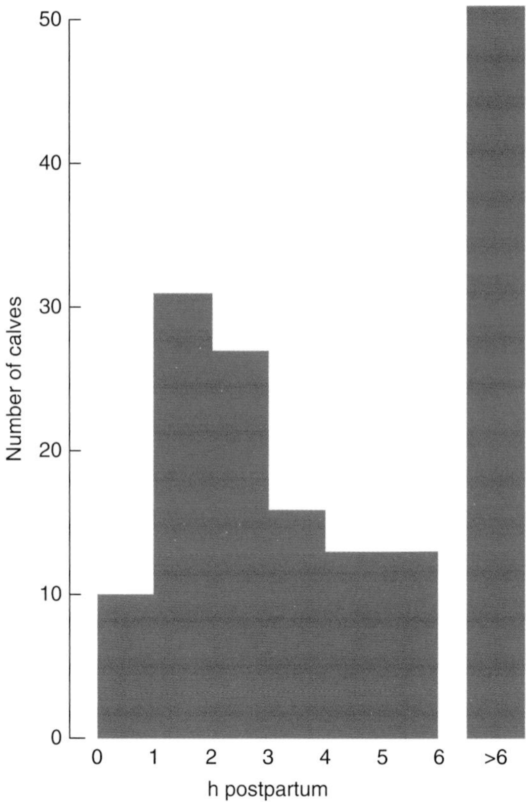

Fig. 19.10. The time to first suckling for 161 dairy calves (after Edwards,1982).

nursing time; other nursing bouts are centred around midmorning, late afternoon and about midnight.

Inter-sucking is a common problem among bucket-fed calves raised together in crowded groups. Such inter-sucking can occur very frequently and can cause skin inflammation. Prolonged sucking of the ears, the umbilical region or the prepuce is typical. Calves that indulge in inter-sucking are often unthrifty, especially if urine drinking occurs. Calves with wet ears from sucking by others may have them frozen in extremely cold weather. The provision of a teat supplying water can substantially reduce inter-sucking in group-housed calves (Vermeer *et al.*, 1988). The persistence of inter-sucking behaviour in adult stock is not uncommon, especially in dairy herds: in a problem dairy herd, as many as one-third of the calves and one-tenth of the cows may be affected. Social facilitation by other milking cows may increase the incidence of the behaviour. The fitting of preventive devices to the cow that sucks may be effective. If not, culling of offending animals

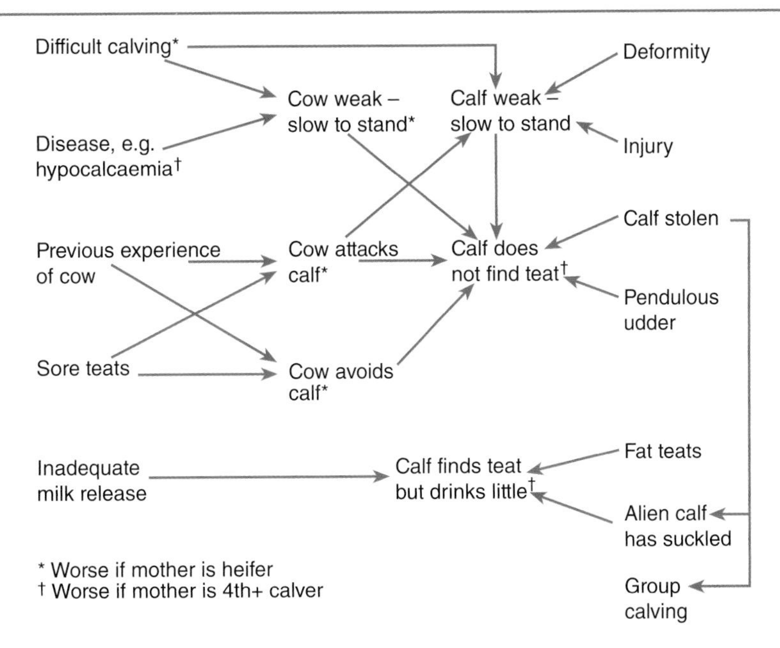

Fig. 19.11. Summary of factors leading to inadequate colostrum intake by calves (after Broom, 1983a).

is necessary and it therefore represents a serious economic problem.

How can calves fostered on nurse cows feed naturally? When cross-fostering is attempted, the normal procedure has been to present a cow already in milk with several young calves, perhaps newly born. Research on fostering has shown a much higher degree of success when the young calves to be fostered are presented to the nurse cow immediately following her parturition, before she has adopted her natural calf but while she is still in the sensitive period of maternal awareness. At this time such cows readily adopt numbers of fostered calves and continue to facilitate their suckling subsequently so that these calves grow better. In selecting calves for fostering it is necessary to realise that calves which have not suckled naturally at all in the first 6 days of life are unable to suckle later on a lactating cow.

Lambs

Lambs can stand, walk and suckle within the first hour, although many of lighter birth weight (e.g. under 3 kg) may require 2 h. Their senses of sight and hearing are evidently well developed. After the first successful suckling they become recumbent and sleep.

Vision is important in the sensory development of young lambs, since they are followers and characteristically move with their mothers throughout the postnatal period. In following the ewe very closely, the lamb often puts its head down close to the dam's head while she grazes. Young lambs recognize their own mothers very rapidly and sometimes will not respond to strange ewes when still only 1 day old. Since this recognition and following response occurs at a distance from the ewe, vision must be involved (Morgan and Arnold, 1974).

The earliest sensory experiences received by lambs are tactile and olfactory, followed by auditory and visual signals, all of which the lamb learns very rapidly. The odour of the inguinal region is used by the lamb when seeking the udder, and the temperature and tactile characteristics of the udder result in the lamb spending some time exploring its immediate vicinity (Vince, 1983, 1984). Sound and vocalizations influence lamb behaviour very strongly. It has been shown experimentally that lambs can hear maternal bleating and throat rumbling sounds before they are born (Vince and Armitage, 1980). The sounds *in utero* are attenuated, particularly in the higher frequencies. Since the rumbles have no high frequencies, they may be familiar sounds when the lamb is born, particularly since different ewes have different rumbles. After birth, the lamb first hears a low rumbling sound made by the ewe as she licks the lamb dry. The structure of rumble sound is very different from that of bleats. These rumblings made by ewes seem to help the lamb orientate to the ewe. This sound also keeps the newborn lamb near to the ewe, particularly in the dark, when mixing with other ewes and lambs (Shillito, 1975).

When a ewe is separated from her young she and the lamb will 'baa' until they are brought together. Even adult members of the flock that become disengaged from the others will 'baa'. They increase in vocalization and become more animated in attempts to locate the main flock. Increased vocalization has been shown to be accompanied by increased mobility. Vocalization in a young lamb separated from its dam, or an adult sheep separated from the main flock, tends to be fairly intense initially. It declines, however, after about 4 h of continuous separation.

Most newborn lambs are able to stand within the first half-hour following birth, and nearly all are able to stand within the first 2 h. The lamb's first attempts to suck are usually unsuccessful; it often seeks out the teat by nosing between the forelegs of the dam or any nearby object which the lamb may feel has maternal properties. If, at this point, the newborn fails to find the teat or is prevented from doing so by the behaviour of the mother, it may cease its attempts to suck.

Within about 1 h of birth, approximately 60% of newborn lambs have begun to suck and, in normal cases, nearly all lambs have sought out the udder within the first 2 h. Once the newborn lamb is able to stand, it sucks and nibbles at any object at hand; this is usually the coat of the dam. While the dam is removing the placenta from the lamb, the latter finds its way to the region of the teats and udder. Sometimes, the lamb is prevented from sucking by the diligent efforts of the mother to remove the placenta. Again, if the udder is too large, the newborn may find the teat difficult to grasp. However, once the newborn can facilitate milk let-down by its nudging movements pushing the teat cistern upwards into the udder, progress in locating the teat again and sucking becomes very rapid. In the first week following birth, lambs suckle very frequently, sometimes on 60–70 occasions during the 24 h period. The duration of suckling at this time is usually from 1 to 3 min, but later on the young are seldom allowed by the dam to suckle for periods of over 20 s.

Sometimes, the dam facilitates suckling by lifting her hind leg on the side at which the newborn is attempting to suck. One aspect of behaviour characteristic of the newborn lamb is the vigorous wagging of its tail when involved in nursing. It has been postulated that this is a mechanism that entices the dam to smell the anal region and so recognize her young, for some ewes do not discourage the approach of another newborn if it is similar in appearance to her own.

Piglets

Piglets stand very quickly and move around the sow soon after birth, following the edge of her body. Those that do not, perhaps because of body deformities or lack of oxygen during the birth process, have reduced chance of survival until weaning (Baxter *et al.*, 2008). Their first response is to any nearby object, which is then investigated by the nose. Piglets have extensive innervation in their snouts, the senses of touch and smell being very important in piglets. The smell and taste of the piglet's amniotic fluid expelled postpartum by the sow may help to keep the piglet near the sow. As with lambs, pre-natal sound could be an influence on piglet postnatal behaviour. Sows are generally vocal and the piglets are likely to have heard the sow's voice before birth. Newborn piglets learn to run to the sow's vocalizations, especially the low-pitched grunts produced before milk let-down (McBride, 1963; Whittemore and Fraser, 1974), and they learn to communicate by vocalizing themselves.

Although piglets can walk, see and hear within a few minutes of birth, certain physiological mechanisms such as temperature regulation are not fully developed at this time, and temperature conservation by huddling is therefore a prominent feature of neonatal behaviour in the pig. Newborn piglet mortality is high, commonly about 10 or 15% before weaning. Most of this mortality is due to crushing by the mother on commercial pig units. Normal healthy and well-fed piglets show positive orientation to their littermates and the sow's mammary region. Disorientated piglets and those that are weak are more liable to become crushed or chilled. Many young piglets fail to obtain sufficient colostrum or milk and hence are weakened or more susceptible to disease (see Fig. 19.12). Fostering of extra piglets on to sows is relatively easy in the first day or two after parturition, because sows generally accept piglets coming from their own nest. Hence, the alien piglets may be accepted. However sows often show aggression to piglets fostered at more than 2–4 days of age, and piglets fostered at more than 7 days of age do not gain weight at the normal rate (Horrell and Bennett, 1981; Horrell and Hodgson, 1985).

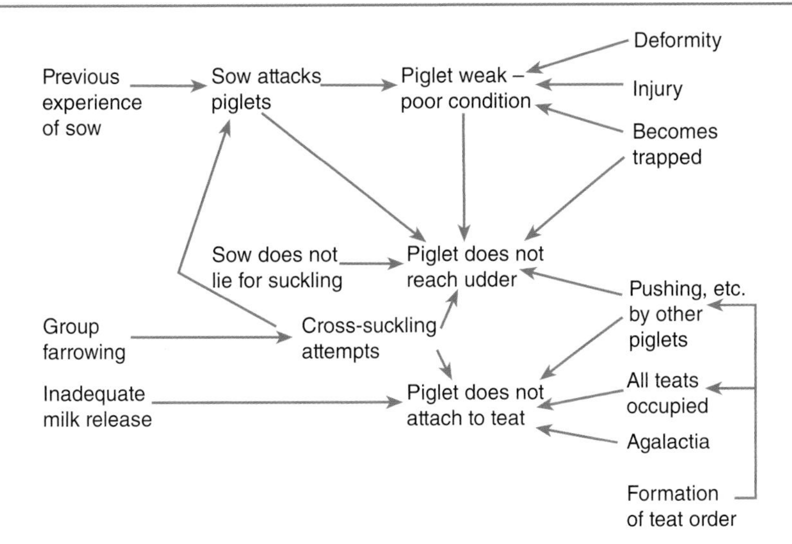

Fig. 19.12. Summary of factors leading to inadequate colostrum or milk intake by piglets (after Broom, 1983a).

One consequence of the acceptance of alien piglets by sows is that group farrowing can lead to the strongest piglets from several litters getting much milk and the weakest being ousted from their own mother's udder (Bryant *et al.*, 1983).

Much study on suckling behaviour in pigs has been focused on the preference of young piglets for specific teats (D. Fraser *et al.*, 1979). Piglets generally suck the same teats, forming a teat order. This order develops among piglets within their first day of life, often within an hour or so of birth. The order is altered when the sow first rolls over to feed the litter from the other side. Many observers have noted the singular preference piglets show for the anterior (pectoral) teats, but the real preference is for a productive teat wherever it is on the udder (McBride, 1963). Piglet birth weights alone do not determine the outcome of these teat competitions, but very light piglets are at risk in their competitive world, especially if functional teat places are fewer than or equal to the total number of piglets.

Each suckling by piglets begins with a massaging operation in which the piglets rub their snouts forcefully in upward and circular directions into the mammary region of the sow. The area massaged by an individual piglet is equivalent to a single segment of the mammary gland. The massaging process lasts approximately 1 min. This is followed by milk ejection from the dam, a process that lasts an average of 14–20 s, during which time the piglets suck vigorously. A final massaging operation follows, which occupies variable periods of time, being frequently protracted.

Sometimes, extended association occurs between the sow and her litter, and another feature of piglet nursing can be observed then. After feeding, somnolent, recumbent piglets may retain attachment to teats. As a result, adherence can persist between litter and dam from one feeding period to another. When this occurs it will tend to ensure that the teat order is preserved and that the entire litter feeds at each let-down of milk. Such adhesive behaviour would appear to have great survival significance by reducing agonistic behaviour and by operating against inanition.

Undue aggressive behaviour within the litter eliminates an increasing number of piglets from the established teat order. Disturbing factors, such as frequent movements by the sow, can lead to increased agonistic behaviour in the litter, as a result of which places in the teat order are lost and mortalities increase. The initial sign of loss of teat position is disorientation in the piglet's suckling behaviour, and such disorientated piglets are prone to being crushed beneath the sow, as is the case also in disorientation due to inanition. In litters larger than 14, where disturbing factors are great because 14 is the average number of teats on a sow, there is a significant increase in the numbers of piglets fatally crushed by the sow.

Further Reading: Marchant-Forde, J.N. (ed.) (2009) *The Welfare of Pigs*. Springer, Berlin.

Juvenile and Play Behaviour

Juvenile Behaviour

Calves

When calves are weaned they begin to show the clearly defined maintenance activities characteristic of adults. Adolescent calves that have acquired behavioural schedules learn to anticipate feeding times and show restlessness as these times approach. Calves at pasture with their mothers and others in a suckler herd form complex social relationships (Kiley-Worthington and de la Plain, 1983; Benham, 1984). More nearest-neighbour associations occur in groups formed of calves reared in a group than in calves reared in isolation. When they are grouped together for only a few days of contact, calves in pairs create a bond, and consequently contribute towards the stability of any group into which they are put. Calves grouped in a lot with no previous social contact form an affiliate group within a week (Kondo *et al.*, 1983).

Lambs

For the first few weeks of life, lambs stay quite close to their own mothers. By 1 month of age young lambs spend two-thirds of their time in the company of other lambs. By this age play is well developed. The gambolling form of play in this species is typical. The play involves upward leaps, little dances and group-chasing. Play is reduced as lambs grow and becomes rarer by 4 months.

Lambs reared in isolation do not respond to other lambs or ewes. Exploratory behaviour in isolated lambs is adversely affected. Such lambs will not inspect a new object or run and chase other lambs. They withdraw from novel environments, vocalize less and are slow to initiate movement and investigate any new object or run and chase other lambs (Zito *et al.*, 1977). Lambs reared in peer groups without mothers behave similarly to lambs reared with them. The main effect on isolated lambs is not, therefore, the result of maternal deprivation but of social deprivation. Twin lambs develop a strong bond with one another and learn to recognize each other's voices and appearance. They stay nearer to each other than to other lambs in the flock when given a choice.

Lambs can find their dams when they are out of sight, by identifying their voices (Shillito-Walser *et al.*, 1981). This ability improves with age, up to 3 weeks, and seems to vary between breeds. A lamb answers bleats more quickly when they are made by its own dam. When a ewe hears her lamb bleat she may help her lamb to find her by bleating promptly in reply. Ewes are inclined to bleat when they hear a lamb bleating, but become more specific in replying as their own lambs get older. As a result, lambs of 6 weeks can use vocalizations to find their ewes very easily.

The distance of the lamb from its dam depends on her activity. If the ewe is lying or walking, the lamb is usually within 1 m of her; if the ewe is grazing the lamb has greater freedom of distance and activity. The distance between ewe and lamb when both are grazing increases rapidly over the first 10 days of life, reaching a mean social distance of 20 m between dam and offspring when grazing on pastures of abundant feed.

Lambs are strongly influenced by the food the ewes eat, and this later affects their own choice of food. The pattern of grazing in lambs of 6 months of age has been determined by the breed of the ewe that reared the lamb. When Welsh Mountain lambs were cross-fostered on to Clun Forest ewes, they grazed in a restricted area, as Clun Forest sheep do, whereas Clun lambs reared by Welsh Mountain ewes grazed in a wide area over the hills, as is normal for the mothers' breed. Lambs reared by mothers that ate grain while the lambs were very young and taking only milk were more likely to eat grain when older than were lambs whose mothers ate no grain (Keogh and Lynch, 1982).

Foals

The maturing foal gradually undertakes sorties of increasing distance from its mother, and progressively spends more time playing with other foals as it grows. The equine bond, however, persists over 1–3 years and can remain strong even when later siblings are born. A behavioural expression of this bond is the head overlay position, in which the foal puts its head over the neck of the mother or the mother puts her head on the neck and forequarter of the foal.

As the foal matures it suckles less frequently, perhaps only about eight to ten times per day at 6 months. From 1 week onwards foals gradually begin to eat grass. They graze for about 15 min per daylight hour by about 3–4 months of age. By 1 year of age foals spend about 45 min per daylight hour in grazing activity. At first, the young foal must spread and flex its forelegs to reach grass and, for a week or two, it is unable to walk and graze simultaneously. Foals spend most of the day resting during the first week. In the next 2 or 3 weeks they rest about half of the time. Foals typically rest in lateral recumbency until they are over 6 months, when physical and physiological maturity requires them to lie flat less often and to lie in sternal recumbency more frequently. Groups of foals may lie down together, and social facilitation of resting behaviour is very evident. In a herd of 15 mares and their foals, the foals, of 7–16 weeks of age, were found to associate in fairly fixed groups of two or three.

Piglets

Piglets suckle frequently when they are young. As piglets mature, the number and duration of suckling bouts gradually decreases from about once every 50 minutes in the first few days of life to about six per day at 2 months of age (D. Fraser, 1974). Piglets attempt to eat solid food by 7–10 days of age. They are particularly attracted to solid food that is sweetened and formed into small pellets. Solid food intake, however, does not become substantial until about 3 weeks of age unless the piglets are milk-deprived. Young piglets quickly learn to eat the same solid food as their mothers and often attempt to share the sow's food with her. If 15–20 day-old piglets observe their mother eating a certain food, they show preference for it 8–10 days later, even if they were not able to eat it when their mothers did (Oostindjer *et al.*, 2011).

Piglets show a normal discipline of defecation, which develops rapidly over the first few days of life. Normal piglet defecatory behaviour develops by 4 days of age. The young pig shows clear preferential use of a communal and restricted dunging area, which is usually the corner of the premises furthest removed from the sleeping area.

Piglets sleep for about 26 min/h and for the first 5 weeks of life, on average 10.5 h/day. This contrasts with pigs of 3–4 months of age, which sleep about 8 h/day. The duration of paradoxical, or rapid eye movement (REM) sleep decreases as piglets develop. Young piglets commonly sleep in a crouched position, with all four legs folded under the body (Fig. 20.1).

Vocal activity is a prominent characteristic of young pigs. The division of piglet vocalization into five different classes has been defined as in Box 20.1 by Jensen and Algers (1983).

Fig. 20.1. Duroc piglets resting in typical juvenile sleeping position (photograph D.M. Broom).

Box 20.1. Categories of piglet vocalizations (after Jensen and Algers, 1983).

1. 'Croaking'. Tonal, short sounds, poor in formants; frequently uttered at the beginning of a nursing. The most typical feature was that the starting and finishing frequencies were approximately equal, while the maximum lay higher and was reached at the middle of the short duration.
2. 'Deep grunt'. Non-tonal, short grunts, rich in formants; uttered throughout nursing; no measurable pitch change. Its most typical feature was the low frequency, making it the only class of piglet calls with a basic frequency <1 kHz.
3. 'High grunt'. Grunts resembling 'deep grunts' except for the basic frequency, which lay approximately 1 kHz higher.
4. 'Scream'. Long, tonal calls with a considerable positive pitch change. Often uttered in within-litter aggressive interactions. Negative pitch change occurred only in a few instances.
5. 'Squeak'. Calls resembling the 'screams' except for the moderate positive pitch change. Like the 'scream', it was often uttered in aggressive interactions between littermates.

Piglet calls during nursing do not form a continuum of sounds, but consist of five discrete classes. The within-class variability was very low and few intermediate patterns occurred, giving these vocal patterns stability. Part of the variation is due to differences between individuals, but a few differences also occur within individuals. Some of the variation within the piglet call-classes relates to frequency and duration. The specific part of a piglet's vocal signal is rather constant, however, and variations only alter the communication specificity to a small extent. Piglets that are hurt show screams of higher and more variable pitch (Wemelsfelder and van Putten, 1985).

Poultry

The domestic chick is active while within the egg. Before hatching it attains an upright position, with the head and neck elevated. It gives various calls within the shell: calls of distress and satisfaction have been identified. When the chick breaks through the shell it quickly seeks a source of heat, making characteristic calls as it does so. The natural heat source is the broody hen, with which the chick makes very close physical contact after hatching. Thereafter, close company with the hen is maintained as the chick matures, and this provides opportunities for learning to refine the behaviour of maintenance. The clutch of chicks with a hen maintains a close association with her and readily recognizes her physical characteristics and calls. When the process of physical development of the chick causes the down to be lost from the head, the hen rejects it. The clutch then becomes dispersed and more dependent on self-maintenance activity, while still associating with the flock in general.

The turkey poult, like the domestic fowl chick, is active within the shell before hatching. After hatching, the poult is very mobile. Great social cohesion within poult groups is notable from the first day after hatching, but attachment to the hen turkey is also evident. Vocal and visual signals are used in the maintenance of close contact. As with the chick, this affinity facilitates the learning of certain critical activities, particularly feeding. Some artificially incubated poults are unable to initiate feeding or drinking and may die as a consequence of the lack of maternal association and the learning facility this provides. After 3 months of age, social hierarchies become formed in turkey groups.

Behavioural Aspects of Weaning and Puberty

Weaning and puberty are the two major events in the course of development of the young mammal. Successful weaning marks the survival of the animal beyond the immature period of dependence on its mother's milk, and the passage of puberty puts the subject into the ranks of reproducing adults. Both events are often associated with temporary but acute turmoil and changes in behavioural orientations.

Weaning

Conditions of domestication often ensure that weaning is not naturally determined. Dog and cat breeders may separate puppies and kittens from their mothers a little earlier than the time that would occur if left to mother and offspring. Most systems of livestock husbandry enforce relatively early separation between calf and cow, piglet and sow, lamb and ewe. In dairy cattle management, the young animal is taken away from the mother soon after birth or no later than that required for colostrum intake. This leads to some signs of disturbance, in the form of increased heart rate and vocalizations (Hopster *et al.*, 1995). In other farm species the young are left with their mothers until the dam's peak of lactation has passed and until the young animal has developed feeding activities alternative to suckling.

These systems of artificial weaning are usually carried out abruptly. The separated dam and young vocalize continuously, calling for one another. The normal forms of behaviour become disturbed, and grazing and resting may virtually cease. After a few days the separated parties readopt normal behavioural activities and the divisive process is concluded. This sudden withdrawal of nursing facilities usually hinders, temporarily, the growth of the young animal, but there appear to be no other untoward sequelae to sudden weaning after a long suckling phase. In dairy calves, the term weaning is applied to the cessation of milk feeding. In a group-feeding situation, abrupt weaning may cause fewer problems than gradual weaning (Barton, 1983a,b).

Although it is not often that sheep and goat mothers are left with their young long enough for natural weaning to occur, this may occasionally happen. Such weaning occurs when the young are between 3 and 6 months old, and the natural weaning process is a progressive one. Mothers tend to lose interest in their young quickly after natural weaning and the young are then liable to become separated from the flock. Lambs of 4–6 weeks of age suckle about six times per day. Evidently, by 4 weeks of age, the gradual weaning process of the young sheep has begun.

Two points regarding natural weaning are worth noting:

1. Before weaning, whether this takes place suddenly or gradually, the young animal must adopt the feeding behaviour it will retain into adult life. Much of this adult feeding behaviour is acquired by imitation of the parent in the selection of feed and manner of eating. It is likely that the roughage of adult feedstuff, in contrast to milk, will better satisfy hunger and so condition the young animal to its ultimate feeding behaviour.

2. The aggressive behaviour sometimes shown by the mother to her own young at the time of the latter's weaning seems in sharp contrast to the general character of maternal behaviour, although an aggressive facet is often apparent in the mother's responses to alien young. The ewe normally allows only her young to suckle and she vigorously drives away and avoids others. This acceptance of her own young and active rejection of strange young depends on the bonding process of the early sensitive period during the first few hours after parturition. In dramatic forms of weaning, what appears to be happening is a reversal of the cognitive processes of the critical sensitive period as a result of which the mother's own young become regarded more like aliens. Whatever the nature of the developments of any counter-critical period might be in the mother, it would appear that the fairly sudden undoing of these developments is involved in weaning aggression.

The age of weaning of piglets in extensive, semi-natural conditions is 12–17 weeks. Since commercially reared piglets are often weaned at 4 weeks, and sometimes as early as 2–3 weeks, all of these are early-weaned. As a consequence, they show various abnormalities of behaviour. Especially when weaned before 4 weeks of age, the piglets deprived of their mother are likely to suck one another. This is obvious as belly-nosing behaviour. Early weaning can also lead to excessive snout-rubbing and increased aggression between piglets. A combination of separation from the mother and lack of space may lead to abnormalities in piglet defecation.

Puberty

Puberty has been defined by various authors as follows:

- The period when the sexes become fully differentiated.
- The period when secondary sexual characteristics become conspicuous.
- The stage when there is an ability to elaborate gametes.
- The ability and desire to effect sexual congress.
- The time when reproduction first becomes possible.
- The period of activation of the neural tissues that mediate mating behaviour.
- The termination of infant sterility.

Puberty is taken here to mean the period when effective mating can first occur, such as the age when oestrus is first noted in the female.

The contribution of experiential influences to the development of hormone-induced behaviour is a complex process and not one of simple learning. Puberty in the dog depends on photoperiod and exposure to sexual pheromones produced by the same and the opposite sex, as well on the age and weight of the animal. The age at which puberty occurs is also affected by a range of variables in the cat, and commonly occurs at 5–9 months of age (Manteca, 2002). Mounting behaviour by young male animals is not indicative of the emergence of puberty, for pre-pubertal mounting is a common occurrence. Mounting can be seen in very young lambs, pigs, calves and kids. Young male domestic chicks show copulation behaviour as early as 2 days of age (Andrew, 1966). In lambs and kids, mounting occurs in the first few weeks of life and in kids, for example, it is quite evident that active mounting has been a frequent activity by 6 weeks of age. This is greatly in advance of puberty, which in the goat has been estimated to occur at about 155 days of age. A similar picture is presented by the pig, young boars being capable of mounting long before they are capable of copulation. About 50% of young male pigs exhibit mounting behaviour before 2 months of age although puberty in the boar is commonly estimated at 7 months of age.

In cattle also, it is clear that active mounting behaviour occurs so far in advance of puberty as to bear little or no relationship to it. Prior to puberty, the majority of bull calves can show good mounting orientation. Improvement occurs in the mounting orientation of calves as they acquire further sexual experience. Most bull calves at the rear of the stimulus animal orientate themselves appropriately on the first breeding occasion, but 30% mount inappropriately. The very young and those in early puberty have a proportion of individuals with poor mounting orientation and will frequently attempt mounting the stimulus animal from the side. Such lateral mounting in an adult animal is indicative of immaturity in its reproductive responses.

Observations have been made on the comparative aspects of behaviour in young bulls and castrates in free association in feeding yards. Young bulls and castrates begin to show differences in behaviour as early as 6 months of age. Young bulls show more general activity, in mounting and riding predominantly, while castrates do not exhibit these features but show, instead, more nosing and licking. Evidently, there is a pre-pubertal differentiation in behaviour between both groups. Masturbatory activities in young bulls are usually not noted until about 11 months of age, the normal age of puberty in the bull.

The age of puberty may vary little from one individual to another, but can be affected by various environmental factors, particularly nutrition. The pig is exceptional, however, since young females receiving a low level of energy intake reach puberty at approximately the same age as those receiving a high level of energy intake. Male pigs that have been undernourished also appear to reach sexual maturation in the usual time. Although boars with slow rates of growth take longer to acquire an ejaculatory capacity than fast-growing boars, no effect of restricted feed intake on sexual behaviour is apparent after 7 months of age.

With cattle the picture is different. High planes of nutrition tend to accelerate puberty, while poor nutrition delays it somewhat. Inbred bulls take longer than others to reach puberty. Even when reared on different planes of nutrition, identical twin bulls develop their sexual responses in very similar fashion.

Experimentally, other factors of environmental origin can exert some influence on the time when puberty is attained in some animals. Ninety per cent of young gilts subjected to transportation exhibited early oestrus while, among control gilts of the same age but not subjected to the same experience, only 28% showed such evidence of puberty (du Mesnil du Buisson and Signoret, 1962). It is concluded that external stimuli can lead to the early manifestation of first oestrus in pigs, and this tends to occur 4–6 days after the stimulating experience; the most effective experience of this nature is exposure to the boar. In sheep, puberty is determined more by seasonal factors than by age alone, and this is most evident at latitudes furthest from the tropics.

The reproductive responses of both sexes are not usually fully developed with the first manifestation of puberty. In males, libido continues to develop for some time after puberty has been reached and, in the female, oestrus is often of less duration and intensity at puberty than in the sexually mature age groups.

When young and old ewes are run together with the same ram, the older ewes show longer and more intense oestrus than the younger ones, and older ewes do not show short, weak oestrus in the way that young ewes frequently do. Heifers show variable characteristics of oestrus.

Play behaviour

All young animals must learn a variety of skills in order to survive in the wild (Bateson and Martin, 2013). In this book, such learning has been emphasized in those chapters and sections on feeding, predator avoidance, body care, and so on. Animals living socially have a much more complex array of skills to learn. Each individual must learn about how to communicate, how to act during social interactions and how to assess its social role, as well as about every other individual in its social group and about all their relationships. Many of the activities that may help in such learning are referred to as play. Some of these are intellectual rather than involving physical activity. The term play has been used for activities that may immediately, or ultimately, be involved in any one of the functional systems in animals. Broom (2014b) discusses the function of play and uses this definition: *play is carrying out a movement or intellectual process, either in the absence of its usual objective, or by using an inefficient means of achieving a goal solely in order to engage in that movement or process.*

The vigorous activities of young animals include: (i) movement or manipulation of objects; (ii) chasing; (iii) fighting without causing injury; (iv) advancing towards then retreating from another individual without contact; (v) acrobatics of various kinds; and (vi) many variations. The idea that these activities might have a function in the improvement of physical fitness and social skills was proposed by Brownlee (1954) as a result of his work with cattle, and was reiterated by Dolphinow and Bishop (1979) and Fagen (1976). The idea is developed further by Bateson and Martin (2013).

The term 'play' will be used for convenience, but the usage of the word does not imply that the examples quoted are similar in motivational state or in function. A wide range of actions, carried out principally by young animals and which may provide useful practice or exercise, are included under this general heading. Actions with an obvious and immediate function are not normally referred to as 'play'.

Activities referred to as play occur most often in healthy young animals, and its absence may be an indicator of poor welfare. Social play, such as chasing and mock fighting, is common (see Fig. 20.2). Solitary play can take the form of manipulatory or locomotor movements. Elements of sexual behaviour appear in the infantile play, of lambs for example, as early as 2 weeks of age. On these occasions lambs will mount each other, clasp and perform pelvic thrusts. Mounting and biting at play feature commonly in foals. This may become aberrant biting if the infantile behaviour is encouraged to persist into adulthood. Male dogs show 'pseudo-sexual' behaviour by week six (Pal, 2008).

Many play movements are truly patterned, for example by the regularity of gaits employed, and these patterns are similar for every individual member of the species concerned.

Fig. 20.2. Cats play-fighting while goat watches. No injury occurred during the play-fight (photograph D. Critch).

In their accompanying emotions and in the duration of action, the various play activities differ from the non-play activities. In a competition situation, when an animal has fled beyond the reach of its opponent, flight ceases. When an animal has repelled its opponent, fight ceases. Such cessations are not observed in play, which may continue for extended periods. The emotions of anger, fear, etc., are absent in play.

Very young but mobile animals show bursts of sudden capricious behaviour as spontaneous acts of locomotor play, in the form of leaping, and infant play appears to be important in the development of motor ability and early social organization. Solitary play is a form of exercise. It appears that the system of neuromuscular play is satisfied when the system is activated in exercise. It would appear that animals are reinforced in playing, as they play repeatedly and spontaneously. When play is denied, as in chronic confinement, even in the adult, an outburst of play activity is usually seen in these animals on being released. The fact that some exercise of muscles is involved in active play does not necessarily mean that maintaining body fitness is a major function of play. Byers (1998) concludes that it is not a significant function of play but that acquisition of social and other skills is more important.

The play movements in the young often simulate those seen in the adult activities of fighting, avoidance, predation and sex-mounting. Many activities in play occur as flight and hiding. Typically, these adult activities are not functional in the young animal. Since it is fed and protected by its dam, it normally is not required to fight nor is it required to initiate flight. As the sex organs are still immature, true sex-mounting is not seen in the young. Yet, although these activities are not required to be carried out by the young, the neuromuscular mounting requirements are in existence and are included in the agenda of practice.

The simulated nature of play movements is well illustrated in the play forms peculiar to each of the different species of domesticated mammals. Young horses use their teeth but do not head-butt in play; adult horses use their teeth but do not head-butt their opponents in fighting. In contrast, cattle do not use their teeth but use head-butting, both in play and in real fighting. Pigs show upthrusts of the head both in play and in true fighting. Thus, in all these species, similarly patterned muscular activities are to be seen both in directly functional adult activities and in the corresponding simulated play activity. Real injury is seldom inflicted in a play-fight, and no escape is truly achieved in a play-fight.

It appears that the system of neuromuscular play is satisfied when the system is activated in exercise. It would appear that animals are reinforced in playing, as they play repeatedly and spontaneously. Enriched environments are conducive to play. Factors encouraging play include environmental objects (Fig. 20.3), dietary enrichment, social stimulation, weather improvement,

Fig. 20.3. Cats using plastic hiding tube during play (photograph A.F. Fraser).

cessation of stress and release from confinement. The tendency to play in young cattle is increased by good feeding and decreased by poor feeding. Calves gallop-play more vigorously on bedded cement flooring than on bare cement flooring.

The characteristic aspects of behaviour displayed in play activities by cattle are prancing, kicking, pawing, snorting, vocalizing and head-shaking. These are seen particularly in young calves, although adults do occasionally indulge in playful activities. Further manifestations of play include trotting, cantering and galloping with the tail at various angles of elevation; bucking, kicking one hind foot, head-butting, pawing loose soil or bedding, and mounting. Calf play also includes butting each other or inanimate objects; they paw the ground, goring bedding, threatening attendants and making snorting noises in the course of playful actions. Playful mounting behaviour is also commonly seen in calves. Male calves do more mounting and pushing than female calves, and overall spend more time in play (Reinhardt *et al.*, 1978). Bursts of play often occur when calves are released from confinement, when freshly bedded, when introduced to other calves and with other changes in their routine or environment. When play-pushing is typically initiated between two calves, the pushing pair may attract a third calf. As all three push one another, other couples in the common group may also be induced to start pushing.

The most common form of play between foals involves nipping of the head and mane, gripping of the crest, rearing up towards one another, chasing, mounting and side-by-side fighting. There are sex differences in foal play, with colts mounting more frequently and engaging in general play more vigorously than fillies. Play in foals tends to be more frequent in males than females. The response of fillies to colt play is often withdrawal or aggression. Most play involves nipping or biting various parts of the body, but also typical are running about alone or in groups, or chasing with much head tossing, sudden stops and starts, and kicking of the hind legs in the air. Foals of 3 and 4 weeks of age often have play bouts lasting 10–15 min. Such bouts are usually initiated by one foal, developing the bout from a mutual grooming episode by changes to acts of nipping. A bout may be ended mutually by the two foals separating or, more often, by one foal simply turning away from the other. Most foals play with foals that are of similar age, but occasionally they play with foals differing in age by 2–3 months.

Piglet play develops in the second week of life and is a prominent feature of their total behaviour; it is mainly in the form of play fights. Cheek-to-cheek fighting, in which each piglet bites and roots at the other's face, neck and shoulders, is standard practice. By several weeks of age, chasing and gambolling are the common forms of piglet play; the chases are usually brief. Playful individual behaviour involves rooting and mouthing of novel items. Oral manipulation and exploration of the environment is a major feature of pig behaviour, extending into adult life. Play in piglets is arrested during illness, as it is in other sick, young animals. In this respect, indices of play can provide indicators of health that could be put to greater use in the practice of preventive paediatric veterinary medicine.

Play is costly behaviour; it seems aimless, capricious and inconsequential, and its sequences include behavioural acts or sequences of behaviour also occurring in high-risk adult activities, but the products of these high-risk activities are absent from play. The risk of death by predation may be increased as a consequence of play, and play does not result in conflict over a contested resource being settled. Because play is so risky and requires so much time and energy, the question arises about beneficial effects that might result from play and compensate for its apparent cost to the animal that is playing.

Benefits to the animal that is playing, both immediately and ultimately (Fagen, 1981) are listed in Box 20.2.

Further Reading: Bateson, P. and Martin, P. (2013) *Play, Playfulness, Creativity and Innovation.* Cambridge University Press, Cambridge, UK.

Box 20.2. Benefits and characteristics of play.

- Play develops physical strength, endurance and skill, particularly in those acts or combinations of acts used in social interactions having potentially lethal consequences.
- Play promotes and regulates developmental rates.
- Play experience yields specific information.
- Play develops cognitive skills necessary for behavioural adaptability, flexibility, inventiveness or versatility.
- Play is a set of behavioural tactics used in intraspecific competition.
- Play establishes or strengthens social bonds in a pair or social cohesion in a group.
- There is a play appetence; the animal that is ready to play is actively looking for an opportunity to play.
- Social inhibitions exist, particularly avoidance of injuring the partner.
- Use of inanimate objects or individuals of other species as substitute playmates indicates lack of stimulus specificity.
- Interruption in every stage by stronger stimuli, such as loud noise or intruder, may occur and indicates that play is not imperative once started.
- Transmission of playing mood to other individuals, particularly to playmates, shows social facilitation.
- Inventing new individual or experimental play, sometimes leading to new nervous and muscular coordinations, can occur.
- Play patterns may be in the form of exaggerated or uneconomical motility.
- Play patterns may be relatively unordered in sequence from one time to another.
- Short sequences and repetitious motor patterns are characteristic of play units.
- Animals repeatedly return to the stimulus source.
- Rapid alternation of behaviour is a common feature.
- The same behaviour may be directed at different stimuli.
- It occurs in a relaxed situation in which there is no imperative maintenance.
- It lacks a consummatory act as an end point.
- The play bout is typically preceded by a signal that indicates 'what follows is play'; these signals may recur during the bout to keep it continuing.
- Actions repeated and performed in an exaggerated manner are very characteristic.
- The activity appears pleasurable to the participants by subjective deduction.
- Social play is characterized by the exaggerated and uneconomical quality of the motor patterns involved. This is most pronounced under the effects of social facilitation.
- The individual movements making up the sequence may become more exaggerated within the play bout.
- Certain movements within the sequence may be repeated more often than they would usually be in non-play situations.
- Movements may be both exaggerated and repeated.
- Individual movements within the sequence may never be completed, and this incomplete element may be repeated many times, indicating that behavioural units in play are not essentially linked as a chain.
- The normal temporal groupings of functionally related actions can break down; elements of a number of different types intermingle in the same behavioural sequences, showing permutation of units.
- Play lacks immediate, biologically adaptive consequences.
- Play partners can change their roles frequently, for instance during play fights.
- Play behaviour is practically repeatable *ad libitum* and lacks the reaction-specific fatigue that is characteristic of much behaviour.
- Play behaviour appears to be of vital importance to animal life; the phenomenon of play, which may have its roots in various fetal activities, could become a basic factor in the determination of good welfare, particularly as it relates to normal development.

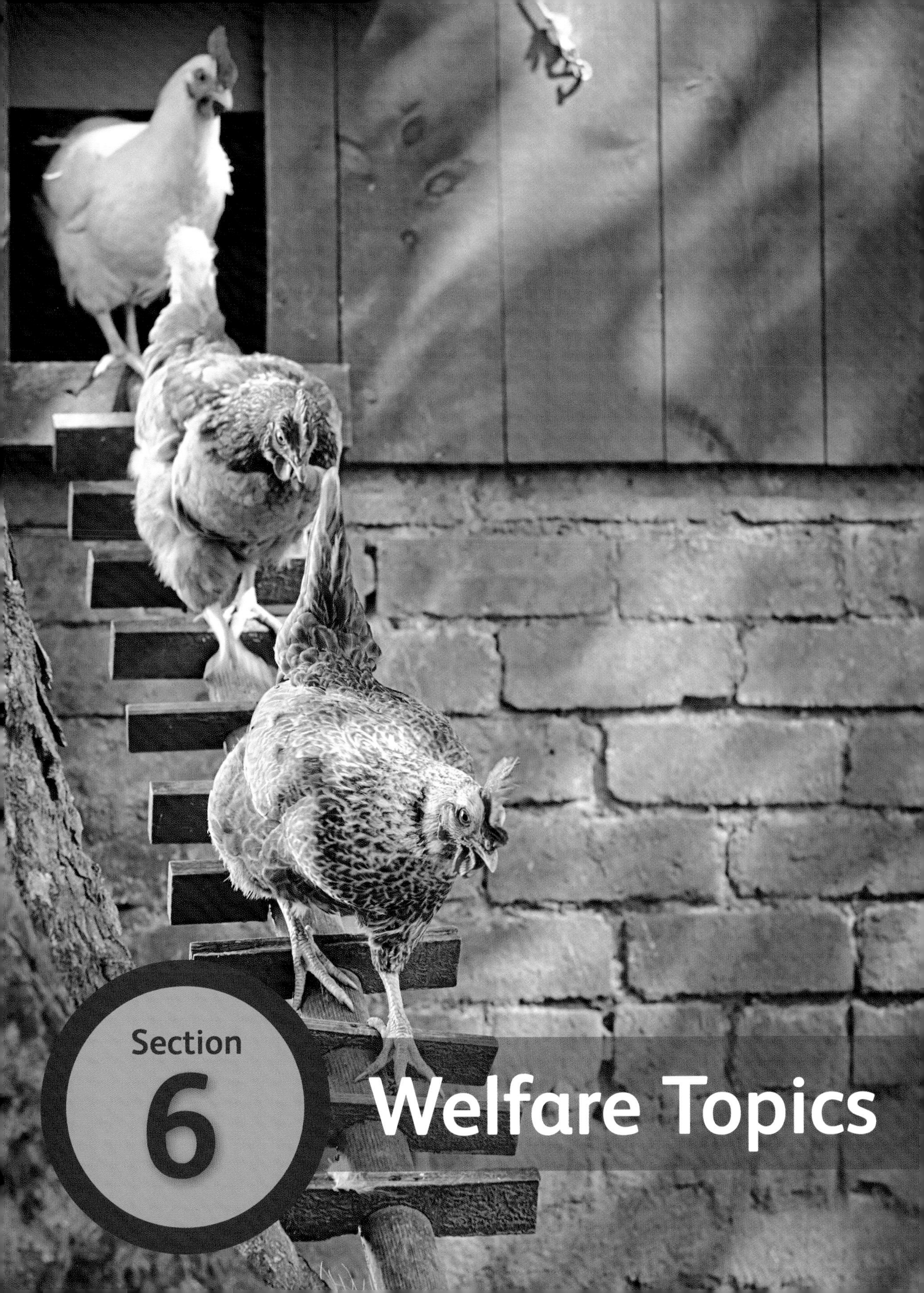
Section
6
Welfare Topics

21 Handling, Transport and Humane Control of Domestic Animals

There is a great variety of ways in which the behaviour of domestic animals can be controlled. Many of these are time-proven and well known to experienced stockmen. Anticipating an animal's movement is the best means of exercising some control over its behaviour.

Introduction

The handling, loading, transporting and unloading of animals can have very substantial effects on their welfare. By far the most frequently transported animals are chickens reared for meat production. With transport occurring at least twice, in 2005 more than 48 billion went as very young chicks and 48 billion went to slaughter, so the total number of journeys in the world was about 96 billion (FAOSTAT, 2006). Figures for several farmed species are shown in Table 21.1.

Animals going from farm to slaughter may pass through a market. Based on data from the Meat and Livestock Commission (UK), Murray *et al.* (2000) reported that, in the UK in 1996, 56% of cattle, 65% of sheep and 5% of pigs passed through a livestock auction market. In a study of 16,000 sheep travelling to slaughter in south-west England, the median journey duration was 1.1 h, with very few journeys of more than 5 h, when the sheep went directly from farm to abattoir, but was 7.8 h with over one-third of journeys 10–17 h when they went via a livestock market (Murray *et al.*, 2000). Online sales are reducing market use.

The first stage of a journey is the selection of animals that will travel and the inspection of these animals to check that they are fit to travel. Animals are prepared for the journey, and this preparation will depend on the species and the length of the journey envisaged. For shorter journeys, preparation can include fasting before collection and possibly movement away from the main herd to protect its health status. For longer journeys, where watering and feeding will be necessary on the vehicle, it can be an advantage to collect the animals involved 2–3 days before the transport, so that they can be prepared for the journey and become accustomed to the feed that will be offered en route.

The animals then have to be loaded on to the transport vehicle. This may be a road or rail vehicle or a boat. During transport, animals will gradually relax and recover from the stress of loading. Animal comfort during transport is highly dependent on vehicle design, driving technique and the roads being traversed. During transport, loading and offloading will increase stress levels and the risk of injury. Moreover, there will be a risk of spreading disease. A vehicle of such a standard that animals can remain on it during the resting, watering or feeding period is required.

At the end of the journey to slaughter, animals are unloaded at the slaughterhouse. Good facilities are required for this. A period of lairage may occur before slaughter. For all species, welfare is better if the animals are not held in lairage before slaughter, or if any lairage period is very short.

Transport

The ships and vehicles used

Although most animal transport is by road, many animals are also transported by ship, air and rail. Ship journeys may involve animals kept in well-designed accommodation or in makeshift conditions that can themselves cause problems for the animals. Similarly, some air-freight and rail travel is in animal housing that is fit for purpose but this is not always provided. On short ferry journeys, road vehicles full of animals may be transported. On roads, horses and cattle are normally carried in one-tiered vehicles, but pigs and sheep may be carried in two- or three-tiered vehicles. Poultry and rabbits are usually carried in crates. These may be stackable crates of such a size that a person can lift one of them, or may be modular units that have to be lifted with a fork-lift vehicle. On the vehicle, there are air spaces between the rows of crates or modules.

Table 21.1. Numbers of selected farmed animals[a] produced in the world in 2005, most of which would be transported to slaughter (from FAOSTAT, 2006).

Species	Total (million)
Chickens (for meat production)	48,000
Hens (for egg production)	5,600
Pigs	1,310
Rabbits	882
Turkeys	689
Sheep	540
Goats	339
Cattle (for meat production)	296
Cattle (for milk production)	239

[a]Fish, often not transported, are excluded.

Farmed fish are not often transported but salmon have to be moved from their freshwater tanks on land to sea cages. This may be done by means of a well-boat, with one or more large tanks on the boat or by use of a helicopter.

Some vehicles are air-conditioned, some have well-designed, openable areas on the sides for ventilation, some have poorly designed possibilities for ventilation and some of the simplest vehicles have open bars on the sides. The latter can provide good ventilation. All but the worst vehicles provide some shade and shelter from rain or other inclement weather. The suspension system in animal transport vehicles varies from very good to negligible. The provision for loading animals also varies greatly. Some vehicles are adapted for use with well-designed ramps, while others have hydraulic lifts on tailgates or floors (see Fig. 21.1). Many have very steep loading ramps that cause poor welfare in all animals loaded.

Fig. 21.1. Vehicle with hydraulic floor lift (photograph D.M. Broom).

The following factors that can result in poor welfare during animal handling and transport are summarized, after Broom (2005, 2008). There is a wide range of attitudes to animals, and these have major consequences for animal welfare. During handling and transport, these attitudes may result in one person causing high levels of stress in the animals while another person doing the same job may cause little or no stress. People may hit animals and cause substantial pain and injury because of selfish financial considerations, because they do not consider that the animals are subject to pain and stress or because of lack of knowledge about animals and their welfare. Training of staff can substantially alter attitudes to, and treatment of, animals.

Laws can have a significant effect on the ways in which people manage animals (Broom, 2014a). Within the European Union (EU), the Council Regulation (EC) No 1/2005 'On the protection of animals during transport and related operations' takes up some of the recommendations of the EU Scientific Committee on Animal Health and Animal Welfare Report, 'The welfare of animals during transport (details for horses, pigs, sheep and cattle)' (March 2002) and of the European Food Safety Authority 'Report on the welfare of animals during transport' (2004), which deals with the other species. Laws have effects on animal welfare provided that they are enforced. Codes of practice can also have significant effects on animal welfare during transport. The most effective of these, sometimes just as effective as laws, are retailer codes of practice, since retail companies need to protect their reputation by enforcing adherence to their codes (Broom, 2002).

Some animals are much better able to withstand the range of environmental impacts associated with handling and transport than are others. This can be because of genetic differences associated with the breed of the animal or with selection for production characteristics. Differences between individuals in coping ability also depend on housing conditions and on the extent and nature of contact with humans and conspecifics during rearing.

Since physical conditions within vehicles during transport can affect the extent of stress in animals, the selection of an appropriate vehicle for transport is important in relation to animal welfare. Similarly, the design of loading and unloading facilities is of great importance. The person who designs the vehicle and facilities has a substantial influence, as does the person who decides which vehicle or equipment to use.

Before a journey starts, there must be decisions about the stocking density of animals on the vehicle and the grouping and distribution of animals on that vehicle. If there is withdrawal of food from animals to be transported, this can affect welfare. For all species, tying of animals on a moving vehicle can lead to major problems and, for cattle and pigs, any mixing of animals can cause very poor welfare.

Both the behaviour of drivers towards animals while loading and unloading and the way in which people drive vehicles, are affected by the method of payment. If people are paid more for rapid loading or driving, welfare will be worse. Hence such methods of payment should not be permitted. Payment of handling and transport staff at a higher rate if the incidences of injury and poor meat quality are low, improves welfare. Insurance against bad practice resulting in injury or poor meat quality should not be permitted.

All of the factors mentioned so far should be taken into account in the procedure of planning for transport. Planning should also take account of temperature, humidity and the risks of disease transmission. Disease is a major cause of poor welfare in transported animals. Planning of routes should take account of the needs of the animals for rest, food and water. Drivers or other persons responsible should have plans for emergencies, including a series of contact emergency numbers for veterinary assistance in the event of injury, disease or other welfare problems during a journey. The methods used during handling, loading and unloading can have a great effect on animal welfare. The quality of driving can result in very few problems for the animals, or possibly in poor welfare because of difficulty in maintaining balance, motion sickness, injury, etc. The actual physical conditions such as temperature and humidity may change during a journey and require action on the part of the person responsible for the animals. A journey of long duration will have a much greater risk of poor welfare, and some durations inevitably lead to problems. Hence, good monitoring of the animals with inspections of adequate frequency, and in conditions that allow thorough inspection, are important.

Animal genetics and transport

Domestic animals have been selected for particular breed characteristics for hundreds of years. As a consequence, there may be differences between breeds in how they react to particular management conditions.

For example, Hall *et al.* (1998b) found that introduction of an individual sheep to three others in a pen resulted in a higher heart rate and salivary cortisol concentration if it was of the Orkney breed than if it was of the Clun Forest breed. The breed of animal should be taken into account when planning transport.

Farm animal selection for breeding has been directed towards maximizing productivity. In some farm species there are consequences for welfare of such selection (Broom, 1994, 1999). Fast-growing broiler chickens may have a high prevalence of leg disorders, and Belgian Blue cattle may be unable to calve unaided or without the necessity for caesarean section. Some of these consequences of breeding may affect welfare during handling and transport. Some fast-growing beef cattle have joint disorders resulting in pain during transport, and some strains of high-yielding dairy cows are much more likely to have foot disorders. Modern strains of dairy cows, in particular, need much better conditions during transport and much shorter journeys if their welfare is not to be poorer than that of the dairy cows of 30 years ago.

Rearing conditions, experience and transport

If animals are kept in such a way that they are very vulnerable to injury when handled and transported, this must be taken into account when transporting them, or the rearing conditions must be changed. An extreme example of such an effect is osteopenia and vulnerability to broken bones, which is twice as common in hens in battery cages than in hens that are able to flap their wings and walk around (Knowles and Broom, 1990). Calves are much more disturbed by handling and transport if they are reared in individual crates than if reared in groups, presumably because of lack of exercise and absence of social stimulation in the rearing conditions (Trunkfield *et al.*, 1991).

Human contact prior to handling and transport is also important. If young cattle have been handled for a short period just after weaning, they are much less disturbed by the procedures associated with handling and transport (Le Neindre *et al.*, 1996). All animals can be prepared for transport by appropriate previous treatment.

Mixing social groups and transport

If pigs or adult cattle are taken from different social groups – whether from the same farm or not – and are mixed with strangers just before transport, during transport or in lairage, there is a significant risk of threatening or fighting behaviour (McVeigh and Tarrant, 1983; Guise and Penny, 1989; Tarrant and Grandin, 2000). The glycogen depletion associated with threat, fighting or mounting often results in dark, firm, dry (DFD) meat, injuries such as bruising and associated poor welfare. The problem is sometimes very severe, in welfare and economic terms, but is solved by keeping animals in groups with familiar individuals rather than by mixing strangers. Cattle might be tethered during loading but should never be tethered when vehicles are moving, because long tethers cause a high risk of entanglement and short tethers cause a high risk of cattle being hung by the neck. Mixing of pigs on vehicles causes a substantial increase in aggression (Shenton and Shackleton, 1990), and cortisol levels in transported pigs were higher if there was mixing of pigs from different origins (Bradshaw *et al.*, 1996). Pigs observed to be more than averagely aggressive in their living conditions were more likely to cause skin lesions in other pigs after mixing before transport (D'Eath *et al.*, 2010).

Handling, loading, unloading and welfare

Many studies have shown that loading is the most stressful part of transport (Hall and Bradshaw, 1998). Unloading may also be stressful. The physiological changes indicative of stress occur at loading and last for the first few hours of transport. Then, the stress response gradually disappears as the animals become accustomed to transport. Therefore, provided transport conditions are good and the journey is not prolonged, the main welfare problems caused by transport result from loading (Knowles *et al.*, 1995; Broom *et al.*, 1996).

The large effect that loading may have on the welfare of the animals results from a combination of several stressors that impinge upon the animals in a very short period of time. One of these stressors is forced physical exercise as the animals are moved into the vehicle. Physical exertion is particularly important when animals have to climb steep ramps. Second, psychological stress is caused by the novelty of being moved into unknown surroundings. Also, loading requires close proximity to humans, and this can cause fear in animals that are not habituated to human contact. Finally, pain may result from mishandling of animals at loading. For example, beating or poking animals with a stick or catching sheep by the fleece will cause pain. This is especially true for sensitive areas such as the eyes, mouth, anogenital regions or belly. The use of electric goads will be painful as well.

Chickens are very much disturbed by close contact with people, for man is a large and dangerous animal to a chicken. Hence, it is not surprising that the handling preceding transport to slaughter has a considerable effect on birds, including considerable physical damage in some birds (Ebedes *et al.*, 2002). The nature of that handling depends upon the conditions in which the birds are kept, but it normally involves one or more birds being picked up and put into a crate that holds about 15 birds for the duration of the journey. If the hens are in battery cages, the sequence of actions normally involves one person opening the cage door, grabbing one or more hens by the legs, pulling them out of the cage, passing them to another person who collects two to five birds held by one leg in each hand, carrying the birds upside down to the end of the row of battery cages or to the door of the house where a vehicle is waiting, and putting the birds into a crate. Such handling may result in broken bones, especially if the birds have lived in a battery cage.

In the study by Gregory *et al.* (1990), 24% of hens had broken bones after handling. Fewer leg bones were broken if two legs were held (Gregory and Wilkins, 1992). The effects on welfare of the rough handling that hens normally receive prior to transport can be assessed by monitoring their adrenal and other emergency responses. In a study in which normal handling and gentle handling were compared (Broom *et al.*, 1986), the plasma corticosterone levels were much higher after the rougher handling, i.e. being carried upside down, (see Fig. 21.2, Table 21.2), indicating that the emergency response elicited in the hens was greater. In this study, a journey of 1 h in a lorry resulted in a smaller increase in corticosterone than that following normal rough handling. The significant effects of handling and transport on plasma corticosterone, plasma glucose and hypothalamic noradrenaline are shown in Fig. 21.3. Glucose production is increased following corticosterone secretion, but the increase is not large because it is used up during transport. Noradrenaline is a neurotransmitter that is used up during handling and transport. These studies suggest that handling of hens has a greater adverse effect on their welfare than does a short journey.

Broiler chickens, turkeys and ducks are transported twice, once as a very young chick and once to the slaughterhouse. Newly hatched chicks require careful handling. Broilers that have reached 2 kg are collected up from the floor of broiler houses, put in crates, taken by vehicle to the slaughterhouse, removed from the crates and hung by their feet on the shackling line,

Fig. 21.2. Hens transported for slaughter are normally carried upside-down by their legs, in this case from a battery cage to a transport crate on a vehicle (photograph D.M. Broom).

Table 21.2. Levels of plasma corticosterone in hens.

Treatment	Corticosterone (ng/ml)
No handling, sampling within 60 s	0.40
5 min after gentle carrying to crate	1.45
5 min after normal carrying to crate	4.30
After gentle handling as opposed to after normal handling	$P = 0.008$

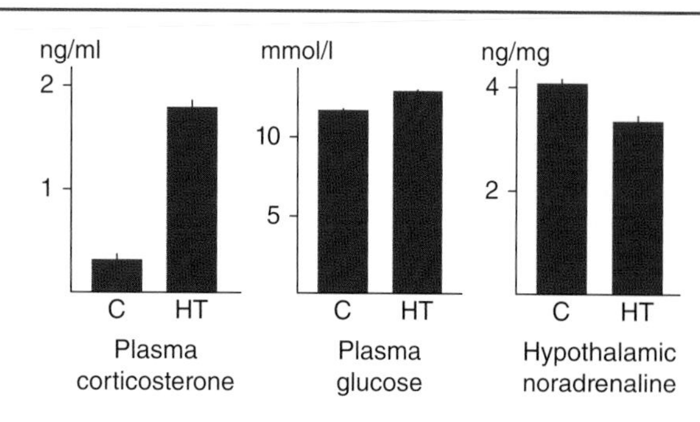

Fig. 21.3. The levels of corticosterone (ng/ml) and glucose (mmol/l) in blood plasma were higher in hens that had been handled and transported (HT) than in hens that had not (C). The level of the transmitter substance noradrenaline (ng/mg) was lower after handling and transport (data from Broom *et al.*, 1986).

stunned and slaughtered. Again, physical conditions on the journey are important in relation to welfare, but even if these are adequate there can be problems with limb breakages, bruises or impaired meat quality.

There is a range of physiological responses to handling and transportation. Freeman *et al.* (1984) reported that untransported control birds had corticosterone levels of 1.0–1.6 ng/ml, whereas levels after handling and 2 h transport were 4.5 ng/ml and, after 4 h transport, 5.5 ng/ml. The normal catching procedure in which a catching gang goes into the house and birds are gathered up by their legs leads to a substantial response. Duncan (1986) studied the use of a broiler-catching device, which is driven through the broiler house, moving birds by rotating rubber flails on to a belt and into a crate. Birds collected up by such a broiler-catcher showed a much briefer increase in heart rate than birds collected up by people. The use of a broiler-catcher also reduced the frequency of bruising and bone breakage (Knierim and Gocke, 2003). The better welfare when the broiler-catcher is used emphasizes the adverse effect on chickens of handling by people.

The slope of ramps is an important aspect when loading or unloading animals. This can be measured in degrees (e.g. 20°) or as percentage gradient (e.g. 20%). The percentage gradient indicates the increase in height in metres over 100 horizontal metres. For example, a gradient of 20% means a slope of 20 in 100 (i.e. 1 in 5) and is equivalent to 11°. The loading of pigs up a suitable ramp is shown in Fig. 21.4. There are important differences between species in their response to handling and loading, and these should be taken into account when choosing appropriate loading procedures. For example, pigs have more difficulties than sheep or cattle in negotiating steep ramps.

Despite all these inter- and intra-species differences, several general recommendations can be made. For example, even illumination and gently curved races without sharp corners facilitate the movement of the animals. Non-slip flooring and good drainage to prevent pooling of water are also important. As animals prefer to walk slightly uphill rather than downhill, floors should be flat or slope upwards. On the other hand, however, ramps should not be too steep (Grandin, 2000, 2014). If the floor of the loading ramp is not slippery, there still remain differences between species in the steepness of slope that they can climb or descend safely.

Well-trained and experienced stockmen know that cattle can be readily moved from place to place by human movements that take advantage of the animal's flight zone (Kilgour and Dalton, 1984; Grandin, 2000). Cattle will move forward when a person enters the flight zone at the point of balance and can be calmly driven up a race by a person entering the flight zone and moving in the opposite direction to that in which the animals desire to go. Handling animals without the use of sticks or electric goads results in better welfare and less risk of poor carcass quality.

The handling and loading of poultry and rabbits is very different from that for the larger mammals. Chickens reared for meat production are often collected by human catching teams and sometimes by broiler collector machines. The welfare is worse when human catching is involved (Duncan *et al.*, 1986). Laying hens are usually collected and put into crates or modules by humans only, and show substantial adrenal responses when caught. Bone breakage is common in hens during the catching, especially if the birds have had insufficient exercise because they have been kept in small cages.

When salmon are transported by means of a well-boat, they have to be moved from the tank on land to the boat and from the boat to the sea cage. As with land animals, it is the handling and movement of the animals that causes the worst welfare, provided that the water in which they are transported is at an adequate temperature and level of oxygenation. Gatica *et al.* (2010b) found that salmon plasma cortisol levels were highest after the fish had been pumped through a pipe and when loading was occurring and relatively low during the journey itself. An open well-boat caused fewer problems than a closed well-boat (Gatica *et al.*, 2010a).

(a)
(b)

Fig. 21.4. (a) A loading ramp for pigs should have no slope, or a slope of up to 14°, and should have solid sides; (b) the ramp should be wide enough for two pigs to walk easily side by side (photographs D.M. Broom).

Good knowledge of animal behaviour and good facilities are important for good welfare during handling and loading.

Temperature and other physical conditions during transport

Extremes of temperature can cause very poor welfare in transported animals. Exposure to temperatures below freezing has severe effects on small animals, including domestic fowl. However, temperatures that are too high are a commoner cause of poor welfare. When sheep or cattle are transported by boat from Australia to the Middle East, high temperature is a major cause of poor welfare (Caulfield *et al.*, 2013). Phillips and Santurtun (2013) found that heat stress was a major problem for *Bos taurus* cattle, somewhat less for Zebu (*Bos indicus*) cattle and less again for sheep, but that it was a major cause of poor welfare in all livestock carried by boat across the tropics. On road journeys, poultry, rabbits and pigs are especially vulnerable to high temperature. For example, de la Fuente *et al.* (2004) found that plasma cortisol, lactate, glucose, creatine kinase (CK), lactate dehydrogenase and osmolarity were all higher in warmer summer conditions than in cooler winter conditions in transported rabbits. In each of these species, and particularly in chickens reared for meat production, stocking density must be reduced in temperatures of 20°C or higher, or there is a substantial risk of high mortality and poor welfare. The impact of temperature on animal welfare is different according to humidity. Villarroel *et al.* (2011) calculated the air enthalpy time derivative (kg water per kg of dry air per s) during pig journeys and found that it was up to ten times higher during the journeys than on-farm or in the abattoir. The air enthalpy time derivative proved to be a good predictor of pig welfare.

A period of rest during a journey can be important to animals (Broom, 2014a), especially those that are using up more than the usual amount of energy during the journey because of the position that they have to adopt or because they have to show prolonged or intermittent adrenal responses that mobilize energy reserves. One way of judging how tired animals become during a journey is to observe how strongly they prefer to rest after the journey. Another way is to assess any emergency responses or adverse effects on their ability to cope with pathogen challenge. For example, Oikawa *et al.* (2005) found that horses on a 1500 km journey showed less adrenal response and less sign of harmful inflammatory responses if they had longer rests and had their pen on the vehicle cleaned during the journey.

Vehicle driving methods, space allowance and welfare

When humans are driven in a vehicle, they can usually sit on a seat or hold on to some fixture. Cattle standing on four legs are much less able to deal with accelerations such as those caused by swinging around corners or sudden braking. Cattle always endeavour to stand in a vehicle in such a way that they brace themselves to minimize the chance of being thrown around, and avoid making contact with other individuals. They do not lean on other individuals and are substantially disturbed by too much movement or too high a stocking density. In a study of sheep during driving on winding or straight roads, Hall *et al.* (1998a) found that plasma cortisol concentrations were substantially higher on winding than on straight roads. Tarrant *et al.* (1992) studied cattle at a rather high, at an average and at a low commercial stocking density and found that falls, bruising, cortisol and CK levels all increased with stocking density. Careful driving and a stocking density that is not too high are crucial for good welfare.

The amount of space allowed for an animal during transport is one of the most important factors affecting animal welfare. In general, smaller space allowances lead to lower unit costs of transport, since more animals can be carried in a vehicle of any particular size. Space allowances have two components. The first component is the floor area available to the animal to stand or lie in. This equates to what is usually referred to as stocking density. The second component is the height of the compartment in which the animal is carried. With multi-decked road vehicles, this may be especially important because there are practical constraints on the overall maximum height of the vehicles, for example, to enable them to pass under bridges. There is thus a commercial pressure to reduce the vertical distance between decks (deck height), and therefore the volume of space above the animals' heads. This reduction may adversely affect adequate ventilation of the inside of the compartment in which the animals are held.

Absolute minimum space allowances are determined by the physical dimensions of animals. However, acceptable minimum allowances will be dependent on other factors as well. These include: (i) the ability of the animals to thermoregulate effectively; (ii) ambient conditions, particularly environmental temperature; and (iii) whether the animals should be allowed enough space to lie down if they so wish. Horses choose to stand in moving vehicles; cattle will lie if they can after 6–10 h in a vehicle; sheep lie after 2–4 h; pigs, poultry, dogs and cats will lie immediately. In all cases, animals continue to stand if they are disturbed or if there is excessive vehicle movement resulting from cornering and braking, or if there is not enough space to lie safely (see Fig. 21.5).

Whether or not animals want to lie down may depend on journey length, comfort in relation to transport conditions, the care exercised in driving the vehicle and vehicle suspension characteristics in relation to the quality of the road surface. One very important consideration in establishing practical minimum space requirements is that the animals need to be rested, watered and

Fig. 21.5. These pigs on a transport vehicle have too little space to lie down (photograph D.M. Broom).

fed on the vehicle. Resting, watering and feeding on the vehicle will require lower stocking densities to enable the animals to access feed and water. Space allowances may need to be greater if vehicles are stationary for prolonged periods to promote adequate ventilation, unless this is facilitated and controlled artificially.

When four-legged animals are standing on a surface subject to movement, such as a road vehicle, they position the feet outside the normal area under the body in order to help them to balance. They also need to take steps out of this normal area if subjected to accelerations in a particular direction. Hence, they need more space than if standing still.

When adopting this position and making these movements on a moving vehicle, cattle, sheep, pigs and horses make considerable efforts not to be in contact with other animals or the sides of the vehicle. Provided that vehicles are driven well, the greater the space allowance the better the welfare of the animals. This is correct up to a maximum space allowance larger than that used in animal transport. However, if vehicles are driven badly and animals are subjected to the substantial lateral movement that results from driving too fast around corners, or to violent braking, close packing of animals may result in less injury to them. The best practice is to drive well and stock in a way that gives space for the animals to adopt the standing or lying position that is least stressful to them.

A separate problem, linked to space allowance, is aggression or potentially harmful mounting behaviour. Pigs and adult male cattle may threaten one another, fight and injure one another. This results in poor welfare and DFD meat. Rams and some horses may also fight. Such fighting is minimized or avoided by keeping animals in the social groups in which they lived on the farm, or by separating animals that might fight. Groups of male animals may mount one another, sometimes causing injuries in doing so. At very high stocking densities, fighting and mounting are more difficult and injuries due to such behaviour may be reduced. However, such problems can be solved by good management of animals, and keeping animals at an artificially high stocking density in an attempt to immobilize them will result in poor welfare.

Floor space allowances need to be defined in unambiguous terms. In particular, stocking densities must be defined as square metres of floor area per animal of a specified live weight, e.g. $m^2/100$ kg or kg live weight per m^2 floor area (kg/m^2). Stocking rates such as m^2 per animal (m^2/animal) are not an acceptable way of defining floor space requirements, since these take no account of variation in animal weight. Definitions of acceptable space allowances must consider the whole range of animal sizes (live weights) to be encountered. One problem is that information applicable to very small or very large animals is sometimes not available. Moreover, the relationship between minimum acceptable space allowance and animal weight is often not linear.

Determining appropriate minimum acceptable space allowances for transported animals relies on several types of evidence. These include evidence based on: (i) first principles using measurements of the dimensions of animals; (ii) behavioural observations of animals during real or simulated transport conditions; and (iii) the measurement of indices of adverse effects of transport. An example of the latter kind of evidence would be the amount of bruising on the carcass (see Fig. 21.6) or the activity of enzymes such as CK in the blood.

For an animal of the same shape, and where body weight is W, linear measurements will be proportional to the cube root of W ($^3\sqrt{}$). The area of a surface of the animal will be proportional to the square of this linear measure ($^3\sqrt{W^2}$). Algebraically, this is equivalent to the cube root of the weight squared ($^3\sqrt{W^2}$), or weight to the power of two-thirds ($W^{2/3}$ or $W^{0.67}$). The minimum acceptable area for all types of animal is:

$$A = 0.021W^{0.67}$$

where A is the minimum floor area required by the animal in m^2 and W is the weight of the animal in kg. The constant in the equation (0.021) depends on the shape of the animal, in particular the ratio of its body length to its body width.

As a result of a review of the literature on the effects of space allowance on welfare, the EU SCAHAW (2002) recommended equations for calculation of this for pigs, sheep and cattle, and examples of the results of such calculations are as shown in Table 21.3. Similarly, the EFSA Scientific panel AHAW (EFSA, 2004) recommended equations to be used for the calculation of space allowances for other farm animal species.

Feeding and watering during transport

Drinking is stimulated when extracellular body fluid is diminished. Animals vary according to species in how often they drink in a 24 h period, and horses may drink only once or twice per day. It is difficult to provide water continuously, and many animals will not drink

Fig. 21.6. The bruising on carcasses, such as these calf carcasses, can be quantified. Poor handling methods, poor driving so that animals fall and too high a stocking density – so that animals have difficulty in standing after falling – can all increase bruising (photograph D.M. Broom).

Table 21.3. Recommended minimum floor space allowances – examples.

Species	Body weight (kg)	Travel duration (h)	Floor space allowance (m^2)
Pigs	100	<8	0.42
		>8	0.60
Sheep – Shorn	40	<4	0.24
		4–12	0.31
		>12	0.38
Sheep – Unshorn	40	<4	0.29
		4–12	0.37
		>12	0.44
Cattle	500	<12	1.35
		>12	2.03

during vehicle movement, so frequent stops may be necessary if adequate drinking is to occur when water is provided on the vehicle.

A study aiming to characterize progressively dehydration, stress responses and water consumption patterns of horses transported long distances in hot weather and to estimate recovery time after 30 h of commercial transport has been performed. It was concluded that transporting healthy horses for more than 24 h during hot weather and without water will cause severe dehydration; transport for more than 28 h, even with periodic access to water, will probably be harmful due to increasing fatigue (Friend, 2000).

Brown *et al.* (1999) compared constant transport for 8, 16 and 24 h without resting periods or watering/feeding and observed the need for pigs to drink and feed during a 6 h lairage period. The results showed that, even though the environmental temperatures were relatively mild (14–20°C), all pigs drank and ate during the lairage period and that, in particular, pigs transported for 8 h ate and drank immediately after arrival before they rested. Sheep often do not eat during vehicle movement. Nevertheless, after 12 h of deprivation, sheep become very eager to eat (Knowles, 1998). As for water deprivation, sheep seem to be well adapted to drought, as they are able to produce dry faeces and concentrated urine. In addition, their rumen can act as a buffer against dehydration. The effects of water deprivation seem to be largely dependent on ambient temperatures, as would be expected. For example, Knowles *et al.* (1993) found no evidence of dehydration during journeys of up to 24 h when ambient temperatures were not >20°C. However, when ambient temperatures did rise above 20°C for a large part of the journey, there were clear indications that animals became dehydrated (Knowles *et al.*, 1994).

If resting periods within the journey are considered as a means of preventing the effects of food and water deprivation, several points have to be taken into account. First, resting periods of 1 h are insufficient and may even have detrimental effects on welfare. Hall *et al.* (1997) studied the feeding behaviour of sheep after 14 h of deprivation and concluded that the extent to which sheep obtained food and water within the first hour was generally low and the composition was sometimes unbalanced in that they sometimes ate food first and were then more dehydrated when there was no opportunity to drink water. Dehydration problems after short resting periods are often worse if the sheep are given concentrates (Hall *et al.*, 1997). Knowles *et al.* (1993) found that recovery after long journeys took place over three phases and that after 24 h of lairage sheep seemed to have recovered from short-term stress and dehydration. It has been suggested that at least 8 h of lairage are needed to gain any real benefit (Knowles, 1998). One further problem is that sheep will not drink readily from unfamiliar water sources, even after prolonged periods of water deprivation (Knowles *et al.*, 1993). A second problem is that feeding during resting periods may cause competition between animals, and the stronger individuals may exclude the weaker ones (Hall *et al.*, 1997). It is therefore important that feeding and drinking space is enough for all animals to have access to food and water simultaneously. Recommended trough space for sheep is $0.112W^{0.33}$ m (Baxter, 1992). This means 30 cm for sheep of 20 kg bodyweight and about 34 cm for sheep of 30 kg bodyweight.

Finally, sheep can be reluctant to eat during lairage, particularly adult animals that are unfamiliar with the feed. Hay has been found to be the most widely accepted form of feed (Knowles, 1998), although Hall *et al.* (1997) found that only small amounts of hay were eaten by sheep after 14 h of food deprivation.

One important decision to be made concerning resting periods is whether the animals are fed and watered on the vehicle or after being unloaded. Feeding and watering sheep in a normal commercial load may not be practicable, mainly because it would be difficult for all animals to have access to food and water simultaneously (Knowles, 1998). However, unloading the animals raises several problems, including an increased risk of disease transmission and the fact that loading and unloading are reported to be the most stressful parts of transport.

Chickens and turkeys have high metabolic rates, especially those bred for meat production. Newly hatched chicks develop very rapidly, faster than those of 20 years ago did. They use up all of their yolk and albumen and so need food and water by 48 h of age. Birds going to slaughter will not eat during a journey and often use up all food reserves within 3–5 h, so journeys have to be short.

Journey duration and welfare

For all animals except those very accustomed to travelling, being loaded on to a vehicle is a particularly stressful part of the transport procedure. However, as journeys continue, the duration of the journey becomes more and more important in its effects on welfare. Animals travelling to slaughter are not given the space and comfort that a racehorse or show-jumper is given. Hence, they are much more active, using much more energy than an animal that is not transported. As a result they become more fatigued, more in need of water and of food, more affected by any adverse conditions, more immunosuppressed, more susceptible to disease and sometimes more exposed to pathogens on a long journey than on a short journey. The extent of fatigue varies with species; sheep can walk on a treadmill for 5 h without substantial fatigue effects (Cockram *et al.*, 2012), so this helps to explain the greater resistance to fatigue of sheep as compared with other farm animals.

In a survey of records of the transport of 19.3 million broilers killed in four processing plants in the UK, Warriss *et al.* (1990) found an average time from loading to unloading of 3.6 h, with a maximum of 12.8 h. Comparable average times for 1.3 million turkeys killed at two plants were 2.2 and 4.5 h, with maxima of 4.7 and 10.2 h (Warriss and Brown, 1996).

Because poultry held in crates or drawers cannot be effectively fed and watered during transport, journeys must be considerably shorter than for red meat species. Mortality is increased progressively with longer transport times (Warriss *et al.*, 1992). These authors recorded the number of broilers dead on arrival in a sample of 3.2 million birds transported in 1113 journeys to a poultry processing plant. Journey times ranged up to 9 h, with an overall average time of 3.3 h. Total time, from the start of loading birds on to the vehicle to the completion of unloading at the processing plant, ranged up to 10 h, with an average of 4.2. The overall mortality rate for all journeys was 0.194%. However, as journey time increased, so did mortality rate. In journeys lasting for more than 4 h the prevalence of dead birds was 0.16%, while for somewhat longer journeys the incidence was 0.28%. In all journeys of more than 4 h, mortality was therefore, on average, 80% higher than in all journeys shorter than this. Liver glycogen, which provides a ready source of metabolic fuel in the form of glucose, is very rapidly depleted after food withdrawal. Warriss *et al.* (1988) found depletion to negligible levels within 6 h. Broilers transported 6 h had only 43% of the amount of glycogen in their livers compared with untransported birds (Warriss *et al.*, 1993). As a consequence, these rapidly metabolizing birds are so short of energy by 4 h of transport that some die. Birds suffering painful traumatic injuries such as broken bones and dislocations, which are not uncommon, will suffer progressively more on longer journeys. Spent hens often travel very long distances to slaughter because of the very small number of plants willing to process them. Their metabolism is slower than that of broilers but this long transport must cause poor welfare.

Rest periods are impracticable and counterproductive for poultry since, as mentioned above, birds cannot realistically be offered food and water. Neither can they be effectively inspected by drivers or veterinary authorities because of their close confinement in the transport receptacles. Moreover, with current systems of passively ventilated transport vehicles, the reduction in airflow likely if vehicles stop without unloading the birds is likely to lead to an increase in temperature within certain parts of the load, and possibly cause the development of hyperthermia in the birds.

Horses stand during transport and have to make balance correction movements throughout any vehicle movement. In a study comparing the effects of road transport ranging from <50 to 300 km, the levels of T_4 and fT_4 lymphocytes were increased in horses after journeys of 150–300 km. Plasma concentration of myocardial depressant factor (MDF) peptide fraction was significantly lowered by road transport in journeys exceeding 100 km. It has been reported that road transportation of Sanfratellani horses over distances of 130–200 km resulted in significant elevations in serum creatinine and CK. Similar changes were recorded after journeys of 130–350 km in 16 untrained horses of various breeds in aspartate amino transferase (AST), lactate dehydrogenase (LDH), alanine aminotransferase (AAT) and serum alkaline phosphatase (SAP) (Ferlazzo, 1995).

It is well known that an increased incidence of equine respiratory disease follows prolonged transport. Predisposition to respiratory disease after transport may be due to a marked increase in the numbers and, in virally infected horses, the activity, of pulmonary alveolar macrophages. Therefore, it is evident that transits of 8–12 h or more tend to be more measurably stressful, and consideration should be given to monitoring welfare and pathology indicators.

A number of experiments have investigated the effects of journey length on cattle welfare. The majority of authors state that, with increasing duration of the transport, the negative effect on the animals increases as well, as represented through various physiological parameters such as body-weight CK, NEFA, BHB, total protein, etc. A period of food and water deprivation of 14 h results in vigorous attempts to obtain food and water when the opportunity arises, but deprivation must be for 24 h before blood physiology changes in calcium, phosphorus, potassium, sodium, osmolarity and urea are apparent (Chupin *et al.*, 2000).

However, food and water deprivation during a journey are likely to have much greater and more rapid effects. The extent of energy deficit when cattle were transported for two successive journeys of 29 h, with a 24 hour rest between, was quantified by Marahrens *et al.* (2003). After 14 h of transport, a break of 1 h for feeding and watering of the animals does not give ruminants enough time for sufficient food and water intake, but merely prolongs the total duration of the journey. Cattle become more fatigued as journeys continue and there are more frequent losses of balance.

The journey length during some sea transport can be many weeks. If pen conditions and sea conditions are good, welfare can also be good during such voyages, However, there are often problems during prolonged sea transport because of adverse temperature and humidity (Caulfield *et al.*, 2013), too high a stocking density, an unaccustomed diet that the animals will not eat, insufficient supply of water, spread of disease among the animals and rough treatment of the animals by inexperienced or uncaring staff. As a consequence, during sea transport from Australia to the Middle East or Indonesia, poor welfare on the ship is frequent (Phillips and Santurtun, 2013). The most frequent indicators of poor welfare were measures of heat stress, inappetance, unloading injuries and high mortality. Periods of prolonged high temperature and humidity on land, similar to those on cross-tropics sea voyages, affect the welfare of sheep but recovery is rapid (Stockman *et al.*, 2011a) so it would seem that the combination of negative effects during voyages causes the problems that do occur during the journeys. If the methods used for handling and for stunning and killing the animals, during transport or at the destination, are inadequate, these can cause major welfare problems. Long-distance transport lasting for several days, except in very good conditions such as those used for valuable animals, would seem to have too high a risk of poor welfare and so public opinion is likely to force it to be replaced by transport of frozen or chilled meat.

Disease, welfare and transport

The transport of animals can lead to increased disease, and hence poorer welfare, in a variety of ways. There can be tissue damage and malfunction in transported animals, pathological effects that would not otherwise have occurred resulting from pathogens already present, disease from pathogens transmitted from one transported animal to another and disease in non-transported animals because of pathogen transmission from transported animals. Exposure to pathogens does not necessarily result in infection or disease in an animal. Factors influencing this process include the virulence and the dose of pathogens transmitted, route of infection and the immune status of the animals exposed (Quinn *et al.*, 2000).

Enhanced susceptibility to infection and disease as a result of transport has been the subject of much research (Broom and Kirkden, 2004; Broom, 2006b). Many reports describing the relationship between transport and incidence of specific diseases have been published. As an example, 'shipping fever' is a term commonly used for a specific transport-related disease condition in cattle. It develops between a few hours and 1–2 days after transport. Several pathogens can be involved, such as: (i) *Pasteurella* species; (ii) bovine respiratory syncytial virus; (iii) infectious bovine rhinotracheitis virus and several other herpes viruses; (iv) parainfluenza 3 virus and a variety of pathogens associated with gastrointestinal diseases such as rotaviruses; (v) *Escherichia coli*; and (vi) *Salmonella* spp. (Quinn *et al.*, 2000).

Commercial transport has been shown to result in increased mortality in calves and sheep (Brogden *et al.*, 1998; Radostits *et al.*, 2000) and salmonellosis in sheep (Higgs *et al.*, 1993) and horses (Owen *et al.*, 1983). In calves, it can cause pneumonia and subsequent mortality associated with bovine herpes virus-1 (Filion *et al.*, 1984), as a result of a stress-related reactivation of herpes virus in latently infected animals (Thiry *et al.*, 1987). Four hours of transport in cattle increased plasma cortisol concentration, decreased phagocyte response, increased CD3+ T cells and increased the ratio of CD4+ to CD8+ cells, all of which would make respiratory disorder more likely (Ishizaki *et al.*, 2004). In some cases, particular aspects of the transport situation can be linked to disease. For example, fighting caused by mixing different groups of pigs can depress antiviral immunity in these animals (de Groot *et al.*, 2001). The presence of viral infections increases the susceptibility to secondary bacterial infections (Brogden *et al.*, 1998).

Transmission of a pathogenic agent begins with shedding from the infected host through oronasal fluids, respiratory aerosols, faeces or other secretions or excretions. The routes of shedding vary between infectious agents. Stress related to transport can increase the amount and duration of pathogen shedding and thereby result in increased infectiveness. This is described for salmonella in various animal species (Wierup, 1994). The shedding of pathogens by the transported animals results in contamination of vehicles and other transport-related equipment and areas, e.g. in collecting stations and markets. This may result in indirect and secondary transmission. The more resistant an agent is to adverse environmental conditions, the greater the risk that it will be transmitted by indirect mechanisms.

Many infectious diseases may be spread as a result of animal transport. If animals from different sources are mixed at lairage (Fig. 21.7) pathogens may spread from one group to another. Outbreaks of classical swine fever in Holland and of foot-and-mouth disease in the UK were much worse than they might have been because animals were transported and, in some cases, transmitted the disease at staging points or markets. Schlüter and Kramer (2001) summarized the outbreaks in the EU of foot-and-mouth disease and classical swine fever and found that, once this latter disease was in the farm stock, 9% of further spread was a result of transport. In a recent epidemic of Highly Pathogenic Avian Influenza virus in Italy, it was found that the movement of birds by contaminated vehicles and equipment created a significant problem in the control of the epizootic.

Major disease outbreaks are very important animal welfare as well as economic problems, and regulations concerning the risks of disease are necessary on animal welfare grounds. If stress is minimized and the mixing of animals and their products is minimized, disease and hence poor welfare can be prevented or made less likely.

Inspection of animals before and during transport

Animals that are to be transported could be unfit to travel because they are injured or diseased, or could be fit to travel only in conditions that are better than those that are the minimal ones permitted by law (EU SCAHAW, 2002).

The person responsible for animals at the point of origin, the driver of the vehicle, and any other person responsible for the animals during the journey should have the ability to evaluate animal welfare. The minimal ability that they should have is to distinguish an animal that is dead, injured or obviously diseased. A person driving a vehicle containing livestock will need to check the animals in the vehicle at regular intervals during a long journey. That person should also check animals after any situation that might cause problems for the animals, such as a period of excessive vehicle movement, a period when overheating might have occurred or a road accident. The intervals between regular checks correspond to the intervals between rest periods that are prescribed by law for drivers.

The checking of animals involves visual inspection and awareness of auditory and olfactory cues that

Fig. 21.7. Sheep at a lairage (staging point) during long-distance journeys. The welfare is affected by: (i) the potential for disease resulting from acquiring new pathogens; and (ii) the effects of extra loading and unloading. In other species, welfare may be poor because of threat or fighting (photograph S.J.G. Hall).

the animals have problems. It is necessary that each individual can be seen, so the design of vehicles, distribution of animals in the vehicle and stocking density must allow for this. If animals cannot be inspected, for example poultry in stacked crates or sliding drawer units, long journeys should not be permitted.

A veterinary surgeon is the person best able to declare an animal fit or unfit for travel. The Federation of Veterinarians of Europe's Position Paper on the Transport of Live Animals (FVE, 2001) provides useful information on those conditions that render an animal unfit for travel. Basically, pregnant animals in the last 10% of the gestation period, animals that have given birth during the preceding 48 h and newborn animals in which the navel has not completely healed are considered unfit for travel in all cases. Animals that are unable to walk unaided onto the vehicle, perhaps because of disease or injury, are considered to be unfit for travel in commercial conditions.

For cattle, sheep, pigs and horses, inspection facilities for the person on the vehicle who is responsible for the animals are needed. It is often possible to check each individual transported horse inside the vehicle without danger to the person inspecting or undue disturbance to the animals. However, other animals, such as adult cattle in groups, cannot be inspected from inside the vehicle without danger to the person inspecting. In this case, external inspection facilities are necessary that allow each individual to be seen. Inspection of sheep and pigs can normally be adequately performed from outside, provided that every individual can be seen. If inspection from inside the vehicle is required, the height of the deck cannot be so low as to render effective inspection too difficult. If sick, injured or dead animals are found, the person responsible needs clear knowledge of, or instructions about, what to do. It is important that records are kept and made available to the competent authority, for example to veterinary inspectors, of all sick, injured or dead animals, including any disposed of during a journey. Where the animals are transported to slaughter, the abattoir as well as the owner of the animals will need a copy of the record. If an animal is found to be sick or injured on a journey, humane killing on the vehicle will sometimes be required. Hence, the responsible person on the vehicle will need to carry, and be trained in the use of, equipment for humane killing of the species carried. Where the injury or sickness is such that the animal cannot complete the journey, for example if it cannot stand unaided, the animal should be killed or unloaded at an appropriate place.

When animals die or are killed during transport, the journey can continue for a time and to a place that is appropriate for the disposal of the carcass. In many cases of injury, sickness or death, it is important to inform the competent authority of the region. This is especially important if any important infectious disease is suspected. Journey plans include the addresses, e-mail addresses and telephone numbers of the competent authorities in each of the regions passed through during the journey.

The breed of animal, its temperament and the type of environment in which it was raised can affect behaviour during handling. Animals raised on open range or away from people will have a large flight distance, and may panic and become agitated when a handler approaches within 15 m. The problems posed by this involve serious disturbance for the animals and sometimes danger for the stockman. For example, Zebu cattle in Queensland, Australia, are not only difficult to round up after spending a long period in extensive paddocks or range areas but may attack the stockman. Similar animals that have been handled regularly are much more docile. Animals that have been raised in close confinement on either solid concrete or slatted floors can also be difficult to control when they are being moved on to other surfaces or manipulated, either individually or collectively, in unfamiliar premises. The animal may be disturbed either by changes in physical conditions or by human presence.

Facilities for moving animals

In order to be designed appropriately, handling facilities on the farm and at lairage should be based on the known behavioural characteristics of farm animals. Given adequate opportunity, cattle, sheep and pigs readily learn about their immediate environment. For example, cattle learn quickly about electric fences and can be contained by them in a field, in a collecting area or in a passageway (McDonald *et al.*, 1981). If, however, they were confronted with an electric fence for the first time when being moved in a strange area, the presence of the electric fence might hinder the process of moving the animals. The animals need time to learn about the fence and they also need the opportunity to avoid the fence. Moving electric fences, like the 'electric dog', are sometimes used in dairy parlour collecting yards. The 'electric dog' is a row of electrically live wires hanging downwards and moved towards the cows in the rear of the yard. It has a considerable adverse effect on some cows, such that their milk let-down may be prevented and they may become extremely unwilling to move towards the parlour.

Every dairy farmer has to be able to move dairy cows in milk to and from the milking parlour. If the races and collecting yards that are used, or the methods of moving the animals are inadequate and disturbing to some or all of the cows, there will be welfare problems. Such welfare problems will often be associated with reduced milk yield. Cows may be reluctant to enter a milking parlour because of the behaviour of the stockman or because of design faults in the parlour that result in uncomfortable milking stalls or stray voltages. Such problems can lead to the use of excessive force by stockmen in the collecting yard.

The problems associated with the design of races for moving cows to the milking parlour are very similar to those of designing races used for other purposes such as movement towards vehicles prior to transport. The most extensive study of how to design good races is that of Grandin (1978, 1980, 1982, 2014). She reported that cattle often baulk if they encounter dark areas or areas of extreme lighting contrast. Races with sharp angular turns may also pose problems for cattle that are being driven, and long, straight races may also result in animals being either reluctant to move or moving too fast. As a consequence of these observations, Grandin recommends that races should be evenly lit, have solid walls if animals unfamiliar with them have to use them, and should be gently curved rather than having sharp corners or long straights.

Negative stimuli, such as intense sound, should not be able to reach collected animals. In no event should disturbing sounds be located at a source which is in the direction that collected animals will be required to go. Shouting represents a disturbing sound and induces negative reactions, ranging from avoidance to flight.

When such reactions are impeded, the extreme fear reaction that we call panic is liable to ensue.

The same principles apply when animals are being moved in slaughterhouses. Grandin (2014) points out that cattle waiting to be sorted can be held in a wide, curved race with a radius of 5 m. From the curved race, the animals can either be sorted into diagonal pens or they can be directed to the squeeze chute, dipping vat or restraining chute at the abattoir. The handler should work from a catwalk located along the inner radius of the race. This facilitates the movement of the animals that tend to circle around the handler in order to maintain visual contact. The curved holding race terminates in a round crowding pen, which leads to a curved single race. Sharp contrasts of light and dark should be avoided. Single-file races, forcing pens and other areas where cattle are crowded should have high, solid walls. This prevents the animals from observing people, vehicles and other distracting objects outside the facility. Where space is limited, the desired effects of positive movement can be obtained using a compact serpentine race system. Holding lanes should be at least 3 m wide. Pigs move most readily in single-file races with solid side fences and open barred tops. Crowd pens for pigs should have level floors and ramps should be avoided. Pigs have 310° panoramic vision, and puddles and shadows should be eliminated in order to avoid negative visual stimuli. Pigs tend to move from a darker area to a brightly illuminated area under artificial lighting and lighting should be so arranged as to give an increasing gradient of illumination in the direction of movement. Pigs that have been raised in dimly illuminated confinement buildings refuse to move towards direct sunlight if this enters the premises. The pigs in Fig. 21.8 were reluctant to enter a dark vehicle.

Most dogs and cats have sufficient exposure to humans to be able to respond to the transport situation with much less fear than is commonly shown by farm animals, since they have seldom had such exposure. However, a few dogs and a much larger number of cats find some aspects of the handling, confinement or vehicle movement associated with transport very aversive. The pet animal, which hides whenever it can predict that a journey is imminent, has learned from previous aversive experience. Most of the information in this chapter refers to farm animals and horses. However, any animal might be affected by the factors mentioned, especially if the conditions are extreme in relation to the regulatory abilities of the animals.

Pharmacological control

The use of drugs to influence or to modify the behaviour of an animal is sometimes necessary in order to avoid excessive force. The range of modern chemical agents of varied pharmacological activities makes it possible to alter an animal's behaviour by tranquillization, sedation or immobilization (Tranquilli *et al.*, 2013). Today, tranquillizers are commonly used when individual animals have to be subjected, for a period, to forms of total restraint or handling to which they are unaccustomed, and which would probably induce in them a state of panic that would be detrimental to their health were it attempted by other means. Tranquillization during transportation, or mixing, for example, has much to commend it in circumstances when animals would react adversely under these circumstances.

Tranquillizers can produce psychological calming of anxiety without substantial physiological depression or clouding of consciousness. However, if tranquillizers are used to produce manageability, the high doses usually necessary may result in ataxia, depressed response to stimulation and respiratory depression.

Tranquillizers do not exert hypnotic or analgesic effects. Increasing the dose does not produce a greater anxiolytic effect, even though the psychological depressant effects are magnified. The psychological state of the animal prior to administration of tranquillizers may markedly affect the degree of tranquillization achieved. Animals that are vicious, intractable and in a state of excitation may not become manageable, except with very high doses, which would be totally incapacitating. Neuroleptanalgesia is a state of sedation and analgesia produced by the combined use of a tranquillizer (neuroleptic) and a narcotic in which, although the patient remains arousable and responds to certain stimuli, various manipulations, including minor surgical interventions, can ordinarily be performed.

Veterinary anaesthesia continues to develop each year. Control of the animal ranges from mild tranquillization, sedation, temporary taming and short-term immobilization to general anaesthesia. In most cases the agents can be given by small intramuscular doses, which makes administration comparatively simple. Oral or intravenous administration of drugs is technically simplified by knowing the dose–response relationship. Only in a few cases and in certain species do unfortunate side effects develop. For example, opioid drugs may have manic effects on a small number of horses, particularly when given in light doses.

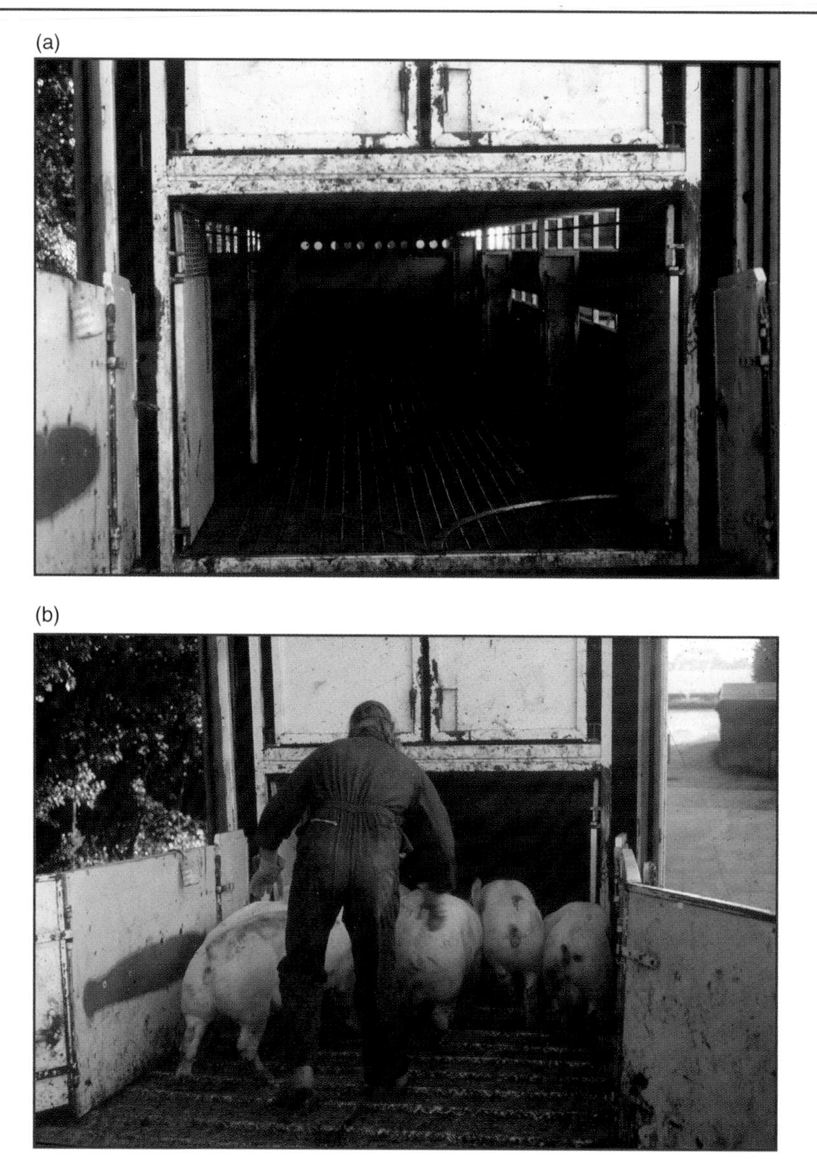

Fig. 21.8. (a) The entrance to this transport vehicle, encountered by pigs at the top of a ramp, is dark; (b) pigs would not enter, so had to be driven in (photographs D.M. Broom).

The administration of these drugs, although simple, requires a sound knowledge of the animal's behaviour and physiology. Certain types of animals react strenuously in fear or rage when control is attempted. There is also variation among individuals in the nature and degree of response. Violent responses are stressful to the animal and chemical restraint of stressed animals is unacceptable on welfare grounds except in emergency. Even then, the physiology of exertional stress must be borne in mind, since the animal will be significantly at risk from the procedure even though the controlling agent has minimal lethal potential. During exertion there is increased production of carbon dioxide from glycolysis in the vigorously contracting musculature. This carbon dioxide is taken up by the red blood cells and transported to the lungs. If this increased carbon dioxide content is not removed by the lungs, the hydrogen ion concentration increases in the blood and the pH decreases, resulting in acidosis. Profound hypoxia also occurs in all tissues.

Exertional stress is, in fact, a generalized state of acidosis, which is a broad condition that cannot be characterized by any single physiological sign. Instead, clinical behavioural signs must be relied upon to determine the condition. With exertional stress, the affected animal hyperventilates. This hyperventilation may result in small pulmonary haemorrhages, and frothy, blood-tinged fluid appears around the nostrils. The head is held down, the mouth is held open and the tongue typically hangs inactively out of the mouth. In addition, there is profound muscular incoordination. Finally, the animal falls and is unable to rise. In association with these signs other clinical features exist relating to cardiac function and loss of swallowing capacity. Few animals survive extreme exertional stress. Some, which appear to survive, are likely to die subsequently from acute heart failure or other pathophysiological developments. Among those that survive this secondary crisis, a number develop localized necrosis of skeletal muscles. This latter condition is commonly termed 'capture' or 'exertional' myopathy.

In spite of these hazards, it is occasionally necessary to treat or handle fractious animals. Using modern drugs it is feasible to do this through chemical control. The potential in this method to complicate physical exhaustion and stress must be considered. The following drugs are used internationally for the chemical tranquillization and restraint of large animals. They should be used only by qualified veterinary surgeons or with veterinary advice.

Acepromazine

This neuroleptic is a potent phenothiazine derivative frequently used for tranquillization and restraint. It should not be trusted on its own as an agent to calm fractious animals, since they can still respond to any threatening stimulus while under the influence of the drug. Since the drug causes hypotension, animals may suffer hypoxia in situations where they are frightened and metabolize large quantities of oxygen. In such an event, shock and hyperthermia may result (Galatos, 2011).

Buprenorphine

Buprenorphine is an opioid that has been found to be useful to calm horses when used in conjunction with xylazine (Cruz *et al.*, 2011).

Butorphanol

When combined with diazepam, butorphanol is used for sedation of young calves, goats or sheep (Abrahamsen, 2013).

Chlorpromazine

This agent keeps the animal conscious and makes it more amenable to handling for a short period. It is an anti-psychotic phenothiazine that creates a dramatic increase in dopamine activity. Phenothiazines without anti-psychotic potency do not have this characteristic (Tranquilli *et al.*, 2013).

Diazepam and midazolam

These agents markedly eliminate or diminish states of anxiety. They can be given to the animal over a prolonged period of time in order to obtain enduring results. Although they reduce anxiety, they do not provide analgesia. Diazepam allows the handling of animals that are only lightly restrained. Given in high doses it can cause respiratory problems (Galatos, 2011).

Ketamine and tiletamine

These drugs are principally used as short-term anaesthetics, but may not provide adequate analgesia in some species such as pigs, goats and poultry. Their use is now restricted because of human abuse of the drugs. The usual effect of ketamine and other phencyclidines is restraint characterized by catatonic immobility but convulsions may occur. To compensate for their deficiencies, these drug are usually administered in combination with another such as diazepam, phenothiazine or a narcotic (Coetzee *et al.*, 2010; Coetzee, 2013).

Xylazine, medetomidine and dexmetomidine

The main desired effects of xylazine and similar drugs in fractious animals are sedation and muscular relaxation. They are best given in combination with other sedatives, since high doses have undesirable side effects such as long recovery periods associated with struggling. They are, nevertheless, very useful drugs for the prompt control and immobilization of large animals (Kreeger *et al.*, 2011). Their anxiolytic properties are valuable.

Effective 'antagonists' to reverse the effects of these drugs are yohimbine, atipamezole and tolazoline (Galatos, 2011). Many animals have been killed as a result of unwarranted chemical restraint. Malignant hyperthermia has been observed in response to anaesthetics, muscle relaxants and to stress in pigs. Predisposition to such hyperthermia has a quantified genetic component in pigs and is commonest in the Landrace, Pietrain and Poland China breeds.

The behaviour of the species in question, the condition of the individual animal, the physiology of exertion and the circumstances of capture and control should all be thoroughly evaluated before chemical restraint is used. The mortality rate in such a procedure is much reduced when these factors are all taken into account and the procedure is carried out with the availability of equipment and materials for intensive care (Abrahamsen, 2013; Coetzee, 2013).

Further Reading: Grandin, T. (ed.) (2014) *Livestock Handling and Transport*, 4th edn. CAB International, Wallingford, UK.

22 Stunning and Slaughter

Euthanasia and Humane Killing

There are ethical questions about whether or not animals should be killed. These are not questions about welfare because welfare is a characteristic of living animals and ends at death (Broom, 1999b) so such matters are not considered in this chapter. Animals are killed in order to: provide a human resource such as food, prevent destruction of a resource by an animal that we might call a pest, prevent spread of disease, provide human entertainment, or benefit the animal itself by preventing suffering. Where the killing is under human control we have an obligation to avoid causing pain, suffering or other poor welfare to the animal prior to death. We use the word humane for such killing. Humane is an absolute term so the action is either humane or not and cannot be somewhat humane. Humane means *treatment of animals in such a way that their welfare is good to a certain high degree.* According to generally accepted principles, for example EU legislation and American Veterinary Medical Association guidelines (commercial slaughter, disease control, veterinarian), humane killing implies that: (i) the treatment of the animals in the course of the killing procedure does not cause poor welfare; and (ii) if there is stunning, the stunning procedure itself results in instantaneous insensibility or, if the agent causing insensibility or death is a gas or injectable substance, no poor welfare occurs before insensibility and then death. This may be achieved because the stunning or killing agent is not detectable by the animal. If the gas used is aversive, the stunning and killing is not humane but the negative effect on welfare is limited in duration. The effects of a carefully conducted injection procedure are usually considered slight enough that the stunning and killing is considered humane.

The word *euthanasia*, which means 'a good death', should be used solely to mean *killing an individual for the benefit of that individual and in a humane way*. If the benefit is for someone else, it should be called killing or humane killing but not euthanasia (Broom, 2007). When a companion animal is suffering, the owner may decide that it would be better for that animal to be dead than to continue suffering. It is assumed that the animal is sentient and would prefer not to suffer further (see discussion in Broom, 2014). The owner believes that it is a benefit to the animal to be killed and the duty of the owner to kill it humanely. In that case, it is euthanasia, but if the animal is killed for the convenience of the owner, for food, because it is considered a pest or because it may spread disease, it is not euthanasia.

Humane killing in the slaughterhouse

In order to kill animals humanely, they must either be killed instantaneously, or killed by a method that they do not detect or is not aversive; or rendered unconscious and then killed without recovering consciousness. Initiating unconsciousness is often called stunning and this should also be instantaneous or not detected or not aversive. Any handling prior to stunning should not cause poor welfare.

The method of killing used for most animals in slaughterhouses is to cut a major artery so that the blood loss results in death. For mammals and birds, the throat is usually cut to achieve this. Without prior stunning, this cut causes extreme pain, fear and other poor welfare. For fish, such as salmon, the artery to the gills is cut. If the animal is unconscious at the time of cutting, there will be no negative effect on welfare. The animal is usually hoisted so that its head hangs down prior to cutting so that bleeding-out is efficient. If the animal has its throat cut but is not stunned, the duration of very poor welfare before unconsciousness is: 20–40 s for pigs, 14–20 s for sheep, up to 120 s for chickens and up to 126 s for cattle. The reason for the variation is the extent to which there is alternative blood supply to the brain via the vertebral artery

(Daly *et al.*, 1988; Gregory, 2007). For fish in cold water, the time before unconsciousness after the artery to the gills is cut can be more than 20 min. Land animals that are not in the slaughterhouse may be killed by shooting.

The method of stunning used in the past, but not now legal in most countries, was to hit the animal on the head with a club or pole-axe. This could be inaccurate. The captive-bolt gun is now widely used for cattle and some other large animals. Smaller mammals and poultry are usually stunned electrically. Gas stunning is described below. The positioning of the captive-bolt gun, or of the electrodes, has to be accurate for effective stunning to occur (Fig. 22.1). Also, the gun must be properly maintained and the electric current must be sufficient for stunning without too rapid a recovery. After stunning, the delay before recovery of consciousness can be as little as 30–40 s so the cutting of the throat, also called sticking, has to occur rapidly.

The efficacy of stunning can be assessed by means of several indicators of unconsciousness. Gibson *et al.* (2009c) recorded electroencephalogram (EEG) in calves stunned with a captive-bolt gun. The total power of the EEG (P_{tot}) decreased sharply at the point of the stun, verifying that stunning had occurred.

Religious slaughter without stunning

Before the advent of modern stunning methods, when animals were often killed for human consumption using very inefficient methods, the Jewish and Muslim communities instituted the much better method of cutting the throat with very sharp implements by skilled persons, the procedures being called shechita and halal, respectively. The religious principles underlying halal slaughter are explained by Aidaros (2014) using quotations from the Qur'an. There are also statements of what the prophet Muhammad said that are very anatomically and technologically detailed so are presumably a modern interpretation. The religious slaughter methods also had the objective of maximizing the amount of blood loss from the carcass before it was used for human consumption. Indeed, after the Jewish slaughter, some people will not eat the hindquarters of the animal because of doubts about whether it is clean if it still has some blood in it. With current knowledge, we know that cutting the throat, even with a very sharp knife, causes much pain, fear and other poor welfare and the duration of this pain is up to 2 min in cattle and poultry, as described above. The reason why some cattle can remain conscious for 2 min after the cut in shechita or halal slaughter (Daly *et al.*, 1988) is that there can be false aneurysms and collateral routes to the brain (Gregory *et al.*, 2008). Gibson *et al.* (2009b) recorded EEG while the throat was cut by a ventral-neck incision and found that the response was consistent with there being a noxious stimulus. The EEG indications of pain were also present if the neck tissues were cut but not the blood vessels (Gibson *et al.*, 2009a) so it is clear that much of the pain results from the neck tissue damage. If stunning with a captive-bolt gun was carried out within 5 s of the throat being cut, the EEG flattened, indicating unconsciousness (Gibson *et al.*, 2009d). We also know that, with normal stunning procedures, the stunned animal is still alive at the time that the throat is cut and that most of the blood is exsanguinated whether the animal is conscious, unconscious or recently dead. However, no animal is completely exsanguinated from any part of its body. The forequarters and hindquarters still have a little blood in them. This small amount of blood does not cause any disease problems so, scientifically, the carcass is not unclean because of it.

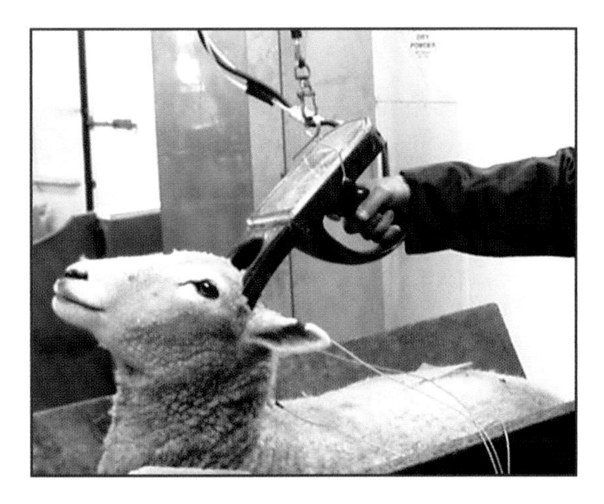

Fig. 22.1. Position of electrodes for electrical stunning of sheep (photograph A. Velarde).

As a consequence of this information, many Muslims and some Jews now accept that stunning should occur close to the time of cutting the throat in the ritualized and efficient way. However, others do not accept this. Most of the public in communities other than the more orthodox Jews and Muslims find it entirely unacceptable that animals should be caused to suffer to the extent that they do because of these traditions about methods of killing. Several countries have made killing without stunning illegal for all animals in slaughterhouses. In countries where religious slaughter is allowed, it is possible for members of the public to avoid eating meat in restaurants or other places where the animals have been killed in this way. However, it is necessary to label meat from animals killed in these inhumane ways without stunning at the time of slaughter, in order that the public can make this choice. At present this is seldom done. It is particularly important to label the hindquarters of animals killed by shechita.

Gas stunning and killing in the slaughterhouse

Another method of stunning involves the use of gas and is sometimes referred to as controlled atmosphere stunning (CAS). Key factors for effective stunning are the gas used, the concentration of the gas or gases, the period of exposure to the gas and the interval before killing if the gas is not being used to kill the animal. Gas stunning can be the best way to stun animals as far as animal welfare and meat quality are concerned. The factors that must be considered are: whether the design of the animal movement system results in stressed or calm animals, whether or not the gas is aversive and whether the stunning and killing are effective.

In a study of broiler chickens, Coenen *et al.* (2009) recorded their behavioural responses, heart rate or electrocardiogram (ECG), and electroencephalogram (EEG) when the birds were rapidly introduced to the controlled atmosphere. A mixture of 98% nitrogen and 2% oxygen, which results in insufficient oxygen in the lungs and hence in the blood and brain (anoxia), led to a median of 1.5 headshakes, 0.5 respiratory disruptions, and 11.5 s before loss of posture and cessation of movement without recovery. The best measure of unconsciousness is onset of isoelectric EEG. A second measure, described by Coenen *et al.* (2009) is the correlation dimension, a measure of complexity of the EEG which indicates unconsciousness when it drops to 60% of the baseline value. These two measures gave delays to unconsciousness of 49 s and 47 s, respectively. Death was defined in these chickens as when the isoelectric EEG was irreversible and was identified as when the heart rate dropped to less than 180 beats per minute. This occurred after 194 s. Vigorous muscle contractions, most obviously wing-flapping in chickens, can occur before loss of consciousness but are frequent after loss of consciousness and sometimes after death when nitrogen anoxia is used for stunning. A second controlled atmosphere was nitrogen with 30% carbon dioxide and 2% oxygen. Chickens in this gas mixture showed 1.0 headshakes, 2.0 respiratory disruptions, 7.5 s before loss of posture, unconsciousness in 34 s and death after 124 s. Hence this gas mixture was a little better for welfare. Other gas mixture conditions included more carbon dioxide. The gas mixtures with more than 30% carbon dioxide resulted in convulsions, so worse welfare. Another approach to investigate what gas mixtures are humane was used by Sandilands *et al.* (2011b) who observed chickens when they had to put their heads into a gas mixture in order to obtain food. A concentration of carbon dioxide of more than 20% in argon or nitrogen was aversive.

A wide range of species of animals, including all mammals, birds and fish that have been adequately tested, have now been shown to show substantial aversion to carbon dioxide (e.g. Kirkden *et al.*, 2005). Carbon dioxide causes dyspnoea (i.e. breathing difficulties) that are highly distressing to humans, fear behaviour and some degree of pain. For example, Thomas *et al.* (2012) found that mice showed the escape response of jumping when in a chamber with a flow of carbon dioxide at 20% of chamber volume per minute. They also showed escape responses with 20% carbon dioxide and 60% nitrogen but not with 20% carbon dioxide and 60% of the analgesic nitrous oxide. The presence of nitrous oxide also led to a reduction in the time before the righting reflex was lost and hence time to unconsciousness.

Low pressure killing

A system using low atmospheric pressure to kill poultry, in cylindrical chambers 6.1 × 2.1 m into which two poultry transport crates could be put, has been developed. The effects of the low pressure are not instantaneous but McKeegan *et al.* (2013b) reported no signs of aversion or escape responses when broiler chickens were exposed to it. The heart rate of the birds dropped and EEG slowed with increased delta wave activity, typical of sleep, starting after 10 s. However, it is not certain that the birds became unconscious, as assessed by EEG and body movement responses (EFSA, 2014). All birds had isoelectric EEG, so were dead on removal from the chamber in which they had been exposed to low atmospheric pressure. Because of the doubts about the time of unconsciousness, it is not yet clear that the killing method is humane.

Stunning and killing method in relation to carcass quality

Handling and pre-stunning can have major effects on the extent of bruises and other injuries and on meat quality problems such as pale soft exudative (PSE) and dark firm dry (DFD) meat. Both of these meat measures indicate poor welfare prior to slaughter. In one study, when stunning of pigs was by gas instead of electrical, the incidence of PSE meat dropped from 35% to 6% (Velarde *et al.*, 2001). The reason for this is probably that the handling procedures prior to stunning are much less stressful for the pigs when they are kept in a group and lowered into a gas than when driven into a chute for electrical stunning. Major advantages for meat quality, for example reduction in blood spots within muscles, have also been found when chickens are gas-stunned in the crate in which they were transported rather than removed from the crate and hung by their feet on a shackling line in order to stun them electrically.

Mass killing for disease control reasons

When large numbers of animals have to be killed for disease control reasons, methods of killing that would not normally be acceptable are sometimes used. For example, there was public outrage at television footage of chickens suspected of having avian influenza in Indonesia, being put into plastic bags alive and burned on a bonfire. A widespread method of killing poultry on farm has been the use of carbon dioxide injected into the building after attempts to seal it. Sparks *et al.* (2010) reported that when a house full of pullets was being filled with carbon dioxide, the mean time for the concentration to reach 45%, at which some birds would die, was 5 min 20 s. However, they did not make any measurement of the welfare of the birds. In a study by McKeegan *et al.* (2011) hens started to show responses to carbon dioxide gradually filling the building when it reached them and continued to do so for 6–10.5 min until loss of consciousness. Death took 12–22 min. Since carbon dioxide is aversive, the procedure involved a prolonged period of poor welfare for many birds. A non-aversive gas such as argon would not cause poor welfare but would take longer before the death of all birds. Sandilands *et al.* (2011b) found that 80% or 90% argon and 90% nitrogen, each with carbon dioxide, were the least aversive gas mixtures for chickens but if there was 30%, 40% or 60% carbon dioxide in the gas mixture it was very clearly aversive.

Another method of killing large numbers of poultry is to use a gas-filled foam. McKeegan *et al.* (2013a) measured heart rate and EEG in poultry and found no major adverse effects of air-filled foam. However, if the foam contained carbon dioxide, there were more head-shakes, foam avoidance and escape attempts. These responses were brief because the time to unconsciousness after immersion in foam was 16 s for broilers, 15 s for turkeys and only 1 s for ducks. For nitrogen-filled foam, time to loss of consciousness was 30 s for hens and 18 s for broilers. It would seem that nitrogen-filled foam, or a mixture of nitrogen or argon and up to 20% carbon dioxide, are the best mass-killing methods for poultry.

Further Reading: Gregory, N.G. (2007) *Animal Welfare and Meat Production*, 2nd edn. CAB International, Wallingford, UK.

Welfare and Behaviour in Relation to Disease

Behaviour and Disease

Behaviour is an important way for individuals to adapt to disease, and selective pressures resulting from disease have had major consequences for the evolution of behaviour. Behavioural responses, adrenal and other physiological responses, immunological responses and brain activity all help in coping with disease (Broom, 2006b; Hart, 2010). The great efforts made by humans to avoid and minimize disease have parallels in all other species (Hart, 2011), including domestic animals.

There are many links between behaviour and health. Behaviour has an important role in the transmission of a wide variety of diseases. For some diseases, an aspect of the normal behaviour of an infected animal, for example mating, is necessary for effective transmission of the pathogen. In other cases the behaviour required for transmission may not be normal, and pathogens and parasites change host behaviour so that transmission is more likely. For example, fish with an eye fluke swim near the surface where the next host, gulls, may catch them (Crowden and Broom, 1980). Many pathogens depend for transmission upon the behaviour of a vector of a different species from the main host.

Domestic animals may become infected with disease because the pathogen is transmitted to them by a non-domestic animal. Wild ducks may be infected with avian influenza and transmit it to domestic poultry. However, this occurs extremely rarely except where domestic poultry are kept in a place where they may contact the wild ducks directly or contact their faeces. Occasional ducks flying over pose almost no risk. A knowledge of behaviour helps to determine the probability of transmission. Another example is bovine tuberculosis. Although cattle-to-cattle transmission is the major route of infection, wild badgers and deer can occasionally be a source of infection. Badgers avoid cattle when they can and cattle avoid badger urine and faeces on pasture but infected badgers might visit cattle food troughs and cattle sometimes rub their heads on earth around badger setts (Fig. 23.1, see Benham and Broom 1989, 1991). Hence food troughs for cattle should be badger-proof and badger setts in cattle fields should be fenced. Studies using electronic proximity monitors on badgers and cattle show that badgers in fields can occasionally be close to cattle (Böhm *et al.*, 2009). However, this may not result in transmission of *Mycobacterium bovis*. Infected and uninfected badgers, monitored by Drewe *et al.* (2013), were close to cattle in only four out of 500,000 animal to animal contacts. Infected badgers did not show different behaviour from uninfected badgers, for example they did not visit cow accommodation. These observations account for the great rarity of transmission of tuberculosis from badgers to cattle. However, it appears that some individual cattle are atypical in behaviour and may therefore be more at risk. Disease control by killing badgers is risky because it is very difficult to kill a sufficiently high proportion of badgers. Attempted killing will often cause badgers to move to a new area and, if badgers are removed from an area, other badgers are likely to move in from the surrounding area. In consequence, unless badgers are eradicated from a large area, killing some of them is likely to increase the spread of the disease.

The changes in behaviour brought about by disease are used by veterinarians and medical doctors in the diagnosis of disease (Broom, 1987a). For example, a dog with abdominal pain may arch its back, sheep with foot-rot may kneel and a bull with traumatic reticulitis may walk with a characteristic stiff-legged gait. Some examples of changes in brain and behaviour associated with sickness or malaise are considered later in this chapter. Some behaviour has evolved because of selective pressures that are principally a consequence of those parasites and diseases that seriously reduce inclusive fitness. One example is preference for bright plumage during mate selection by female birds, because sick or parasitized males would often be less bright and less

Fig. 23.1. Cattle sometimes rub their heads on earth banks and holes. In this case, the rubbing is occurring at the entrance to a badger sett. If tuberculosis in badgers or cattle is suspected, all badger sett entrances should be fenced to prevent this possible route for infection (photograph D.M. Broom).

suitable mates (Hamilton and Zuk, 1984). Another example is behaviour resulting from feelings of malaise and consequent inappetence that allows energy to be retained for immune system function when fighting infection.

Disease and welfare

Diseased animals very often have difficulty in coping with their environment, or fail to do so, hence their welfare is poorer than that of a healthy animal in otherwise comparable conditions. The effects on an animal of laminitis, mastitis, pneumonia or severe diarrhoea are easy to appreciate. Whether the disease causes pain or other kinds of discomfort or distress, veterinary treatment that reduces the effects of the disease is clearly improving the welfare of the animal. It is important to emphasize, as have Jackson (1988) and Webster (1988, 1994), that it is not the diagnosis of the disease that improves welfare but the consequent treatment. As Webster (1988, 1994) points out: 'The fevered pneumonic calf, shivering in the corner of a damp draughty barn, feels rotten, and is in no way comforted by the fact that its condition has been diagnosed by a trained veterinarian.'

If the consequence of disease diagnosis in a pig is preventive measures in the whole pig unit, the welfare of the animals already diseased is not improved. One important moral question for all veterinarians to ask themselves is whether or not they put the welfare of the individual farm animal first when confronted with an animal that is diseased or injured. Is the patient considered before the client, who may or may not wish to pay for treatment? When veterinary surgeons report on the treatment of animals for disease or injury, they sometimes describe the measures used to evaluate the welfare of the animal after treatment, but often do not undertake those measures (Christiansen and Forkman, 2007). A range of welfare indicators should be used in clinical investigations of the effects of veterinary treatment. One key moral question for a farmer is whether or not to allow an animal to suffer when the suffering could be reduced or prevented by seeking veterinary advice and treatment. Most of the public would say that if a farmer takes on the responsibility of keeping the animals, that farmer should pay for veterinary treatment when it is needed by the animal.

One of the consequences of the poor welfare associated with disease is that the consequences of poor welfare reduce resistance to other disease. This has been known for a long time in the medical and veterinary professions and is part of the more general process whereby poor welfare, whatever its cause, can lead to increased susceptibility to disease. The relationship can account for the downward spiral towards death that has often been described in animals initially affected mildly by disease or difficult conditions. This positive feedback effect, which may or may not go as far as death, is shown in Fig. 23.2. Also shown in Fig. 23.2 are the simple relationships: disease always means poor welfare and, whenever welfare is poor for any reason, there will often be a greater susceptibility to pathogen replication and adverse effects.

Pathology is the detrimental derangement of molecules, cells and functions that occurs in living organisms in response to injurious agents or deprivations (Broom and Kirkden, 2004, modified after Jones *et al.*, 1997). Any kind of pathology involves some degree of poor welfare. If an individual has a parasite or pathogen within its body it may be that there is no effect upon that individual, so no pathology or effect on welfare. However, as soon as there is 'detrimental derangement', as described in the definition of pathology, the individual will have more difficulty in coping with its environment and some harmful effect on its functioning, and so there will be worse welfare because of the pathology. It might be that the individual is aware of the consequences of infection and the pathology might lead to feelings of pain or malaise. In that case, the pain or malaise plus the other effects of the pathology will make the welfare worse than when there is pathology alone.

Consider a sheep infested with the sheep scab mite, *Psoroptes ovis*. The mite can be present without affecting the sheep, so there is no pathology and no effect on welfare. However, mite activity on the skin leads to an inflammatory response, lesions, biochemical changes in the blood, irritation and rubbing responses, changes in blood cell counts and hormones, mouthing stereotypies, changes in pain sensitivity and often death (Corke, 1997; Corke and Broom, 1999; Broom and Corke, 2002). Welfare can be assessed during changes in the extent of pathology in sheep with sheep scab (see below). Some of the variables mentioned here will always result in some degree of poor welfare, while others can indicate good or poor welfare. Animals with pathology – whether from infectious disease, metabolic or other disorders – can have very poor welfare, which can be evaluated scientifically.

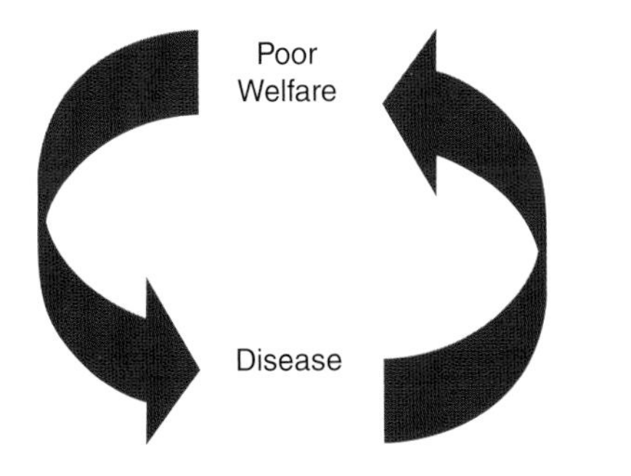

Fig. 23.2. The interaction between poor welfare and disease over time (modified after Broom, 1988c).

The criteria listed in Box 23.1 can be used in assessing the welfare and of sheep with sheep scab (Broom and Corke, 2002).

In some animal housing and management systems, the most important causes of poor welfare are disease conditions. For example, the major welfare problems of dairy cows are leg disorders, mastitis and reproductive disorders (Webster, 1993; Greenough and Weaver, 1996; Broom, 1999c, 2001c; Galindo *et al.*, 2000). In high-producing herds there are about 40 cases of leg disorders and lameness per 100 cows per annum. These disease conditions will vary with the environmental conditions, but are also considerably affected by the metabolic pressures on the individual. Cows producing very large quantities of milk are much more likely to be overtaxed metabolically and hence to have such problems (Pryce *et al.*, 1997).

Another example, which involves the largest numbers of animals kept by humans, is the broiler chicken. In chickens kept for meat production, the major causes of poor welfare are pathologies resulting from selection for fast growth. Some of the problems that have arisen in the very fast-growing birds are those associated with

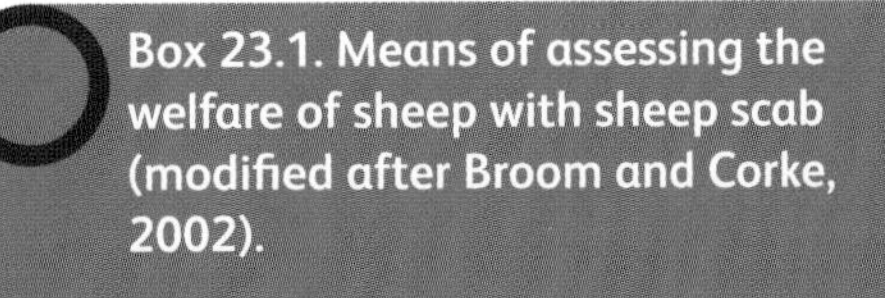

Box 23.1. Means of assessing the welfare of sheep with sheep scab (modified after Broom and Corke, 2002).

- Clinical examination.
- Behavioural assessment.
- Nociceptor response.
- Physiological assessment.
- Immune system function.
- Variables affecting welfare.
- Body condition score.
- Lesion dimensions.
- Mite numbers.
- Skin sensitivity.
- Body temperature.

cardiovascular disorders, which lead to ascites. A larger proportion of the problems are those leg disorders that result in reduced ability to walk, for example tibial dyschondroplasia, femoral head necrosis and valgus-varus syndrome (Broom, 2001a; Bradshaw *et al.*, 2002). Reduced ability to walk or stand often results in breast blisters and hock burn, because the bird has to spend a long time crouching on poor-quality litter. In a study in supermarkets, 82% of 'grade A' birds, rated as being of suitably high quality for sale as whole birds as opposed to chicken portions, had some degree of hock burn (Broom and Reefmann, 2005). The dermatitis seen in such birds is painful in itself, but the effects of inability to walk are much more severe.

Welfare and disease susceptibility

Poor welfare resulting from a wide variety of different causes may make disease more likely, often by initiating immunosuppression (Kelly, 1980; Broom, 1988a; Broom and Kirkden, 2004). Depression can be a consequence of an environment that is difficult, perhaps because the individual has little control over it, and this condition, which certainly involves poor welfare, has further pathological consequences (Irwin, 2001). It is also possible for injurious behaviour, which is caused to one individual by another and is often associated with poor welfare in both, to increase the likelihood or extent of pathology.

On the other hand, good welfare, sometimes facilitated by the social support provided by conspecifics, can help to protect individuals against disease (Lutgendorf, 2001; Sachser, 2001). Positive behavioural and mental responses can increase the likelihood that the individual will succeed in coping. Indeed, general clinical experience suggests that pathologies, such as tumour growth and proliferation, can be reduced or sometimes prevented in humans if they are happy and thinking positively. This may also be true in other species (Broom and Zanella, 2004). Welfare has important and complex interrelationships with pathology.

The evidence linking welfare with susceptibility to disease is of three kinds (Broom, 1988d): (i) clinical data concerning which individuals show signs of disease; (ii) experimental studies and surveys comparing levels of disease incidence in different husbandry systems or after different treatments; and (iii) studies of immune system function after different treatments. Each of these will now be discussed.

Every veterinary surgeon can give examples of situations in which a number of animals live in apparently similar conditions but only one or two show signs of disease, or most show signs of disease but only one or two die. The individuals affected more by disease are those that, using physical or behavioural signs, had looked weaker and less well able to cope with the environment, for example in calves (Morisse, 1982). In group-housing situations, the more susceptible animals are often those obviously at the bottom of a social hierarchy, with the consequence that they are chased a lot, injured by others, excluded from favoured places and sometimes prevented from obtaining an adequate diet. There is little well-documented scientific evidence concerning this effect in farm animals, but much clinical evidence suggests that research is desirable in this area. For example, what is the reason why runt piglets are more likely to develop chronic enteritis than their larger siblings? Studies of man and of laboratory animals have concentrated much more on the question of why it is that certain individuals succumb to disease while others do not.

Some housing systems for farm animals or treatments such as handling, transport or farm operations lead to more welfare problems than do others. Hence, there is the possibility of relating variation in welfare to variation in disease incidence. In some studies, an experimental treatment can be related directly to disease effects; for example, Pasteur (Nichol, 1974) found that chickens whose legs were immersed in cold water became more susceptible to anthrax. In other studies, it has been noticed that changes in husbandry methods are associated with changes in disease incidence; for example, Sainsbury (1974) reported a gradual increase in chronic infections of poultry over a period when the frequency of intensive production practices was increasing.

Direct comparisons of disease incidence levels in different housing systems are also possible, but any apparent relationship between poor welfare and disease incidence must be interpreted with care, as other factors which vary with conditions may affect disease incidence. Ekesbo (1981) has emphasized that environmentally evoked diseases are caused by a combination of factors. Examples of studies investigating the effects of housing conditions on disease incidence are those of Bäckström (1973) and Tillon and Madec (1984) on pigs and a range of studies by Gross, Colmano and Siegel, who carried out disease challenge on chickens treated in ways that increased plasma corticosterone concentrations in blood. When chickens were introduced to strange birds they displayed, fought and showed increased adrenal cortex activity. Frequent social mixing of this kind resulted in reduced resistance to *Mycoplasma gallisepticum*, Newcastle disease, haemorrhagic enteritis or Marek's disease (Gross, 1962; Gross and Colmano, 1965; Gross and Siegel, 1981). In contrast, such social mixing led to increased resistance to *Escherichia coli* and *Staphylococcus aureus* (Gross and Colmano, 1965; Gross and Siegel, 1981). When antibody activity was measured, it was clear that chickens subjected to social mixing showed less activity against both viral antigens such as Marek's disease and particulate antigens such as *E. coli* (Gross and Siegel, 1975; Thompson *et al.*, 1980). The social mixing leads to increased adrenal cortical activity, and this can help in counteracting inflammatory responses.

Pathogens elicit a range of responses in animals. The set of immunological responses includes proliferation and activation of: (i) B cells synthesizing antibodies that bind to antigens, thus facilitating antigen ingestion by granulocytes and macrophages or destruction by complement, which is a set of enzymes; (ii) T cells that destroy cells with foreign antigens on their surfaces; (iii) T-helper cells that assist (i) or (ii); (iv) NK (natural killer)-cells that destroy cells that lack normal antigens, such as tumour cells or those containing viruses; and (v) memory cells that increase humoral- and cell-mediated responses. These and other defence mechanisms against pathology interact with other coping systems, as explained below (Broom, 2006b).

Glucocorticoids play an important part in many body-regulating functions and facilitate important adaptive brain processes (Broom and Zanella, 2004). However, cortisol or corticosterone may be produced in emergency situations where coping is difficult. High levels of cortisol have been shown to: (i) decrease synthesis of both interleukin-1 by macrophages and interleukin-2 by T-helper cells, causing poorer B-lymphocyte and cytotoxic T-cell activity; (ii) decrease interleukin-beta that regulates T-helper cell-1 and -2 balance and action; and (iii) result in generally worse effects of respiratory pathogens, *Toxoplasma*, *Salmonella* and tumours (Dantzer, 2001; McEwen, 2001; Broom and Kirkden, 2004).

Some of the situations that elicit glucocorticoid production have rapid and serious effects on the health of animals. In a report of clinical cases by Madel (2005), a group of elderly sheep had been gathered, transported for 30 km and then left on an exposed field during a stormy night. At least 13 of the sheep became hypocalcaemic, and three had died by the following morning. The treatment would have led to increased glucocorticoid production, as well as to the severe loss of calcium that had caused their sickness or death. There are many other examples of stress, in the sense of an environmental effect on an individual which overtaxes its control systems and reduces its fitness or seems likely to do so (Broom and Johnson, 1993), leading to pathology. Adrenal activity can be monitored by arteriosclerosis, myocardial lesions, gastric ulcers and various other forms of organ damage (Manser, 1992).

As explained in detail by Broom and Kirkden (2004), the relationship between the chronic activation of physiological coping mechanisms, immunomodulation and susceptibility to infectious disease has been explored in the field of psychoneuro-immunology. Environmental challenges leading to poor welfare, whether they threaten mental or bodily stability, activate these coping mechanisms. However, the relationship is not a simple one. The response of the neuroendocrine system varies according to environmental challenge (Mason *et al.*, 1968a, 1968b, 1975), species (Griffin, 1989) and how individuals perceive the challenge, i.e. whether it elicits investigation, freezing, active defence, etc. (Broom, 2001c). Glucocorticoids and other hormones modulate the immune system in various ways, but a given change in the immune system may affect an animal's susceptibility to different pathogens in different ways (Gross and Colmano, 1969).

Glucocorticoid effects on the immune system may include the following (Griffin, 1989): (i) reduction in the number of circulating lymphocytes (lymphopenia); (ii) increase in the number of neutrophils (neutrophilia); (iii) in many species, reduction in the number of eosinophils (eosinopenia); (iv) a drop in the total number of circulating leucocytes (leucopenia); or (v) in species with relatively low numbers of lymphocytes, an increased leucocyte count (leucocytosis). The differential effects of glucocorticoids upon different leucocyte populations may explain the observation that a given stressor can increase the susceptibility of chickens to some pathogens while reducing their susceptibility to others (Gross and Siegel, 1965, 1975; Siegel, 1980). Not only do glucocorticoids reduce the number of circulating lymphocytes, they also suppress the activity of B cells and cytotoxic T cells, by interacting with macrophages and T-helper cells. For example, glucocorticoids decrease the synthesis of interleukin 1 (IL-1) by macrophages

(MacDermott and Stacey, 1981) and the synthesis of interleukin 2 (IL-2) by T-helper cells (Gillis *et al.*, 1979). These cytokines increase the activity of B cells and cytotoxic T cells, as well as that of other leucocytes, including macrophages and T-helper cells.

Glucocorticoids are very important mediators of the immune system, but they are not the only means by which stressors influence immunocompetence (Griffin, 1989; Biondi and Zannino, 1997; Yang and Glaser, 2000). Other hormones produced when animals are trying to cope with difficulties in life include beta-endorphin, vasopressin and oxytocin. It is known that beta-endorphin can promote T-cell responses and that vasopressin and oxytocin stimulate T-helper cells to produce interferon-gamma which, in this circumstance, activates NK cells and macrophages. Oxytocin is produced during mammalian nursing of young and various other pleasant experiences (Carter, 2001). Hence, it seems that pleasure can sometimes lead to better defence against pathogens (Panksepp, 1998).

Both the synthesis of β-endorphin by the anterior pituitary gland (Haynes and Timms, 1987) and the release of vasopressin and oxytocin from the neurohypophysis (Wideman and Murphy, 1985; Williams *et al.*, 1985) are increased in response to environmental challenges. In humans at least, catecholamines suppress the cell-mediated immune response while enhancing the humoral immune response (Yang and Glaser, 2000). Furthermore, the lymphoid organs – including the bone marrow, thymus, spleen and lymph nodes, where lymphocytes are produced and stored – are all innervated (Felton and Felton, 1991; Schorr and Arnason, 1999), permitting the CNS to influence lymphocytes directly. Vasopressin and oxytocin (Gibbs, 1986a, 1986b; Gaillard and Al-Damluji, 1987) and catecholamines (Axelrod, 1984) also stimulate the secretion of ACTH while β-endorphins, secreted in parallel with ACTH, form their mutual precursor pro-opiomelanocortin (Guillemin *et al.*, 1977; Rossier *et al.*, 1977).

Environmental conditions that elicit physiological coping responses in animals, thus involving poor welfare, alter animals' susceptibility to infectious agents and hence their health (Peterson *et al.*, 1991; Biondi and Zannino, 1997). The wide range of responses to pathology includes behavioural changes, physiological changes in the body – such as the production of acute-phase proteins in body fluids and production of cytokines in the brain – as well as immunological changes. Short-term responses to pathological effects include vomiting, which gets rid of some toxins and is promoted by certain interferons, and diarrhoea, which also helps to get rid of toxins and is promoted by interleukin-2 (Gregory, 2004). Longer-term responses include malaise or sickness behaviour, which is linked to immunological changes (Hart, 1988, 1990, 2010). Immune system responses may need much energy while pathogens may take energy directly from their host (Forkman *et al.*, 2001). Hence, some sickness behaviour results in energy saving, some promotes body defence mechanisms and all is adaptive (Broom, 2005). Sickness behaviour is initiated when cytokines are released by infected cells, endothelial cells, phagocytes, fibroblasts and lymphocytes, so there are many peripheral sources as well as brain-mediated (Gregory, 1998, 2004). However, the importance of the brain in relation to responses to pathogens is clear from brain lesion studies. Lesions to the hypothalamus and reticular formation reduce cellular immune responses, while lesions to the locus coeruleus reduce antibody responses (McEwen, 2001).

Some adaptive cytokine responses to pathology

At the individual level, adaptation is the use of regulatory systems, with their behavioural and physiological components, to help an individual to cope with its environmental conditions (Broom, 2005). Welfare may be good or poor while adaptation is occurring. Some adaptation is very easy and energetically cheap, and therefore, during this period welfare can be very good. Other adaptation is difficult and may involve emergency physiological responses or abnormal behaviour, often with unwanted consequences such as pain or fear. In that case, welfare is poor or very poor, even if complete adaptation eventually occurs and there is no long-term threat to the life of the individual. In some circumstances, adaptation may be unsuccessful, the individual is not able to cope, stress occurs and welfare is ultimately very poor.

One important part of trying to adapt to, or cope with, pathology is the acute-phase response (Kushner and Mackiewicz, 1987; Gregory, 1998). This is a set of body defences initiated and largely controlled by chemical mediators: the cytokines. Cytokines can affect leucocyte adhesion, alter capillary permeability, stimulate production of neutrophils, break down muscle proteins to allow production of acute-phase proteins and initiate fever (Gregory, 2004). Fever is an effect promoted by the cytokines interleukin-1 and interleukin-6 (Gregory, 2004). Fever helps recovery, for example from *Pasteurella multocida* infection, and helps the infected individual combat pathogens such as *Babesia* (Hart, 1988; Gregory, 1998). As explained by Hart (2010) the behavioural and physiological changes associated with fever involve heat-loss reduction. The animal may curl up, huddle, seek a warm place, erect the fur and shift blood flow from the skin to the interior parts of the body. In order to produce heat there may be muscle contraction, including shivering, and other muscular and biochemical activities initiated by interleukin-1. There is an energetic cost associated with fever estimated as 13% extra metabolism per degree C of fever, but, in many cases, the response is lifesaving. Interleukin-1 also induces excessive sleep.

A quite different adaptive effect of cytokines in sick animals is that of causing loss of appetite. This saves energy in the short term because gut activity does not occur. Fasting has been shown to reduce the *Listeria* population in the body of mice and to reduce mortality rates (Wing and Young, 1980). Failure to eat with consequent reduction in energy availability can, however, harm immunological defences if prolonged (Dallman, 2001). Brain processing efficiency can be impaired by some cytokine activity, perhaps because of reduction in energy availability.

A wide range of other effects are caused by cytokines in response to pathology (Gregory, 2004; Broom, 2006b). Wound healing is promoted by TGF-beta-2, and the anti-inflammatory cytokines interleukin-4, interleukin-10 and TGF-beta help to control septic shock caused by internal poisons that might be produced by pathogenic action. Interleukin-1 reduces grooming behaviour (Hart, 2010).

The most dramatic sickness behaviour when animals have some degree of pathology is often reduced activity (Gregory, 1998, 2004). The individual may feel tired, rest and sleep more. This is adaptive because sleep deprivation and too much activity when fatigued can lead to a reduction in NK cell activity and a reduction in interleukin-2 response to antigen challenge (Irwin *et al.*, 1994). It is also known that an increase in interleukin-1 level leads to more non-REM (rapid eye movement) sleep. A second kind of reduced activity is that caused by pain, promoted by interleukin-2. The feeling of pain itself can be mediated via products of damaged tissue such as nitric oxide, which promote the action of inflammatory agents such as bradykinin and prostanoids (Dray, 1995). A third type of activity reduction is that which results when the individual socially isolates itself. This will be adaptive, because an isolated, infected individual whose activity is reduced is less likely to transmit infection or have other infections transmitted to it. All three kinds of activity reduction can help an individual recover from disease.

Behaviour and disease diagnosis

Altered behaviour is usually the first indication of illness and, indeed, animal behaviour and veterinary diagnosis have long been closely associated. There are numerous references to the behaviour of sick animals in the classical literature of Ancient Greece. Practising veterinarians rely heavily on behavioural observations in diagnosis of illness. Examples include: deficiency diseases such as aphosphorosis, metabolic diseases such as hypo-magnesaemia and hypocalcaemia, and the infectious conditions such as encephalitis. Animal illness is often first manifested behaviourally by loss of appetite, altered activity, diminished body care or lack of responsiveness. Examples of what the veterinary clinician using behaviour in diagnosis may find are: the hyperexcited state in a cow is due to hypomagnesaemia; the stiffened bull's gait is the result of traumatic reticulitis; the dirty coat and nose of a steer, from arrested body care, is the effect of a septicaemia; the aggressively prancing mare has an ovarian tumour; the depressed steer is toxic; the asymmetric forelimb posture of the horse is due to navicular disease; the subdued sheep has toxaemia; the pig that has ceased eating has an infection; the calf with abnormal reactions has a neural impairment; the horse walking stiffly has tetanus.

Some diseases are known by behavioural descriptions that assume that the abnormal behaviour occurs in response to an abnormal physical state. Examples of these include staggers, swayback, nymphomania, louping-ill (leaping), star-gazing, gid (giddiness), wobbler, wanderer, circling disease, daft lambs, doddler calves and other more vague clinical syndromes such as the downer cow.

In conditions of severe pain, animals may distend the nostrils, roll the eyes in the head, extend the head and neck vigorously and groan. Some horses lie on their backs in a position of dorsal recumbency, with all four legs held in the air for up to 15 min. More violent manifestations of pain are: throwing itself down, rolling from side to side or walking into objects oblivious to the surroundings.

Diagnosis from locomotion

When the locomotor behaviour of animals is to be examined, animals should be observed singly, in good light, moving about using different gaits on a clean, dry and level surface. *Lameness can be defined as impaired locomotion or deviation from normal gait.* More frequently in the veterinary context, lameness refers to abnormal gait caused by painful lesions of the limbs or back or to mechanical defects of the limb. Neurological deficits producing lameness are usually defined separately after their differential diagnosis. Signs indicative of intracranial or brain disease include changes in mental state, seizures or convulsions, abnormal head posture, incoordination of head movement and cranial nerve deficits. Frequently, some cranial signs will accompany abnormalities of both posture and gait. Alterations of mental state such as depression, disorientation, coma or hyperexcitability may accompany locomotor signs.

In most animals, a progression through disorientation to seizures or fits is seen in many inflammatory conditions of the brain and its meningeal sheath, such as in bacterial meningitis. Seizures are characterized by a markedly diffuse increase in muscle tone, falling into lateral recumbency and rhythmic clonic convulsions simulating running or pedalling movements of the limbs. Depression or even coma may follow a seizure. Opisthotonus (dorsiflexion of the neck) with extension of the limbs is also seen with meningitis. Several metabolic and other systemic disorders may lead to confusion, coma, syncope, collapse, seizures, tremors, vision disturbances, quadriparesis, paraparesis and episodic weakness (Palmer, 1976). Hypoglycaemia, though not a specific disease, is frequently the main metabolic manifestation of starvation in piglets during the first week of life. Confusion and ataxia progress to quadriparesis, sometimes seizures and to death in coma. In the ataxic stages piglets may rest their noses upon the floor, apparently to gain further support. Ataxia means incoordination of muscular action or gait. Signs include wide-based stance, swaying movements, falling, rolling, imprecise gait, crossing of the limbs and exaggerated abduction of the limbs on turning.

Paralysis is defined as inability to move because of loss of motor or sensory nerve function; paresis is partial paralysis. A general muscular weakness may appear as apparent quadriparesis and, in practice, may be indistinguishable from true neurogenic paresis. In such cases, when a neurogenic component of the 'paralysis' can be ruled out, the term muscular weakness is thought to be more appropriate than paralysis.

Assessment of gait should take account of length of stride. Dysmetria may take the form of movements that are too long (hypermetria) or too short (hypometria). Hypermetria, called 'goose-stepping', is a relatively common sign of locomotor dysfunction in the pig. Goose-stepping is sometimes due to hock joint abnormalities preventing its movement. A dysmetric gait often includes a long stride with a prolonged supporting phase. Painful skeletal lesions give a short stride with a reduced support phase.

Recumbency in animals may also indicate paralysis or muscular weakness. Pain in a limb will produce abnormal limb positioning: often, flexion or abduction to avoid weight bearing. Shifting of weight from one leg to another in the standing position is seen in polyarthritis. When there is pain involving all four feet, as in laminitis in the acute form, a posture is adopted with the back arched and the feet bunched together under the abdomen. Such animals are reluctant to move.

Diagnosis from feeding behaviour

Failure to feed and abnormal feeding behaviour have long been used in clinical diagnosis. Automatic recording of feeding has opened up possibilities for aiding disease diagnosis in some species. For example, González *et al.* (2008) describe methods of recording feeding behaviour in dairy cows that can facilitate monitoring of health disorders. Ketosis is associated with a rapid decrease in intake and in rate of eating. Acute lameness leads to food intake and feeding time dropping but rate of feeding increasing. Chronic lameness causes identifiable changes in feeding pattern over the long term.

Diagnosis from other behaviour

The following list of specific features can be given as indications of the scope for clinical applications of ethology:

- Cattle with hypocalcaemia adopt a very characteristic recumbent posture and many of them show an equally characteristic 'S' bend in the neck while recumbent.
- In hypomagnesaemia, it is common to observe greatly increased excitability in the behaviour of the affected animal. This excitability is evident in

such behavioural features as unusual and excessive flicking of the eyes and ears and in an unusual style of walking.

- Urolithiasis occurs quite commonly in rams that are housed and heavily fed. The condition is associated with characteristic behaviour including grating of the teeth, straining and arching of the back.
- In gangreneous mastitis, affected ewes characteristically draw one hind leg behind the other while walking. An animal's general posture indicates a toxic state and the head is often held low.
- Hyperexciteability is the initial feature of toxicity from an organoarsenical feed additive.
- In about 80% of clinical cases of cystic ovarian disease in cattle, restless behaviour is very conspicuous. There is a typical feminizing behaviour, with increased production of vaginal mucus and frequent acceptance of mounting by other cattle. There may be a deepening of the voice and an increase in vocal activity. Among the various forms of masculine behaviour, digging with the forefeet is often observed.
- During milk fever, the cow is disinclined to move and has a depressed facial expression with staring eyes. She grinds her teeth and often makes intermittent paddling movements with her hind legs. Often, the cow produces exaggerated abdominal efforts to defecate, to little or no avail. As the condition progresses, the cow may show some degree of incoordination in her movements. As her powers of balance recede she is inclined to wander stiff-legged, swaying a little and subside to a position of sternal recumbency. Once the cow is recumbent she becomes more placid, although sometimes showing hyperexcitability and sweating if approached. As coma supervenes, she either remains in sternal recumbency and lowers her head and neck to the ground or rolls over onto lateral recumbency.

Throughout veterinary literature, among the identifying signs of many clinical conditions in animals, mention is made of depression. For example, depression is described as a major diagnostic clinical finding in such conditions as shipping fever in cattle, the mastitis-metritis-agalactia (MMA) syndrome in sows and Newcastle disease in poultry. By the clinical use of the term is meant a marked reduction in general activity, diminished responsiveness to exteroceptive stimuli and an appearance of reduced awareness. Head-pressing is a notable example. The depressed behaviour of an animal can therefore be recognized by both a decrease in the frequency of certain classes of maintenance behaviour and an increase in the frequency of certain anomalous forms of behaviour. Depression and disorientation are features of Aujeszky's disease and encephalomyelitis.

Further Reading: Hart, B.L. (2010) Beyond fever: comparative perspectives on sickness behavior. In: Breed, M.D. and Moore, J. (eds) *Encyclopedia of Animal Behavior*, Vol. 1. Academic Press, Oxford, UK, pp. 205–210.

24 Abnormal Behaviour 1: Stereotypies

What is Abnormality?

In order to recognize that behaviour is abnormal, the person observing must be familiar with the range of normal behaviour of that species. For some abnormalities, indeed, recognition depends upon a knowledge of the behaviour of that particular individual. One of the qualities of a good stockman is the ability to identify abnormal behaviour using knowledge acquired as a result of looking carefully at the animals. A difficulty for the stockman arises if many of the animals kept show the same kind of abnormal behaviour, since such a situation can lead the stockman to believe that behaviours like bar-biting in sows are normal.

In order to obtain knowledge of the behavioural repertoire of animals and establish what is normality, it is necessary to study the animals in a relatively complex environment where they have the opportunity to show the full range of their behaviour. This would not be the wild environment, but it should include those components of it that are important for the animal. An extensive knowledge of the biology of such animals and a detailed ethological investigation are therefore needed in order to be able to decide what behaviour is abnormal.

The most common abnormalities are those where the frequency of the movements, the intensity of the actions or the context in which the behaviour occurs is different from the norm. The animal may show the behaviour in an attempt to cope with some aspect of its environment. In some cases, that abnormal behaviour may help the individual to cope but, in other cases, it may confer no beneficial effect. Abnormal behaviour is behaviour that differs in pattern, frequency or context from that which is shown by most members of a species in conditions that allow a full range of behaviour.

Some abnormal behaviour has an obvious detrimental effect on either the animal showing it or on other animals. Examples are horses eating wood and pigs tail-biting. The word 'vice' is sometimes used to refer to abnormal behaviour. However, when used in a human context, this word implies that blame should be attributed to the individual showing the behaviour. Since almost all of the so-called vices shown by domestic animals have been shown to be a consequence of the ways in which the animals are housed or managed, it is an illogical term and will not be used in this book. The use of such a term can have an effect on the people who manage, own or advise about the animals. If they think that the abnormal behaviour concerned is not their responsibility but, rather, is a fault of the animal, they may fail to rectify the bad practice or inadequate housing. Such attitudes have been an important factor in the perpetuation of many systems resulting in poor welfare.

In this chapter, stereotypies are described. There follow chapters on other abnormal behaviour classified according to what it is directed at: the individual's own body or the inanimate surroundings (Chapter 25), or other individuals (Chapter 26). Chapter 27 concerns behaviour in which there are inadequacies of normal function, and Chapter 28 deals with abnormal reactivity. Each of these chapters includes a description of the abnormal behaviour and comments on the causative mechanisms, where this is possible. Some common abnormal behaviours in livestock are listed in Table 24.1 with indications of factors that lead to them. It should be emphasized, however, that while some kinds of abnormal behaviour are a direct result of some specific problem for the animal, the actual abnormality shown is often very individual in its characteristics. The level and quality of the abnormality can vary greatly from one individual to another.

Stereotypies

It has long been known that some caged animals in zoos and some human prisoners in isolation cells will pace out the same route over and over again. Similarly, birds

Table 24.1. Ten common abnormal behaviours in livestock.

Behavior	Confined in small space	Lack specific stimulus	Social isolation	Frustration	General lack of stimulation
Stereotyped pacing and weaving in horses	✓	–	✓	✓	✓
Crib-biting in horses	✓	–	✓	✓	✓
Bar-biting in pigs	✓	–	–	✓	✓
Excessive grooming in calves	✓	–	✓	✓	✓
Tongue-rolling in cattle	✓	–	–	✓	–
Sham-chewing in sows	✓	–	–	✓	✓
Wool-pulling in sheep	–	✓	–	✓	✓
Inter-sucking in calves	–	✓	–	✓	–
Tail-biting in pigs	–	✓	–	✓	–
Snout-rubbing in pigs	✓	–	–	✓	–

in small cages will fly or hop from perch to perch, following a route, and both monkeys in cages and autistic children will rock backwards and forwards for long periods. Hediger (1934, 1950) and Meyer-Holzapfel (1968) gave many examples of such behaviour in zoo animals, and Levy (1944) described examples of head-shaking in battery hens and various movement patterns in children. Brion (1964) described crib-biting and -sucking by horses, and D. Fraser (1975) described bar-biting by pigs. *A stereotypy is a repeated, relatively invariate sequence of movements which has no obvious purpose.* Its occurrence and causation are described in detail by Broom (1981, 1983b) and Mason and Rushen (2008).

A stereotypy is usually recognized because a sequence of movements is repeated several times with little or no variation. However, the behavioural repertoires of animals include many examples of repeated action patterns, such as walking, flapping flight and various displays, that would not be called stereotypies (Broom, 1983b). Hence, it is necessary to include in the definition of a stereotypy some reference to its apparent lack of function. Does it form part of one of the normal functional systems of the animal (see Chapter 1)? Detailed studies using video recording show how much variation there is in stereotyped behaviour. Figure 6.5 illustrates a pig drinker-pressing, a behaviour sometimes shown in stalls. In the same study (Broom and Potter, 1984), the sham-chewing stereotypy occurred in bouts of mean length 17.5 min. Below in Box 24.1 is a typical sequence of events seen in drinker-pressing.

Box 24.1. Example of sequence of events during drinker-pressing stereotypy.

- Sow standing.
- Presses drinker with snout (5–8 s, no water drunk).
- Pauses (1–2 s).
- Repeats above 7–15 times.
- Presses drinker as above.
- Swings head to left and puts nose in neighbour's pen.

Just as action patterns, which are part of normal behaviour, are seen to be somewhat variable when analysed in great detail (Broom, 1981; Chapter 2, this volume), so the repeated movements in stereotypies show some variation. When the descriptions of behaviour are subjected to analysis using information theory, however, the stereotypies are found to include much more redundant information, i.e. the same sequences occur, than do non-stereotyped behaviours. When Cronin (1985) analysed sequences of the behaviour of tethered sows he found that some sows chewed on their tether chain in a rigid sequence of actions whereas there was more variation in the sequences shown by other sows (Broom and Potter, 1984; Fig. 24.1). These other sows, however, showed action patterns that were themselves repetitive and repeated each of these action patterns even

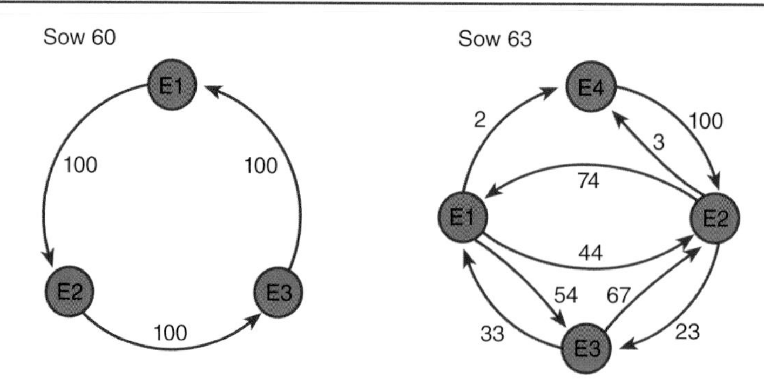

Fig. 24.1. Quantitative description of stereotyped sequences in two pregnant, tethered sows. Each element corresponds to a specific motor act (e.g. E3 is sham-chewing in both sows). Numbers along arrows indicate the percentage of occasions that individual elements succeeded other elements (from Cronin, 1985).

if the order was not constant, so the behaviour would still be called a stereotypy. These examples emphasize that stereotypies are relatively invariate rather than absolutely invariate. The repetition may be regular, but it need not be, and the sequence of movements may be very short, as in the head-shake of a hen, or long and complex, as in some route-tracing by bears in zoos or in the elaborate weaving sequences by some pigs in stalls or mink in cages.

As a way of trying to understand the significance and motivational basis of stereotypies, various physiological investigations have been carried out. Several different psychostimulant drugs that interfere with the metabolism of the catecholamine neurotransmitters dopamine and noradrenaline affect the incidence of stereotypies in various animals, including farm animals. The possible mechanisms involved were reviewed by Dantzer (1986), who concluded that: 'There is good evidence that performance of stereotyped behaviour depends upon brain dopamine systems involved in the control of movement.' Opiate peptides in the brain may also be linked with stereotyped behaviour in some way, for Cronin *et al.* (1985) found that when naloxone, which blocks the mu-receptor sites for opioids such as beta-endorphin, is administered to stereotyping sows they cease the behaviour. However, it is unlikely that the tethered sows use stereotypies to induce the action of the analgesic opiate peptides in the brain, as Cronin *et al.* suggested.

Zanella *et al.* (1996) found that sows that showed much stereotypy had lower mu and kappa receptor densities and lower dopamine in the frontal cortex, while McBride and Hemmings (2001) reported that horses showing more stereotypies had more dopamine (DI) receptors in the nucleus accumbens than low stereotypies. The roles of opioids and dopamine in stereotypies are reviewed by McBride and Hemmings (2009).

Stereotypies occur in situations where the individual lacks control of its environment. In some cases, the animal is obviously frustrated and in other cases the future events are rather unpredictable. Frustration about food inadequacy is one factor leading to increased likelihood of stereotypies (Lawrence and Rushen, 1993; Mason, 1993; Vinke *et al.*, 2002). Many examples of such situations are included in the rest of this chapter. Ideas about the causation of stereotypies and their possible function for the animal have been complicated by the fact that some situations where stereotypies are shown are barren environments, but others include disturbing or threatening factors. Hence, the stereotyped behaviour might increase the total sensory input in the barren environment but produce a more predictable and familiar input in the disturbing situation. McBride and Hemmings (2009) consider that horses are most at risk of showing stereotypies if they lack opportunities for sufficient social contact, or locomotion, or food intake. Housing in individual stables is a widespread cause of stereotypies and hence of poor welfare.

The behaviour sequence that becomes stereotyped is sometimes an incomplete form of a functional behaviour pattern (van Putten, 1982). It might arise from direct attempts to remedy some problem, such as to remove a bar that is preventing escape or to obtain the last available particle of food. By the time that the stereotypy is established, no simple function is served. The observation that animals showing stereotypies are often difficult to disturb is of interest, for they may have their brain state modified in a way that reduces responsiveness.

There is no clear evidence that stereotypies alleviate the effects of adverse conditions. However, whether or not they are of any help to the animal, they are clearly an indicator of poor welfare. Mason *et al.* (2007) explain how the poor welfare associated with the occurrence of stereotypies can be alleviated or eradicated by appropriate environmental enrichment.

The remainder of this chapter includes descriptions of a wide range of stereotypies occurring in domestic animals and is organized according to the nature of the movement. The first few stereotypies involve the whole body. There follow stereotypies involving much of the body or part of the body. Finally, those in which the oral region is used are detailed. The link between stereotypies and poor welfare is very well established and, since there are almost always alternatives, the use of a housing or management method that causes stereotypies involves unnecessary suffering and distress and hence is grounds for prosecution in most countries.

Pacing or route-tracing

The repeated action patterns during pacing or route-tracing are those used in walking or other locomotion, but the animal follows a path that returns to its point of origin and which is often repeated with only minor modifications. The route-tracing of zoo animals in cages, of some confined domestic animals and of confined or disturbed people has often been described. Some obvious frustration is normally evident, most frequently that the animal cannot escape from confinement in a cage or pen, but occasionally that access to a social partner, a sexual partner, food or some other resource is impossible.

In the horse, stereotyped pacing is shown under conditions of minimal exercise in chronic confinement. The condition closely resembles weaving, with the precision and repetition in which the animal performs its rhythmic movements. Stereotyped pacing also occurs in poultry; the condition can be induced, by thwarting birds that are very hungry and which have a high expectation of food provision. Duncan and Wood-Gush (1971, 1972) trained hens to feed from a dish in a particular position and then thwarted feeding by putting a transparent perspex cover on the dish. Stereotypies were among the responses elecited (see Table 24.2). In poultry, the stereotyped pacing resembles escape movements. Affected birds typically show repetitive pacing movements occupying the full range of one side of the pen or cage. When the condition has become established in the bird's behavioural repertoire it shows a strong tendency to persist, although it may reduce in frequency if thwarting circumstances are eliminated.

Table 24.2. Conditions leading to stereotyped pacing by hens (Duncan and Wood-Gush, 1972).

	Mean number of stereotyped pacing routines in 30 min
Deprived, fed	13.3
Not deprived, not fed	18.7
Deprived, frustrated (food under perspex cover)	161.0

Hens also pace before oviposition if no nest material is available. If hens have nest material they will build a nest before egg-laying, and their frustration if this is not possible is probably the major factor leading to stereotyped pacing at this time (Wood-Gush, 1969, 1975; Wood-Gush and Gilbert, 1969; Brantas, 1980).

Circling and tail-chasing

Animals of various species may sometimes turn in tight circles and dogs may do so, apparently trying to catch their own tails (Burn, 2011). Occasionally, this is a result of a neurological disorder. More often it is a consequence of a dermatological or other disorder, or an environmental inadequacy in which a key need is not fulfilled. Tail-chasing is most likely to occur when a dog is excited and frustrated, for example, because there is a possibility that it will be taken for a walk, but it cannot control when or whether or not this happens. The behaviour may stop when the frustrating situation is resolved but in some cases there is never adequate resolution and the stereotypy is frequent.

Rocking, swaying and weaving

The individual remains in one place when carrying out this stereotypy, but *the body is moved backwards and forwards or from side to side, with or without head-swinging.* Monkeys in captivity, especially those deprived of their mother or of companions for some time, show rocking behaviour; so too do autistic children and other children in very disturbing circumstances. Horses, calves and adult cattle that are tethered or in small pens will sometimes rock and sway.

Weaving in horses involves swinging the head and neck and anterior parts of the body from side to side, so that the weight rests alternately on each forelimb. In most cases the forefeet remain on the stable floor during the behaviour but, in extreme cases, each foot is raised as the weight passes on to the other foot. Lack of variety in the environment is a likely cause (Houpt, 1981), as it occurs most commonly in riding horses that have been stabled for long periods in idleness. Many animals exhibiting weaving become physically exhausted and lose weight progressively. Once the condition is acquired it is extremely difficult to control, and it is believed that the anomaly can be induced in other horses in a stable through mimicry. To some extent, weaving can be controlled by tying the horse with cross-reins so as to limit the lateral movement of its head, but this is probably of no benefit to the horse. Ideally, affected animals should be turned out to pasture but, when this is not possible through lack of space, enforced exercise can be provided by lunging or the use of a mechanical exerciser.

Rubbing

Some part of the body is moved back and forwards against a solid object and the movement is repeated so many times that it could not function merely to alleviate a local irritation.

Cattle confined to stalls for extended periods, such as in winter, may rub their heads repeatedly against some part of the stall. This behaviour is more noticeable in horned breeds and more in bulls than in other stock. Head-rubbing in pigs is sometimes observed in animals subject to chronic restriction within narrow, single stalls. In this behaviour, the upper snout region of the sow is rubbed repetitively and vigorously along the underside of a bar across the front of the stall to excess. The horses may back into a post, tree, fence or a portion of a building and move the hindquarters rhythmically from side to side while the tail is pushed into the perineal region.

Pawing and stall-kicking

Although pawing is a normal behaviour of four-legged animals, it can be shown in abnormal form when it is performed with vigour in a persistent, stereotyped fashion. Dogs may show pawing in certain frustrating situations, and pawing may occur when horses are frustrated in obtaining food. The anomalous condition is shown in pawing that is so frequent and vigorous that holes may be dug in the stall floor and the hoof worn down severely. Continual pawing on a hard floor can result in various forms of leg strain and injury. Attempts to control this problem through negative conditioning have not been successful. It occurs most frequently in confined and isolated horses, so may be alleviated by turning the affected animal out to pasture in the company of other horses.

Some stalled horses repeatedly lower the head, pull back the ears, arch the back and kick with a hind leg against the wall or door behind them so that a loud bang is produced. The noise may initially attract attention, which the horse may be seeking, but the action can lead to injury and damage. The splintering of woodwork can result in exacerbation of injuries and the production of material that the horse may later ingest. Attempts to prevent stall-kicking usually involve the use of hanging mats or barriers that might reduce injury and nuisance, but do not address the cause of the problem. As with other stereotypies shown by stalled horses, putting the animal out at pasture or providing frequent exercise and more companionship are the real answers.

Head-shaking or head-nodding

The head is repeatedly moved vertically, laterally or with a rotary movement of the neck. Head-shaking occurs in the domestic fowl and takes the form of a rotary movement of the head, with a series of rapid side-to-side turns ending with a slight downward movement (Levy, 1944). These spasms of movement last only for a second or so but may be repeated in succession for several minutes. They are also shown by jungle fowl (Kruijt, 1964).

Kruijt (1964) and Forrester (1980) suggest that head-shaking is linked to attentional mechanisms and the preparation for making a response. Hence, the behaviour may have a function when shown occasionally but should be regarded as abnormal and a stereotypy when shown often. Increased head-shaking sometimes results from the close presence of an observer from which the bird cannot escape. For example, it has been found that, in certain strains of birds, the incidence of head-shaking increases fivefold in the presence of an observer in an obvious position (Hughes, 1980). There appears to be more head-shaking in caged birds than in floor-housed hens, and it is affected by breed, space allocation, group size, transfer to novel conditions and social rank (Hughes, 1981; Bessei, 1982). Chickens show head-shakes when they encounter a noxious gas or deep foam (Coenen *et al.*, 2009; McKeegan *et al.*, 2013a).

Head-nodding in the horse occurs as stereotyped behaviour in various forms. The most common is repeated 'bobbing' up and down of the head. As with some other stereotypies, such as weaving, it has been suggested that there may be a self-hypnotic component in this behaviour. When animals show this anomaly they appear to be in a light somnolent state, showing little attention to their environment.

Wind-sucking

Movements during which air is sucked in and expelled, sometimes combined with head-nodding or crib-biting in horses, or tongue-rolling in cattle. In some forms of wind-sucking, the horse nods its head and neck several times before making the intake effort. In the initial act of wind- sucking, the head may be jerked upwards, the mouth opened taking in air and the floor of the mouth raised as the neck is flexed. The characteristic wind-sucking sound is made as air is expelled. The air taken into the mouth is not swallowed and passed into the gut (McGreevy *et al.*, 1995). It is known that horses are more likely to show wind-sucking if those stalled nearby do so, and foals may acquire the habit if their mothers do it. The behaviour is not normally seen among horses at pasture.

A common method for preventing wind-sucking is use of the wind-sucker strap. This is a strap fastened tightly around the throat, with a heart-shaped piece of thick leather held between the angles of the jaws with the pointed end protruding towards a pharyngeal area. With this device in place, difficulty and apparent discomfort are caused to the horse when the neck is flexed in the attempt to suck wind. Some horses will continue to practise the abnormal behaviour in spite of this device, so that they eventually acquire pressure sores on that part of the neck where the strap presses.

Various surgical methods have been attempted in the prevention of wind-sucking. One of these involves the creation of fistulae on each side of the mouth between the buccal cavity and the outer cheek. Such fistulae prevent the formation of a vacuum in the mouth that is required in the act of swallowing air. In addition, a wind-sucking prevention operation has been devised in which the small muscles surrounding the pharynx are partially removed. None of the operations prevents stereotypy completely. The use of straps and the surgical methods make the welfare worse. They are harmful to the horse and it would clearly be much better to improve the environment of the animal.

Eye-rolling

The eyes are moved around in the orbit at a time when no visible object is present, and the head moved in such a way as to lead to such eye movement. Young calves confined in crates sometimes stand immobile for extended periods and do not show the normal variations between lying and upright positions. During some of these episodes, the head is held motionless and the animal rolls its eyes within the orbits so that only the white sclera is shown, such eye-rolling being frequently repeated.

Sham-chewing

Jaw movements such as those shown when chewing food are shown at a time when the animal has no food in its mouth. This condition is typically seen in sows that are tethered or kept singly in stalls in which no litter is provided. The animal chews vigorously at a time when all food available has been eaten and, since pigs are not ruminants, there can be no oral content except saliva. Constant features involve periodic chewing and mouth-gaping, while the chewing motion causes frothing and foaming of saliva. This foam may collect on the outer edges of the lips and the corners of the mouth and drops to the ground, where such material can remain in portions, for some time, as evidence of this activity. Sham-chewing occurs most often while the sow is lying in a prone position or on its haunches in a dog-sitting position. It can be maintained as a prominent activity enduring throughout consecutive days. Broom and Potter (1984) reported that sows spent up to 90 min sham-chewing (median 26 min) during the 8 h of daylight, and Sambraus (1985) describes sows that were sham- chewing for many hours, day after day.

Sham-chewing may be reduced if the sows are given straw or other fibrous material to chew and root. When sow diet was supplemented by the same weight of oat hulls, the total frequency of stereotyped behaviour was not altered but sows lay down for longer and, as a consequence, showed more sham-chewing but less of the stereotypies shown while standing (Broom and Potter, 1984). A change to a group-housing system is the best way to alleviate the adverse effects on sows resulting in sham-chewing.

Tongue-rolling

The tongue is extruded from the mouth and moved by curling and uncurling outside or inside the mouth with no solid material present. This stereotypy has been described in detail for cattle and it includes components of the movements involved in the prehension of forage plants during grazing. The tongue is typically extruded and rolled back into the open mouth, after which partial swallowing of the tongue and gulping of the air takes place. Tongue-rolling may be associated with air intake, together with frothing at the mouth. Durations of tongue-rolling episodes range from a few minutes to several hours. It occurs most commonly immediately before and after feeding. Tongue-rolling might have an origin in forms of calf-feeding in which suckling is deficient. In horses, the tongue may be repeatedly extended and withdrawn without rolling movements. This stereotypy is called tongue-drawing.

Attempts to control the condition have been only partially successful. Wind-sucking straps have sometimes been fixed to affected animals. Other control methods include the insertion of a metal ring through the frenulum of the tongue. Therapeutic success has been reported through the provision of diets improved by salt mixtures. The provision of freedom of movement is also helpful in the control of this condition.

Licking or crib-whetting

In stereotyped licking, the tongue is applied repeatedly to an area of the animal's own body or to an object in the surroundings, with the same pattern of movement. This action may result in injury to the tongue, a wearing away of the area licked or ingestion of substantial quantities of hair or other materials (see Chapter 25). Stereotyped licking occurs in situations where animals have inadequate quantities of food, no teat from which to suck or insufficient total sensory input.

Some horses subject to chronic confinement show a form of anomalous oral behaviour in which the body of the tongue is slowly, but repeatedly, drawn across the edge of some part of the stall such as the crib or manger. The animal keeps the tongue still and firm during this action, so that the behaviour does not represent true licking.

Bar-biting, tether-biting or crib-biting

The animal opens and closes its mouth around a bar, tether or stable door, engaging the tongue and teeth with the surface and performing chewing movements. Bar-biting has been described for pregnant sows housed in stalls, or tethers that are very restrictive and do not allow the animal to turn around. The crate front and sides are made of metal piping. Tethers are commonly metal chains that the sow can bite and move up and down. Floors may be solid concrete or slats.

When engaged in bar-biting (see Fig. 24.2), the sow takes into its mouth one of the cross-bars at the front of the crate and bites it, rubs it with the body of the tongue or slides the mouth across the bar in rhythmic side-to-side motions (whetting). While biting the bar, the sow may take a firm grip on it with its jaws or may press the body of the tongue against the bar. In some instances the sow disengages from bar-biting and rubs its nose, above the snout, underneath the bar in side-to-side motions. Tether-biting occurs in much the same way but movements after the tether chain is taken into the mouth are more variable, as the tether can be moved more than can a bar. The sequences of movements include series of elements that are repeated exactly and others that are more variable. Breaks in these activities occur so that they are produced in episodes of activity. Trauma to the sow is not usually observed as a result of this condition.

Bar-biting and tether-biting can be partially controlled by improving the husbandry condition so as to provide the animal with oral occupation. This can be done by providing straw or other fibrous material that the animal can chew or in which it can root. Bar-biting, crib-biting and other stereotypies are more frequent if straw or other manipulatable material is not present (D. Fraser, 1975). Bar-biting and other stereotypies were not reduced by eating straw or oat hulls (D. Fraser,

Fig. 24.2. Bar-biting stereotypy by a stall-housed pregnant sow (photograph D.M. Broom).

1975c; Broom and Potter, 1984), so it appears that the possibility of manipulating is important rather than the bulk content of the diet. Substantially increased food rations do reduce its incidence, however (Appleby and Lawrence, 1987). Further details of such studies are reported in Chapter 31.

Bar-biting may also be shown by cattle kept in close confinement. Horses crib-bite by grasping the edge of the manger or some other convenient fixture with the incisor teeth. The upper incisors are most often used alone. The subject presses down, raises the floor of the mouth, the soft palate is forced open and there is usually a distinct grunt. In some cases, horses may rest their teeth against the bottom of the manger, the lower edge of the rack or the end of a shaft or pole. In rare cases, the mouth may be placed against the knees or cannons. Some horses that have been crib-biters may change to being wind-suckers when remedial measures are attempted. Some horses engage in these activities when alone in the stable but others will show it when in the company of other horses. Some affected animals will never show any signs of the disorder when under close supervision, but most disregard the presence of humans.

In horses that are crib-biters the incisor teeth, particularly of the upper jaw, show signs of excessive wear. This tooth wear may progress to such an extent that the incisors no longer meet when the mouth is shut, and grazing then becomes impossible. The muscles of the throat increase in size. In the advanced condition, the animal becomes physically unfit. Control of this condition is difficult except by change to less confined housing conditions. The most common measure is to fasten a strap round the throat, sufficiently tight to make arching of the neck uncomfortable, but not tight enough to interfere with respiration. Such straps usually require removal during feeding. In some types there is a metal 'gullet-piece', which has a recess into which the windpipe fits and which allows the device to be worn without danger. Another preventive device consists of a hollow, cylindrical perforated bit that prevents the animal from making its mouth airtight so long as it is worn.

Physical prevention of crib-biting does not reduce the behaviour overall and results in poor welfare (McGreevy *et al.*, 1995; McGreevy, 2004). A thick rubber or wooden bit that prevents the jaws from closing is sometimes used, but entails movement deprivation, acute discomfort and hence suffering. Surgery is sometimes performed: this is a highly specialized procedure involving a section of the throat muscles essential to the behaviour. It is harmful to the horse to use surgery to prevent an abnormality that is induced by inadequate management and housing conditions. The suffering caused is unnecessary.

Drinker-pressing

This stereotypy (see Fig. 6.5), *pressing an automatic drinker repeatedly without ingesting the water*, is shown by pregnant sows kept in stalls or tethers and provided with a nipple drinker. The drinker is one of the most interesting items in the animal's surroundings, and some individuals spend long periods manipulating it (see above). In a study by Broom and Potter (1984), sows spent from 2 to 74 min pressing their drinkers during 8 h of daylight. The median time spent was 10 min, which is considerably longer than necessary for drinking.

Further Reading: Mason, G.J. and Rushen, J. (eds) (2008) *Stereotypic Animal Behaviour: Fundamentals and Applications to Welfare*. CAB International, Wallingford, UK.

25 Abnormal Behaviour 2: Self-directed and Environment-directed

The general principles of what constitutes behavioural abnormality are discussed in the previous chapter. Domestic animals show some behaviour that is, for the most part, normal in its pattern but is abnormal in respect of the object to which it is directed or the extent to which it occurs. In this chapter, those behaviours, other than stereotyped behaviour (Chapter 24), that are directed towards some part of the animal's own anatomy or some inanimate feature of the animal's surroundings, are discussed. Chapter 26 concerns abnormal behaviour directed towards other animals.

Self-mutilation

Dogs may chew their limbs or other parts of the body because there is an injury or irritation in that area, or in the absence of any detectable localized injury. Such chewing can continue to the point of significant self-mutilation. Repeated rubbing of a part of the body against a solid object until injury results can also be shown by dogs.

Mink in fur-farm cages remove fur from the tail quite frequently and, much less frequently, chew the tail itself so that it is shortened (de Jonge and Carlstead, 1987; Mason, 1994).

Self-injury through vigorous body friction or flank-biting is a serious behavioural anomaly in horses. Animals affected with this disorder may bite at their sides or rub their neck crests damaging the coat, the mane, the skin and occasionally causing flesh wounds. It is a form of behaviour characterized by its intensity and is sometimes accompanied by vocalization. This disorder appears to occur more commonly in stallions than in mares or geldings.

In dogs, horses and other animals showing self-mutilation, the condition typically occurs in circumstances of confinement and isolation. Tranquillization can sometimes temporarily terminate an episode. The provision of a companion or a more extensive and complex living condition can be helpful in controlling the disorder. While affected animals are not usually found to have any pathological skin condition, parasitism or gastrointestinal clinical condition, these matters should be taken into account in the assessment of the case.

Occasional individuals of all domestic animal species show rubbing behaviour resulting in the development of a wound and, in certain circumstances, animals will peck, bite or kick at themselves to an extent that results in injury. This behaviour is often associated with some localized infection, parasitism or pain (see Chapter 28), but extreme self-mutilation, like that shown by monkeys which are confined and deprived, may sometimes occur. The removal of hair, wool or feathers is considered in the next section.

Licking, plucking and eating of own hair, wool or feathers

Excessive licking of the coat can be shown by dogs, cats and many other domestic animals. Many young calves housed in individual crates spend long periods of each day licking those parts of their bodies they can reach. This behaviour, which may be stereotyped in form, results in the ingestion of large quantities of hair, which aggregates into hairballs or bezoars in the rumen. Balls as large as 15 cm in diameter have been found in the rumens of calves (Groth, 1978), and these hairballs clog the rumen and openings to it. Digestive problems and even death can result from this. This excessive licking occurs mostly in early-weaned calves: all dairy calves would come into this category and more in individually housed than in group-housed calves. Ingestion of such material is occasionally shown by young lambs and by poultry, but this is more frequently a different kind of behaviour addressed to other individuals.

Some caged birds, such as parrots, pull out their own feathers, with or without eating them. Parrots in situations in which there is insufficient stimulation and those in which there can be disturbance that the parrot cannot control are more likely to show feather-plucking. The behaviour may result in large areas of the body being denuded of feathers. An improvement in the quality of the parrot's environment by provision of much more for the parrot to do, or by removal of the uncontrollable disturbance, is the way to improve the welfare of the parrot and prevent the feather-plucking. Efforts to reduce feather-plucking are seldom successful unless they involve removal of the cause of the poor welfare that is causing the problem. The welfare of most parrots in small cages is very poor so a move to an aviary large enough for them to fly easily is needed. Parrots can be given control of their environment by having an opportunity to conceal themselves from humans or other disturbance and by carrying out operant actions that modify aspects of their surroundings. Social contact benefits almost all parrots, as they are species with elaborate social structure and communications.

Sucking and eating solid objects

Recently weaned mammals will often suck and lick the walls and bars of their pens in a non-stereotyped way. Such behaviour is particularly frequent in young calves and piglets that are weaned at a much earlier age than would occur naturally. Calves separated from their mothers in the first few days after birth will nibble, chew and suck at any object in their environment, but they suck more on teat-shaped objects, especially artificial teats and the appendages of other calves (Waterhouse, 1978; Broom, 1982; van Putten and Elshof, 1982). Such behaviour will be discussed further in the next chapter.

The propensity for dogs and, to a lesser extent, cats to eat various solid materials is well known and often a source of concern to pet-owners. A dog may gnaw a destructible material such as wood or may pick up fibrous or other objects and then chew and eat these materials. This may indicate a nutrient deficiency, but some individuals show the behaviour when they are not deficient in any nutrient. Chewing furniture and human clothing can occur when dogs are left without companions, either canine or human. The object may have the smell of a companion. Providing companionship is often the best solution to the problems that lead to the abnormal behaviour.

Among older animals of all grazing species, chewing at or eating solid objects is occasionally recorded. The seeking out and eating of wood, cloth, old bones and other objects by cattle, sheep and horses is sometimes referred to as pica (A.F. Fraser, 2010). In some circumstances such behaviour is a result of phosphorus deficiency, and it is frequent among free-range animals on phosphorus-deficient land. This deficiency can be remedied by ingesting some materials mentioned above. Control of this abnormal behaviour is to supply phosphorus compounds to affected animals. It has been found that merely offering a phosphorus-rich supplement, such as bone meal, may not be sufficient to rectify a serious deficiency. Some deficient animals fail to ingest a sufficient quantity of the supplementary feed to attain satisfactory body levels of phosphorus. In such instances, phosphorus can be given by injectable solution.

Abnormal chewing and eating of wood, or lignophagia, is not uncommon in horses in confined quarters or paddocks. It is not restricted to stalled horses, since it can be observed in horses in outdoor enclosures. In pastures, wood-chewing may take the form of debarking tree trunks. Wood-chewing can lead to serious intestinal obstruction (Green and Tong, 1988). Although wood-chewing horses do not usually ingest most of the wood they chew, some splinters may be consumed and some can cause damage within the mouth. Excessive tooth wear also occurs. Affected animals may transmit the habit to associating horses. This can lead to the destruction of wooden fences, partitions and doors. Lack of roughage in the diet undoubtedly predisposes a horse to wood-chewing. Horses fed on concentrate diets with a low supply of roughage show the condition much more frequently than horses fed hay in abundance. A wood-chewer may chew 0.5 kg of wood per day from stall edgings. It has been found that ponies confined to stalls and fed a high-concentrate diet spend 10% of their time wood-chewing. When a high roughage diet was given, wood-chewing dropped to 2% of eating time. Access to extensive pasture is helpful in treating this condition, but the habit can persist and trees can be ringed by debarking.

Eating litter, earth or dung

Some animals are kept on bedding that is a potential food source for them so it is not surprising that pigs, cattle and horses will eat some of their straw bedding.

This is not abnormal behaviour unless carried to an extreme (A.F. Fraser, 2010). However, some litter used for animals is almost or completely non-nutritive and yet animals that are confined in a small space will eat their bedding, even after it has become soiled. Litter-eating, as seen in chicks and turkeys, occurs most commonly when they are reared on chaff or wood litter. The incidence of this behaviour is highest within flocks that are not provided with sufficient feed-trough space. The incidence is also higher in some breeds and strains of birds than others, and this indicates a genetic predisposition to the condition. The condition can sometimes be alleviated by supplying an abundance of grit to birds, and it may be that the abnormal appetite component of litter-eating represents a search for mineral material. Birds that practise litter-eating are liable to develop impaction of the gizzard or other alimentary regions and this, in turn, causes death in many cases. In the control of litter-eating in poultry, it is important to ensure that there is abundant feed-trough space so that birds in low positions within the peck order have an opportunity to find secure space somewhere at the feed trough. Without adequate trough space they are likely to be driven away and start to eat litter in compensation.

When horses eat litter they become increasingly indiscriminate with regard to the nature of the litter eaten and may eat mouldy, contaminated bedding. Colic is liable to occur in such cases, with severe illness and death as possible consequences. In the horse, several causes of litter-eating are recognized. Imbalanced rations, feeding at the wrong time of day and heavy worm burdens have all been found to contribute to this condition. Horses kept outdoors graze most of the day, and eating is clearly their major occupation. Within stables, this occupation is curtailed and grain or compounded food is often consumed quickly. If such food is not followed up with the provision of clean hay, for ingestive occupation as much as balanced nutrition, horses are likely to seek other available materials to consume. Control of this abnormal behaviour requires close attention to all aspects of the diet. Appraisal of feed is necessary to ensure adequate quantity and variety. Supplementary feed should be provided in the form of salt licks and mixtures rich in minerals and vitamins. Fresh feed, such as grass, greens or carrots, should be offered regularly. Feeding times should be observed on a precise timetable with late night or early morning feeding being included in the schedule. Horses found to have a worm burden should receive effective, appropriate anthelminthic treatment.

Dogs, horses and cattle sometimes eat soil or sand. Animals that do this are susceptible to alimentary dysfunctions. The condition has been termed geophagia, and has been thought to be the result of mineral-deficient diets. Phosphorus and iron deficiencies are known, in some cases, to be responsible for soil-eating, but other affected animals do not appear to have a nutritional deficiency. Close confinement and lack of exercise appear to be the most common causal circumstances. Excessive eating of sand can result in sand impaction of the caecum and colon in the horse. Sand impactions have also been encountered in cattle in the abomasal region as a consequence of this habit.

Coprophagia, the eating of faeces, is normal behaviour in rabbits. A double passage through the gut, when an individual eats its own faeces, leads to more efficient digestion in rabbits and can also be functional in other species. Coprophagia is not uncommon in in dogs, pigs and horses. The faeces of other species may provide useful nutrients, so they are part of the food of some pigs. Dogs may be able to eat the faeces of other species without any harm resulting but most dog owners attempt to prevent the behaviour. The habit is so common in foals under conventional management that it is generally considered as normal behaviour in these animals. It may help to establish an adequate gut flora initially but otherwise would seem to be maladaptive. Coprophagia in the adult horse is anomalous and is induced by particular circumstances such as chronic enclosure in loose boxes or a change in management from regular exercise to no exercise.

Overeating

Overeating, or hyperphagia, and rapid eating are habits observed in dogs, horses and, occasionally, in cattle. In the course of bolting their food some of these animals may choke. Since the food consumed is not fully masticated, digestive disorders can occur. When grain is consumed in excessive quantity in cattle it can be fatal. One way of controlling hyperphagia is to make it difficult for the animal to consume the food rapidly. Some competitive feeding situations make rapid eating the most successful strategy for the animal to obtain sufficient food so these feeding methods should be changed.

Polydipsia

As mentioned in the previous chapter, some confined sows spend long periods showing stereotyped drinker-pressing, but this is not polydipsia, or excessive drinking, because little of the water is ingested. Polydipsia nervosa is seen in some horses that are isolated and confined in stalls with water supplied *ad libitum*. Some horses will consume about 140 l/day, or about three to four times the normal quantity. This excessive consumption can be spread over a period or may be concentrated within a relatively short time of 2–3 h. Excessive drinking is also encountered in other species that are subject to close confinement and, in these cases too, the water consumption by the individual usually represents a two- to fourfold increase in the normal water intake. It has been observed in sheep subject to chronic, close confinement in stalls and in metabolism crates.

While it is difficult to be precise about the adverse effects of this anomaly on the animal, the constant flushing of ingesta probably reduces the nutritional value of the ration. In some instances, polydipsia has been noted in sows with the 'thin sow' syndrome. It could be that the sudden intake of an abnormally large quantity of water may allow a segment of the alimentary canal to become heavily loaded and liable to twist. This could explain cases of gut-twist in the horse. Polydipsia is most common in closely confined animals given little exercise, and the habit can be controlled and broken by providing better housing and regular exercise.

Further Reading: Fraser, A.F. (2010) *The Behaviour and Welfare of the Horse*, 2nd edn. CAB International, Wallingford, UK.

26 Abnormal Behaviour 3: Addressed to Another Individual

Introduction

Animals that are kept with other members of their own species, or that have an opportunity to interact socially with members of other species, sometimes direct abnormal behaviour to those other animals. Much of this behaviour involves behaviour patterns that are in the normal repertoire but which are inappropriately directed. The behaviours described in this chapter are grouped according to the apparent motivational state of the animal. Animals sometimes treat other animals, or parts of their anatomy, as if they were objects to be investigated, obtained or eaten, just as were the objects described in the last chapter. Other types of abnormal behaviour are directed inappropriately towards other animals as if they were a sexual partner, a mother or a rival. The extreme artificiality of many farm animal management and housing systems has resulted in much more abnormal behaviour and more kinds of such behaviour being shown by farm animals than by companion animals.

Animals Treated as Objects

The behaviour preceding activities described in this section is often indistinguishable from that which precedes activities reported in Chapter 20. The animal approaches another individual, or more often a particular part of that individual, as if it were exploring its environment or looking for food. In circumstances where the animal approached is unable to move away because of lack of space or the close proximity of other animals, an action may be completed that is damaging to that animal. If the animal that is showing the behaviour can readily obtain the resource that it seeks, such behaviour is less likely; for example, provision of straw and other enrichment for pigs results in less behaviour directed at pen-mates (Beattie *et al.*, 2000).

Egg-eating

Egg-eating is a habit found in chickens kept in pens and cages. It appears to occur more on wire mesh floors than among flocks on litter. The behaviour begins with a bird pecking at an egg until it is broken. The contents of the egg are then partially ingested. When a bird acquires this habit it is likely to increase the practice, and other birds may also acquire the habit through mimicry. In some cases, significant amounts of eggshell are eaten and this leads to the suspicion that the diet of affected birds may be deficient in grit.

Control of this condition involves the elimination of affected birds, but this may be difficult in a large flock as the perpetrators are difficult to identify. It is sometimes found possible to inject strong food dye into the substance of an egg and have this egg left lying on the ground. An egg-eating bird choosing this egg will be marked by coloration about the head. It is advisable to provide a supply of grit or oyster shell chips in dealing with problems of this nature. It is important to lay out the grit in long troughs so that all birds can have occasional access to it. In cages, the problem is reduced if eggs can roll away out of reach of the birds. The provision of nest boxes in larger cages reduces egg-eating, as floor-level eggs are eaten most frequently.

Wool-pulling and wool-eating

Wool-pulling is a form of abnormal behaviour that occurs in sheep within restrictive enclosures and indoor management systems. It is clear that crowding within pens is a contributing factor, but it is also believed that a deficiency of roughage in the diet may contribute to it. In addition to pulling wool from other members of a group, the individual sheep also ingests some of the wool.

Wool-pulling in adult sheep is usually practised by one individual within the group. In time, the anomaly may be induced in others in the group. The sheep receiving most wool-pulling are usually those that are lowest in the social hierarchy within a group. The condition is therefore related to social dominance. The wool-pulling animal is usually identifiable through having an intact fleece. When wool-pulling first begins, affected animals are observed to pull with their mouths on the strands of wool on the backs of others. As afflicted sheep receive more attention from the wool-puller, the long wool becomes denuded from the back area. Over this region the fleece may be reduced to wool fibres of approximately 3 cm, while fleece of normal length is still borne elsewhere on the body. As the anomaly intensifies, afflicted animals can lose wool so extensively over the entire body that they begin to appear semi-naked, as a result of pink skin showing through the sparse remaining wool fibres.

Since this condition is clearly associated with overcrowding within indoor pens, control of the condition is possible through reduction in pen densities. If pens of 20 m^2 can contain ten mature sheep, wool-pulling is likely. A reduction to 50% of this density is effective in controlling wool-pulling. At this lower level of population concentration the anomaly can be eliminated, especially if there is also the provision of a regular supply of quality roughage. Hay is ideal, but straw can also be useful for this purpose. While nutritional deficiencies have been suspected as a cause of wool-pulling no link has been demonstrated, and it would appear that any nutritional need associated with this anomaly relates to an inadequacy of structured feed rather than to any specific nutrient factor. Control can also be effected by releasing animals into outdoor, extensive husbandry conditions for long periods.

Young animals sometimes remove parts of the coats of their mothers while in close contact with them, by licking and sucking on parts of the maternal body other than the mammary gland. Young lambs may begin wool-eating as early as 1 or 2 weeks of age. The lamb sucks, chews and ingests the wool from parts of its mother's fleece on such regions as stomach, udder and tail. The accumulation of ingested wool in the lamb's stomach (abomasum) leads to the formation of compact fibrous balls (bezoars). Lambs with such wool balls may suffer severe colic attacks resembling fits. Affected lambs harbouring bezoars become anaemic, unthrifty and progressively lose bodily condition. Affected lambs stand for long periods in a stationary posture displaying distended stomachs with their backs arched. Complete alimentary obstruction, for example, in the region of the pylorus or small intestine, can cause death.

Feather-pecking, body-pecking and eating pecked matter

Feather-pecking is a form of anomalous behaviour common in poultry. Under conditions of intensive management it can occur in all ages and many species, including chicks, adult hens, turkeys, ducks, quail, partridge and pheasants. The normal exploration and food investigation behaviour of such birds involves pecking, so it is not surprising that in a barren environment they investigate the feathers of other birds in this way. Hens crowded together on wire floors have few objects at which to peck. In these conditions birds peck on the backs, tail, ventral region and cloaca of associate birds. Mutual pecking, in which chicks in close parallel and opposite positions peck at each other, is common. In other cases, several birds may be involved so that chains of peckers may form. Young birds have been observed to show no resistance or other response when their feathers are pecked, but adult birds try to avoid being pecked.

Feather-pecking is especially prevalent in intensive husbandry systems and is seen more often in some breeds, such as light hybrids. Environmental factors considered to initiate this behaviour include poor ventilation, high temperatures, low humidity, excessive population density and excess illumination. Feathers pecked from other birds may sometimes be eaten. Feathers may be picked out from preferred sites of other birds, such as the tail and pinions, which are the largest feathers in the body. In smaller chicks, feathers are pecked mainly from the back and ventral region of the body.

Birds that feather-peck may subsequently start to peck and remove blood, skin and flesh from other birds (Brantas, 1975; Blokhuis and Arkes, 1984). None of this behaviour involves aggression but is motivated by investigation, need for dust-bathing materials or desire to eat. The causation of feather-pecking and body-pecking with consequent cannibalism is described in Chapter 32. Body-pecking and consequent cannibalism can begin when wounds arise when blood-filled new quills from the wings or tail are pecked and start to bleed (Sambraus, 1985). The outlet of the uropygial gland, which protrudes slightly, and the protruded cloaca after egg-laying elicit body-pecking. The most severe effects often ensue after the cloaca has been pecked. Wounds in the cloacal region can rapidly become severe, and the intestines can extrude through a cloacal wound. These are likely to be the subject of more pecking and, in due course, be pulled out and ingested. Mortality is therefore frequent once a wound has been produced.

Feather-pecking does not lead inevitably to body-pecking, and many birds show feather-pecking but not body-pecking. Neither feather-pecking nor body-pecking is preceded by threatening behaviour, and both are preceded by body orientation and movements that are typical of investigatory behaviour. The bird pecked usually has little opportunity to escape, but the failure of the pecker to respond differentially to an inanimate object and to another bird that responds, even by submission, is clearly abnormal. Pecked birds too are abnormal in their behaviour, for they cease to show much escape behaviour when pecked often, presumably because they have learned that previous escape attempts have proved fruitless. One bird usually initiates the body-pecking but other birds are likely to join in, so that pecked birds may be subjected to a barrage of pecks. Body-pecking is shown by domestic fowl, turkeys, pheasants, quail and ducks.

Other parts of the body are also subjected to pecking. Head-pecking is an aggressive behaviour and occurs in older birds confined together in cages, while pecking at toes and back is sometimes widespread among younger birds. In toe-pecking, the active bird pecks at the toes of associate birds and, on rare occasions, their own. While toe-pecking may not lead to significant wounds in the case of young chicks, wounds on the toes result from this behaviour in adult hens. Following injury there is bleeding, and portions are then picked off the wound. The resultant open wound is liable to infection and further haemorrhage. Birds so afflicted show depression behaviour, retreating to a corner of the pen, refusing to eat and losing weight. In the absence of appropriate husbandry intervention, death may occur. Head- and tail-pecking are often observed in the young in overcrowded conditions. Injury and haemorrhage of the caruncle region of the head is not uncommon among male turkeys after fighting. Toe-pecking may occur, especially in deep-litter pens. It is believed that hybrid strains of birds show this behaviour more often than others.

The control of feather- and body-pecking is most commonly effected by beak-trimming, also called debeaking. Beak-trimming involves the removal of the anterior part of the upper mandible. By removal of this portion of the beak, pecking becomes inefficient but, as noted in Chapter 32, this is a painful procedure for birds. Beak-trimming does not eliminate aggressive pecking entirely, or prevent the development of the peck order, but treated birds are less able to pull feathers. Another method used to control the condition is to limit the vision of birds. This can be done by darkening poultry pens and changing the light to a red hue through the use of infrared lamps or painting window panes red. The vision of each individual bird can be restricted by fixing aluminium rings to the upper beak or applying 'poly-peepers', although the use of such devices is banned in some countries. None of these procedures can be carried out without some adverse effects on bird welfare and are illegal in some countries. Changes in the bird's environment that minimize the likelihood of this behaviour are preferable.

Anal massage

Young pigs rub their noses on other pigs and, while some of this behaviour is similar to teat-searching and udder-massage (see later section on belly-nosing), other behaviour appears to be of a more general investigatory nature. The anomalous behaviour of anal massage by snout-rubbing and ingestion of faeces seen in pigs occurs typically among growing pigs kept in crowded conditions. It is more noticeable where tail-docking at an early age is used for the control of tail-biting. Within dense groups, affected animals move from one animal to another, nosing the anal regions with upward massaging motions of the snout.

Although some animals approached in this fashion avoid the contact, others do not. Anal massage is carried out with considerable pressure, so that the snout of the active animal is pushed deeply into the perineum of the associated animal. Such an animal frequently responds to such pressure by reflex defecation. When faeces are expelled as a result of this activity, faecal ingestion may occur. This coprophagia may be carried out by the snout-rubbing animal alone or it may be joined in the activity by other pigs in the group (Sambraus, 1979). Some pigs do not avoid snout-rubbing by showing any resistance, and animals that tolerate such attention often acquire swollen wounds of the anus or adjacent perineal area. Such afflicted pigs become weak, have difficulty in standing, and lose appetite and physical condition. Badly afflicted animals may die.

Anomalous snout-rubbing may be reduced by supplying pigs in pens with objects to occupy them by chewing and rooting. Both this behaviour and anal massage are signs that, if the strong motivation of pigs for rooting and manipulation behaviour cannot be satisfied, perhaps because the animals are housed on concrete, welfare is always poor and abnormal behaviour is a consequence. The control of anal massage and associated coprophagia can be attempted by easing crowded conditions and providing rooting and manipulation opportunities.

Tail-biting

Of all the abnormal behaviours of farm animals, tail-biting in pigs has attracted most attention due to the problem it has created in the pig industries. The behaviour is seen among growing pigs grouped in pens, but it is sporadic in its occurrence. Tail-biting was recognized as a problem in pig-rearing for many years, but was not considered as a serious matter until the modern pig industry became established following World War II (Dougherty, 1976; EFSA, 2007b). The behaviour first appears with a pig taking the tail of another crossways into its mouth and chewing on it lightly. The animal receiving this attention usually tolerates it. In due course, the tail-biting attention becomes more severe, with resultant wounds on the tail and haemorrhage. It is believed that haemorrhage encourages more active tail-biting, and other pigs in the group begin to chew on the damaged tail. The injured tail becomes progressively eaten away to its root. At this point, tail-biting pigs may begin to bite the afflicted animal on other parts of the body such as the ears, the vulva and parts of the limbs. All of this behaviour is associated with much unrest in the pen-mates.

A pig injured as the result of excessive biting becomes submissive and then depressed in its behaviour, reacting only slightly to being bitten. Wounds may become contaminated with infection, resulting in abscessation of the hindquarters and the posterior segment of the spinal column. Secondary infection may occur in the lungs, kidneys, joints and other parts as a result of pyaemia.

Conditions that predispose to tail-biting include: breed type such as Landrace; dense grouping of rapidly growing pigs; insufficient trough space; insufficient drinking facilities; and adverse environmental features including high levels of noise, noxious gas, humidity and temperature. In general, an impoverished environment with little opportunity for the satisfaction of some needs – such as the need to manipulate material with the mouth or to dig in earth with the nose – increases the risk of tail-biting. Comparison of systems showed from two to ten times more tail-biting in barren systems (Hunter *et al.*, 2001; Moinard *et al.*, 2003).

Oral activity is greater in the pig than in other farm species, and pigs try to explore items in the environment with the snout or mouth by rooting and chewing. It is noticed that under extensive husbandry systems pigs engage in considerable mouthing activities, including such acts as picking up and carrying sticks in the mouth and chewing up material for bedding. Much activity within pig groups occurs during the morning and when temperatures are high. The cannibalistic behaviour of tail-biting is independent of a social hierarchy and it is frequently the smaller animals in a group that develop the habit (van Putten, 1978).

In the control of tail-biting, amputation of the distal half of the tail (tail-docking) has become a widespread practice. It has been found that it is not necessary to remove the entire tail to prevent the anomaly developing, but that the docked tail is sufficiently sensitive that pigs react effectively when a tail-biting attempt is made on them. When the tail is cut, a neuroma may form (Simonsen *et al.*, 1991; Done *et al.*, 2003). Neuromas usually result in frequent pain, apparently emanating from the severed or damaged appendage or other tissue, and increased sensitivity in the area in which the neuroma is situated. Hence, it is likely that pigs with docked tails have frequent pain and that they respond rapidly to tail-biting attempts because of enhanced pain when the tail stub is manipulated.

Attempts to control tail-biting by removal of tail-biters from a group or changes in atmospheric factors have little effect on tail-biting. Reduction in stocking density can reduce tail-biting, but a combination of lower stocking density and provision for needs to root and manipulate materials such as straw are necessary to obviate tail-biting. Animals having the opportunity to root in earth seldom show tail-biting, so the provision of earth may help to reduce the incidence of this and other abnormal behaviours in pigs.

Animals Treated as Sexual Partners

Many farm animals are kept in single-sex groups and seldom or never encounter a member of the opposite sex. The same is true for some dogs and cats. Hence, it is likely that sexual behaviour will often be directed to individuals of the same sex. Such behaviour may be considered abnormal in that it cannot result in offspring production, but the normal development of sexual behaviour often involves parts of adult sexual behaviour being directed towards various individuals of either sex. Therefore, some homosexual acts or sexual behaviour towards other species are to be expected among young animals, and may also have a function when shown by older animals.

Homosexual interactions are very frequent among groups of cows. Cows and heifers in oestrus are mounted by other cows, and this behaviour is used as a sign of oestrus by stockmen (see Chapter 17). Occasionally, such mounting can lead to injury but, partly because of its use to the stockman, it is not usually considered to be a problem. Hall (1989) found that female–female mounting was not shown by Chillingham wild cattle in the UK except by animals found to be partially masculinized. Hence, this behaviour must be considered to be abnormal, especially the high frequencies of mounting seen in the absence of a bull.

Sexual mounting of males by other males occurs frequently when young or mature male animals are held together in monosexual groups (Stephens, 1974). When this is a continuing arrangement as, for example, in the husbandry of large numbers of young bulls or rams being raised for breeding, it is found that the sexual orientation of these animals is frequently directed to their own kind. When exposed subsequently to female stock in conventional breeding arrangements, some of these animals are incapable of altering their sexual orientation. Homosexual rearing may lead to impotence in stud animals, which will persist for a variable period in rams, bulls, boars and male goats, although it is not usually permanent (Price and Smith, 1984).

In beef cattle production, it has been known for many years that some bullocks, or steers, will stand to be mounted by other steers under conventional pasturage conditions. In extensive grazing the condition does not represent a serious problem, and these animals, or 'bullers', do not suffer from excessive mounting. Under commercial feedlot conditions the incidence of bulling increases greatly, and the result is poor welfare in excessively mounted animals, reduced growth rates, increased frequency of injury and serious economic losses. It is recognized that these bullers characteristically stand to be mounted, but will also engage in mounting other bullers. In addition, some steers, despite the fact that they are castrated, are active 'riders' on bullers. In cattle feedlots some riders repeatedly mount a few specific bullers. This may continue until the bullers become exhausted, injured or collapse. Death may also occur following injury to bullers. It has been estimated that the economic loss in feedlot systems in North America due to bulling is second in importance only to respiratory disease.

Although the buller syndrome can occur in feedlots where synthetic hormones are not used, the incidence of the condition rises in such feedlots following hormonal administration, such as injection with progesterone and oestradiol or oral administration of DES (stilboestrol) (Schake *et al.*, 1979). A common denominator in the buller steer syndrome is that both bullers and riders are castrated with a growth-promoting female sex hormone compound. Physiologically, buller steers have been found to have more creatinine, 17-hydroxycorticosterone and oestrogen in their urine and blood than normal steers.

Other factors contribute to the aetiology of the syndrome and, increasingly, it has been recognized that a further common denominator in affected groups is a high population of animals within a group. Both crowding and total population size within a feedlot appear to be important. It would appear that crowding stress, together with a mixture of other husbandry factors such as hormone administration, vaccination and dipping, constitute the causal circumstances. Repeated introduction of new steers into feedlots increases the buller incidence. The vast majority of steers in cattle pens, however, avoid potentially harmful social interactions in a manner typical of normal bovine social behaviour.

More aggression occurs when the amount of space for a given quantity of animals, in number or bulk, decreases. In other words, as animal volume per unit of group space increases, so the quality of social exchange declines. There is no doubt that the buller steer syndrome is a welfare problem as well as an economic one. The behaviour is part aggressive and part sexual. Control of this condition really requires reduction of population pressure and withdrawal of artificial hormonal stimulation. Animals housed indoors are sometimes deterred from mounting by bars placed over the pen that physically prevent mounting or by electric wires over the pen. Both of these measures fail to deal with the root of the problem, and the electric wires may lead to severe effects on some individuals.

Animals Treated as Mother

Young mammals that have been separated from their mother at an earlier time than the normal weaning age often show teat-seeking and -sucking behaviour that is directed towards inanimate objects (see Chapter 24) or pen-mates. Sometimes this behaviour, which is called belly-nosing and inter-sucking in piglets and calves, persists into adulthood.

Belly-nosing

Belly-nosing is a behaviour shown by piglets. It is an up-and-down movement of the snout and the top of the nose on the belly of other pigs, on the soft tissue between their hind legs and between their forelegs. This behaviour was described by van Putten and Dammers (1976) and Schmidt (1982), and it is similar in form to the massaging movements directed by piglets towards the udder of the sow. Belly-nosing is not shown before weaning, but is shown by piglets that have been weaned much earlier than the normal weaning age (D. Fraser, 1978b).

Since many piglets are weaned at 3–5 weeks of age, which is 2–3 months before weaning would occur in the absence of human influence, the behaviour is very widespread. Some piglets kept in flat-deck cages are pursued and belly-nosed for long periods and their nipples, umbilicus, penis or scrotum may become inflamed. In some cases, male piglets urinate when belly-nosed and the bellynoser ingests the urine. High levels of belly-nosing were negatively correlated with weight gain in the study by Fraser (1978b). Later weaning is associated with less belly-nosing. The incidence of the behaviour is also reduced by the provision of straw, which the piglets can manipulate (Schouten, 1986). In the study by Schouten, belly-nosing was seen most frequently during the transition from an active to a resting phase. Piglets in a bare crate spent several minutes in this transition and often showed belly-nosing at this time. Those provided with straw usually lay down simultaneously and started chewing straw. As a consequence, the duration of belly-nosing was significantly greater in the piglets without straw.

Inter-sucking by calves

Calves separated from their mothers suck and lick at their own bodies, at objects in their pens (see Chapter 25) and at parts of the bodies of other calves. They commonly suck on the navel, prepuce, scrotum, udder and ears of other animals. Sucking the scrotum is very common among calves. The testes are pushed up by the nose of the sucking calf, which then sucks on the empty scrotal sac. The posture and position of the sucking animals is that of the naturally suckling calf and includes the pushing movements the calf normally directs at the cow's udder. The calves that are sucked usually show a passive response. Such calves may suck others in their turn, so several animals may be involved. If the coat of the other calf is sucked significant quantities of hair may be ingested, leading to formation of hairballs (see Chapter 24). If the penis is sucked, urine is often produced and may be drunk by the sucking calf. This can lead to liver disorders and to reduced nutrient intake. Another adverse consequence of inter-sucking is that the part of the calf sucked may become inflamed, damaged and infected (Kiley-Worthington, 1977).

Experimental studies have shown that injurious inter-sucking is more common in bucket-fed calves than in those fed from the mother or an artificial teat (Czako, 1967). The presence of the teat is important, but the time taken to feed is also a factor. Calves suckling their mothers spend 60 min/day doing so, but bucket-fed calves spend about 6 min/day drinking. Therefore, in the control of this anomaly, the best results are obtained by providing feeding conditions that resemble those of normal ingestive behaviour in young animals.

It is found that feeding calves with milk through automatic nursers with teats, whose aperture is such that the sucking time of each feed period is sufficiently prolonged, will reduce the likelihood that other objects or calves will be sucked. Sucking periods lasting approximately 30 min appear to eliminate inter-sucking. Another remedy is tying up calves for 1 h following bucket-feeding, for this allows time for the desire to suck to diminish somewhat in order that inter-sucking is not shown so often after that time period. The supply of supplemental roughage, such as 150 g of straw, can also result in a reduction in inter-sucking. Urine-drinking can be reduced by the plentiful supply of water from a normal drinker for the calves. Mees and Metz (pers. comm.) provided calves with water trickling down a wall or pipe, and the animals licked that water rather than showing urine-drinking.

Inter-sucking or milk-sucking by adult animals

This behaviour involves a cow or bull sucking milk from the udder of a cow. Among adult milking cattle, sucking the udders of other members of the dairy herd is seen periodically. Such inter-sucking by adult cattle involves the withdrawal of milk from a lactating animal, and is a behavioural abnormality of occasional occurrence under a variety of circumstances of husbandry. Among 'suckler' herds of cows with nursing calves, 'sneak sucking' may occur by alien calves or by an adult animal such as a bull. On rare occasions, milking cows are discovered with the habit of sucking milk from their own udders. Cattle that suck milk from herd-mates may choose the same lactating animal, and this leads to a paired arrangement.

The loss of milk from inter-sucking can become significant, and frequent sucking by an adult can lead to teat damage, pathological changes and deformation of the udder. Within some very large dairy herds this behavioural anomaly can represent a serious problem, with as many as 10% of lactating cattle in the herd becoming involved. The anomaly is worsened by the influence of social facilitation and by the large numbers in modern herds.

In contrast with the condition in the young calf, inter-sucking behaviour in the adult animal is more common in open husbandry systems. Wood *et al.* (1967) report that those calves showing inter-sucking may go on to inter-suck milk as adults, and other authors have suggested that various forms of experience with or without teats in calfhood may increase or decrease inter-sucking as an adult. The behaviour is sufficiently unpredictable, however, as to render it a difficult subject for experimentation. Waterhouse (1978, 1979) reared calves individually with a non-nutritive teat on the wall of the pen for 3 months, but neither those calves nor any without such a teat showed inter-sucking as adults.

Attempts to control anomalous milk-sucking in the past have taken the form of applying devices carrying pointed prongs to the face and nose region of the sucking animal. The attempt is to ensure that the animal seeking to suck will cause an avoidance reaction in any animal approached. Unfortunately, some of these devices can hinder the affected animal's natural feeding. Furthermore, if the affected animal is persistent, it can inflict wounds on other animals. An electrical device secured to the forehead and giving an electrical shock to the wearer when the circuit is closed by head pressure is reported to give good results. Since the shock is received by the sucking animal, the method is more appropriate than methods in which the aversive stimulation is directed at the receiving animal.

Animals Treated as Rivals

Some aggressive behaviour shown towards members of the same or other species is normal. It serves a function in defence of the individual, defence of the young or establishment of ownership or social position. Problems arise in farm animal husbandry where a situation that is conducive to aggressive behaviour is created and the aggression is particularly severe in its effects or too frequent. Aggressive behaviour, which leads to welfare problems, is discussed in Chapters 29 to 39. Some of this behaviour would be considered to be abnormal because of its intensity or frequency. Threats or attacks directed towards man are a cause for concern, but many of these are not abnormal behaviour. A cow with a calf or a bull that attacks a stockman is not behaving abnormally. The process of domestication has not led to the elimination of all defensive or competitive behaviour directed towards man, although it has reduced it considerably. Certain forms of husbandry make such behaviour much more likely to occur than do others.

Threat and attack directed towards humans by dogs are a matter of widespread concern. Whether or not the dog is a pet deliberately trained to attack people or a guard dog, police dog, army dog or pet in which such behaviour has been encouraged, each person needs to be able to recognize the prelude to an attack (see Fig. 26.1). This will be different according to the situation and training of the dog. Most of these attacks are defensive responses. A dog defending territory, a resource such as food, or another individual – whether this is a dependent dog or a human – will usually show a threat display before attacking. This will often consist of standing erect, raising the hair round the neck, raising the ears, pulling back the cheek area to expose the teeth and growling.

If this threat display successfully repels the intruder or challenger, there may be no attack. An attack may involve preliminary movements towards the human subject with continued growling and other display. However, the attack may occur suddenly. If the human is approaching the dog, for example to take food or something else from the dog, the attack may occur only when the human is at close range. A dog trained to attack, or for any other reason tending to treat the human subject as prey, will not threaten at all, but will run at the human and attempt to bite. If the sole motivation is predatory – for example, in the case of dogs attacking a small child – the face and neck will often be bitten. Most dogs, even those that will attack humans, have received training that discourages them from biting in the most vulnerable areas, so an arm or leg is usually bitten.

Cats attack humans in ways that cause less severe injuries, but attacks or defensive responses are not uncommon. The threat shown has similarities to that of dogs in that raising of fur, exposure of teeth and vocalizations are often involved. The sound is higher pitched than a dog growl, but serves the purpose of threatening and hence reducing the likelihood of continued approach or other adverse action by the human subject. The cat is more likely to use its claws than to bite but, in more extreme circumstances, may do both.

Fig. 26.1. This dog is attempting to initiate aggression towards another individual (drawing A.F. Fraser).

The threat display of a bull is a sequence of actions (Box 26.1) that all of those who might encounter a bull should know.

The horned ungulates defend or attack by striking aggressively with the head. In polled animals the blow delivered is in the form of a knock, while in the horned animals the effect may be one of goring, i.e. horning. Some individuals may show heading behaviour. These animals use their heads in threat displays where the head is worked into the ground and the earth is loosened. Butting by bulls and male goats is a very well-known behaviour problem. It would appear that the habit occurs most often in animals confined by themselves and that have previously experienced some degree of socialization through close human contact. In some instances the behaviour is directed at physical structures such as doors or gates.

In a horse threatening a human there is muscular tension, extending the head, laying back the ears and exophthalmos, which causes the sclera of the eyeball to become visible. This latter feature has been termed by horsemen as 'showing the white of the eye'. In other animals such as boars, tenseness of posture through muscular rigidity is a prominent feature of threatening.

Box 26.1. Behaviour of a bull that may attack.

- The animal turns towards the source of stimulus, with the long axis of the body acutely angled.
- The animal's head and neck are slightly deviated, more obliquely from this line. In this position, the animal presents the lateral aspects of the head, neck and shoulder closer to the stimulus than the hind parts, which are turned away slightly.
- The line of the back, from the withers to the tailhead, drops as a result of the hind legs being drawn slightly forward below the abdomen.
- There is fixation and protrusion of the orbit.
- Hair on the dorsal aspect of the neck is erected.
- Muscular rigidity is demonstrated by the tenseness of the posture.
- Deep, monotonal vocalizations or snorts may accompany the display.

One version of a single hind limb kick is the 'cow kick', so named because it is the common natural method of kicking in cattle. In the cow kick, one hind limb is briefly projected forward, outward and backwards. Another hind limb kick involves the extended projection of the limb forward, sideways and backwards at full stretch. Such kicks often are delivered with precise aim.

Biting is exhibited occasionally by horses. Stallions are particularly prone to this, but young horses and chronically enclosed horses may also exhibit the behaviour (Houpt, 1981). The biting is usually in the form of snapping and nipping with the incisor teeth on the body of any person coming within reach or approaching the animal. This biting behaviour is aggressively directed in some instances to other horses. The typical biter exhibits the behaviour with the warning signals of ears laid flat, lips retracted, teeth bared and tail often switching. Biting attempts are usually very sudden.

Rearing and striking with the forefeet is a dangerous habit of some horses, more commonly seen in stallions and in light horses than in others. Striking out with one forelimb may be done without rearing. Animals having this tendency or disposition in their behaviour may exhibit it when first approached or may show it after they have been held by the head in restraint for a period.

In kicking with the hind feet together, the feet can be used to strike out explosively to reach another animal or person within 1.5–2.0 m. Such extended hind limb kicking is a feature of mule behaviour, and this type of kick is sometimes termed a 'mule kick'. This is a natural method of self-defence among horses, donkeys and mules but some individuals may use it habitually and aggressively. The head is lowered, the body is lifted behind the withers and both hind limbs are vigorously extended backwards. Kicking with one hind foot is again a natural defensive action in ungulates when the individual space of the animal, in the region of its posterior pole, is invaded. It is usually delivered sharply in one downwards or backwards direction, without full extension of the limb.

The control of all of these aggressive actions is not easy. In some cases, avoidance of the specific eliciting circumstances is the answer. In others training procedures are necessary, and many books have been written about this. For most animals early experience is of importance, and housing conditions can have a considerable effect. Dogs trained or allowed to attack humans are potentially dangerous, as are bulls and stallions that have been confined for a long period. Long-term isolation of dairy bulls during rearing and during adulthood often results in animals being dangerous to man and to other cattle, whereas bulls kept in groups are much less aggressive.

Further Reading: EFSA (2007a) The risks associated with tail biting in pigs and possible means to reduce the need for tail docking considering the different housing and husbandry systems. *The EFSA Journal* 611, 1–13.

Abnormal Behaviour 4: Failure of Function

The conditions imposed on domestic animals lead to some abnormalities of sexual behaviour, parental behaviour and basic body movements, which are inadequacies of function rather than stereotypies or active misdirection of behaviour. Individual variation within a species also accounts for some anomalous behaviour. In this chapter such abnormalities are described, factors affecting their occurrence are discussed and possible control measures are considered.

Inadequacies of Sexual Functioning

Failure to reproduce or delay in reproduction is an important economic problem and such failure often involves behavioural inadequacy. It is an occasional problem in dogs and cats and a major problem in several farm animal species. Such abnormalities in behaviour can be most conveniently dealt with according to the phase of the reproductive cycle that is affected. Oestrus and libido are, of course, the important prerequisites for reproduction and it is in these areas that some of the principal anomalies of reproductive behaviour occur (A.F. Fraser, 2010).

Silent heat

Silent heat, which has been discussed in Chapter 17, is a term that emphasizes the fact that normal oestrus is principally a behavioural phenomenon. Among dairy cattle an overall incidence of 20–30% has been recorded by various observers in a variety of countries, and it occurs more among high-producing cows. The incidence in cattle is highest among heifers low in position in the social hierarchy of the herd. Silent heat is not uncommon in mares, ewes and sows.

The anomaly relates to absent or weak manifestation of overt behavioural oestrus in animals that have all the other, physical, characteristics of the phenomenon, including uterine turgidity and congestion, with follicular ripening leading to ovulation. The physical characteristics in this condition are so complete that these animals are capable of being fertilized if bred artificially. Mares with this condition, for example, are found to have a normal level of fertility if forcefully bred. Similarly in cattle, if the condition can be satisfactorily identified, artificial insemination at an appropriate time can be associated with normal levels of fertility. Control of this condition must relate to improved management of breeding animals to minimize the amount of metabolic and social stress (see Chapter 29).

Male impotence

During the breeding of farm animals under conditions where male and female are together for a short period only, it is sometimes found that the male animal does not respond to the breeding female. Among cattle there sometimes occurs a condition of 'somnolent impotence' in bulls having an abnormally protracted reaction time at breeding. In this condition the bull lays its chin on the hindquarters of the cow and directs little attention to her. The animal maintains an inactive stance with which there is an appearance of somnolence, created by the eyes being partially or periodically closed much of the time. This condition has been recognized as a breeding problem of considerable significance. It appears to occur more in beef breeds than in others and, to date, has been most often recognized in the Hereford and Aberdeen Angus breeds. In these breeds the condition is refractory to treatment and, in such circumstances, becomes a state of behavioural sterility that usually leads to the animal being culled.

Coital disorientation

During the controlled breeding of the larger farm animals, it is standard practice to have the female secured and the male led to her on a lead rope. This action initiates reaction time, which is the time lapsing between initial contact and mounting. Immediately following contact it is normal for the male animal to align himself in the same long axis as the female. This active alignment is normally carried out positively and briskly. Once assumed, alignment is generally fully maintained up to the time of mounting. In a number of cases male animals show positional disorientation such that the animal's long axis is markedly deviated from the female axis. Such deviation is maintained throughout the reaction time which, under these circumstances, is normally so protracted that mating fails to occur. In the truly anomalous condition the animal is inactive and is found to be behaviourally impotent in most cases. Each subject, when deviating, does so constantly to a given side, some always to the left and others always to the right. This form of deviated posture is maintained for lengthy spells within the reaction time, or throughout it, in contrast to any accidentally created misalignment which, in other circumstances, is quite quickly corrected.

Intromission impotence

Another form of male impotence involves intromission failure. In this condition, which has been observed in bulls and rams, the subject can mount but fails to achieve intromission in spite of active pelvic thrusting movements. The bull mounts readily but only partially covers the cow. The bull's hind feet are not brought forward close to the hind feet of the cow. As a result, close genital apposition does not occur and the bull thrusts in vain for a time before dismounting. Such episodes can be repeated many times without effect. The condition exists in several dairy breeds of cattle and in the Suffolk breed of sheep. In the case of bulls, the condition can be found in animals that have previously had a satisfactory history of normal copulatory function. In the case of rams, the condition has been encountered in young animals within certain breedlines. The anomaly may not be of permanent duration. It frequently persists for 6–12 months, but often becomes resolved spontaneously. Some aspect of previous experience is often the cause of inadequate sexual behaviour in male domestic animals. A lack of interest in receptive female animals may occur because sexual behaviour is directed towards males. Control normally necessitates a change in management practice. Some rams are slow to attain puberty and are reproductively inactive when others of the same age are active, but no hormonal deficiency is detectable. General social deprivation, or specific deprivation of contact with and odours of females, could reduce the ability of rams to respond to the smell of oestrous ewes (Zenchak and Anderson, 1980) or their behaviour. Silver and Price (1986) showed that orientation for mounting by beef bulls is more frequently correct after juvenile mounting experience. However, Orgeur and Signoret (1984) showed that rams isolated in early life still showed normal copulatory behaviour provided that they had had contact with females during adolescence. Boars reared in isolation with solid-walled pens reacted inadequately in mating tests, but contact with other boars through wire mesh during rearing improved later mating behaviour markedly (Hemsworth *et al.*, 1978a; Hemsworth and Beilharz, 1979). Remedies for male sexual inadequacy involve avoidance of prolonged isolation rearing and giving animals complete or partial contact with both sexes during development.

Inadequacies of parental behaviour

Parental behaviour in domestic animal species is largely maternal, in practical terms, for the only behaviour of males that benefits the young is defending the group against attack, and such behaviour is relevant in only a very limited range of farming situations. Inadequacies or abnormalities of maternal behaviour are often of very great importance to the welfare of the young and the economics of farming enterprises. Dogs and cats seldom have inadequacies of parental behaviour. Selection of farm animals for breeding, however, has been based on the production characteristics of the progeny, on milk output, on reproductive output and on absence of handling problems. A characteristic that has been largely neglected is quality of maternal behaviour. As a consequence there are breeds of farm animals, such as Merino sheep, in which the mothers often neglect or desert their young. High mortality in young calves and piglets is partly a consequence of inadequate maternal behaviour. Some of this inadequacy is a consequence of the housing system, but there are genetic differences between good and bad mothers. Genes resulting in poor maternal behaviour would have a very low incidence in wild populations.

Neonatal rejection

Various forms of neonatal rejection can occur on the first day postpartum. Notable among these cases are active desertion or persistence of aggressive reactions by the mother towards the newborn. Together with maternal failure, neonatal rejection is the principal form of abnormal maternal behaviour among farm species. In the sow this condition can take the form of cannibalism, to be discussed later. While most cases of neonatal rejection occur spontaneously, some are due to short-term separation from the newborn animal early in the postpartum period. Immediate separation of kids or lambs or other farm species from their mothers at parturition, for periods as short as 1 h, can lead to a rejection of the young by the mothers.

Desertion of lambs by ewes, involving the ewe walking away from the lambs, is more frequent in many of the fine-wooled breeds (<22%) than in other breeds. When fine-wooled Merinos have twins they may move off after giving birth to the second lamb, and are followed by only the firstborn, so that the second lamb dies unless found by a shepherd (Stevens *et al.*, 1982). Ewes of British breeds do not move off unless both lambs follow them. Nutritional factors are also likely to be important and it might be considered adaptive, rather than abnormal, for a ewe on a low-protein diet to desert her offspring if she would not be able to feed it (Arnold, 1985).

In a number of cases the reaction of the mother to the newborn is very aggressive, and includes attacks on the latter in the form of butting, striking, driving away or biting. Since the newborn animal will persist in orientating itself towards the maternal figure, the repeated approach of the young animal to the aggressive mother worsens the condition and makes the affected maternal subject hyper-reactive. Neonatal rejection occurs more frequently in mothers with their first young than with experienced mothers, indicating that maternal inexperience may be a predisposing factor to this anomalous behaviour. Aggressive responses towards older offspring are shown by some mothers.

Broom and Leaver (1977) and Broom (1982) found that isolation-reared heifers turned away from their calves more than did group-reared heifers after the initial licking period, and drew parallels with the inadequacies in social responsiveness to peers shown by the isolation-reared heifers. The finding that reduced early social experience can lead to rejection of young has been reported for monkeys (Harlow and Harlow, 1965; Chamove *et al.*, 1973). Surveys of foal rejection by mares emphasize the importance of previous experience with young, in that most problems occurred with primiparous mares (Houpt, 1984). Disturbance at the time of foaling, either by stallions or by humans, seemed to be a contributory factor.

Maternal failure

Some animals that do not actively desert or aggressively reject their young fail to show adequate maternal responses to them. The failure to supply maternal attention towards the newborn is frequently first shown in a delay or a failure to groom and clean it immediately following its birth. When such failure occurs, the neonate is left in a wet condition and the mother fails to acquire the olfactory and gustatory stimulation from the neonate that should initiate bonding and assist in identification of her progeny. The next feature that is deficient in the behaviour in this syndrome is the reluctance by the maternal animal to accommodate the suckling attempts by the neonate. As the newborn commences teat-seeking behaviour, the mother shows persistent, negative reactions, moving the udder away from the neonate. Such turning has the effect of keeping the young animal in front of the mother's head and prevents it from suckling. Another inadequacy involves movement whenever the young tries to grasp the teat.

Such behaviour, like desertion and aggression to young, is most frequently shown by primiparous mothers and is often only temporary. Such young mothers are often very attentive to their young, but older mothers may sometimes not restrict the movements of the young adequately or not return to them often enough, so that the young are more likely to become associated with mothers that are not their own (see next section). Careful stockmanship is necessary in order that problems of maternal failure can be detected and remedied.

Stealing of young

Pre-parturient ewes, cows and mares often approach, sniff and remain close to the newly born young of other members of their group. This is not abnormal behaviour, but it can lead to problems if the mother of the young animal approached does not retain close contact with it because of weakness, maternal failure or social subordination. An alien female can dispossess the mother in these circumstances, and this is sometimes facilitated by movement of the young animal away from the mother to other adults nearby. Young ungulates commonly attempt to suckle from alien adults, but these attempts are usually rejected. The intense interest shown by many ewes in alien lambs (Welch and Kilgour, 1970) and cows in alien calves (Edwards, 1983) can be followed by acceptance of suckling attempts and the young being stolen from its mother. Edwards (1983) found that 33% of calves born in group housing suckled from an alien female during the first 6 h of life and older cows were more likely to steal calves.

The resulting problems are common in cattle, horses and sheep, wherever large numbers of pregnant females are enclosed together. For example, lamb-stealing leads to confusion over lamb ownership. Sometimes, newborn lambs with disputed ownership find themselves with foster mothers not yet in lactation. At other times the disputed or stolen lamb fails to acquire its due share of colostrum. Even when a newborn lamb is adopted through lamb-stealing, the foster mother may later reject her own lamb when it is born or may have no colostrum left for it (Edwards, 1982, 1983). In these various circumstances lamb deaths frequently occur. Some lamb-stealing is done by ewes that have lost their own lambs in stillbirth. While the stealing may not be entirely detrimental to the stolen lamb, it may affect the maternal behaviour of the deprived ewe if this is her first lambing; in the following season, such a ewe will be inexperienced and more likely to show anomalous behaviour at lambing.

The control of these problems can be effected by separating the cow, mare or ewe from the group before or very soon after parturition. Animals giving birth indoors can be put in separate calving boxes, preferably in sight of other animals, shortly before giving birth; the separation can be achieved by erecting a hurdle pen around the mother.

Killing of young and maternal cannibalism

Violent behaviour is sometimes shown by parturient domestic animals, as it is by many wild animals. The original function of such behaviour may have been to defend the offspring, or to sacrifice one group of young in order to maximize long-term breeding success. The control of such puerperal aggression against people or offspring involves prompt restraint and the provision of antipsychotic therapy, such as chlorpromazine. Alternatively, given appropriate therapy with a neuroleptic agent such as azaperone, most cases are found to be normal in behaviour when they recover consciousness.

Maternal chewing of their own newborn is seen in pigs and sheep. The most dramatic form of maternal cannibalism occurs in the sow and involves the biting, killing and eating of newborn piglets by sows. Within the general anomaly, three subtypes of behaviour, varying in degree of expression, can be recognized in affected sows (Sambraus, 1985):

1. In the simplest form of the condition, the sow is hyper-reactive following the birth of the piglets and responds with agitation to their activities, including their vocalizations. Piglets become crushed by the sow's agitated movements. The accidentally killed piglets may be eaten or partially eaten.
2. The second syndrome of cannibalism resembles the anomaly of neonatal rejection occurring in other species. In this the sow shows persistent active avoidance of her piglets. Avoidance leads to aggression directed towards piglets approaching the sow closely. Such piglets are then likely to be bitten and killed. Dead piglets may be eaten or partially eaten.
3. The third cannibalistic syndrome in the sow resembles the general condition of puerperal aggression occurring in other species. The sow is hyperactive following parturition and shows aggression towards humans or piglets coming within her range. The affected sow snaps aggressively at any intruding piglet. This aggressive biting usually leads to the death of the whole litter. As with the other two allied syndromes of cannibalism, piglets killed by biting may be eaten, partially eaten or left.

All these forms of cannibalism in the sow are associated with hyperexcitability and are limited, chiefly, to sows with their first litters, although they can first appear in experienced breeding sows. Although the anomalous behaviour is usually shown soon after the birth of the piglets, the sow may accept the piglets normally at first and exhibit cannibalism 1 day later. More usually, the condition develops immediately following the parturient process. Once cannibalism has started, the anomalous behaviour is likely to continue until the litter is lost entirely. On rare occasions, a single piglet will be killed and eaten with the remainder of the litter untouched. Although some killed piglets may be left uneaten, the consumption of piglets by sows, under any circumstances, must be considered as cannibalism since the normal behaviour of sows encountering dead piglets is to leave them uneaten.

Certain husbandry conditions are found to be associated with maternal cannibalism in the sow. For example, if a sow is placed in a novel environment at the time of parturition it appears that cannibalism is more likely to occur. This implies environmental maladjustment as a factor. In this regard it has also been widely observed that the supply of straw to pre-partum sows, which allows them to engage in nest-building, seems to lessen the likelihood of cannibalism.

One form of maternal cannibalism occurs where ewes lamb in crowded indoor conditions and takes the form of persistent nibbling of the lamb's appendages. Such ewes may eat off the tails and feet of their newborn lambs. While this may not lead to the death of the lamb, those with severely damaged feet require to be destroyed. In addition to serious hoof injury, tails that have been chewed down to the stump can become seriously infected, with resultant inflammation of tissues in that region.

Control of this condition can be effected simply by adequate provision of outdoor space. Parturient ewes indoors require quarters in which a lambing pen is incorporated. Since this condition, like all others affecting some aspects of abnormal appetite, creates grounds for suspicion about nutritional factors, it is important to provide well-balanced rations to lambing ewes for this and other reasons of perinatal health.

Abnormalities of Basic Movements

Some types of animal housing prevent certain movements from occurring or make normal sequences of

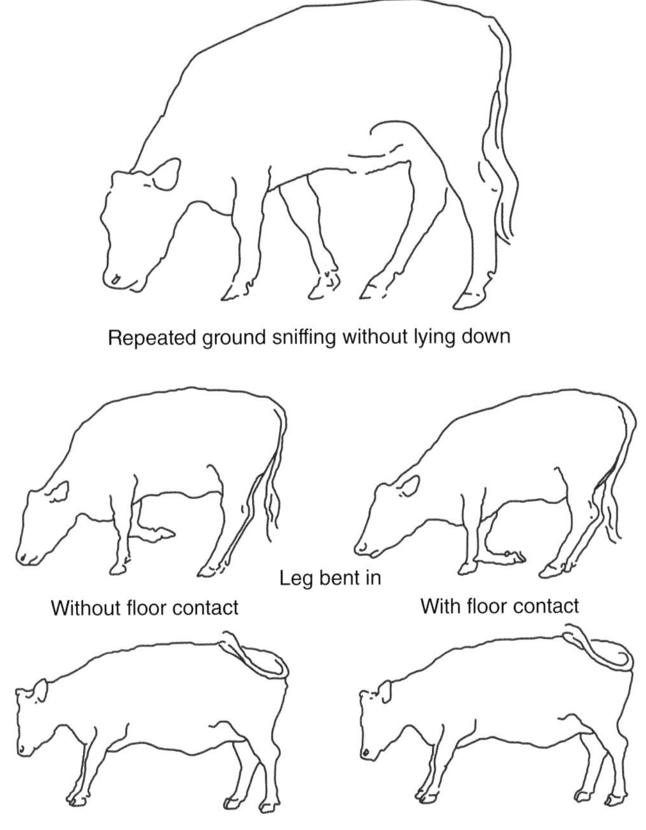

Fig. 27.1. Young cattle on slippery slatted floors show alterations in behaviour that inhibit, delay or prolong lying. For comparison with normal lying, see Fig. 10.2 (from Andreae and Smidt, 1982).

movement difficult to carry out. Movements which are prevented include wing-flapping and flying by hens in battery cages; walking by calves in crates, sows in stalls or tethered animals; and running by many housed farm animals. Abnormalities in grooming by calves in crates or confined sows, which cannot reach the back of the body, have been described in Chapter 25. Hens in battery cages have insufficient room for normal preening with associated stretching (see Chapter 32).

Abnormal lying and standing

Hooved animals kept on slippery slatted floors have especial difficulty in lying down and standing up again. Such problems are described in detail for cattle (Andreae and Smidt, 1982). Cattle normally lie after a brief period of sniffing the ground, then lowering the forequarters (see Fig. 27.1). The sequence of movements shown by fattening bulls on slatted floors is shown in Fig. 10.1. The period of ground-sniffing before lying is prolonged, presumably, because the animal is apprehensive about the slipperiness of the floor.

There are also many interruptions in the sequence of movements once the body starts to be lowered. The ground-sniffing intention movements occurred at a mean of seven or eight times before lying was completed when bulls were put on to slatted floors and 40% of the lying sequences were interrupted. On deep straw, intention movements were followed directly by lying on most occasions, and <5% of lying sequences were interrupted. Young bulls did adapt to slippery floors to some extent, but many of them did so by lying down hindquarters first (Fig. 27.2).

Fig. 27.2. In the same situation as that in Fig. 26.1, some young cattle lie down rump first, presumably to minimize painful events when trying to lie on the slippery floor (see also Fig. 10.2) (from Andreae and Smidt, 1982).

This very unusual way of lying may be less hazardous for the animals in these difficult circumstances. The lying behaviour of sows in stalls and in farrowing crates is different from that of sows in a larger area, because lateral movements are not possible. The sow normally moves her body to the side in the course of lowering her body to the ground but, if bars prevent this from happening, she is forced to drop down from a greater height. Such movements are more likely to result in sow injuries and much more likely to lead to piglets being squashed by the sow. Another factor contributing to this abnormal lying behaviour is weakness of leg and other muscles consequent upon lack of exercise (Marchant and Broom, 1996). Sows that have been in stalls or tethers for a long time may be unable to lie down slowly and carefully because of their inactivity during the non-farrowing period. Other abnormalities of lying behaviour are a consequence of lameness or other localized body pain.

Among breeding sows restrained for most of their pregnancy in narrow single stalls devoid of bedding, the condition of dog-sitting is observed. This anomalous behaviour is, however, not restricted to stalls, since it can be observed in mature pigs kept in high-density groups within pens, again in the absence of bedding. The animal gives the overall appearance of somnolence, and the term 'mourning' has been applied to this behaviour. This can lead to cystitis and nephritis and, in due course, these inflammatory conditions can result in wider systemic infection leading to abortion in some cases and, in other cases, to sudden death associated with pyaemia.

One form of dog-sitting is sometimes observed in veal calves permanently confined in narrow crates. Here, it is observed that the behaviour of such a calf has been influenced negatively by its husbandry circumstances. Again, occasional adoption of a dog-sitting posture can be seen in heavy livestock such as bulls, but this may not be abnormal. The control of anomalous dog-sitting requires changes in housing systems. Dog-sitting is one of several indicators of poor welfare that call for husbandry alteration. The control of abnormal lying and standing involves removal of the causes. Slippery slatted floors should be avoided and conditions leading to leg weakness should not be used. Farm animals need exercise on adequate flooring and suitable lying conditions.

Further Reading: Fraser, A.F. (2010) *The Behaviour and Welfare of the Horse*, 2nd edn. CAB International, Wallingford, UK.

28 Abnormal Behaviour 5: Anomalous Reactivity

Very low or very high levels of activity and responsiveness are also abnormalities of behaviour. Just as in certain circumstances people can become very lethargic or hyperactive, so too can domestic animals. The causes are occasionally specific neurological disorders but, most frequently, they are an inadequacy in rearing or housing conditions, including especially lack of social contact.

Prolonged Inactivity

Prolonged inactivity by companion or farm animals shows obvious parallels with the behaviour of many human depressives. The inactive behaviour of cats in cages, for example in a veterinary hospital, has long been remarked upon by veterinary nurses, veterinary surgeons and cat owners. The cat often lies down at the back of the cage, as far as possible from humans moving past and shows little reaction to stimuli, often keeping the eyes closed for long periods. It is now clear from the work of McCune (1992) and others that this is abnormal behaviour associated with poor welfare. A cat that is well adapted to its environment is active and alert for longer than those cats that are disturbed by the confinement conditions.

Dogs also show abnormal inactivity in some circumstances. A dog deprived of social contact, perhaps when it has no canine companion and its human companions are away from it for many hours, may show great, sometimes destructive, activity or may be inactive and relatively unresponsive.

Motionless sitting, standing or lying is reported as abnormal behaviour in various farm animals by Wiepkema *et al.* (1983). There is much variation among species and among individual wild animals in the proportion of time that they spend active. Clubb and Mason (2003) describe the extent of the time and distance in locomotion and report that locomotor stereotypies in confinement are greater when ranging is greater. Any estimate of the degree of abnormality to ascribe to animals that seem inactive must involve a comparison with animals of that genetic type in conditions that allow a wide variety of activities. Prolonged inactivity has been reported for sows in stalls and tethers; for example, Jensen (1980, 1981) recorded that tethered animals were lying for 68% of the daytime period, while the pigs in an area of woodland and field studied by Wood-Gush and Stolba spent 50% of the daytime rooting and only a short period lying (Wood-Gush, 1988). Various factors must affect the level of activity, but it is frequently found that confined animals are less active. Prolonged lying in sows can lead to urinary tract disorders (Tillon and Madec, 1984), and this is discussed further in Chapter 31.

When calves are kept in small crates such that they are unable to turn around, lying down is sometimes difficult and it reduces sensory contact with events in the building. Calves often stand for long periods, lean against the side of the crate or adopt a semi-seated posture against the rear of the crate. During this chronic standing they may show some stereotyped behaviour (see Chapter 24) or may remain completely immobile for very long periods.

A posture indicating depression in horses is described by Fureix *et al.* (2012). The 'withdrawn' posture involves standing with the neck at approximately the same height as the back. The nape–withers–back angle is approximately 180° and the neck is stretched out. When horses are looking at their environment, the neck is higher and when a horse is resting, the neck is rounder. Chronic standing in horses and the withdrawn posture are more common in horses kept in separate stalls than in those in groups. A slightly different condition is sometimes encountered in horses that have acquired orthopaedic conditions of the hindquarters and hind legs, but such animals are aged. In consequence of their localized clinical or subclinical conditions,

difficulty is experienced by these animals in rising and lying. Since the horse, in rising, finally gets to an upright stance by a forceful extension of the hindquarters, chronic orthopaedic lesions in these parts are likely to be the seat of pain during such sudden movement. Experience of such pain would condition the horse against lying.

This problem of prolonged standing and avoidance of lying has long been recognized by horse-keepers; it was once a particular problem among heavy horses.

The only way to prevent prolonged lying in pigs or chronic standing in calves is to provide conditions that allow more movement and the expression of a greater variety of normal behaviour. For horses with orthopaedic lesions, it was a not uncommon practice in former times for horsemen to provide such animals with a strong chain or timber across the rear posts of a horse stall for the animal to lean its hindquarters against, in order to permit rest and sleep.

Tonic Immobility

Tonic immobility as a behaviour problem is lying or freezing in protracted recumbency. It is characterized by an abnormally low level of reactivity to such stimulation as would otherwise be effective in making the animal change position or posture.

The freezing response is a normal response of birds to close contact with a predator, and in young domestic chicks its duration is affected by the treatment received by the birds and by their previous experience (Broom, 1969a,b; Rose *et al.*, 1985). If a domestic chicken or adult fowl is handled by a person and then laid down on the ground and its head covered for a few seconds, it will show tonic immobility for a period of seconds or minutes (Ratner and Thompson, 1960). It seems that this behaviour is the normal response to capture by a dangerous predator. It is not, therefore, abnormal for the species but it is a response that a wild jungle fowl is unlikely to show very often during its lifetime. Hence, it is an unusual response for a bird to show and it gives some indication of how dangerous the bird judges the situation to be.

It is not clear whether the tonic immobility shown by cattle and other large farm animals is homologous with that shown by poultry but there is some similarity in the conditions in which it occurs, as well as in the response itself.

Certain situations are seen to be closely associated with the appearance of this behaviour. The subject, in virtually all cases, is closely restrained or limited in movement when some stressful circumstances occur. The duration of the phenomenon varies considerably and may be greater in individuals whose level of fear is higher at the time of initiation of the response (Jones, 1996). There may be parallels between the state, in some cases, of the 'downer' and 'tonic immobility' or catalepsy. When the downer cow is maintaining recumbency in the face of varieties of stimulation but is apparently physically sound, a transfer to a new situation – e.g. from indoors to outdoors or otherwise modifying the environment by, for example, removing a dog – may make the subject rise immediately. The condition is therefore not so much an inability to rise as a strong unwillingness to try to rise. This unwillingness leads to a pathological bodily state.

Unresponsiveness

Measures of activity level can be obtained accurately, but it is difficult to know whether reduced activity means poor welfare. In descriptions of abnormal behaviour, Wiepkema *et al.* (1983) emphasized that confined sows may be unresponsive to events in the world around them, in addition to being inactive. Such behaviour is sometimes called apathetic. In studies of sows in stalls, Broom (1986d,e, 1987a) measured their responsiveness to three different stimuli. All animals video recorded were responsive to stimuli associated with the advent of food, but they showed little response to a stranger standing in front of them or to 200 ml of water at room temperature tipped onto their backs while they were lying awake. Group-housed sows, in contrast, were much more likely to take notice of strangers and to sit or stand and carry out other activities when the stimulus was presented. This work shows that stall-housed sows are abnormally unresponsive to such stimuli (see Table 28.1). The results of such work are likely to depend upon the precise nature of the stimulus presented, for a very frightening stimulus might elicit a maximal response in all sows. The behaviour of head-pressing is shown by animals that may be in pain and are unresponsive to most stimuli. Studies of this kind indicate parallels with the behaviour of human depressives (Broom and Johnson, 2000; Goodyer, 2001).

Table 28.1. Responsiveness of stall-housed and group-housed sows; behaviour in the 20 min after stimulus presentation (from Broom, 1988d).

	Stall-housed	Group-housed
Median time to sit or stand[a] (s)	27.5	349
Median number of other activities[b]	2.5	6.5

[a]P = 0.096; [b]P = 0.004.

Hyperactivity

Animal handling is made difficult when animals show freezing responses or startle responses, but neither of these is necessarily abnormal behaviour. Shying, jibbing, or baulking by horses can sometimes be extreme in the extent to which they are shown and may necessitate the use of blinkers and the avoidance of potentially startling situations. Other domestic animals may also show extreme flight responses. However, this behaviour is within the normal range of responses to danger.

Problems arise when individuals injure themselves because of their high reactivity or if they influence others to behave similarly. High-density housing of animals and the presence of dense flocks or herds at pasture can make such socially transmitted hyperactivity dangerous to the animals and to humans. Grazing animals suddenly disturbed, even by an innocuous object such as blown paper, may stampede. They are more likely to be injured due to collision or falling during a stampede. Primitive man exploited this behaviour in order to catch large herbivores; for example, Native American peoples in North America caused bison to stampede over cliffs. The behaviour is present in wild populations but it can be maladaptive. Stampedes of cattle, horses or sheep can be very damaging to the animals.

Hyperactivity can also be a problem in pet dogs. Young dogs might be regarded as being too active by relatively sedentary owners, but pathological hyperactivity can also occur and sometimes necessitates drug treatment (Manteca, 2002).

Hysteria

The occurrence of the extensive alarm reaction in poultry is often termed hysteria. Flightiness in the domestic chicken appears in different types of nervous and hysterical behaviour occurring in differing environments and age groups. Hysteria in the caged laying hen is characterized by sudden flying about, squawking and trying to hide. The incidence of hysteria in penned poultry is higher at greater flock density. Flocks of 40 have been found to have 90% incidence of hysteria while flocks of 20 had an incidence of 22%. Even in cages hysteria can occur, but it is less of a problem in multiple-hen cages containing three to five rather than the larger numbers of birds.

Keeping in smaller groups controls hysteria in poultry to some extent, although it can spread throughout a flock kept in battery cages. Hens in a cage adjacent to hysterical birds may or may not be affected, but individual birds apparently trigger an episode within a cage leading to hysteria among all the individuals within that cage. Following episodes of hysteria within caged birds, traumatic sequelae occur in the form of torn skin over the back. In addition, there is a drop in feeding and egg production following hysteria (Craig, 1981).

In broiler chickens, and particularly in turkeys, hysteria may result in pile-ups of birds under which some die. Good stockmanship prevents hysteria responses; for example, the good stockman entering a poultry house always knocks to warn the animals that human entrance is imminent. This minimizes the escape response that might otherwise occur and be magnified as it moves in a wave down the house. Another method of controlling hysteria is to place baffles in the poultry house so that birds do not move too far before reaching a baffle. The consequent pile-ups are smaller and mortality is reduced.

Further Reading: Fureix, C., Jego, P., Henry, S., Lansade, L. and Hausberger, M. (2012) Towards an ethological animal model of depression? A study on horses. *PLOS ONE* 7 (6), e39280, 9 pp.

Section 7

Welfare of Various Animals

The Welfare of Cattle

Public Perceptions of the Dairy and Beef Industries

During the last 50 years, some aspects of cattle management have changed considerably but, at the same time, our knowledge of cattle physiology and behaviour has been improving. It is clear that cattle have complex brain mechanisms regulating their behaviour processes, an elaborate social structure and sophisticated learning ability (Kiley-Worthington and de la Plain, 1983; Phillips, 2002; Hagen and Broom, 2003, 2004). Data establishing this have made many animal scientists reconsider the effects of conditions and procedures on farms, both in terms of their efficiency as regards production and with respect to the welfare of the animals.

Most members of the public asked about the dairy and beef industries think of cattle grazing in fields and cows living for many years while a series of calves are born and milk is produced. Milk products and beef are considered by the public in relation to their effects on human nutrition and health, animal welfare and the environment. If production is perceived to be bad in relation to any of these aspects, sales of the products could be severely affected. Bovine spongiform encephalopathy (BSE) resulted in a substantial but brief decline in beef consumption in several countries. Some people may limit their intakes of milk products because of a desire to reduce cholesterol intake while certain aspects of the dairy industry, such as methane production, may be criticized in relation to pollution, but animal welfare also has significant effects on purchasing by the public.

For many years, the welfare of the dairy cow was not often perceived to be poor, and it has been only in calf rearing that dairy production systems have been regularly criticized. However, the dairy industry has been changing (Webster, 1993). Evidence of poor welfare in cows has accumulated and has had influence on public opinion in several countries. It is important to the dairy industry that welfare problems should be addressed before there is any widespread public condemnation of breeding and management practices. Similarly, beef cattle production is not often criticized on welfare grounds. However, a few critical newspaper articles or television programmes that appear well founded can be very damaging to producers, processors and retailers.

The general range of welfare problem areas is similar for cattle and other farm animals. This chapter is concerned with the effects of housing and management systems. Possible causes of cattle welfare problems are shown in Box 29.1.

In many aspects of farm animal management, improved welfare leads to improved production. If the welfare of a dairy cow is improved there is often a greater milk yield and, if the welfare of very young calves is improved, the resulting increases in growth rate and survival chances lead to economic advantages for the farmer. In other situations, however, improving welfare leads to reduced profits, for example when high stocking density is detrimental to welfare. Modern cattle husbandry systems do lead to some welfare problems, as discussed below. Genetic selection for high yield and a general change in cattle management methods has led to an increase in production pressure. Nutritional expertise has increased to the point where animals now convert feed to meat and milk very efficiently. If animals are pushed hard energetically there need not be welfare problems and management difficulties, but these are more likely and should be taken into account when deciding or advising about which system to choose. Welfare problems that should be taken into account in cattle practice have been reviewed by Broom (2004).

Box 29.1. Some causes of poor welfare in cattle.

- Breeding procedures and consequent difficulties.
- Ill-treatment.
- Neglect: calculated, accidental or due to lack of knowledge.
- Inadequacies in design of housing, including pens, etc.
- Inadequate management system or poor husbandry on the farm.
- Mutilations of the animals, including those that are unnecessary or poorly executed.
- Poor conditions and procedures: (i) during moving or loading; (ii) during transport; (iii) at market; or (iv) at slaughterhouse.
- Inadequate provision for emergencies.

Ill-treatment and Neglect

Human actions that can lead to poor cattle welfare include ill-treatment, neglect and inadequate production systems. Ill-treatment occurs most frequently when animals are being moved around the farm, when they are being loaded into or out of vehicles, or when they are at market or lairage. Those who do ill-treat animals can be advised of likely economic effects of their actions as well as being told about the laws on the subject. Neglect includes failure to provide an adequate diet, failure to treat disease and the lack of normal husbandry procedures. The diet may be inadequate in nutrient composition or in quantity. Cattle are sometimes undernourished for a period while food is scarce or expensive, with the expectation that compensatory growth will occur when more food is provided. If the undernourishment amounts to starvation, and this is clear from the condition of the animal, then this is serious neglect. Lack of knowledge on the part of the farmer may result in the provision of a poor diet or in failure to treat disease. This is poor husbandry, a very important cause of welfare problems. Advice on good husbandry methods can be an important veterinary service. If animals are diseased and require treatment there is a moral obligation upon the farmer to obtain adequate advice and on the veterinarian to treat them.

General Housing and Management: Feeding Trough Barriers

Certain general points are relevant to the management of all cattle and so will be made before considering problems specific to calves, beef cattle and dairy cows. Feeding of housed cattle may lead to difficulties for the animals because the acquisition of food in housing conditions is very different from that when grazing. Physical difficulties may occur, as described by Cermak (1987), but social factors are also very important. Cattle synchronize their feeding to a large extent (Benham, 1982a; Potter and Broom, 1987) so, where group feeding is possible, enough feeding places for each animal are required (Wierenga, 1983; Phillips, 2002). Those animals that cannot find a feeding place may not get sufficient food and it is likely that there are adverse effects on their welfare.

The precise effects of the frustration occurring when food is inaccessible because of competition remain to be determined. Competitive feeding situations where there are no individual feeding places pose extra problems for cattle. The subordinate individual has to attempt to obtain food despite the attacks or threats of other individuals. Bouissou (1970) found that the greater the extent of the barrier between feeding places for cows, the fewer the attacks that occurred (see Fig. 29.1). A trough that requires subordinate cows to come close to dominant individuals results in those subordinates walking greater distances and taking longer to obtain a meal (Albright, 1969; Fig. 29.2). Calves of low social rank obtain less of the favoured food if trough space is restricted (Broom and Leaver, 1978; Broom, 1982).

In order to minimize such welfare problems that are often associated with poor weight gain, farmers should provide feeding spaces for all individuals, preferably with barriers between the individual places. Adaptation to a single food source is possible for cattle, however, as a transponder-operated feeding stall can be successful (Albright, 1981), but certain individuals in a herd may have difficulties in such systems.

Another general problem for housed cattle is being required to stand on floors that are wet, slippery, uneven or hazardous because of sharp edges (Galindo *et al.*, 2000). Slippery slats can lead to difficulties in standing or lying (Andreae and Smidt, 1982; see also Figs 27.1 and 27.2). These and other inadequacies of flooring can

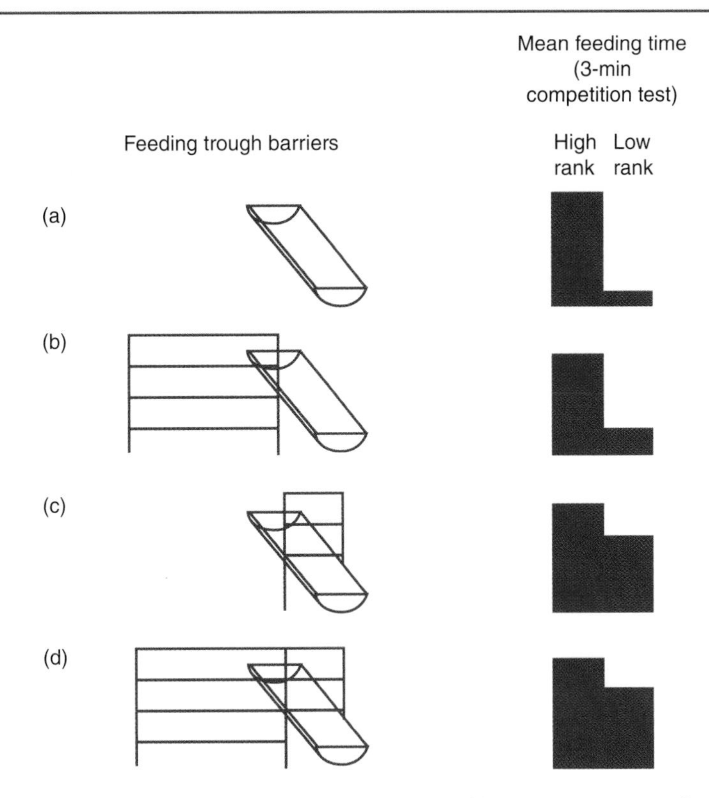

Fig. 29.1. Physical barriers affected feeding times by cows ranking high and low in a competitive order. With no barrier (a), the low-ranking cows were scarcely able to feed. A body barrier (b) improved the situation slightly for the low-ranking cows but a head barrier (c) and a complete barrier (d) had a much greater effect (redrawn after Craig, 1981; data from Bouissou,1970).

result in limb injuries, foot lameness, tail-tip necrosis and various diseases. Hock lesions resulting in ulceration were found to occur in 16% of the hocks of 4000 UK dairy cows on 77 farms and 81% had hair loss that might lead on to hock lesions (Potterton *et al.*, 2011). Ninety per cent of farmers thought that hock lesions cause poor welfare in cows. In fact, lameness is widely regarded as the greatest welfare problem of housed dairy cows. Barker *et al.* (2010) visited 205 UK dairy farms and found that the prevalence of lameness was 37% (range 0–79%). Factors influencing its occurrence include poor floor quality and drainage, and also high milk production, as discussed later in this chapter.

Welfare of Calves

In the first few days after birth, the major calf welfare problems are enteric and respiratory diseases. The calves of dairy cows may fail to obtain sufficient colostrum for a variety of reasons (Edwards, 1982; Edwards and Broom, 1982; Broom, 1983a). Management practices that maximize the chance that colostrum will be obtained and minimize contact with pathogens have important beneficial effects on calf welfare. If calves of dairy cows are normally left with their mother for the first 24 or 48 h, the risk that the calf will not suckle early enough to obtain and absorb the immunoglobulin from colostrum can be minimized by the stockman placing one of the mother's teats in the mouth of the calf as early as possible after the calf stands. Group-calving situations, where several cows calve during a short period, can lead to a cow's colostrum being drunk by a calf other than her own or to calves being rejected by their own mothers (Fig. 18.5 and Chapter 18). Such occurrences can be prevented by providing separate calving boxes that should, ideally, allow the cows some visual contact with other cows. The provision of soft bedding for the calf is also desirable and is easier where special calving accommodation is available. Dairy calves are deprived of their mother from an early age and

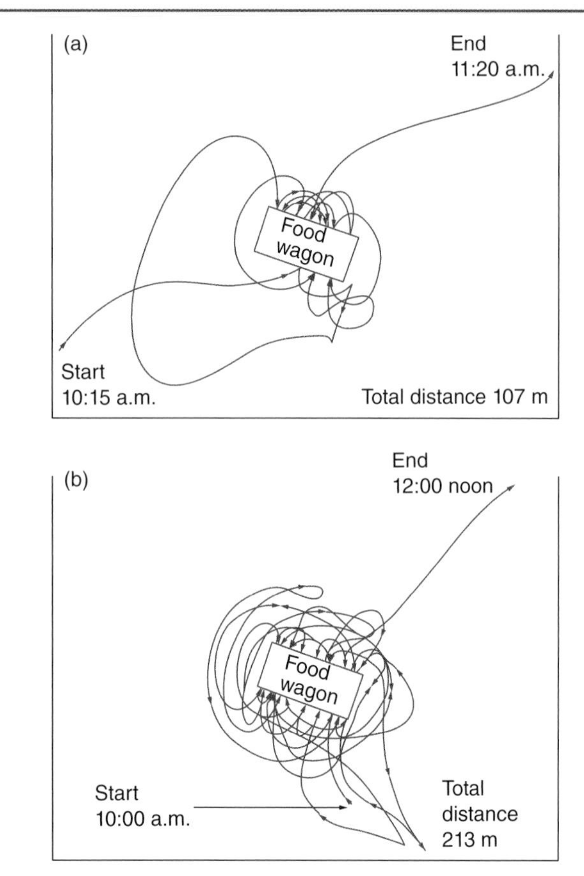

Fig. 29.2. The paths of two cows in a herd after food is provided in a food wagon are shown. Animal (a) was found to be high in a competitive order whereas animal (b), which was low in that order, walked further because of displacement at the food wagon and took longer to feed (after Broom, 1981; modified after Albright, 1969).

many are individually housed, so that they are confined in a small space and deprived of all or most social contact.

A need is a requirement, which is a consequence of the biology of the animal, to obtain a particular resource or respond to a particular environmental or bodily stimulus (see Chapter 1). The needs of calves are described in detail by Broom (1991, 1996) and by EFSA (2006a), and examples are given here. Calves need the following: (i) to breathe in good air conditions; (ii) feed and drink, including sucking, manipulating food and ruminating; (iii) normal gut development; (iv) rest and sleep; (v) exercise; (vi) lack of fear; (vii) ability to explore and have social contact; (viii) minimal disease; (ix) ability to groom; (x) ability to thermoregulate; (xi) avoidance of harmful chemicals; and (xii) avoidance of pain.

The calf needs to ingest colostrum very soon after birth, and milk thereafter. It also needs to show sucking behaviour and, if a calf is not obtaining milk from a real or artificial teat, it sucks other objects (Broom, 1982, 1991; Metz, 1984; Hammell *et al.*, 1988; Jung and Lidfors, 2001). A variety of nutrients are needed by calves. Sufficient iron is needed to allow normal activity and to minimize disease. A blood haemoglobin level in calves of 4.5–5.0 mmol/l is generally considered a threshold below which iron deficiency anaemia occurs (Lindt and Blum, 1994a,b), since at this concentration there is substantially reduced immunocompetence. When calves were forced to walk on a treadmill, those with a mean haemoglobin level of 5.5 mmol/l consumed more oxygen and exhibited higher cortisol levels after walking than calves whose haemoglobin level was 6.6 or 6.9 mmol/l (Piquet *et al.*, 1993). Hence, the level at which there is clearly no harm to the calf is 6.0 rather than 4.5mmol/l.

Normal calf anatomical, physiological and behavioural development occurs only if the calves have some fibre-containing material to eat (van Putten and Elshof, 1978; Webster *et al.*, 1985; Webster, 1994), so it is clear that they need fibre in their diet after the first few weeks of life. The lack of appropriate roughage is a major determinant of abnormal oral behaviours in veal calves (Veissier *et al.*, 1998; Cozzi *et al.*, 2002; Mattiello *et al.*, 2002).

Calves need to rest and sleep in order to recuperate and avoid danger. They need to use several postures including one in which they rest the head on the legs and another in which the legs are fully stretched out (de Wilt, 1985; Ketelaar de Lauwere and Smits, 1989, 1991). Calves that have more rest in comfortable conditions grow better (Mogensen *et al.*, 1997; Hänninen *et al.*, 2005).

Exercise is needed for normal bone and muscle development, and calves choose to walk at intervals if they can, show considerable activity when released from a small pen and have locomotor problems if confined in a small pen for a long period (Warnick *et al.*, 1977; Trunkfield *et al.*, 1991).

Exploration is important as a means of preparing for the avoidance of danger and is a behaviour shown by all calves (Kiley-Worthington and de la Plain, 1983). Calves need to explore, and it may be that higher levels of stereotypies (Dannemann *et al.*, 1985) and fearfulness (Webster *et al.*, 1985) in poorly lit buildings are a consequence of inability to explore. Play is a form of exploration, and calves show more play in environments that meet their needs (Jensen *et al.*, 1998).

The needs of young calves are met most effectively by the presence and actions of their mothers. In the absence of their mothers, calves associate with other calves if possible and they show much social behaviour. This is not possible in individual pens (Fig. 29.3). The bond between dam and calf is likely to develop very soon after birth: calves separated from their dam at 24 h can recognize the vocalizations of their own dam 1 day later (Marchant-Forde *et al.*, 2002). In their review of the effects of early separation of dairy cows and their calves, Flower and Weary (2003) conclude that, on the one hand, behavioural reactions of cows and calves to separation increase with increased contacts but, on the other hand, health and future productivity (weight gain for the calf, milk production for the cow) are improved when the two animals have spent more time together. Calves reared by their dam do not develop cross-sucking, while artificially reared calves do so (Margerison *et al.*, 2003). The need to show full social interaction with other calves is evident from calf preferences and from the adverse effects on calves of social isolation (Broom and Leaver, 1978; Dantzer *et al.*, 1983; Friend *et al.*, 1985; Lidfors, 1993, 1994; Gygax *et al.*, 2009).

In order to minimize disease, calves have a wide range of immunological, physiological and behavioural mechanisms. One example of a behavioural mechanism is the behaviour that maximizes their chances of obtaining colostrum; another is that calves show preferences to avoid grazing close to faeces. They also react to some insects of a type that may transmit disease. If infected with pathogens or parasites, calves will show sickness behaviour that tends to minimize the adverse effects of disease (Broom and Kirkden, 2004). Young calves, <4 weeks of age, are not well adapted to cope with stressful events. An inability to mount an effective glucocorticoid response, which is adaptive in the short term, may be a contributing factor to the high levels of morbidity and mortality occurring in young calves (Knowles *et al.*, 1997), as may neutrophilia (Simensen *et al.*, 1980; Kegley *et al.*, 1997), lymphopaenia (Murata *et al.*, 1985) and suppression of the cell-mediated immune response (Kelley *et al.*, 1982; MacKenzie *et al.*, 1997).

Grooming behaviour is important as a means of minimizing disease and parasitism, and calves make considerable efforts to groom themselves thoroughly. Calves need to be able to groom their whole bodies effectively.

Calves need to maintain their body temperature within a tolerable range. They do this by means of a variety of behavioural and physiological mechanisms. When calves are overheated, or when they detect that they are likely to become overheated, they move to locations that

Fig. 29.3. These calves have straw and adequate food but are individually housed so are deprived of social contact (photograph D.M. Broom).

are cooler. If no such movement is possible, the calf may become disturbed, thus exacerbating the problem, and other changes in behaviour and physiology will be employed. Responses to a temperature that is too low will also involve location change if possible. Overheated, or potentially overheated, calves adopt positions that maximize the surface area from which heat can be lost. Such positions often involve stretching out the legs laterally if lying, and avoiding contact with other calves and with insulating materials. If too cold, calves fold the legs and lie in a posture that minimizes surface area. Overheated calves will attempt to drink in order to increase the efficiency of methods of cooling themselves.

Comparisons of calf housing and management systems

The major housing systems that have been compared in studies of calf welfare are individual crates, group housing on slats and group housing on straw. Where calves are housed individually, the size of the crate and whether or not the sides are solid have been varied. Aspects of diet are important in relation to welfare. For example, if inappropriate proteins or carbohydrates are fed, the calf may be unable to utilize them and, if milk is acidified too much, calves may find if very unpalatable. However, the two aspects of diet that have been of greatest concern in relation to calf welfare have been the amount of fibre and the amount of iron, as discussed above.

Some feeding systems for young calves involve the use of a bucket, while others use teats. The provision of milk through a teat, a long milk meal and the possibility of sucking a dry teat can decrease non-nutritive sucking in artificially reared calves, but do not abolish it (Veissier *et al.*, 2002; Jensen, 2003; Lidfors and Isberg, 2003).

Calves reared in individual crates show various signs of poor welfare: (i) the occurrence of stereotypies; (ii) difficulties in standing, lying and grooming; (iii) excessive grooming of the front of the body with the ingestion of much hair and the formation of hairballs in the gut; and (iv) substantial adverse reactions to walking and to transport (as described by Broom (1991, 1996) and in several chapters of this book). All calves are motivated for social contact, which is not possible in individual crates. Such a motivation was shown using operant conditioning by Holm *et al.* (2002). Individual housing can be stressful to calves as measured by adrenal responses to ACTH (Raussi *et al.*, 2003) and results in inability to show cognitive skills, such as reversal learning, that calves with social experience do show (Gaillard *et al.*, 2014). Group housing or grouping in a field (Fig. 29.4) can help calves acquire social skills (Boe and Faerevik, 2003). Some experience of mixing is of particular importance: calves that have been reared for a while in a group dominate calves that have always been in individual crates (Broom and Leaver, 1978; Veissier *et al.*, 1994).

Fig. 29.4. Dairy calves in a group in a field (photograph D.M. Broom).

The amount of space available to calves in groups has an effect on their welfare. In dairy calves it has been shown that spatial environment stimulates play: calves in small group pens performed less locomotory play than the ones kept in larger pens (Jensen *et al.*, 1998; Jensen and Kyhn, 2000). Dairy calves kept from birth to 1 month of life in larger stalls (1.00 × 1.50 m) showed a higher percentage of lying behaviour and grooming than calves kept in smaller stalls (0.73 × 1.21 m); in addition, lymphocyte proliferation was significantly higher in calves reared in large stalls (Ferrante *et al.*, 1998).

The incidence of disease in young calves is high; for example, 25% of veal calves had to be treated for respiratory disease in a study by van der Mei (1987). The use of antibiotics to prevent disease is also a problem. It is important for calf welfare and for farm economics that disease levels be lowered. However, the transmission of both respiratory and gastrointestinal diseases in a calf house occurs whether the calves are housed individually or in groups. The key factor affecting this is ventilation (Heinrichs *et al.*, 1994) rather than individual or group housing.

One aspect of management that causes problems is the practice of mixing calves from different sources. Webster (1994) found that calves purchased and brought into a unit were five times more likely to require treatment for disease. A second aspect is hygienic practice by farm staff and a third is early detection of disease. These variables seem to be more important than housing system in exacerbating disease.

As a consequence of the evidence of poor welfare in veal calves, the EU passed a Directive in 1997 which required the following criteria: (i) group housing of calves after 8 weeks of age; (ii) individual pens at least as wide as the height of the calf at the withers; (iii) no tethering of calves except for <1 h at feeding time; (iv) sufficient iron to ensure an average blood haemoglobin of 4.5 mmol/l; and (v) fibre in the diet increasing from 50 g/day at 8 weeks to 250 g/day at 20 weeks. Many EU calf producers have found group housing of calves to be more successful economically than the old crate system, and white veal can still be produced from systems which comply with the new law.

In a comprehensive review of calf welfare, the European Food Safety Authority Scientific Panel on Animal Health and Welfare (EFSA, 2006a) produced the list in Box 29.2 of risks to poor welfare from housing and husbandry in intensively kept calves.

Box 29.2. Risks to poor welfare in intensively kept calves (modified after EFSA, 2006).

- Inadequate colostrum intake – duration.
- Inadequate ventilation, inappropriate airflow, airspeed, temperature for some husbandry systems.
- Exposure to pathogens causing respiratory and gastrointestinal disorders.
- Continuous restocking (no 'all in–all out').
- Mixing calves from different sources.
- Inadequate colostrum intake – quantity.
- Inadequate colostrum intake – quality.
- Insufficient access to water.
- Insufficiently balanced solid food.
- High humidity.
- Indoor draughts.
- Inadequate ventilation, inappropriate airflow, airspeed temperature for some husbandry systems.
- Poor air quality (ammonia, bio-aerosols and dust).
- Poor floor conditions: gaps too large, too slippery, wet floor for lying, no bedding.
- Insufficient light for response to visual stimuli.
- Exposure to pathogens causing respiratory and gastrointestinal disorders.
- Poor response of farmer to health problems, especially necessary dietary changes.
- Lack of maternal care.
- Separation from the dam.
- Iron deficiency resulting in haemoglobin levels <4.5 mmol/l.
- Allergenic proteins.
- Diet too rich (overfeeding).
- Insufficient floor space allowance.
- Inadequate health monitoring.
- Inadequate haemoglobin monitoring.

Farm mutilations

In order to facilitate certain management procedures, many cattle, usually calves, are subjected to mutilations such as castration, disbudding, dehorning and various sorts of individual marking. These are often carried out by people with no veterinary qualification and, in most cases, no anaesthetic is used. The disbudding involves removal of the tissue on the head that will grow into horns by means of a surgical scoop, a hot iron or caustic paste. Castration is carried out by surgical removal of the testicles, application of a constricting elastic band at the base of the scrotum so that the blood supply is blocked and the tissue dies, or external clamping of the base of the scrotum (Burdizzo method). Cutting tissue to remove testicles or horn buds, tightly constricting tissues, the use of caustic materials that remain in contact with flesh for many hours and the use of a hot iron that burns tissue will cause severe pain.

When calves are in pain, the increased cortisol production and increased occurrence of pain-related behaviours (Stafford and Mellor, 2005; Stilwell *et al.*, 2008a,b, 2009, 2010) are readily quantified. Pain-related behaviours in calves include head-shaking, ear-flicking, head-rubbing, inert lying, alterations in gait, amount of walking, licking scrotum, lifting hind leg, abnormal lying, rapid transitions between behaviours and reluctance to go to the food trough. When 4-week-old calves were disbudded with caustic paste, plasma cortisol concentration increased for 60–90 min and pain-related behaviours increased substantially for 3–4 h. Some changed responses were reported up to 24 h and, after castration, responses continued for up to 48 h. If the anaesthetic lidocaine was used by cornual nerve injection, the pain indicators were reduced during the first hour but not during the next two or more hours. If the analgesic flunixine meglumine, a non-steroidal anti-inflammatory drug (NSAID) known to inhibit mainly COX-1, was also given intra-venously, neither physiological nor behavioural indicators of pain were shown on the first day. The use of the analgesic alone did not prevent pain indicators in the first 3 h after disbudding (Stilwell *et al.*, 2009). Treatment with the analgesic carprofen prevented pain indicators for 48 h (Stilwell *et al.*, 2012). Carprofen is an NSAID that inhibits COX-2, and has a half-life of up to 70 h. There are also other measures that indicate pain. Gibson *et al.* (2007) measured EEG changes in Friesian heifers that were dehorned at 6–9 months of age using a scoop. The median frequency (F50) and 95% spectral edge frequency (F95) components of the EEG increased and the total power (P_{tot}) component decreased when there was no anaesthesia but were not affected when a lignocaine ring block was used.

The indicators of pain are not seen if the animals are given a combination of anaesthesia to prevent the immediate pain and long-lasting analgesia to prevent pain when the effects of the anaesthetic ceases. The use of this pain relief has a cost to the farmer. However, there is also a cost for the whole product as a result of members of the public refusing to buy it if such pain has been caused to the animals.

Welfare of Beef Cattle

The housing conditions for calves destined for beef production are sometimes similar to those kept for veal production, so they have similar welfare problems. Older beef animals are kept in small individual pens or are tethered in some countries and they then show much stereotyped behaviour. Riese *et al.* (1977) reported that stereotyped behaviour included tongue-rolling, weaving movements and self-licking. Wierenga (1987) reported that one-third of young, individually housed, bulls spent several minutes in every hour showing tongue-rolling. Physiological responses to confinement also occur. Ladewig (1984) reported that tethered bulls showed more frequent episodes of high blood cortisol levels than did bulls able to interact socially in groups. Such abnormal behaviour and physiology is probably exacerbated by both social deprivation and inability to perform behaviours because of spatial restriction. Tethered animals lack exercise and have different patterns of muscle fibres from those free to walk (Jury *et al.*, 1998) and more osteochondrosis (de Vries *et al.*, 1986). Individual housing of beef animals is more frequent when they are bulls than when they are steers. In Germany almost all beef animals are bulls but, in the UK, most were steers before the ban on growth promoters.

Studies of beef cattle temperament in relation to production and welfare (Turner *et al.*, 2011; MacKay *et al.*, 2013) are described in Chapter 1. A useful test of temperament, described by Core *et al.* (2009), is the amount of eye-white visible when the animal is disturbed by being in a squeeze-chute. Fighting and mounting can lead to welfare problems when beef animals, especially bulls, are kept in groups. The most important way of minimizing such problems is to keep the animals in stable groups, since social mixing leads to much fighting with consequent injuries, bruising and extreme physiological responses (Kenny and Tarrant, 1982). In stable groups, mounting may lead to more injury than does fighting (Appleby and Wood-Gush, 1986).

Animals that are frequently mounted become bruised and may suffer severe leg injuries. Mounting can be greatly reduced by the use of overhead bars that physically prevent it or an electrified grid, which deters animals that wish to mount. Although experience of an electric shock has a less adverse effect on welfare than the serious effects on animals that are repeatedly mounted, mounting is best prevented by management changes (Chapter 26).

The stocking density of beef animals and the flooring provided also have considerable effects on welfare. High stocking densities lead to more aggression, injury and bruising. Beef animals increase rapidly in body weight but they have little exercise if they are housed in small pens, and their leg growth may not be able to keep pace with that of the rest of the body. The final weights reached are much higher now than they used to be, so the legs are scarcely adequate to support the body. The consequence of this is cartilage damage, clear indications of limb pain and obvious difficulties in standing and lying (Dämmrich, 1987). Graf (1984) found that these problems were absent if fattening bulls were reared on deep straw and that such conditions also led to fewer behavioural problems. Beef cattle have a strong preference for straw or other bedding. All of these issues are reviewed in the EU SCAHAW (2001a) report on the welfare of cattle kept for beef production.

Traceability of animals is important for disease control, other welfare reasons and practical management (Broom, 2006d). However, a humane method of marking animals should be used. Hot-iron branding of cattle is illegal in many countries because of the severe pain that it causes. Careful freeze-branding causes little response from cattle so is an alternative. However, the trend towards electronic tagging of all large farm animals will obviate the need for such marking in the near future.

Welfare of Dairy Cows

The major welfare problems of dairy cows are lameness, mastitis and any conditions leading to impaired reproduction, inability to show normal behaviour or emergency physiological responses, or injury. For a review of all aspects described here see EFSA (2009). A mutilation that is allowed in a few countries is tail-docking. This is carried out solely for the convenience of those who milk the cows and there is no evidence that it is necessary for hygiene reasons. Since the tail is used by cows as their main method of responding to insect attack or annoyance, tail removal can cause great hardship to them. Where the insects carry disease, tail-docking can increase risk of disease and this is the reason why using the tail for defence against flies is a very high priority behaviour for animals such as cows (Hart, 2011).

Dairy cow leg and foot problems

Leg and foot disorders are painful and frustrating because many of them prevent normal control of movement. They often cause animals to walk with a limp, reduce walking to a low level or avoid walking whenever possible (see Fig. 29.5). The ulcerated hock lesions reported by Potterton *et al.* (2011) would be painful. The ability of lame cattle to carry out various preferred behaviours is generally impaired, and there may be adverse consequences for various other aspects of their normal biological functioning. Lameness reduces ability to obtain some resources, such as food and preferred lying places. Lameness always means some degree of poor welfare, and sometimes means that welfare is very poor indeed.

Measurements of the extent to which some degree of lameness occurs in dairy cows include 35–83 cases per 100 cows per annum in the USA, the UK and the Netherlands. While there may have been some improvement since the 1990s, a large-scale UK study (Barker *et al.*, 2010) gave a prevalence of 37%. In addition to causing poor welfare, lame cows produce less milk. Archer *et al.* (2010) found that lame cows

Fig. 29.5. A sole ulcer on a dairy cow's hoof is very painful and leads to abnormalities of walking (photograph F. Galindo).

produced about 1 kg less milk per day. EFSA (2009) concluded that 'on farms with a high prevalence of recognizable locomotor difficulties, e.g. approaching 10%, there should be improvement of housing conditions, genetic strain and management practices'. This sets a threshold for codes of practice and legislation that would greatly improve dairy cow welfare.

Dairy cow mastitis

Mastitis in mammals is a very painful condition. The sensitivity to touch of affected tissues is clearly evident and there is obvious diminution of normal function. Mastitis prevalence should have declined greatly with improved methods of prevention and treatment, but it has not declined as much as it should have done. Webster (1993) reports 40 cases of mastitis per 100 cows per year as an average for the UK (see EFSA, 2009 for review).

Reproductive problems

Reproductive problems in dairy cows have become very common in recent years, with large numbers of cows being culled because of failure to get in calf. In a study of 50 dairy herds in England, Esslemont and Kossaibati (1997) found that farmers reported failure to conceive as the predominant reason for culling, with 44% of first lactation, 42% of second lactation and 36.5% of cows in total being culled for this reason. However, mastitis, feet and leg problems, ketosis and other disease conditions can lead to reproductive problems and it is difficult to discover their initial cause from farmers' records. A report by Plaizier *et al.* (1998) concerning Canadian herds indicated that reproductive culling risk varied between 0% and 30%, with a mean of 7.5% (see EFSA, 2009 for review).

Calving difficulty, or dystocia, is a major welfare issue for some cows and has been estimated to affect one in six Holstein calvings in the UK and 50% of primiparous calving in the USA (Barrier *et al.*, 2012). The use of analgesics at this time can alleviate some of the problem and may have beneficial effects on later calf growth (Stilwell *et al.*, 2014).

Indoor and outdoor management systems and welfare

Is the welfare of dairy cows better if they are outdoors or indoors, with access to pasture or without? Studies of the preferences of dairy cows make it clear that, while they prefer adequate food supply to inadequate food supply, if they have access to pasture they will work for it by walking to it. When cows had access to a building where they were milked and where concentrate food was provided, as well as to pasture, they spent time in both places. At night, in several different studies, 80% of time was spent at pasture (Spörndly and Wredle, 2004; Legrand *et al.*, 2009; Charlton *et al.*, 2011, 2012). This preference was independent of distance to pasture when this was 60 m, 140 m or 240 m. In the daytime, however, when food was sometimes available in the building, the time spent outside was 41% if the distance to the pasture was 240 m, but 51% if it was only 60 m; more time was spent on the track as distance to pasture increased (Charlton *et al.*, 2012). A mean of five transitions from indoors to pasture was made.

The incidence of lameness is much worse in housed cows than in cows at pasture. Cows at pasture may have stone damage to hooves if they do not have a suitable place to walk, but wet cubicle houses or poorly maintained straw yards can result in very high levels of lameness. Even the best cubicle housing systems seem to have some lameness problems, which are exacerbated by social factors (Broom, 1997; Fig. 29.6).

Since the best straw yards, with an abrasive area on which normal hoof wear occurs, have little lameness, these may be the best solution for housed cows. Mastitis incidence is affected by hygiene at milking and various other conditions of management. Poorly designed housing systems can result in a variety of welfare problems, and these can be exacerbated by high stocking density. Most of these problems – such as those resulting from cubicles being too short for the length of the cows now occupying them or from poor design of cubicles which do not allow adequate movements in the cow – are well known, so are mentioned only briefly here. In general, it seems that many dairy cow housing systems, and cubicles in particular, do not provide an environment to which cows can adapt easily. The best straw yards seem to be the most successful as they give the cows more opportunity to control their interactions with their environment. EFSA (2009) concludes that the use of straw yards or substantial improvements to cubicle house design is needed in the dairy industry.

In addition to the design of the accommodation for dairy cows, methods of management can have substantial effects on the welfare of the animals. This topic is considered in Chapter 21 and elsewhere in this book.

Fig. 29.6. If cows have to stand in wet places, perhaps because of social pressure, they are more likely to become lame (photograph D.M. Broom).

Cows adapt better to the difficult conditions that we impose on them if they have the same routine each day and if the actions of the people who interact with them are predictable. For example, when entering the milking parlour they often select the same side and may enter with a similar order of individuals on each day (Paranhos da Costa and Broom, 2001).

The size of a dairy herd does not, in itself, result in better or worse welfare. However, there is a tendency for those who run large herds to use systems that cause more prolems (Broom, 2013). Each herd should be evaluated using good measures of welfare.

Milk yield and welfare

The milk production of a beef cow is about 10 l/day and 1000–2000 l/year. In the late 1980s, a single dairy cow in the UK produced about 30 l/day and 6000 l/year. Ten to 15 years later, the comparable figures were 75 l/day and 18,000 l/year for some cows. The dairy animal is producing considerably more than its ancestor would have. This raises questions of whether it is at or beyond its maximum production level and the extent of any welfare problems (Knaus, 2009).

The peak daily energy output of the dairy cow per unit body weight is not very high in comparison with some other species such as seals or dogs, but the product of daily energy output and duration of lactation is very high indeed. Hence, long-term problems are likely to occur (Nielsen, 1998). This is what we see because, although some cows seem to be able to produce at high levels without welfare problems, the risk of poor welfare indicated by lameness, mastitis or fertility problems is greater as milk yield increases.

The steady increase in reproductive problems as milk yields have increased is well known. As Studer (1998) states: 'Despite programmes developed by veterinarians to improve reproductive herd health, conception rates have in general declined from 55–66% 20 years ago to 45–50% recently (Spalding *et al.*, 1975; Foote, 1978; Ferguson, 1988; Butler and Smith, 1989). During the same periods, milk production has greatly increased.'

Studies showing that milk yield is positively correlated with the extent of fertility problems have come from a range of different countries (van Arendonk *et al.*, 1989; Oltenacu *et al.*, 1991; Nebel and McGilliard, 1993; Hoekstra *et al.*, 1994; Pösö and Mäntysaari, 1996; Pryce *et al.*, 1997, 1998; EFSA, 2009; Oltenacu and Broom, 2010). Studer (1998) explains that high-producing cows that are thin and whose body condition score declines by 0.5–1.0 during lactation often experience anoestrus. A loss of condition score of about 1.0 during lactation was normal in the review presented by Broster and Broster (1998). Data on the relationships between milk yield and reproduction measures from two large-scale studies are included in Tables 29.1 and 29.2.

Table 29.1. Positive correlations between milk production level (33,732 lactation records) and indicators of poor welfare (Pryce *et al.*, 1998).

Welfare indicator	Correlation
Calving interval	0.50 ± 0.06
Days to first service	0.43 ± 0.08
Mastitis	0.21 ± 0.06
Foot problems	0.29 ± 0.11
Milk fever	0.19 ± 0.06

Table 29.2. Positive correlations between milk production level (10,569 lactation records) and indicators of poor welfare (Pryce *et al.*, 1998).

Welfare indicator	Correlation
Calving interval	0.28 ± 0.06
Days to first service	0.41 ± 0.06
Mastitis	0.29 ± 0.05
Somatic cell count	0.16 ± 0.04
Foot problems	0.13 ± 0.06

In some studies, effects of health problems on reproduction are evident; for example, Peeler *et al.* (1994) showed how cows that were lame in the period before service were less likely to be observed as being in oestrus. The lameness could be more likely in high-producing cows. Direct links between level of milk production and extent of disease conditions are also evident from a range of studies, positive correlations being reported by Lyons *et al.* (1991), Uribe *et al.* (1995), Pryce *et al.* (1997, 1998), EFSA (2009); see also Tables 29.1 and 29.2. In addition to mastitis and leg and foot problems, which are often measured in such studies, the occurrence of other clinical conditions can also be affected by production level. Modern, high-producing cows with good body condition have a high incidence of milk fever, retained placenta, metritis, fatty liver and ketosis (Studer, 1998).

The high yields of modern dairy cows are a consequence of genetic selection and feeding. Cows are adapted to high-fibre, low-density diets. The ways in which they have been modified genetically do not change these basic characteristics much. Cows do not adapt easily to high-grain diets or to manufactured diets with high protein and low fibre. Genetic selection has not taken adequate account of the adaptability and welfare of cows. Current trends towards ever-greater milk production and feed conversion efficiency should not be continued unless it can be ensured that welfare is good (Broom, 1994; Phillips, 1997, EFSA, 2009, Oltenacu and Broom, 2010). EFSA (2009) reported that 'The genetic component underlying milk yield has been found to be positively correlated with the incidence of lameness, mastitis, reproductive disorders and metabolic disorders' and 'Higher weight should be given to fitness and welfare traits when these may conflict with selection for milk yield'.

Bovine somatotrophin (BST) is a naturally occurring peptide hormone. The recombinant BST produced by genetically modified bacteria has a slightly different structure but similar effects. When injected into cows, rBST results in higher milk yields and higher levels of mastitis, lameness, reproductive disorders and some problems at the injection site (Broom, 1993; Willeberg, 1993, 1997; Kronfeld, 1997; Broom, 1998). While some of the effect of BST is mediated via the increased milk production, whether or not this is so, the substantially poorer welfare caused by the BST is avoidable. EU SCAHAW (1999) concluded as follows:

> BST is used to increase milk yield, often in already high-producing cows. BST administration causes substantially and very significantly poorer welfare because of increased foot disorders, mastitis, reproductive disorders and other production related diseases. These are problems which would not occur if BST were not used and often result in unnecessary pain, suffering and distress. If milk yields were achieved by other means which resulted in the health disorders and other welfare problems described above, these means would not be acceptable. The injection of BST and its repetition every 14 days also causes localised swellings which are likely to result in discomfort and hence some poor welfare.

The EU Committee also made the following recommendation:

> BST use causes a substantial increase in levels of foot problems and mastitis and leads to injection site reactions in dairy cows. These conditions, especially the first two, are painful and debilitating, leading to significantly poorer welfare in the treated animals. Therefore from the point of view of animal welfare, including health, the Scientific Committee on Animal Health and Animal Welfare is of the opinion that BST should not be used in dairy cows.

The use of recombinant BST is banned in many countries because of the poor welfare that it causes in dairy cows.

Welfare Consequences of Other Breeding Procedures

Conventional methods of cattle breeding have changed these animals considerably during recent years, and future changes are likely to be accelerated by new possibilities for genome manipulation. For example, selection for double muscling in beef cattle and the possibility of transferring genes which increase growth rate or modify final body form could both result in animals with larger, faster-growing bodies. Growth promoters such as bovine somatotrophin (BST), are discussed in the section on dairy cows above. In addition, body weight increase without corresponding increase in leg size and strength could result in more lameness. Any modification of animals should be checked carefully using proper scientific measures of welfare to ensure that animals do not find it more difficult to cope with their environment. Such studies should be carried out over a period as long as the maximum farm lifetime of the animal.

The offspring of transgenic animals should also be studied in this way. New techniques should not be licensed for general use unless such welfare checks have been carried out and have shown that there are no adverse effects on the animals. Some modifications of animals could result in improved welfare: for example, if genes were implanted which increased the efficiency with which disease could be combatted by the individual.

The crossing of breeds of animals can lead to welfare problems for cows if a large breed of bull is crossed with a smaller breed of cow, resulting in increased calving difficulty. Similar problems can arise if embryo transfer is used. Multiple implantation of embryos could lead to other problems. Cloning involves transfer of embryos and some placental abnormalities, dystocia and welfare problems for cloned offspring (Broom, 2014b). Any novel breeding procedure should be such that cow welfare is not worse than that of cows undergoing a normal pregnancy.

Welfare Consequences of Feeding Systems

Many dairy cows are fed from a trough or feed face at which the amount of space per animal is an important factor affecting welfare. Insufficient space can lead to increased aggression and some individuals may not be able to obtain enough food. In order to investigate this, Rioja-Lang *et al.* (2012) trained cows to distinguish black bins with high palatability food from white bins with low palatability food. They then offered the cows a choice of high palatability food near another dominant cow or low palatability food. If the distance from the other cow was 0.3 m or 0.45 m, the cows would not take the high palatability food but if the distance was 0.6 m or 0.75 m they sometimes did take it. As a consequence, all cows should have sufficient space at the feeding place, with the head at least 0.6 m from each other cow.

A quite different development area that can have effects on welfare is the development of microprocessors and other electronic control units. Cows can already carry transponders that allow them to be fed individually, and this methodology could be improved to minimize the chances that any individuals fail to obtain food. This system of feeding cows at a single or small number of feeding stalls can lead to problems, because dominant individuals may attack others or deter them from feeding. Some very timid cows might be quite unwilling to approach a feeder when an aggressive individual is near it. Wierenga and van der Burg (1989) suggested the use of auditory signalling devices on the cow's ear that tell each individual when to come to feed. Animals that have already fed receive no food if they enter the feeder, so this system provides a means for distributing feeding times for each individual throughout the day. Such a system should work well for all animals except for those that are stimulated to feed only when another animal is feeding.

Electronic systems could also allow cattle greater control over their physical environment, for example by giving them the opportunity to regulate environmental temperature and air-flow rates. Lack of control is a major cause of welfare problems (Broom, 1985, 2003), so such possibilities could improve welfare. The development of automatic milking systems for cows, which are recognized individually on entry to a milking stall so that a computer, pre-programmed with their udder coordinates, initiates robotic attachment of a milking machine to them, can improve welfare for animals that can adapt to the system (Fig. 29.7). The cow can come to be milked wherever she chooses to do so. However, pasture systems need to be changed to a system of fields radiating out from the milking parlour. If this is not done, there is a tendency for the cows to be kept in buildings continuously, and this leads to increased leg and foot disorders and failure to allow enough exercise.

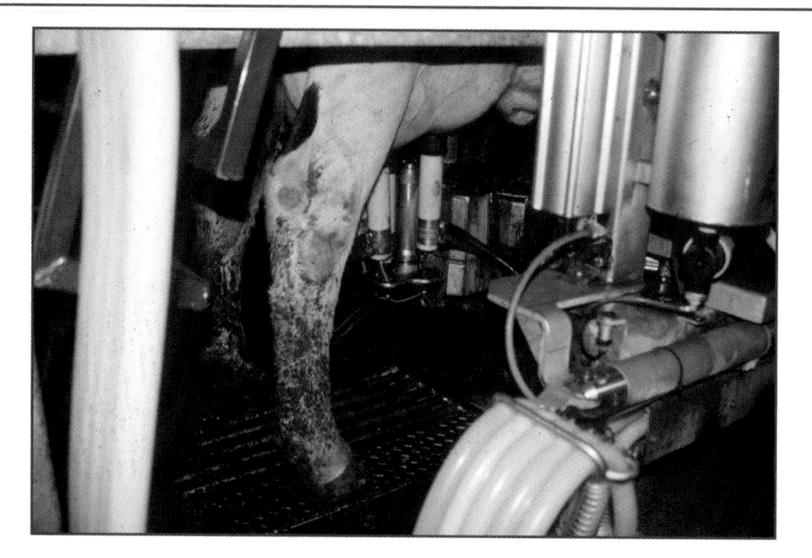

Fig. 29.7. Hind leg of cow with automatic milker cluster attached to her udder. The cow has entered the milking stall voluntarily (photograph D.M. Broom).

Bullfighting

Bulls reared for bullfighting in Spain and a few other countries are normally kept in good conditions before they are put in a bullring. The traditional view of the bullfight is of a contest in which humans attempt to overcome a dangerous adversary who is a willing participant in the competition. However, bulls make substantial efforts to avoid fights with obviously superior opponents. If they encounter a group of armed humans, they seldom fight unless cornered. The bull is a reluctant participant and the bullfight methodology is partly designed to goad the bull into fighting. There are also methods to minimize the risk to humans, for example putting small lances into the neck muscle so that the bull cannot easily toss a man in the air. This action causes significant tissue damage, as do the stabbing blows with swords and other weapons. It has been suggested that endogenous opioids prevent the stabbing of the bull from being painful. However, there is no evidence for this. There must be a substantial amount of fear in many bulls while in the bullring. The scientific evidence is that bullfighting causes very poor welfare during most of the period in the bullring and that there are no mitigating circumstances for this.

Further Reading: EFSA (2009) Scientific report on the overall effects of farming systems on dairy cow welfare and disease. *Annex to the EFSA Journal* 1143, 1–38.

30 The Welfare of Sheep and Goats

Sheep and goats may be similar in size and similar in that they are social ruminants but they differ greatly in behaviour and somewhat in usage (Dwyer, 2009). Goats are frequently kept for milk production and are important for milk, meat and skin in tropical countries. Sheep originate in cold or cool mountainous areas and were kept principally for their wool until the market for wool declined. Now their meat is the main product but a few are used for milk production. Sheep are well adapted to dry, rocky conditions and can tolerate high temperatures. Goats can climb trees as well as rocks, jump well and often browse rather than graze. They are very vocal in a social situation and when disturbed. In behaviour, they are more like cattle than like sheep, for example in oestrous behaviour. Sheep do not adapt well to indoor conditions but both sheep and goats are mainly kept extensively.

Sheep and goats have welfare problems that are common to other farm animals, such as ill-treatment, failure to provide sufficient food, failure to obtain veterinary treatment for disease, mutilations of the animals, and being kept in housing conditions and with procedures that do not meet their needs. Problems for the animals may occur: (i) during moving or loading; (ii) during transport; (iii) at market; or (iv) at the slaughterhouse. Sheep and goats are sometimes left for long periods, in an area where there is pasture deemed to be sufficient for them, without being checked on a daily basis. If the amount of food and water is sufficient they may flourish, but they can also die from starvation and lack of water. They may also be subject to predation, accidental injury and disease. Daily checking can minimize such adverse effects, all of which can result in very poor welfare.

Goats that are milked are less likely to starve or have their diseases and injuries untreated but the likelihood of problems depends on the resources available to the people who care for them, some of whom are very poor. Like cows, milking goats are subject to mastitis and other disorders because of the large amount of energy that has to be used for milk production.

The housing conditions required for goats are similar to those required for growing cattle. Sheep and goats are especially susceptible to disease in humid, poorly ventilated conditions. Too high a stocking density, poorly designed flooring and a barren environment also have considerable effects on the welfare of both species. Sheep kept in a barren environment and handled in an inconsiderate way show indications of negative mood and poorer welfare, when tested later, than sheep kept in a calm and enriched environment (Reefmann *et al.*, 2012; see Chapter 4, this volume).

There have been many studies of cognition and emotion in sheep and goats in relation to welfare. Welfare indicators and behavioural measures in sheep are discussed at length in several chapters of this book. These make it clear that both species have complex, sophisticated behaviour and show emotional responses with consequences for brain activity (e.g. Doyle *et al.*, 2011; Muehlemann *et al.*, 2011).

Welfare of lambs

Lamb mortality in the early days of life is often high and the welfare of those lambs that die is generally poor in the period before death. Some of this early mortality is a consequence of bad weather conditions. Lamb thermoregulation is less good when very young so death due to exposure can occur. The risk of this can be greatly reduced by providing adequate shelter for the lambing ewe and attending the ewe to assist when necessary. Occasionally mortality is a result of inadequate maternal behaviour (see Chapter 19), especially in primiparous ewes (Dwyer and Smith, 2008).

As explained in Chapter 23, weak individuals are more susceptible to disease. However, both weak and strong lambs may become infected by pathogens of various kinds. Respiratory and gastrointestinal disorders account for the deaths of many lambs. The more time that human caretakers give to the ewes and lambs, the better the lamb survival rate. Shelter with adequate ventilation is important if the conditions are adverse. In general, the factors listed in relation to calf welfare are relevant to that of lambs and goat kids.

Farm mutilations

Lambs are sometimes subject to tail-docking, male lambs and kids to castration and fine-wool breeds of sheep that may be at risk of fly-strike may be subjected to mulesing. Sheep and goats have very similar pain systems to those of humans and other mammals (Chapter 4) and pain can be assessed effectively in both species (Chapter 6). When in pain sheep may avoid showing active behavioural responses, including vocalizations, because this is important anti-predator behaviour. Goats are much more vigorous and noisy in their responses to pain. Facial grimacing is a response to pain that is evident in diseased sheep (K. McLennan *et al.*, in preparation). It may not be obvious to predators but is measurable by trained observers using a scale of severity.

Many studies have been carried out to establish the extent of pain in lambs subjected to a variety of castration and tail-docking procedures. It is clear from such work that the use of rubber rings, Burdizzo clamps or surgical procedures all cause pain, but that this can be avoided by the use of appropriate anaesthetics and analgesics. The rubber ring placed around the scrotum of the lamb or kid is easy to use but particularly painful. The cost of the anaesthetic and analgesic is not great in relation to the total cost of production so, given the public pressure to avoid such extreme pain, could be incorporated into normal farm practices. Tail-docking of lambs may cause neuroma formation and hence more prolonged pain. This procedure should not be routine and is not necessary in many circumstances.

As described in Chapter 6, the mulesing operation involves cutting off a 15-cm diameter area of skin around the anogenital area of the young sheep. It is very painful for many days (Figs 30.1 and 30.2). The objective of the procedure is to permanently remove an area of wool that might become damp and be a site where blow-flies (mainly *Lucilia*) lay their eggs. When these eggs hatch, the maggots consume the tissues of the sheep and may cause the death of the animal. The welfare of such a sheep is very poor so the painful operation can lead to a benefit for welfare and an economic advantage to the farmer as fewer sheep die. The risk of fly-strike is reduced if the sheep are regularly checked and treated with insecticide if necessary. Farmers who wish to leave their animals for long periods without checking them developed the use of mulesing. Regular checking of extensively kept sheep and goats is advantageous for other reasons (see below).

Fig. 30.1. Merino sheep upside down on cradle while mulesing is carried out (photograph D.M. Broom).

Fig. 30.2. Merino sheep after mulesing showing the area of skin cut off (photograph D.M. Broom).

Fig. 30.3. Merino sheep being driven to the left by two dogs but some looking apprehensively at the photographer, other people and another dog (photograph D.M. Broom).

Footrot and other disease

Leg and foot disorders are relatively rare in sheep in hot dry countries but are the commonest cause of poor welfare in sheep kept in cool temperate countries.

Sheep standing or walking for long periods in wet conditions are susceptible to the pathogens that cause footrot. This condition is painful to sheep and may cause them to be unable to stand or walk without lameness. Footrot can destroy much tissue on and between the claws of the hoof. Sometimes the sheep will be forced to graze while on its 'knees'. The condition can be treated, and it should be, but some farmers fail to identify it as a problem for which they are responsible.

Many other diseases of sheep have major effects on their welfare. The likelihood of a sheep farmer calling for veterinary assistance to deal with disease in one or more sheep should be the same in all conditions but it is often affected by the price that can be obtained for the sheep. A growing lamb or pregnant ewe might be treated while an old ewe might not.

Outdoor and indoor management systems and welfare

Since most sheep and goats are kept extensively, they have much control over their interactions with their environment and, in general, this results in good welfare. Extremes of weather may lead to problems and responses to very hot and very cold conditions are discussed in Chapter 9. Deep snow and prolonged dry conditions can result in the deaths of adult as well as young animals.

Predators can be a problem for sheep and goats in some countries and the shepherd caring for the flock may be present to act as a deterrent to predators or even to directly defend them. Guarding dogs that identify with and protect the flock are sometimes used to keep predators away (Coppinger and Schneider, 1995; Chapter 7, this volume). The sheep and goats habituate fully to these dogs and, in most cases, have been reared in their presence. In contrast, dogs used for herding are treated as potential predators and cause substantial welfare problems, particularly for sheep (Fig. 30.3). As explained in Chapters 6 and 7, sheep often show a maximal emergency physiological response to the approach of a dog. The management of extensively kept sheep is difficult without using dogs but these dogs should be closely controlled and their use minimized. Their training should be limited to what is essential. The practice of using dogs in sheep-dog trials for public entertainment or as a sport is cruel to sheep and should be discontinued.

Sheep may be kept in buildings but, in general, they have more difficulty in adapting to indoor conditions than goats or cattle. Social conditions, dry floors, material to manipulate and sufficient space for exercise are among the factors necessary for good welfare.

Further Reading: Dwyer, C. (2009) The behaviour of sheep and goats. In: Jensen, P. (ed.) *The Ethology of Domestic Animals*, 2nd edn. CAB International, Wallingford, UK, pp. 161–176.

31 The Welfare of Pigs

The complexity of pig behaviour, and the brain mechanisms controlling it, is evident from the studies described in earlier chapters. Their learning ability is considerable and their social behaviour elaborate. As a consequence, welfare problems arise for pigs if they are unable to control events in their environment, if they are frustrated or if they are subjected to unpredictable situations. For example, inability to prevent attack by another pig, to regulate body temperature or to groom adequately can all lead to poor welfare. Such effects are additional to those resulting from injury, disease or other pain and physical discomfort.

Pig welfare problems (Broom, 1989b; Jensen *et al.*, 2001) include those due to physical abuse, neglect, handling, transport, farm operations and disease (see Chapter 23). Some of the welfare problems of pigs, like those of other domestic animals, are a consequence of the ways in which the animals have been selected by breeders. When a population is strongly selected for high production, fewer resources are available to adequately respond to other demands such as coping with stressors (Rauw *et al.*, 1998). Such effects are shown to include behavioural capacities (Schütz *et al.*, 2004).

This chapter is concerned principally with the effects of widely used housing systems on pigs, in particular dry sow housing, accommodation for farrowing, piglets after weaning and fattening pigs. The handling and transport of pigs is a major welfare issue, in part because pigs have been selected in a way that causes them serious problems when subjected to transport or other difficult conditions, as discussed in Chapter 21.

Dry Sows

Dry sows and gilts are those that are not lactating. They may be pregnant, i.e. during gestation, and they are kept in a variety of different housing systems. These include tethers, stalls, other individual pens, groups with feeding stalls, groups with food supplied on the floor or in a communal trough, groups with electronic sow feeders (Figs 31.1 and 31.2), groups in deep-straw pens, groups in fields or large yards and, experimentally, in the family pen system. There is some variation within each of these categories, particularly with respect to the microclimate, the type of flooring, the use of straw or other bedding materials, the diet and the frequency of feeding. Their welfare is discussed below in relation to several welfare indicators.

Piglet production

It is possible that a gilt that fails to grow or a gilt or sow that produces very small litters of piglets may have a welfare problem, although other factors contribute to the wide variation in individual production. In comparative studies of production systems, however, it is often difficult to discover how many individuals do badly in this way, because the data are presented as a mean of animals in a system. Hence a few pigs producing well could mask a few bad producers. Where average production figures are used, it is clear that well-managed units can do equally well whether the sows are confined or in one of the forms of group housing. In a comparison of matched sets of sows housed in stalls, groups with an electronic sow feeder and groups with individual feeding stalls, Broom *et al.* (1995) found that stall-housed sows had shorter body length but there was no difference in piglet production. In an EU report involving a much larger number of animals, being kept in sow stalls or a group-housing condition was not found to affect production (Jensen *et al.*, 2001).

Reproduction problems

Some sows are culled because they do not become pregnant and others because they have small litters.

Fig. 31.1. This sow is wearing an electronic transponder that will trigger food delivery when she enters the feeding gate. The transponder may be on a collar, on an ear tag or, since it can be very small, implanted under the skin to allow electronic recognition throughout life (photograph D.M. Broom).

Fig. 31.2. Sows kept in groups have access to this electronic sow-feeder that delivers food to them, after their entry, only if they have not received their ration for the day. The sow should have a front exit to the feeder (photograph D.M. Broom).

These reproductive failures or inadequacies can occur because the sow encounters difficult conditions and has difficulty in trying to cope with them. Many factors lead to anoestrus in the pig, including tethering of gilts. Individual penning of sows and bullying can lead to fewer sows becoming pregnant after service, or attempted service. Other measures of reproductive problems may reflect poor welfare during the gestation period and problems at farrowing. Most studies of such problems are complicated by the fact that sow accommodation during both gestation and farrowing may influence the results. Bäckström (1973) compared 1283 sows confined in a crate during pregnancy and farrowing, with 654 sows free in a pen at both times. In crate-housed sows there was a higher incidence of mastitis/metritis/agalactia (11.2% versus 6.7%, respectively) and greater numbers of sows whose farrowing time was longer than 8 h (5.4 versus 2.3%, respectively). Vestergaard and Hansen (1984) studied four groups of sows that were tethered or loose-housed during pregnancy and during farrowing. The duration of farrowing was significantly shorter in those sows that were loose-housed throughout than in those that were tethered at either stage. It seems possible that lack of exercise has some adverse effect that is evident during the parturition process. In the review by EFSA (2007a) it was concluded that reproductive problems were least in sows that were kept in stable groups throughout pregnancy.

Sow disease, bone strength and injury

Animals that utilize their adrenal cortex frequently may have impaired immune system function and greater susceptibility to disease. The effects of housing conditions on the health of sows and piglets were reviewed by Ekesbo (1981), who pointed out that the relative levels of disease in different systems are much affected by differences in the use of antibiotics.

If sows are exposed to very infectious diseases such as Aujeszky's or swine fever, all may become infected, irrespective of their previous welfare. Non-infectious, or less infectious, diseases such as those leading to some foot and leg problems, sores, torsion of the gut or ulcers are more obviously related to environmental conditions and the animals' attempts to cope with them.

Tillon and Madec (1984) noticed that urinary tract disorders had increased in frequency in France during a period when more and more sows were confined. They reported on the relatively high incidence of such disorders in tethered sows, and Madec (1984) suggested that sows might be more prone to urinary disorders if they have to lie on their faeces. They also found that tethered sows drink less and urinate less often than do loose-housed sows, so that urine is more concentrated and bacteria have longer to act within the urinary tract (Madec, 1985). This problem is probably a consequence of low activity levels and consequent infrequent drinking; therefore, while it could be in part a consequence of the effect of the housing system on the animal, it may be reduced within that system by stockmen encouraging the animals to stand and drink. It is unlikely that farm staff would do this with sufficient frequency. There is clearly much variation among sows here, as some inactive sows drink infrequently but other active sows drink very often. Tillon and Madec (1984) also reported that, in one-quarter of tether units, more than 20% of sows showed serious lameness. Several other authors have reported similar findings. Bäckström (1973) found that the number of traumatic injuries caused by pen fittings and flooring was 6.1% in confined sows, but 0.8% in loose housing. Most studies of leg injuries and infections that cause lameness have related their incidence to the type of flooring. Penny *et al.* (1965) attributed the high incidence of foot rot to poor concrete floors, and Smith and Robertson (1971) described how poorly designed or maintained slatted floors resulted in many leg and foot injuries and high culling rates. Bäckström (1973) found that 6.3% of 588 sows on partly slatted floors had foot lesions, but these were shown by only 3.3% of 3520 sows on unslatted floors. It is now clear that good slats cause fewer problems than poor slats, but the incidence of sow lameness is still very high. There remains the probability that confinement and associated lack of exercise cause lameness even on good flooring. Using a precise method for quantifying integumental lesions, de Koning (1983) reported that such lesions can occur with high frequency on farms where sows are tethered.

Sows in stalls or tethers have almost no opportunity to exercise. One consequence of this is that certain of their muscles are weaker and they have substantially weaker bones than sows that can exercise. Marchant and Broom (1996) found that sows in stalls had leg bones only 65% as strong as sows in group-housing systems. The actual weakness of bones means that the animals are coping less well with their environment, so welfare is poorer in the confined housing. If such an animal's bones are broken there will be considerable pain and the welfare will be worse but, in practice, bone breakage is rare.

Injuries resulting from attacks by other sows can be serious in group-housing conditions. A good feeding system and the maintenance of stable groups minimizes fighting and consequent injury, but injury can have a serious detrimental effect on welfare in a poorly managed system. Where sows are attacked by others, the lesion can be quantified in a precise way (Gloor and Dolf, 1985). Any system for keeping sows that results in high levels of fights that cause injury, vulva-biting or tail-biting is clearly bad for the welfare of at least some of the pigs. This topic is considered further, below, in relation to behavioural and physiological measures.

Activity and responsiveness

Abnormally low levels of activity and lack of responsiveness to events in the surrounding world have been proposed as indicators of poor welfare in pigs (van Putten, 1980; Wiepkema *et al.*, 1983). Several authors have reported that sows confined in stalls or tethers are inactive for longer periods than are sows in groups. Sow activity is affected by parity, stage of pregnancy, extent of lameness and stereotyped behaviour. If inactivity – with associated unresponsiveness – and stereotyped behaviour are alternative strategies that sows use to try to cope with adverse conditions, the gross measures of the activity of sows in a particular housing condition are not very useful. It is better to study individual animals in detail and to try to assess responsiveness in a precise way. In a series of experimental studies on the responsiveness of sows (Broom, 1986d,e, 1987a), stall-housed sows were found to be less responsive to stimuli other than food presentation than were group-housed sows. There was, however, considerable variation among the stall-housed sows in this respect.

Stereotypies

Confined sows are not able to groom normally, they may have difficulty thermoregulating, most are fed small volumes of food infrequently, they cannot interact normally with other sows and they cannot move away from people or other potentially hazardous stimuli. One response shown by a variety of animals to such situations where the individual has little control of its environment is to show a stereotypy such as bar-biting, manipulating the tether-chain, or drinker- or sham-chewing (see Fig. 6.5, Fig. 24.2). Such behaviour is occasionally shown by group-housed sows, but the mean frequency is extremely low. Examples of reported total duration of stereotypies in stall-housed sows are generally 10–29% of time active, but some individual tethered sows (Cronin and Wiepkema, 1984), performed stereotypies for a mean of 80% of daytime observation periods. The figures obtained in such studies depend upon the efficiency of the recording method, especially on the use of video recording.

Many reports of low incidence of stereotypy are a result of failure to notice some stereotypies such as sham-chewing, and there may be some reduction in the incidence of stereotypies when human observers are present. In the comparison of sow welfare by Broom *et al.* (1995), the greatest difference between the animals in the different housing conditions was the much higher level of abnormal behaviour in the stall-housed sows than in either of the group-housing conditions. Stereotypies, such as bar-biting and sham-chewing, and behaviour that included a substantial stereotyped component such as drinker manipulation and rooting in the trough, were much more common in the confined sows (Broom *et al.*, 1995).

Diet can have an effect on stereotypies. If added roughage is in the form of a manipulable material such as unchopped straw, there can be considerable reduction in stereotypies. The addition of high bulk material to concentrates caused a redistribution of stereotypies but no net change in total duration in one study (Broom and Potter, 1984). A high level of feeding (4 kg/gilt/day) resulted in much lower levels of stereotypies than a low level (1.25 kg/gilt/day) in a study by Appleby and Lawrence (1987). However, the sows in the study by Broom *et al.* (1995), which were fed at a commercial level of 2.2 kg/day rather than the low level of 1.25 kg/day used by Appleby and Lawrence (1981), showed very little stereotypy in the group-housing conditions but much stereotypy in the stalls. The stereotypies are an indicator of poor welfare and they are frequent in most sow stall and tether units. The relatively low levels at which pregnant sows are commonly fed may be a contributory factor to poor welfare, but the confinement itself must be a major part of the problem for the animal.

Aggressive and other injurious behaviour

The welfare of an animal that is injured by another, or which is often pursued by another or whose movements are severely restricted by the presence of other dominant individuals, is poor. Genetic selection of pigs can affect the amount of aggression shown but has often not been included as a factor taken into account in pig breeding because most focus has been on meat production. However, when 'social breeding value', which includes effects on other individuals such as the consequences of aggression, was included in pig selection indices, the frequency of skin lesions and other injuries in fattening pigs was reduced (Canario *et al.*, 2012).

Aggression is frequent in tethers and stalls, probably because disputes cannot be resolved, and so they escalate (Broom *et al.*, 1995). Some farmers using group-housing systems for sows report serious fighting and extensive wounding. The extent of this fighting is greatest if there is competition for food. A group-housing system with well-designed individual feeding stalls allows all sows to feed all at once, with no aggression being possible during feeding, so this would seem to be a good system on welfare grounds (see Fig. 31.3).

The electronic sow-feeder system has the advantage that sows receive their own individual ration but the disadvantage that only one sow can feed at a time, so sows tend to queue for feeder access. Early versions of

Fig. 31.3. Sows can be fed individually in feeding stalls so as to minimize aggression. Most feeding stalls are closed at the back when the food is provided. The majority of such systems are for sows kept indoors (photograph D.M. Broom).

electronic sow-feeder systems had disadvantages that led to considerable problems of bitten vulvas and other injuries. These included poor design and positioning of the feed crate (e.g. a back gate which did not prevent contact by sows behind), a rear exit-only system, a gate which allowed other sows to follow the first in or a feeder positioned in a lying area. Work on management and studies of behaviour (Hunter *et al.*, 1988) in electronic sow-feeder systems has shown that efficient training, a sufficient number of feeders, once-a-day feeding, stable grouping, front-exit feeders and a good bedded lying area or kennel can result in few welfare problems. Damaging aggression was absent in all conditions in the study by Broom *et al.* (1995), but the level of aggressive behaviour was highest in the stall-housed sows.

In the family pen system in which boars and offspring are present with the sows, there is very little aggression (Stolba, 1982; Wood-Gush, 1983). The sows have a more diverse and interesting environment in this system and there are seldom any indicators of welfare problems. Even more is demanded of the stockman running this system, however, so there might be risks to welfare if an inefficient stockman was involved. In general, when sows are kept in stable groups, aggression is not a major problem (EFSA, 2007a).

Adrenal and other physiological measures

Measurements of plasma glucocorticoid levels and of heart rate are useful indicators of the welfare of animals subjected to short-term procedures such as handling and transport. Assays of glucocorticoids or measurement of heart rate, however, are of little use when the long-term effects of housing systems are being evaluated. The evaluation of responses to an adrenocorticotrophic hormone (ACTH) challenge after various previous housing conditions is of some interest, although precise interpretation is rather difficult. Barnett *et al.* (1984) compared the cortisol responses to loading and transport and to ACTH challenge in sows previously kept in tethers, pairs or groups. The tethered sows showed a greater response to the transport and some sign of a higher response to ACTH challenge, but there were also high responses from some animals in pairs. Barnett *et al.* (1981) had found that tethered sows had higher plasma cortisol levels than stall-housed sows, and a further study (Barnett *et al.*, 1987) showed that this higher level could be reduced by using wire mesh between the tether stalls. These results do not, however, reflect long-term responses to the housing conditions. Measurement of opioid receptors in the brain has provided evidence of differences in coping methods used by sows (Zanella *et al.*, 1996, 1998).

Studies of preferences and the improvement of welfare

In designing systems for keeping sows, studies of how sows preferred to spend their time in an extensive and varied outdoor environment are useful (Stolba *et al.*, 1983). Sows in fields (see Fig. 31.4) spend much time rooting, and Hutson (1989) found that pigs will carry out operant responses many times where the opportunity to root in earth is the reinforcer. The extensive usage of straw as a material to manipulate is clear from many studies on sows, e.g. Jensen (1979). Straw may not be the only material that serves this purpose, and its beneficial effects may not counteract all the effects of an otherwise adverse environment, but its use should be considered in all systems where no comparable material is present.

Farrowing Sows and Suckling Piglets

One major welfare problem in all countries is the high level of piglet mortality within the first few days of life (Marchant *et al.*, 2000). In modern, well-organized pig farms, the proportion of piglets born alive that fail to survive to weaning is typically 11.5% in the UK, 12.5% in Denmark and 13% in the Netherlands (British Pig Executive, 2011; Kirkden *et al.*, 2013b). The most important cause of death in young piglets is overlying by the sow. Piglets may be squashed and killed or may be weakened so that they are less likely to be able to suckle and more likely to succumb to diseases. Selection by breeders has resulted in a substantial increase in the size of the modern sow in relation to her wild ancestor, but much less change in the size at birth of the piglets and some tendency for maternal behaviour to be impaired (Marchant *et al.*, 2001). Selection for maternal behaviour can improve maternal behaviour and piglet survival (Baxter *et al.*, 2011). For example, she may be more responsive to the distinctive call of a piglet that is being crushed (Illmann *et al.*, 2013). However, the selection of pigs in order to increase litter size has

Fig. 31.4. Sows kept in a field with arks to shelter in can walk, run, root and carry out a wide range of other activities. The best fields are those with shallow soil and stones or rock near the surface, for these do not get too muddy. Welfare can be poor in very cold conditions unless bedding is provided (photograph D.M. Broom).

had a negative effect on piglet survival as it results in there being a larger proportion of weaker pigs and a greater likelihood that the number of piglets will be more than the number of functional teats that the sow has (Weber *et al.*, 2009; Andersen *et al.*, 2011).

Diseases are also a major factor in piglet mortality, and the likelihood of serious disease effects is dependent upon the absorption of immunoglobulins from colostrum. A summary of the factors leading to inadequate colostrum or milk intake by piglets is shown in Fig. 19.12. Piglet weakness and mortality is clearly a major welfare problem, as well as being a great economic problem for the farmer.

The use of a farrowing crate (see Fig. 31.5) led to less mortality than that which occurred in a simple pen and various versions of the farrowing crate are now very frequently used on pig farms. Piglet survival is improved by using a warm creep area and bars on the farrowing crate that reduce the chance that the piglet will move under the sow (Vasdal *et al.*, 2010). However, overlying is still a problem. After farrowing, the sow's environment is interesting because she is frequently visited by piglets. However, in a crate, she is very restricted in her movements and she cannot move much towards the piglets. This is a frustrating situation for the sow. Before farrowing, sows will build large nests if they are given the opportunity, and inability to build a nest is also frustrating for the sow (Lawrence *et al.*, 1994).

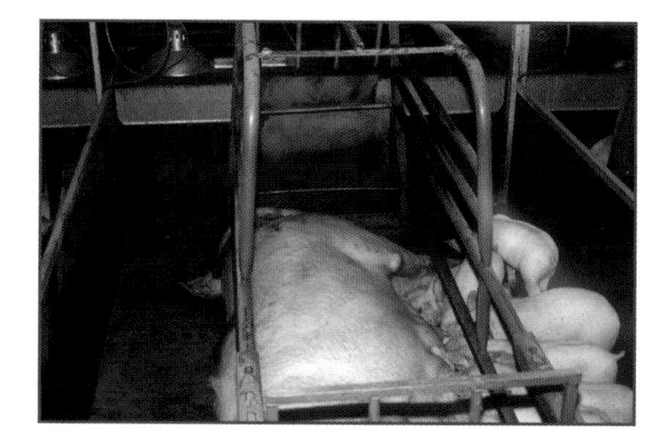

Fig. 31.5. Sow in a farrowing crate (photograph D.M. Broom).

Overall, the widely used farrowing crate is easy to manage but is far from ideal for the sow. Outdoor farrowing in huts on well-drained soil can provide good conditions for the sow and the piglets. Piglet mortality is similar to or lower than that in farrowing crates except in unusually bad weather conditions. However, indoor systems are needed in some climate and soil conditions.

Farrowing crates are widely used in commercial practice, but restrict sow movement and have received an increasing amount of criticism due to evidence that such restriction impairs welfare (e.g. Cronin *et al.*, 1991; Lawrence *et al.*, 1994; Arey, 1997; Edwards and Fraser, 1997). Various alternative systems have, therefore, been developed which are less restrictive for the sow and allow varying degrees of freedom of movement. These systems may be divided into: (i) group-farrowing systems, where sows share a communal area and can build an individual nest (Gotz and Troxler, 1993; Cronin *et al.*, 1996; Wechsler, 1996); and (ii) pen systems, which confine the sows individually, without a common communal area, but allow some freedom of movement. Examples of this latter system include the turn-around crate (McGlone and Blecha, 1987), the 'Ottawa' crate (D. Fraser *et al.*, 1988), the ellipsoid crate (Lou and Hurnik, 1994; Rudd and Marchant, 1995), the slope-floor pen (McGlone and Morrow-Tesch, 1990), the farrowing box (Schmid, 1997; Damm *et al.*, 2003), the Werribee pen (Cronin *et al.*, 2000), the Pigsafe pen (Baxter *et al.*, 2012) and the UMB pen (Andersen *et al.*, 2014). In the review by Baxter *et al.* (2012), the welfare score for farrowing crates was low and that for outdoor farrowing with huts and some larger farrowing pens was substantially higher.

Problems encountered with some systems have included difficulty in management in some of the earlier systems (Arey, 1997) and high piglet mortality due to crushing. Benefits to the sow, such as greater freedom of movement and possibility to build a nest, must be balanced against costs to piglets. Group-farrowing systems have the advantage of allowing the sow considerable freedom of movement and the opportunity to integrate back into the group shortly after farrowing, but introduce added difficulties due to the possibility of social aggression. An individual pen system may therefore provide a solution in providing stockmen with good protection, piglets with an area to avoid sow movements (i.e. being crushed) and the sow some degree of freedom of movement for the performance of behaviours including nesting behaviour. There will be some extra cost of systems that require more space in a building and this is quantified for several systems by Baxter *et al.* (2012).

The UMB farrowing pen (Andersen *et al.*, 2014) comprises two compartments: a nest area covered with a rubber mat, and an area for activity and dunging, separated by a threshold that the sow can step over and removable doors for use if the sow has to be temporarily confined in one compartment (Fig. 31.6). The nest area is at the rear of the pen because sows choose to farrow there. They probably make this choice to avoid disturbance, such as that by stockmen who approach the pen from the front. The front dunging area is surrounded by fences made of vertical stainless steel rods to enable the sow to see her surroundings and provide visual and limited physical contact with neighbouring sows. The nest area has solid side walls to provide a closed cave-like environment for the sow and piglets, affording the sow a visual barrier for privacy from neighbouring sows while in the nest, and hence some sense of isolation from herd mates. There is a hay or straw rack from which the sow can pull out as much material as she is motivated to eat and use for nest-building. In the nest area are sloping panels along two walls and two, independently

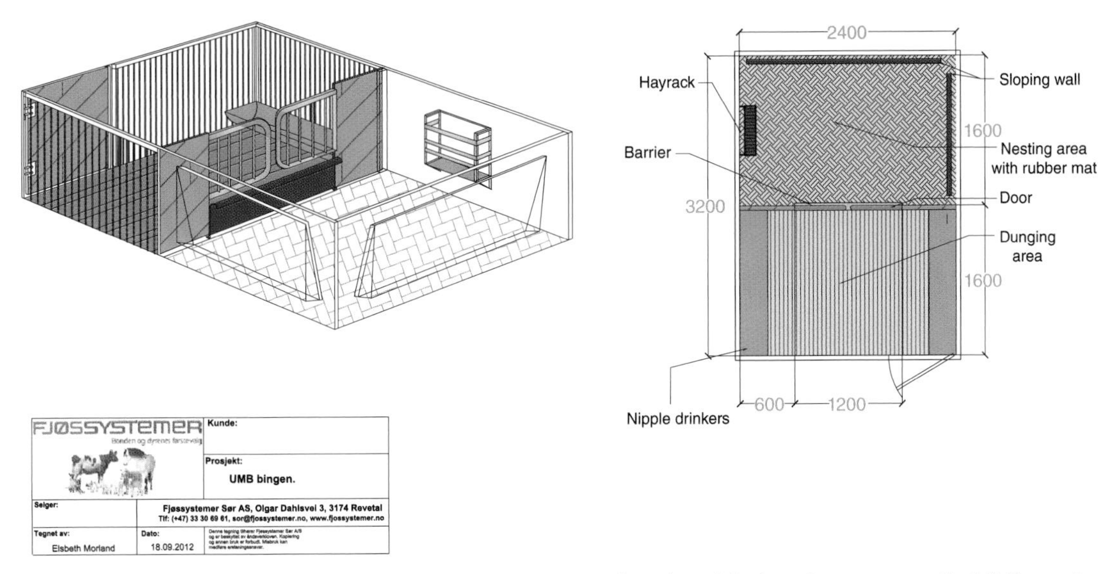

Fig. 31.6. Diagram of UMB farrowing pen with the rear nesting area on the right and the front dunging area on the left. The nesting area has sloping walls and a rack for nesting material. There is a barrier that the sow can step over and doors that are removed or open except when closed for management purposes (diagram reproduced by permission of I.L. Andersen).

controlled, heated-floor zones. Sows prefer to lie against sloping panels when descending from standing to lying posture (Damm *et al.*, 2006). Hence, with the ability to control temperature in different floor zones, it is possible to influence where the sow lies relative to her litter and greatly reduce the risk of overlying. Figure 31.7 shows a sow in the dunging area before parturition and Fig. 31.8 shows a sow with a nest and piglets in the nesting area. In trials of this pen in Norway and Australia, the piglet mortality rate of 12–13% was achieved without birth assistance or any particular surveillance around the time of farrowing. This is better than that often reported for farrowing crates.

Induction of farrowing has become a relatively common practice on pig farms. It is normally carried out using prostaglandin, or a synthetic analogue, sometimes together with oxytocin. Supervision of farrowing and effective fostering, if needed, are facilitated by determining the time of farrowing and these will improve sow and piglet welfare. However, piglet viability is reduced if the birth is too early so induction should be no earlier than 2 days before the expected farrowing date (Kirkden *et al.*, 2013a). Given that the prediction of farrowing date is not always accurate and that oxytocin use increases the risk of dystocia, farrowing induction should not be used unless it has a net beneficial effect on sows and piglets and only if expert care of the animals is available at farrowing.

Fig. 31.7. Sow standing in dunging area of UMB pen before parturition. The rack has nesting material in it (photograph I.L. Andersen).

Fig. 31.8. Sow in nesting area of UMB pen with piglets (photograph I.L. Andersen).

Good care of sows and piglets at farrowing includes assistance of dystocic sows, measures to prevent and treat sow hypogalactia, good hygiene, provision of a warm micro-environment for piglets, early fostering of supernumerary piglets, helping weak piglets to breathe and obtain colostrum, and intervention to prevent crushing or savaging of piglets (Kirkden *et al.*, 2013b). All of these measures are most effective if carried out by trained persons who are aware of the needs of sow and piglets. The actions can substantially improve welfare, piglet survival and hence the economics of pig production. The measures are most effective if the farrowing occurs in an appropriately designed farrowing pen.

Piglets and Fattening Pigs

The young piglet is often subjected to a series of operations within its first few days of life. In many countries the piglet is picked up, injected with iron supplement and sometimes antibiotics, tail-docked and castrated, all without anaesthetic. In experimental studies of the response to castration, this operation leads to increased high-frequency calls (>1000 Hz) or screams, as compared with sham-manipulated or anaesthetized pigs (Weary *et al.*, 1998; Taylor and Weary, 2000; Prunier *et al.*, 2002; Marx *et al.*, 2003). These calls are accompanied by physical resistance movements and an activation of the sympathetic nervous system, as demonstrated by an increase in heart rate (White *et al.*, 1995) and adrenal cortical activity (Prunier *et al.*, 2004). Analysis of the calls during the overall procedure of castration suggests that pain is more acute during extraction of the testes and severing of the spermatic cords (Taylor and Weary, 2000). This is further supported by the observation that local anaesthesia is most effective by reducing behavioural resistance when the cords are cut (Horn *et al.*, 1999). Piglet responses to the pain of tail-docking were increased if the mother sows had been stressed on three occasions during pregnancy (Rutherford *et al.*, 2009). In parallel with these physiological reactions, behaviour is modified. Castrated pigs spend less time at the mammary glands massaging or suckling (McGlone and Hellman, 1988; McGlone *et al.*, 1993; Hay *et al.*, 2003). They remain more inactive while awake, they show more pain-related behaviours (prostration, stiffness, trembling) and tail-wagging, they frequently seek solitude and their behaviour is more frequently desynchronized compared with uncastrated control piglets (Hay *et al.*, 2003). Castration without anaesthesia or analgesia leads to immunosuppression (Lessard *et al.*, 2002).

A traumatic event encountered by each piglet on commercial farms is weaning. As described by Jensen (1986), piglets left with their mother are weaned after 10 weeks of age, but the common commercial practice is to wean at 3 or 4 weeks of age. Such early weaning has considerable effects on the piglets, leading to poor welfare. The absence of the mother and her teats from which milk can be obtained forces the piglet to look for food elsewhere and leads to much nosing and sucking of other piglets. Belly-nosing is an up-and-down massaging movement with the snout placed under the belly of other pigs, often described as being shown frequently by early-weaned piglets (van Putten and Dammers, 1976; Schmidt, 1982). During or after belly-nosing, the genitalia or navel may be sucked and urine drunk. Anal massage may also occur and can lead to lesions with bleeding (Sambraus, 1979). Piglets may be chased around in a pen by those attempting belly-nosing, etc., and it seems likely that their welfare is adversely affected by the chasing, the inability to escape and by injuries which may result. The piglet that is attempting belly-nosing, etc., or snout-rubbing and sucking on floor, wall or bars is manifesting its desire to suck, which would be directed at the sows' teats had these still been available. The persistence of such behaviour gives some indication of the extent to which the piglet feels deprived.

Another welfare problem arising at the time of weaning is fighting. If piglets from different litters are mixed after weaning they interact in various ways, including fighting, as would be expected when they have to establish new social relationships. Such fighting can result in injuries. As with belly-nosing, the major problem arising is probably that of the individual that is continually being chased around the pen by one or more other piglets. If piglets are provided with the opportunity to hide their heads in a 'pop-hole' or to hide behind a barrier in the pen (see Fig. 31.9), belly-nosing and chasing can be reduced (McGlone and Curtis, 1985; Waran and Broom, 1993). Fighting may be reduced by the use of tranquillizers at the time of mixing piglets, but there is no evidence that this improves their subsequent welfare. Fighting when pigs are mixed is a greater problem with older pigs since they can inflict much more serious injuries on one another, so mixing of pigs should be avoided. Lesions on the skin are good indicators that

Fig. 31.9. The barrier in this pen allows piglets to hide if chased (photograph D.M. Broom).

there has been fighting in groups of pigs (Turner *et al.*, 2009). Genetic selection of pigs to reduce the likelihood of fighting has been demonstrated to be possible (Canario *et al.*, 2012) and should be utilized.

Environmental conditions such as the presence of toxic substances or adverse light, temperature or noise conditions, will affect the welfare of pigs. Pigs can adapt to a certain temperature range but freezing conditions or too hot conditions cause problems. In a comparison of pigs kept with straw and objects to manipulate, Parker *et al.* (2010) and O'Connor *et al.* (2010) found that aggression in pigs was increased by an ambient ammonia concentration of 20 parts per million and by a light level of 40 lux as compared with 200 lux. Play behaviour was reduced at the high ammonia concentration and adrenal size was increased. If there was a high level of noise (over 80 dB) the pigs showed less submissive behaviour and a combination of high noise and increased ammonia led to more disease.

In some countries it is legal to treat pigs with beta-agonist drugs in order to increase their growth rate. Beta-agonists result in there being more water in the meat, so that there is less nutrient value per kilo, and they may also change the proportion of fat. The beta-agonist clenbuterol is persistent in the carcass but ractopamine is less persistent. The metabolic fate of ractopamine is similar in cattle, pigs, laboratory animals and humans and it results in increased anxiety in humans. In pigs it leads to tachycardia, peripheral vasodilation, higher circulating catecholamines, increased activity, more biting of other pigs, more chasing of other pigs, greater difficulty in handling by people and changes in transmitter substances in the frontal cortex and amygdala of the brain (Marchant-Forde *et al.* 2003; Poletto *et al.*, 2009, 2010a,b). The substantial negative effects of ractopamine on the welfare of pigs and the small risk of adverse effects on the welfare of human consumers is the reason why ractopamine use as a growth promoter is banned in 160 countries.

The amount of space available to piglets or fattening pigs affects their ease of movement, thermoregulation, defecation, exploration, ability to find an adequate resting place and ability to avoid the undesirable attention of others. It is not just the area of floor that is of importance, however, but the quality of space available. When assessing the space requirements of pigs, all of their movements during lying and standing need to be taken into account (Petherick, 1983a). Other factors that should also be considered are: (i) the provision of adequate separate lying and dunging areas (Baxter and Schwaller, 1983); (ii) the possibility of feeding without there being a high possibility of attack by others; and (iii) the possibility of avoiding serious attack, etc., by others.

In calculating the amount of space needed by pigs, Petherick (1983a), expressed the area (A) in m^2, as a relationship with body weight (W): $A = kW^{2/3}$. The factor k differs according to the posture adopted by the pig and, as explained in EFSA (2006b), to all of the movements that the animal has to perform in order that its needs can be met. As k increases from a low space allowance value, performance such as weight increases (Edwards *et al.*, 1988; Gonyou *et al.*, 2004). Meunier-Salaun *et al.* (1987) argue that behavioural and physiological responses are earlier and more sensitive indicators of adaptation to the environment than productivity. The conclusion reached by EFSA (2006b) was that in order for pigs to be able to have separate dunging and lying areas, to thermoregulate adequately at temperatures that may exceed 25°C and to move around when some are lying, and to interact socially (Arey and Edwards, 1998; Spoolder *et al.*, 2000; Aarnink *et al.*, 2001; Turner *et al.*, 2001), the value for k should be 0.047.

Another major welfare problem for fattening pigs is having to stand on inadequate flooring. Pigs may trap their claws, suffer abrasion injuries or break or strain their legs because the floors are slippery. Old concrete floors and old slats cause special problems, but some of the newer floors for piglets or older fattening pigs are clearly very uncomfortable for the animals to stand and move on. Beattie *et al.* (2000) found that the pigs in an environment that includes straw spent more time in exploratory behaviour and less time inactive or engaged in aggressive behaviours towards pen-mates than pigs in slatted floor housing. The pigs in the barren environments persistently nosed and chewed their pen-mates. Kelly *et al.* (2000) found that the behaviour of weaners weaned at 3 weeks was less directed at pen-mates and the pen and more towards straw if pigs were raised on either deep-straw or straw-flow systems compared with pigs housed in flat-deck systems with expanded metal as flooring. Hence, tail-biting is less likely if straw is provided (Zonderland *et al.*, 2004). A comparison of pigs in typical, rather barren conditions with pigs with straw and a variety of other manipulable materials showed that the animals from the enriched environment showed a positive cognitive bias in a go/no go test with sound pitch as the cue. This is interpreted as indicating positive emotion and good welfare. If animals from the enriched environment were moved to the barren environment, the bias, and hence welfare, became more negative (Douglas *et al.*, 2012).

Young pigs prefer places with straw (Steiger *et al.*, 1979) and prefer solid floors to slatted floors (Marx and Mertz, 1989), especially if the former have straw available. Ducreux *et al.* (2002) found that pigs preferred to rest on litter and not on slats, the more so if they had been reared at lower temperatures. Pigs preferred not to walk on slats if these allowed a claw to go into the slot (Greif, 1982). Where solid floors are used, it is easy to supply pigs with straw or other manipulable materials. The provision of straw leads to reduced bursitis and other pathologies (Pearce, 1993). If floors are slatted, or partly slatted, the manure disposal system can be such that straw, etc., can be disposed of. This is done by incorporating chopping equipment in the manure disposal pipe so that it is not blocked by manure or straw.

Further Reading: Kirkden, R.D., Broom, D.M. and Andersen, I.L. (2013b) Piglet mortality: management solutions. *Journal of Animal Science* 91, 3361–3389.

32 The Welfare of Poultry

Introduction

Human attitudes to poultry often differ from those to domestic mammals. The chicken is less often thought of as an individual and few people would ascribe much intellectual ability to it. These attitudes are partly a consequence of the very large numbers of these animals that are kept in one place, partly to the fact that they are birds and therefore harder for a person to identify with than are the larger mammals, and only slightly because of any real difference in intellectual ability. In fact, when individual chickens are put into experimental learning situations that are not frightening to the birds, they perform quite well (see Rugani *et al.*, 2009; Chapter 3, this volume). Their sensory ability is very good when compared with that of man and their social organization is complex, requiring considerable ability in learning and memory in order to maintain it.

The chicken is too often seen in a situation where its fear of man dominates most of its activities. Hence, the average member of the public or chicken farmer is unable to assess adequately its behavioural complexity and awareness of its environment. These attitudes to chickens, and to a lesser extent to other poultry, have had a considerable influence on man's assessment of what conditions these animals need and what treatment they can tolerate. Those who have studied the behaviour of the domestic fowl in detail (Appleby *et al.*, 2004), especially those who have looked at feral fowl (McBride *et al.*, 1969; Wood-Gush *et al.*, 1978), always acquire much respect for the members of this species. There are few differences in behaviour between the wild Burmese red jungle fowl (*Gallus gallus spadiceus*) and the domestic fowl (*Gallus gallus domesticus*), and the calls, displays and other behaviour are diverse. The concept of rank order, or peck order, in a social group originated from work on chickens (Schjelderup-Ebbe, 1922; Guhl, 1968), and has been developed substantially by further work on this species.

The examples of welfare problems reported in this chapter start with those of the domestic hen, because most work has been carried out on this species and because of its numerical importance. The domestic fowl is the most common bird in the world, with numbers estimated at 9–10 billion. It is a very successful species and it can reasonably be said that it exploits man very effectively. If there are widespread welfare problems in this species, however, then enormous numbers of individuals may be affected. Hence, the welfare of chickens is a very important subject. The two main sections in this chapter concern the effects on welfare of housing systems for hens kept for egg production (see Newberry, 2004) and of housing for broilers kept for chicken meat production (see Weeks and Butterworth, 2004). Welfare problems in breeding hens, in the rearing of young chickens and in rearing turkeys, ducks and geese are then discussed. A key issue for all of these birds is whether welfare is worse in large groups. A review of data by Estevez *et al.* (2007) indicates that in the main, for poultry species, where sufficient space is available larger group size is not associated with increased aggression. However, there is a need for farm staff to look carefully at individuals to check for problems when numbers are large.

The important problems associated with the handling and transport of poultry are discussed in Chapter 21.

Laying Hens

The needs of hens

As explained in Chapter 1, animals have needs for particular resources and needs to carry out actions whose function is to obtain an objective. Needs can be identified by studies of motivation and by assessing the welfare of individuals whose needs are not satisfied. Unsatisfied needs are often, but not always, associated with bad feelings, while satisfied needs may be associated with good feelings. When needs are not satisfied, welfare will be poorer than when they are satisfied. The needs of hens are summarized below (Broom, 1992, 2001b). These needs include basic biological functioning and the means to achieve this by carrying out a variety of activities, responding to certain stimuli and maintaining certain physiological states. High levels of motivation related to these needs are not continuous:

- Obtain adequate nutrients and water.
- Grow and maintain themselves in such a way that their bodies can function properly.
- Avoid damaging environmental conditions, injury or disease.
- Be able to minimize the occurrence of pain, fear and frustration.
- Show certain foraging and investigatory movements.
- Have sufficient exercise.
- Show preening and dust-bathing behaviour.
- Explore and respond to signs of potential danger.
- Interact socially with other hens.
- Search for, or create by building, a suitable nest site.

Management systems

Hens kept for egg-laying were largely maintained in free-range conditions in most countries until 1950. The birds were normally shut up in some sort of house at night, but were able to roam around a farmyard or field throughout the day. There was then a change to indoor housing on deep litter for a period of 10–15 years followed by a further change to the use of battery cages in controlled indoor environments. During the 1990s, concern about hen welfare led to an increase in free-range and barn egg production (see Table 32.1). Free-range systems are those where hens are kept with access to an open area at a stocking density defined by regulation in the European Union (EU). The cost of eggs to a consumer can be twice as much for free-range eggs as for the cheapest eggs, in which case the profit is much higher for free-range because eggs from good free-range systems do not cost twice as much to produce. Battery cage usage has ceased in the UK since the system was made illegal in the EU.

Table 32.1. Different methods of keeping laying hens in the UK (from Ewbank, 1981; DEFRA, 2006).

Year	Free range (%)	Deep litter (%)	Battery cages (%)
1948	88	4	8
1956	44	41	15
1961	31	50	19
1965	16	36	48
1977	3	4	93
1981	2	2	96
1998	15	5	79
2005	30	6	64

The first use of rows of cages for hens in a building where the physical conditions could be controlled occurred in the USA in the 1930s. Such rows could be called a battery of cages, so the term 'battery cage' gradually became widespread as the system developed. Battery houses (see Fig. 32.1) allow separation of the birds from their faeces to a greater extent than some other systems, and this has advantages for disease and parasite control. The faeces may be carried away by a belt under the cages or may fall directly or via deflectors into a pit. Figure 32.2 shows some designs of battery houses. Disease is further reduced by the all-in–all-out management system. Hens are brought into a clean, empty house shortly before they start to lay and they are all removed for slaughter at the same time when they are about 72 weeks old. The number of birds in a cage is commonly five but is often more in North America. The average temperature, humidity and ventilation throughout the house can be controlled with some accuracy. Water is usually available continuously from drinkers, and food is supplied regularly and usually automatically by a moving chain or belt.

There can be modifications of battery cages that reduce the incidence of injuries but do not give the animal more freedom to carry out normal behaviour. Furnished cages have been designed that have a perch, a nest box and a sand bath, where dust-bathing is possible. These cages can have automatic food provision and egg-removal mechanisms (see Fig. 32.3). At present, such cages are the same height as battery cages or a little taller, to allow the perch to be used, but sufficient space for birds to stretch and flap their wings could be

Fig. 32.1. Hens in battery cage (photograph D.M. Broom).

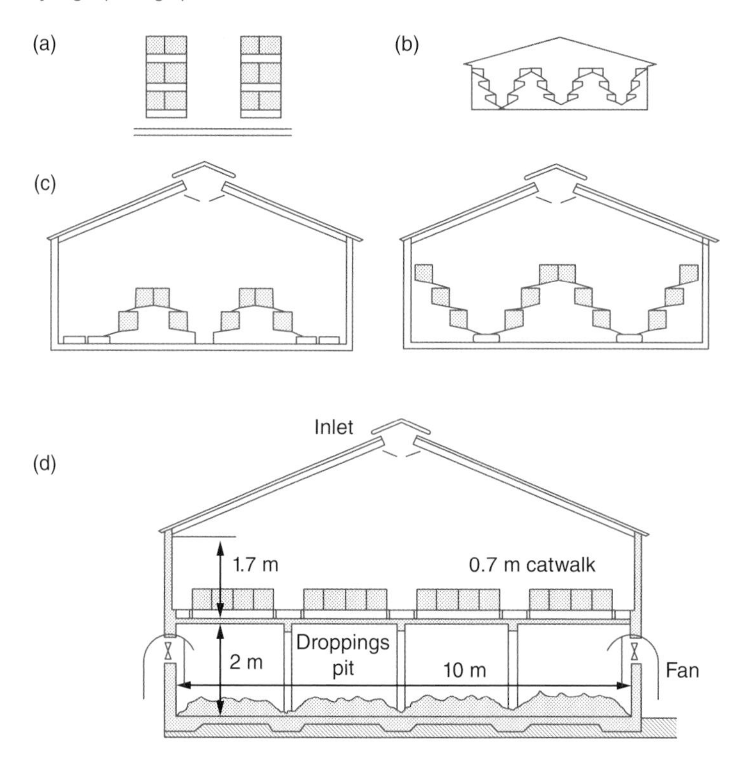

Fig. 32.2. Battery cage houses: (a) vertical cages; (b) semi-stepped cages; (c) fully stepped cages; and (d) flat-deck cages over deep pit for droppings. Some modern systems have more tiers of cages (from Sainsbury and Sainsbury, 1988).

provided in a taller cage with a floor area a little larger than the conventional cage.

Since major problems for a hen in a battery cage seemed to be lack of space and inability to escape from a bird that is pecking it, the get-away cage was designed by Elson (1976). This was developed further in the Netherlands (Brantas *et al.*, 1978) and in Germany (Wegner, 1980). These cages (see Fig. 32.4) housed

Fig. 32.3. Hens in furnished cage with perches, nest box and dust bath (photograph D.M. Broom).

about 20 hens on two levels and they had nests, perches and a sand-bathing area. As explained later, they were not successful.

The deep-litter system in which many hens were housed together on some sort of litter, or partly on slats, in a building with many nest boxes was very widespread in the early 1960s. The space allowance for hens was commonly 0.18 m^2 for heavy breeds and 0.14 m^2 for light breeds, if slats were provided, and 0.27–0.36 m^2 per bird if there were no slats (Sainsbury, 1980). In the deep-litter system the hens have most of the space available to them, whereas in small cages with the same space allowance the birds can take only a few paces before reaching the end of the cage. The deep-litter system is in widespread use for breeding birds so it is generally acceptable to the poultry industry except for the extra difficulties in collecting eggs as compared with a battery house. However, most of the volume of a building is not utilized in contrast to multilevel battery houses or modern tiered floor systems. Hence the building cost per egg produced is high for a deep-litter house.

The perchery is a development from the deep-litter house in which birds are not just on the floor, but they have rows of perches at different levels in the building that they can and do use. One problem arising in houses with many perches is that some young birds come into the house and do not use the perches. Appleby (1985) found, however, that perch use was almost universal if perches had been available to the hens when they were young chicks, so early experience of perching is important. Percheries have now been superseded by aviaries.

The range of intermediates between the get-away cage and the perchery includes aviaries such as voliere, volierenstall, voletage, Rihs Boleg 1, Rihs Boleg 2, Natura, Hans Kier unit and tiered-wire-floor units. The aviary has extra floors of wire or slats with feeders, drinkers and nest boxes. Tiered-wire-floor units have four or more tiers whose floors allow droppings to go through on to a conveyor belt below that removes them. There are usually perches above the top tier, and adjacent stacks of tiers are staggered to facilitate jumping from tier to tier by the birds. The floor of the building has litter on it so hens can scratch in it. There are stacks of nest boxes on the sidewalls (see Figs 32.5, 32.6 and 32.7).

Housing and Management in Relation to the Needs of Hens

Space for movements

EFSA (2005a) reviews hen-housing systems in relation to welfare. If hens need to carry out a range of normal movements, how much space is required for these? Measurements of the space occupied by a hen when carrying out such movements have been made (M. Dawkins and Hardie, 1989; Table 32.2).

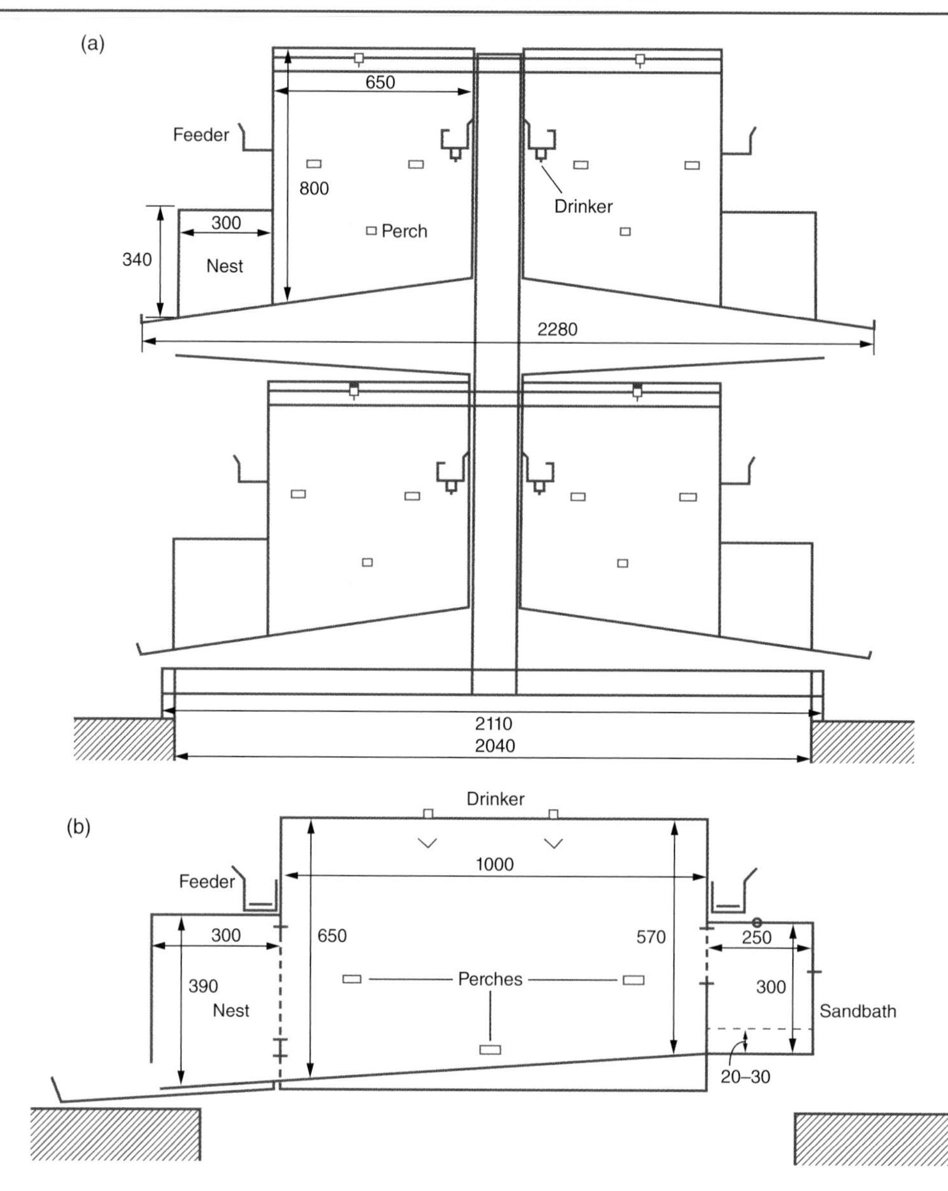

Fig. 32.4. Sectional diagrams of two forms of get-away cage (dimensions in mm). (a) Get-away cage with roll-away nest ± droppings pit; (b) get-away cage with roll-away nest and sand bath (from Wegner *et al.*, 1981).

If there are five hens in a cage these hens will not show all of the different movements simultaneously, and some hens might be relatively inactive while one bird uses more space. Some possible combinations of movement are considered in Table 32.3. It is clear from the calculations presented in this table that a cage for five hens, allowed 450 cm^2 each and hence occupying 2250 cm^2, severely inhibits normal movements. Wing-flapping is not possible with commonly used cage heights of 50 cm or less. If hens are allowed more space than 450 cm^2 per bird, the amount of disturbed behaviour shown is decreased (Nicol, 1987a). Hens will work for a larger space allowance of up to 1125 cm^2 per bird and they continue to space themselves out in cages of 1410 cm^2 per bird but, in much larger space allowances of 5630 cm^2 per hen, they cluster. The effects of space allowance on the extent of injurious behaviour are not a linear relationship in battery cages (Polley *et al.*, 1974; Al-Rawi and Craig, 1975), but depend upon the

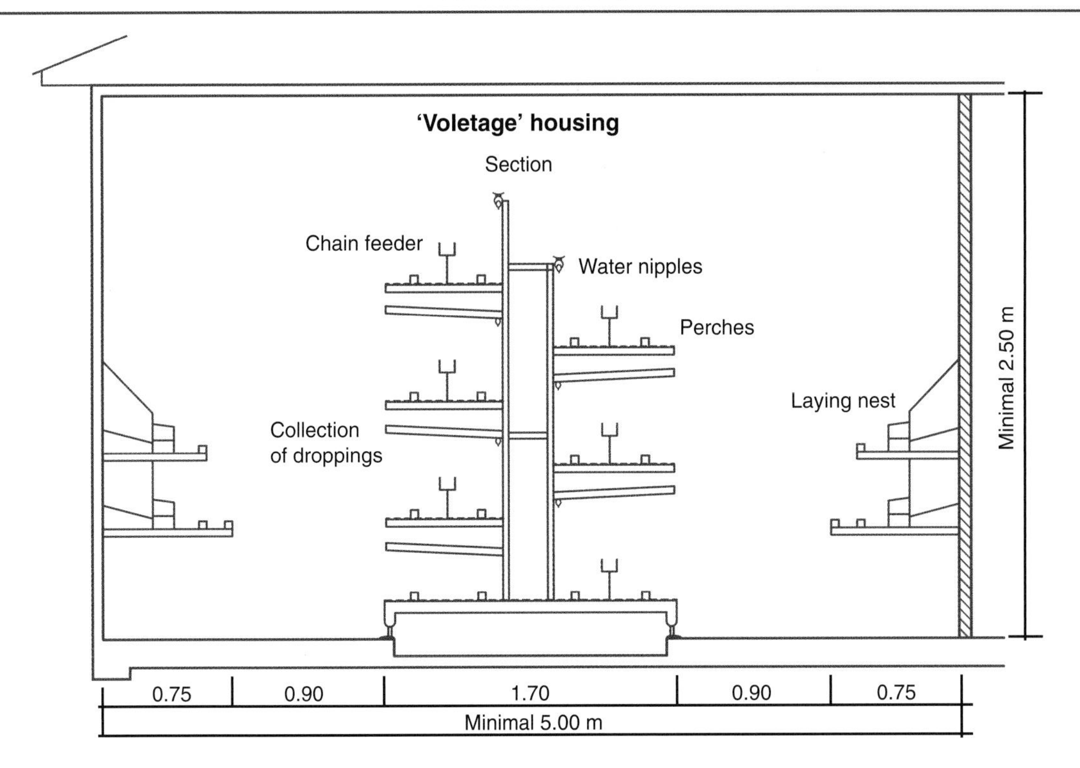

Fig. 32.5. Voletage for hens (after Fölsch *et al.*, 1983).

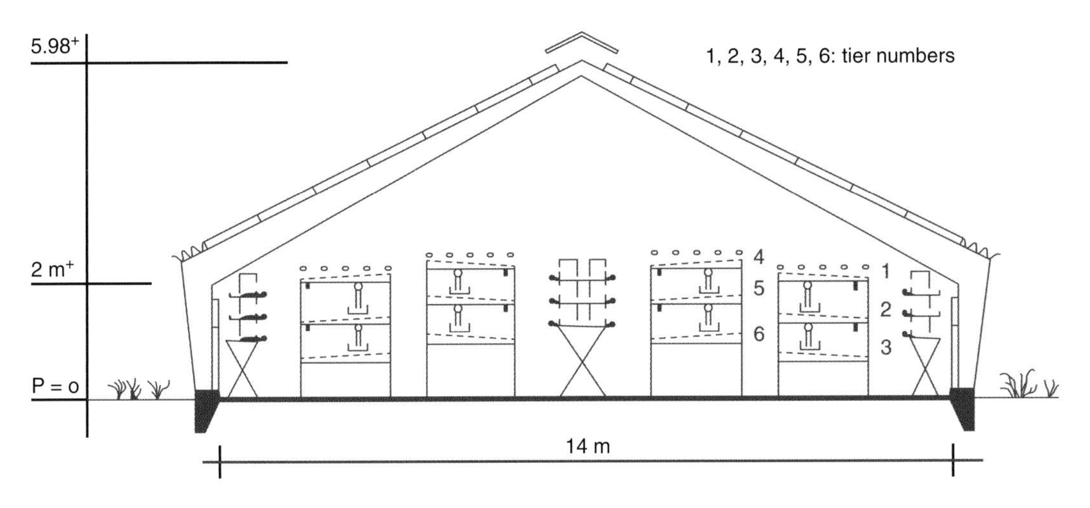

Fig. 32.6. The tiered wire floor system for housing of laying hens, with the following characteristics: (1) three tiers of floors – the upper floor is a resting area; (2) bordering wire floors mounted at unequal heights to create a staircase effect; and (3) litter covering all of the ground floor area (from Dutch Society for the Protection of Animals, 1986).

complexity of the environment. In order to provide opportunities for escape and to hide from birds that tend to feather-peck or cause tissue damage by pecking, more space allowance than that normally provided in a battery cage is needed. Such escape possibilities are important in order to minimize injuries caused by other birds. As long as these are available, the level of injurious behaviour can be low at various aviary space allowances.

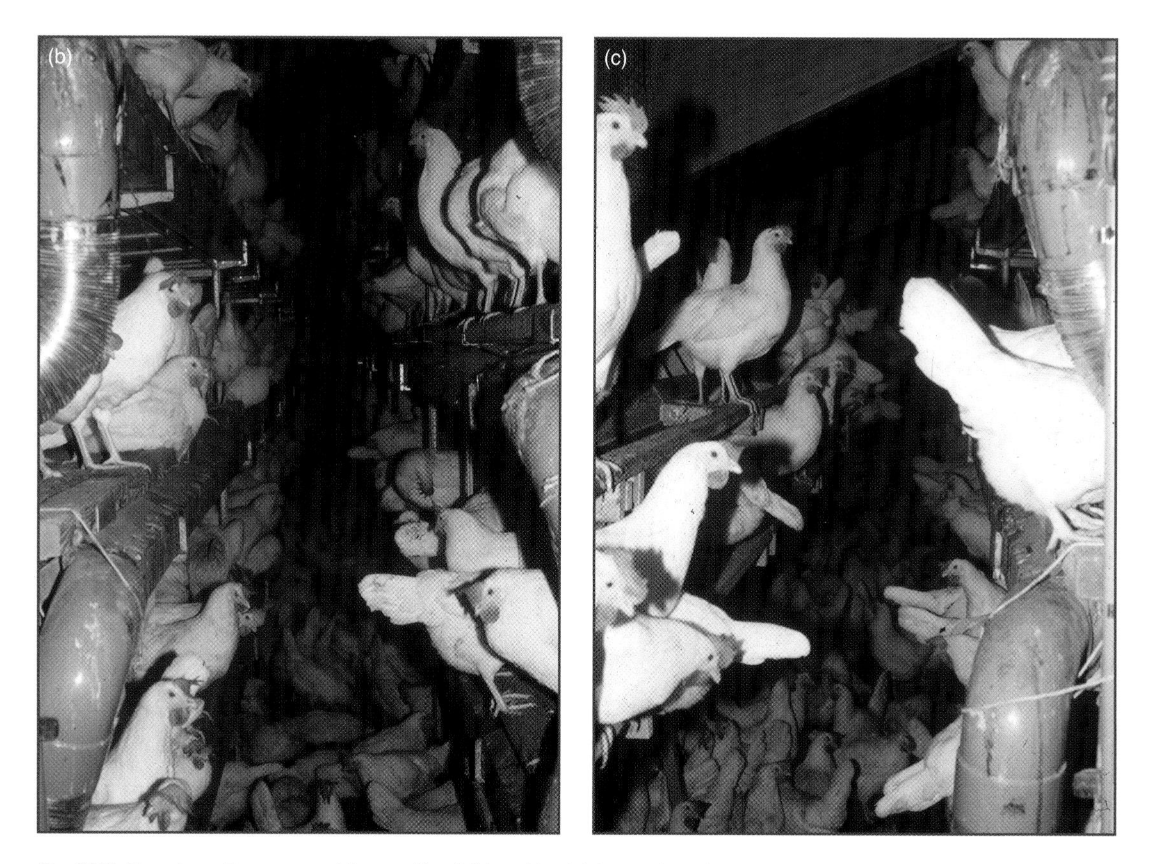

Fig. 32.7. Tiered-wire floor system: (a) ground level; (b) mid-level; (c) upper level (photographs D.M. Broom).

Table 32.2. Area required by hens for different behaviour patterns (data from Dawkins and Hardie, 1989).

Activity	Area required (cm²)
Standing	428–5922
Turning	771–1377
Preening	818–1270
Ground-scratching	540–1005
Wing-stretching	653–1118
Wing-flapping	860–1980

Table 32.3. Space required for hens in a cage holding five birds.

Activity	Total space (cm²)	Space per bird (cm²)
Four hens crowded together plus one wing-flapping	2720	544
Four hens crowded together plus one wing-stretching	2185	437
Four hens crowded together plus one preening	2342	468
Three hens crowded, one turning, one wing-flapping	3469	694
Two crowded, two turning, one wing-flapping	4218	844
Four hens standing, preening	3074	615
Four hens standing, one wing-flapping	3460	692
Two hens standing, two turning, one wing-stretching	4050	810
Two hens standing, two turning, one wing-flapping	4584	917

Bone fragility

The calcium and vitamin D in the diet of hens is adequate for bone development, but the bones of hens from battery cages break easily. In a series of studies, 25–40% of end-of-lay hens from battery cages were found to have at least one broken bone following handling prior to stunning, and 98% of carcasses had a broken bone (Gregory and Wilkins, 1989; Gregory *et al.*, 1990, 1991). Some reduction in bone weakness can be brought about by feeding more omega-3 polyunsaturated fatty acids to hens (Tarlton *et al.*, 2013). However, the numbers of broken bones in hens from percheries and aviaries were much lower than from battery cages. Hens sometimes broke their bones while they were in poorly designed or overcrowded percheries. In the most confined systems, the strength of the bones in wings and legs was reduced because there was insufficient opportunity for exercise. Birds that lived in cages in which they could not flap their wings had wing bones that were only half as strong as those of birds in a perchery that could and did flap (Knowles and Broom, 1990; Nørgaard-Nielsen, 1990). The commonest bone breakage is that of the keel bone and, as might be expected, keel bone fractures result in pain-related behaviour that is reduced by treatment with the analgesics morphine, a mu-receptor agonist, and butorphanol, a kappa-receptor agonist (Nasr *et al.*, 2012).

Investigatory pecking and dust-bathing

Chickens strongly prefer litter floors to wire floors (M. Dawkins, 1982; Appleby *et al.*, 2004). The opportunity to peck at objects on the floor, scratch on the floor and dust-bathe in a suitable substratum reduces the likelihood that injurious behaviour will be shown by hens and broiler breeders (Blokhuis, 1986; Kjaer and Vestergaard, 1999). Studies of the developmental and motivational basis of feather-pecking behaviour indicate links with deprivation of ground-pecking and dust-bathing opportunities.

Avoidance of injurious pecking

Hens may peck other hens in the course of investigation, attempts to obtain feathers, food-finding or aggression. In chickens, aggressive pecks are distinctive, forceful, usually downward pecks aimed at the head or dorsal region of another bird (Kjaer, 2000). Recipient birds tend to move away to avoid being pecked. Feathers are sometimes damaged, but this is usually restricted to the head region (Bilcik and Keeling, 1999). The term 'aggression' is sometimes used within the chicken industry to describe feather-pecking or cannibalism. However, social aggression is quite distinct from these behaviours both in form and origin (Savory, 1995). Pecking of the cloaca is probably motivated in a similar way to food-searching; but cloacal pecking, effectively cannibalism, causes physical damage. It is often also reported to be common in birds showing production diseases and infections such as salpingitis (Tauson *et al.*, 1999). Feather-pecking is discussed later.

Nest boxes and egg-laying

An appropriate nest box is used by almost all hens if readily accessible (Wood-Gush, 1975), and behaviour is clearly disturbed if none is available. The abnormal behaviour most frequently observed when no suitable nest site is present is stereotyped pacing (Wood-Gush and Gilbert, 1969; Fölsch, 1981). This stereotypy is a sign of long-term, intense frustration.

Perching

Perches are preferred resting places for all but the youngest chickens. The design should be right, and early experience of perches facilitates effective use (Appleby *et al.*, 1993). The presence of perches can increase leg strength (Hughes and Appleby, 1989). Where cloaca-pecking is a possibility, the perch should not be sited at such a height that the heads of some birds are level with the vents of others. Injurious pecking has been an important reason for the failure of some 'get-away' cages (Moinard *et al.*, 1998).

Lighting

If domestic fowl are kept in low light levels they are not able to show normal exploratory behaviour. At the lowest levels, eye development is impaired. Chickens show quite good visual discrimination down to light levels of about 5 lux but their ability declines very rapidly below this so normal behaviour would not be possible (Gover *et al.*, 2009). A review by Manser (1996) indicated clear welfare problems at light levels of up to 20 lux. Kristensen *et al.* (2007) have found that chickens prefer certain kinds of fluorescent lighting to incandescent lighting.

Beak-trimming

Mutilations involving tissue damage are painful at the time of the operation and can sometimes cause neuromas that result in lasting pain. The effect on welfare of beak-trimming is substantial, but is much greater if neuromas are present (Gentle, 1986). Beak-trimming also seriously impairs sensory input and pecking behaviour. The method used for beak-trimming that does not seem to cause substantial pain at the time of the operation is infra-red beak treatment (McKeegan and Philbey, 2012). In their study, the beak length 4 weeks after the operation was reduced by 44%, some regrowth of nerves had occurred and there was no neuroma formation.

Problems with Systems

Most hens utilize the opportunity to walk, flap, scratch, peck and interact socially in range conditions. These advantages can sometimes be outweighed by disadvantages. If hens are left in a free-range unit for some time without being moved to a new area, the risk of disease is high. For example, Löliger *et al.* (1981) reported that the incidence of worm infestation and coccidiosis was at least ten times higher in the free-range systems that they studied than in battery cages and there was also predation by birds of prey. Both disease and predation must be considered when welfare is assessed (see Fig. 32.8). In addition, free-range birds are subject to extreme weather conditions, and their welfare may be so poor during extreme winter weather that their egg production is substantially reduced. Hence, welfare can be very poor in free-range units, but these problems can be solved.

If stocking density is high, the risk that parasites and diseases transmitted via faeces will have significant effects will be reduced by moving the animals to fresh ground. The effects of inclement weather can be reduced by providing well-insulated and ventilated houses. Domestic fowl are, by origin, a tropical species, so in winter conditions the hens may seldom venture outside. Therefore, free-range houses need to be as good as the indoor group-housing accommodation. Hens often stay indoors during periods of rain or wind and cold, so the problems of aggression and house design referred to later in this chapter are relevant to free-range hens. The fact that conditions may effectively prohibit free-range hens from going outside is often not appreciated by those who advocate such systems without qualification. It is clear that welfare problems can be reduced to a very low level by free-range accommodation for hens, but it is quite possible that all of the advantages to the hen of free range can be obtained within a well-designed building.

Low disease incidence, good air conditions, constant water supply and a regular supply of a well-balanced diet are all advantages of the battery cage system in the maintenance of good welfare. Against

Fig. 32.8. Free-range hens are vulnerable to predation from predators like foxes and birds of prey, so this unit has a good fence and netting over the outside area (photograph D.M. Broom).

these advantages must be set an array of disadvantages. Some of these are also disadvantages of extended cage systems and, for a few of them, aviary systems.

One important risk for animals in any building with automatic systems is that these may fail. Drinkers may become blocked, food conveyors may fail to convey food to some cages or ventilation and heating systems may cease to work effectively. In each of these situations, and in those where birds become ill or trapped, alarm systems and careful inspection are needed. The inspection of birds in many cages is relatively easy: the good stockman can check regularly on every bird and identify some severe welfare problems. It is well known, however, that inspection is much less efficient if the birds are in the bottom row of cages or in a row the stockman cannot see. As Elson (1988) points out, the bottom row should not be too low for inspection and, if there are many tiers of cages, it is essential that gantries there are positioned so as to allow inspection of the upper tiers. Another risk of poor welfare in all houses accommodating animals is fire. All animals should be kept in such a way that they can be evacuated from their building in case of fire. It would take a long time to open all the cages in a house containing 100,000 birds, and birds that have lived in a cage for a long period might not leave the cage readily even if the building was on fire. This is a serious inadequacy of the system, as there are sometimes cases of poultry houses burning down. Fire alarms and sprinkler systems could help to reduce the risk, but it will always remain a serious problem.

Many of the problems for a hen in a battery cage are a consequence of things that she cannot do. She cannot move freely, flap her wings, perch, build a nest before oviposition, scratch for food, dust-bathe or peck at objects on the ground. The consequences of such deprivations are frustration, pecking at other birds and certain abnormalities of growth and body form. The evidence concerning the resultant poor welfare will now be summarized. Free-range hens walk much more than those kept in cages or in groups on a wire floor (Fölsch, 1981). Wing-flapping is entirely prevented in battery cages and there is a reduction in the total amount of comfort behaviour carried out in a cage as compared with group housing or free range (Hughes and Black, 1974). Battery cages provide little room for hens to stretch their legs or preen normally. One behavioural test of deprivation magnitude involves giving the animal the opportunity to perform the activity and measuring performance after various periods of deprivation: this is the rebound effect. For example, the amount of water drunk can be measured after various periods of water deprivation. Wennrich (1975, 1977) showed that birds removed from a battery cage and put into a larger area showed much wing-flapping. Nicol (1987b) studied this rebound effect further and compared birds kept in cages for different times. Longer periods confined in cages were associated with more prolonged wing-flapping when released. This is clearly a welfare problem, but it is difficult to say how widespread or severe it is.

The long-term effects of lack of exercise are those on bones and muscles. Battery-caged birds have lower bone weight and greater bone brittleness than birds that have more freedom of movement (Meyer and Sunde, 1974), and more caged birds are lame (Kraus, 1978). The levels of osteomalacia and osteoporosis are considerably higher in caged hens than in group-housed or free-range hens (Löliger *et al.*, 1981), and it is suggested by Martin (1987) that there is much more muscle weakness in caged birds. The muscle characteristics of birds kept in different ways are evident from meat quality studies, but these do not necessarily provide information about welfare.

Perhaps the most important evidence of poor welfare is the incidence of broken bones when hens are taken to slaughter. Simonsen (1983) reported that the incidence of broken wing bones on arrival at the slaughterhouse was 0.5% in hens free to move but 6.5% in hens from cages. A further 9.5% of hens from cages had broken bones after slaughter. Gregory and Wilkins (1989) dissected 3115 spent hens from battery cages in the UK and found that a mean of 29% had broken bones before the time that they reached the waterbath stunner at slaughter. Removal of the birds from the cages and hanging on the shackling line were identified as points where damage was most likely to occur. Knowles and Broom (1990) have found that hens from battery cages are much less active than those in a perchery or an Elson terrace system, and their humeral and tibial breaking strengths are less.

Exercise is correlated with bone thickness and strength in hens, as in other animals, and it is generally accepted that lack of exercise in cages is the main cause of bone weakness (Fleming *et al.*, 1994; Michel and Huonnic, 2003). Poor design of percheries and too high a stocking density can also lead to bone breakage (Gregory and Wilkins, 1996). It is not known how much discomfort is associated with osteomalacia and osteoporosis, but there can be no doubt that broken bones are extremely painful. Anything that regularly leads to a high incidence of broken bones is a very serious welfare problem. Handling and transport procedures are a major problem, and these are not quite the same for free-moving and caged birds. However, it is clearly going to be necessary for some substantial change in housing or handling, or both, if this high incidence of broken limbs is to be avoided.

Hens use perches when these are available, especially at night when they often perch close together and are highly motivated to perch (Fröhlich, 1993; Olsson *et al.*, 2002; Oester, 2004), but perches are seldom provided in battery cages. When perches at several levels are provided, hens choose to perch high above ground (Blokhuis, 1984), but they will utilize a low perch in a battery cage (Tauson, 1984). Hen arousal levels are reduced if they can perch, and provision of a perch leads to increased leg bone strength (Hughes and Appleby, 1989; Abrahamsson and Tauson, 1993).

Perches must be properly designed, with no sharp edges and not too large or small, in order that foot disorders do not occur (Moe *et al.*, 2004). Using a flattened-top, hardwood, 38 mm diameter, circular perch was found to give the lowest incidence of bumblefoot (Tauson and Abrahamsson, 1997). Gunnarsson *et al.* (1999) showed that rearing young chicks without access to perches and giving them access only after 4 weeks of age doubled the prevalence of cloacal cannibalism in the adult flocks. Yngvesson *et al.* (2002) showed that rearing without perches impaired laying hen escape behaviour in a simulated cannibalistic attack. The strong preferences that most hens have for using a perch, and the improvements in bone strength and in incidence of injuries and deformities, particularly if the perches are in a solid floor system rather than a wire cage (Moe *et al.*, 2004), show that the provision of perches improves welfare.

The fact that hens in free-range conditions walk considerable distances in good weather conditions is evidence for their preference for being in a larger rather than a smaller space (Hughes, 1975; M. Dawkins, 1976, 1977). Birds housed at 600 cm^2 per bird, or above rather than at 450 cm^2, have a broader and more varied behavioural repertoire and greater freedom of movement (Appleby *et al.*, 2002). Some activities require more than 600 cm^2, if only for a certain proportion of time. This has led to a reinterpretation of some early preference test results where hens selected increased cage sizes for a proportion of the day (Lagadic and Faure, 1990). Rather than indicating a preference for a small cage, these results suggest there is an intermittent preference for a large cage that is context dependent (Cooper and Albentosa, 2003). Recent assessments of spatial preference have demonstrated that hens in functional cages adopt an even spatial distribution, demonstrated most clearly when hens are allowed to move between two linked, furnished cages (Wall *et al.*, 2002, 2004; Cooper and Albentosa, 2004). These preference test results suggest that hens in furnished cages at 600 cm^2 cage floor area per bird are still attempting to maximize their personal space allowance. This preference outweighs any competing preference for additional cage height (Cooper and Albentosa, 2004).

Domestic fowl show elaborate nest-searching, nest-building and other behaviour before egg-laying, which is essentially the same as that of wild jungle fowl (Wood-Gush, 1954; McBride *et al.*, 1969; Fölsch, 1981). Hens in a cage have no nesting material, no nest cup and no quiet, dark place in which to lay. They are often pushed aside from their planned laying place (Brantas, 1974). However, these birds have a strong preference to find and lay in an enclosed nesting place (Freire *et al.*, 1996). There are various indications of the extent of the frustration felt by hens that are about to lay an egg in a battery cage.

Plasma corticosterone levels increase pre-oviposition, but this occurs whether or not a nest is available (Beuving, 1980), so it is probably a preparation for the egg-laying procedure. Hens give a pre-laying 'Gackeln' call (Baeumer, 1962), whose intensity and duration is three times higher if the bird is in a battery cage (Hüber and Fölsch, 1978; Schenk *et al.*, 1984). This may indicate greater frustration, but the clearest behavioural indicator is stereotyped pacing. At the time when a nest would normally be built, the hen walks up and down in the cage in a repetitive way (see Chapter 23). Provision of an artificial nest with a floor of AstroTurf seems to be entirely acceptable to hens (Wall *et al.*, 2002), and hens will use a nest shared by many birds (Abrahamsson and Tauson, 1997).

Another abnormality of the battery cage that prevents certain normal behaviours is the wire floor. The hen cannot scratch on the ground for food or dust-bathe. The preferences of hens for different sorts of metal flooring have been assessed and, while hens preferred a chicken-wire floor to a floor with larger mesh, they spent more time on litter floors, especially around the time of oviposition (Hughes and Black, 1973; Hughes, 1976; M. Dawkins, 1981). When hens are prevented from dust-bathing they may show anomalous behaviour in the cage such as attempting to bathe in the feed (Martin, 1975; Wennrich, 1975; Vestergaard, 1980), and dust-bathing movements on cage-mates or in the air (Vestergaard, 1982; Martin, 1987).

The amount of dust-bathing behaviour was proportional to the duration of the period of deprivation (Vestergaard, 1980). Oden *et al.* (2002) reported significantly fewer birds dust-bathing when the quality of the litter was poor, while Wall *et al.* (2008) showed that birds tend to dust-bathe once every second or third day. If the dust bath is too small, birds may compete for access to it and use it less (Olsson and Keeling, 2003). In order to allow dust-bathing, some furnished cages have a sand bath; this is small, so is rapidly emptied, and the sand thrown out may damage food provision or egg collection machinery. If food only is provided in the sand bath area, this does not damage machinery when used for dust-bathing, but may be wasted or polluted with faeces and then eaten.

Hens spend much time investigating their environment by pecking at objects, and their bills are richly provided with sensory receptors. Birds in cages have little at which to peck so they direct much pecking to substitute objects (Fölsch, 1981). Some pecking stereotypies are also shown in cages (Martin, 1975). The fact that pecking is not just a matter of obtaining food is demonstrated by the finding that birds will work for food reward by pecking at a key even when food is present (Duncan and Hughes, 1972). Hens in cages sometimes peck persistently at cage fittings and at the feathers and vents of cage-mates. Blokhuis (1986) suggests that the major reason why feather-pecking occurs in battery cages is that hens do not have enough other things at which they can peck. Important causes include dust-bathing deprivation, lack of food-investigation possibilities, type of floor, stocking density, flock size, food structure and composition, genotype and light intensity (Blokhuis and van der Haar, 1989; Savory *et al.*, 1999). In order to minimize the occurrence of injurious pecking, Newberry (2004) recommends:

1. That hens be provided with perches, attractive foraging materials and feed in small particle form such as mash or crumbles throughout rearing as well as in the laying hens.
2. The age at first egg should be delayed until the hens are at least 20 weeks old and the birds should be managed to minimize the availability of preferred victims and to prevent the discovery that flock mates represent a highly palatable food source.
3. High perches should be available to act as a refuge, and nest boxes should be designed to minimize visibility of the cloaca during oviposition.
4. Sufficient space should be provided to facilitate access to all resources.

Pecking has rather more severe effects on the individual pecked and often starts in young hens with pecking at the vent and progresses to serious injury or death. Factors affecting the occurrence of injurious pecking involving cannibalism and death are reviewed by Savory (1995) and Newberry (2004). One death is often followed by more in the same group (Tablante *et al.*, 2000), which could be a result of the fact that injured, less fit or 'different' individuals are attractive victims (Savory *et al.*, 1999; Yngvesson and Keeling, 2001; Cloutier and Newberry, 2002), or that birds may learn from one another (Cloutier *et al.*, 2002).

Unchecked growth of bill and claws can be a problem in battery cages. If there is no possibility for birds to wear these down they may grow to the extent that they are seriously deformed. As Tauson (1986, 2003) has shown, the provision of an abrasive strip in the cage can solve this problem. Birds can become trapped in the wire mesh of cages by the head, body, wings, legs or feet. Steep floors (e.g. of 23% or steeper) lead to high levels of foot deformities because birds' feet slip down onto cross-wires, but this can be avoided by using slopes no greater than 12%. Foot deformity incidence can also be reduced by provision of a perch (Tauson, 1980, 1988; Tauson and Holm, 2003).

The number and severity of welfare problems in battery cages are so great that most consumers who know about them are unwilling to accept the use of the system. As a consequence, laws and codes of practice are gradually changing so that the system is beginning to be phased out. Furnished cages are now very much better, but the problem of how to provide material for dust-bathing without the function of machinery such as that for supplying food or collecting eggs being impaired has not yet been solved. Also, few furnished cages are high enough to allow the birds proper exercise and hence to avoid weakened bones. A study by Wilkins *et al.* (2011) showed that furnished cages result in weaker bones in the hens but not necessarily more bone breakages. Present designs of furnished cages are significantly more expensive than the best aviaries, so it is the aviary that is being utilized most to replace battery cages. Beak-trimming is still widely used for both battery cages and aviaries.

In aviaries, the hens have much freedom of movement, although there are some social restrictions on this. A variety of activities is possible, but injurious pecking can sometimes occur and be very serious, affecting many birds. As a consequence, beak-trimming with its consequent pain and other disadvantages (see later section) is often carried out. In aviaries and in other group-housing systems, aggression can be reduced by providing uniform illumination with no bright patches in which hens may accumulate and fight (Gibson *et al.*, 1985). It is also important that there should be no sharp corners from which retreat is difficult when pursued. It seems that neither group size nor stocking density in aviaries has an effect on injurious pecking (Gunnarsson *et al.*, 1999; Green *et al.*, 2000; Oden *et al.*, 2002). There can also be welfare problems because disease and parasitism incidence can be higher. Good stockmanship is necessary to identify birds with health problems or signs of injury and to separate these from the main group.

One economic problem is that some birds do not lay in nest boxes, so their eggs are difficult to find and are often dirty, but the incidence of floor-laying is very low if good nest boxes are used. In commercial systems all egg collection is automatic. In general, however, the best tiered floor aviary systems are the best commercially viable housing systems for hen welfare. Their production costs per egg are similar to those of battery cages, allowing 500 cm^2 per bird. Furnished cages have practical problems and are, at present, more expensive per egg. Free range is more expensive but attracts a higher price from consumers.

In a comparison of six or seven examples of farms with four laying hen-housing systems, Sherwin *et al.* (2010) found more old fractures and birds in poor conditions in barn systems, more bone breakage at depopulation in battery cage systems, more vent-pecking in free-range systems and fewest problems in furnished cage systems. A review by Lay *et al.* (2011) also described weaknesses in each system. It is clear that poor management and variations in system design can lead to problems in any system. However, the difficulties for all birds in a system that does not meet their needs such as the battery cage are a paramount factor. In each other system, the mean welfare will be better than in battery cages but the worst examples of any system can result in widespread poor welfare. As Sherwin *et al.* point out, improvements in the genetic strains of birds are needed to ensure good welfare in the aviary or free-range systems.

Chickens Reared for Meat

The numbers of chickens reared for meat are very large indeed, so any welfare problems would involve very many individuals. The modern breeds of broilers grow very quickly and most are now slaughtered at 5–6 weeks of age, by which time they have reached a weight of 1.5–2.5 kg. The major problems of broiler production, resulting in poor bird welfare, are a consequence of selecting birds for a short, very fast-growing, life. Many of the birds become too heavy for their legs to support. They therefore develop leg disorders and do not have normal locomotion. Knowles *et al.* (2008) studied the walking ability of 51,000 broilers and found that at 40 days of age, 3% were almost unable to walk and 28% had poor locomotion. A biomechanical analysis of the gait dynamics of broiler chickens was conducted by Paxton *et al.* (2013). The leg disorders are disabling, being associated with inflammation of joints, hocks and bone, and causing pain during walking (Kestin *et al.*, 1994; McGovern *et al.*, 1999; Danbury *et al.*, 2000). This pain is severe and, if the broilers are trained to perform an operant for self-administration of the analgesic drug carprofen, they do so (McGeown *et al.*, 1999; McGovern *et al.*, 1999; Danbury, 2000). In terms of how bad the problem is for individuals and numbers of animals affected, this is the most serious animal welfare problem in the world.

A typical broiler house is something like the deep-litter house for hens. It is usually rectangular, with a good ventilation and temperature control system. Before birds are put in, the clean floor is covered with litter such as wood shavings or straw to a depth of about 15 cm. Day-old chicks are introduced into the house at a density that will result in a weight of birds at the end of the growing period of about 30–45 kg/m^2. Typical numbers in a house are 10,000–20,000. At the beginning of the rearing period the chickens have plenty of space, but at the end of the rearing period they are crowded close together (see Fig. 32.9).

One problem with this rearing system is that there are many birds present and there is little or no facility for inspection, so an individual that is weak, injured or sick is often not detected. Most of these individuals die and their bodies remain in the litter. In some cases, weak individuals die because they are trampled on by the other birds, for as the density approaches the final level of 30 or more kg of birds in each square metre, it is essential to be able to stand up in order to survive. Another disadvantage of having very large numbers of birds in a single building arises if the birds are suddenly frightened and hysteria develops. If this happens, many birds may move rapidly to the end of the building where there is a pile-up, under which many individuals may be crushed to death. Hysteria can be minimized by good stockmanship and the effects can be reduced by putting baffles in the house.

The rapid growth rate of a modern broiler is not uniform throughout its body. Muscle grows very quickly but bones, and in particular the leg bones, grow less fast. As a consequence, a point is reached at which the bird's legs cannot easily support its body. There is then a risk of being trampled, as mentioned above, but there will also be prolonged contact with the litter

Fig. 32.9. Chickens in typical broiler house (photograph D.M. Broom).

beneath the bird. The various disease conditions associated with leg weakness in broiler chickens are reviewed by Bradshaw *et al.* (2002) and Mench (2004). These include: bacterial femoral head necrosis, tenosynovitis and arthritis, infectious stunting syndrome, varus valgus disease, tibial dyschondroplasia, rickets, chondrodystrophy and spondylolisthesis, osteochondrosis, degenerative joint disease, spontaneous rupture of the gastrocnemius tendon and contact dermatitis. The welfare of broilers with leg disorders may be impaired due to pain from the condition, an inability to walk leading to frustration and associated problems of being unable to feed and drink due to immobility. In assessing welfare, the individual broiler should be considered.

Faecal matter accumulates very rapidly in a broiler house, so that the litter is covered with faeces well before the end of the growing period. The faeces and their breakdown products are alkaline and have a corrosive effect on skin, so birds that have to sit on the soiled litter for long periods get breast-blisters, hock-burns and footpad lesions (see Fig. 32.10). The dermatitis lesions can develop in less than 1 week (Ekstrand and Carpenter, 1998). These can be widespread in a broiler unit and, while footpad lesions are seldom seen by consumers because the feet are usually cut off before sale, the first two are visible to the customer buying a chicken. One consequence of this is that legs are often cut short on carcasses so that hock-burns cannot be seen. Another consequence is that many carcasses are downgraded further so that they can be used only for chicken pieces.

Broom and Reefmann (2005) investigated the extent of 15 kinds of visible lesions in Grade 'A' broiler carcasses. Six lesions were analysed histopathologically, and this showed that the hock-burn that was seen would have occurred several days before death and would have been painful. Many broiler carcasses in the supermarkets had dermal lesions and 82% had detectable hock-burns. Of these, 18% were >0.3 cm^2 in area

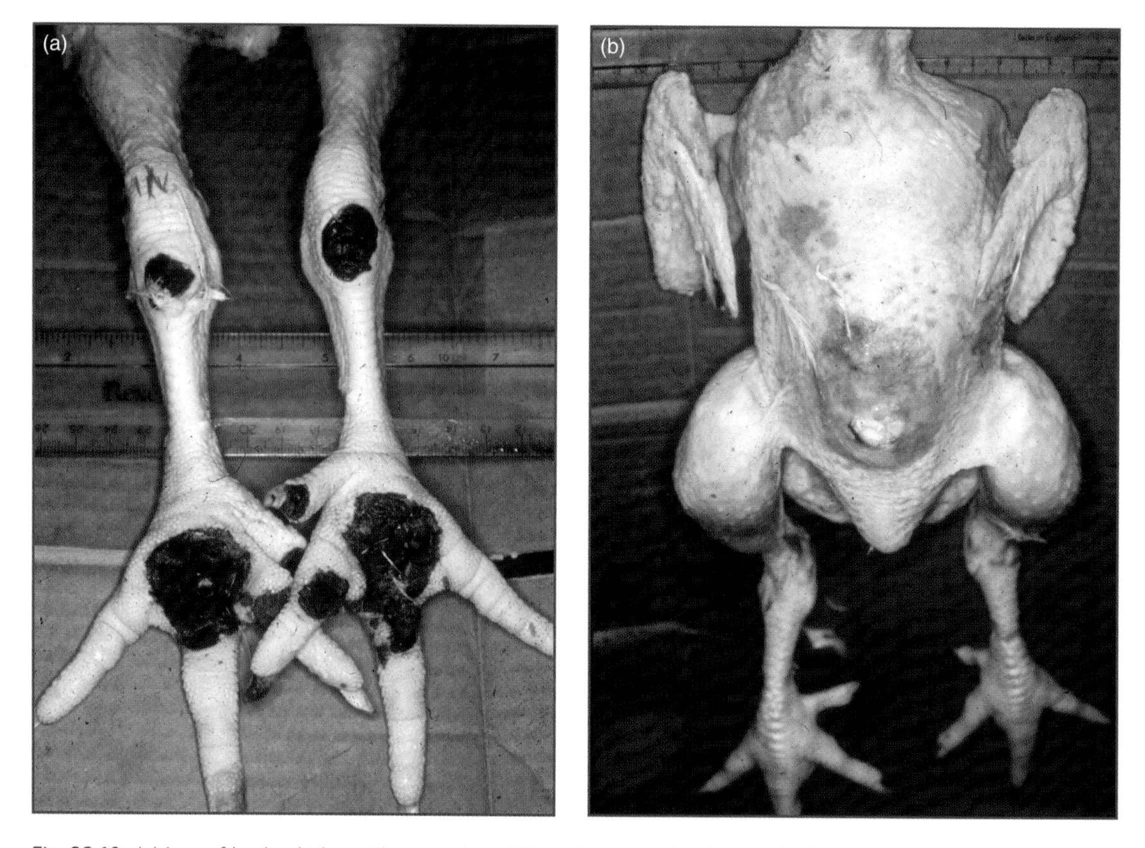

Fig. 32.10. (a) Legs of broiler chicken with severe dermatitis on the footpad and severe hock-burn. A small brown hock-burn can commonly be seen on broiler carcasses and indicates pain in the bird. (b) Breast-blister on broiler chickens (photographs D.M. Broom).

(e.g. 6 × 5 mm). 'Organic' chickens had half as many hock-burns as conventionally reared broilers, perhaps because of differences in litter quality or leg strength. The Grade 'A' chickens observed exclude birds with obvious visible defects, because these birds would have had the blemishes removed and the carcass would have been portioned. Hence, the frequencies of lesions in farmed birds would be higher than those reported. Serious skin abrasions and blisters cause considerable pain and discomfort to the birds so the system of housing, or the breed itself, should be modified so as to avoid pododermatitis, hock-burns and breast-blisters.

Leg disorders in broiler chickens become worse as the birds get older and heavier (Mench, 2004; Sandilands *et al.*, 2011a). Kestin *et al.* (1994) reported that 90% of broiler chickens had some walking ability impairment in the last week before slaughter and 26% had a severe impairment. Sanotra *et al.* (2001), studying a broiler strain used in many countries, found that 30% of birds on commercial farms had severe walking difficulties by market age. As shown by the Knowles *et al.* (2008) study mentioned above, the problem does not seem to have declined over the 9 years from 1999 to 2008. This is true, even though the rate of culling for leg problems has declined, perhaps because the relationship between gait score and the various pathologies is complex (Sandilands *et al.*, 2011a). It is widely known that birds with weak legs sit on litter and, when the litter quality is not good, many chickens, as a consequence, have contact dermatitis visible on carcasses as breast- or hock-burn. A comparison of 1957 and 1991 strains of broilers showed that growth rates and, hence, leg problems, have an origin that is much more a consequence of genetics than of food quality (Havenstein *et al.*, 1994).

When broiler houses become very crowded, as is typical in the days leading up to thinning, i.e. reduction in numbers or to slaughter, locomotion is reduced. This reduced exercise makes leg problems worse. The complexity of the environment also has an effect on the extent to which broiler chickens exercise, as greater complexity leads to more activity. The introduction of more interesting food and materials to investigate, manipulate and climb on will increase activity and hence leg strength. Low light level is associated with low activity (Boshouwers and Nicaise, 1987) and with a greater frequency of leg disorders (Gordon and Thorp, 1994).

Ascites is another pathological condition associated with fast growth in broiler chickens and is a major cause of poor welfare and mortality (Maxwell and Robertson, 1998). It is also known as pulmonary hypertension syndrome, and results in fluid from the blood leaking into abdominal cavities. It affects approximately 5% of young broilers and 15–20% of the larger birds, and it can kill or weaken the birds and result in carcass condemnation. Although originally described as occurring especially at high altitudes, it is now widespread at all altitudes. The main cause of ascites is failure of heart function associated with lack of oxygen supply to tissues. It is extremely rare in old strains of broilers and results from failure of the cardiovascular and pulmonary systems to grow fast enough to keep pace with the demands from the muscles and gut.

As stocking density of broilers increases, the extent and frequency of poor welfare tends to increase. For example, there are various reports of growth rate declining at high stocking densities. Mortality increased with stocking density over a range from 5 to 45 kg/m^2 (Shanawany, 1988). Locomotor activity and locomotor problems are generally found to increase (Kestin *et al.*, 1994), and the extent of wet litter and hock-burn usually increases at higher densities.

Üner *et al.* (1996) compared broiler communal systems stocked at 24–36 kg/m^2 and found that, at the higher stocking densities, there was less walking, running, preening, total activity and calm behaviour and more time concentrated around the feeders. There is less possibility for environment enrichment at high stocking densities. M. Dawkins *et al.* (2004) claimed that stocking density was less important to broiler chicken welfare than other management factors. There is no doubt that litter management and ventilation in the building are important in their effects on hock-burn but the study of Dawkins *et al.* did not consider stocking densities of less than 30 kg/m^2. They did find that locomotor problems were twice as high at 46 as at 30 kg/m^2 and lumped all other variables to compare with stocking density. Their results do not allow the conclusion that stocking density is unimportant, in part because the farmers had notice that the researchers were coming to make measurements.

The poor welfare occurring in broiler chickens as they near the age of slaughter affects a very large number of individuals. However, the problems are soluble. Birds can be bred for stronger legs, and this has been done for many years, but some slowing of growth by genetic selection or management is the only real solution. Leg problems can be reduced if food intake is limited for a period during growth (Classen, 1992). Some problems are exacerbated by high stocking density, so this should be limited to a maximum of 25 or perhaps 30 kg/m^2. Breeding should encourage better leg development or less muscle development. Birds should not be kept on faecally contaminated litter.

Broiler breeders are often beak-trimmed but are usually kept in quite good conditions. A significant cause of poor welfare is that they are prevented from becoming too heavy by restricting their diet. As a consequence, they may be hungry for a substantial part of their lives. The degree of hunger can be assessed by increasing how much they will eat when given the opportunity and how hard they will work for food (Hocking, 2004). This is a problem that cannot be solved in its entirety.

Turkeys

The majority of turkeys raised for meat production are kept in a controlled environment consisting of windowless buildings where heat, ventilation and lighting are precisely regulated. Some parent stock and the majority of Christmas turkeys produced on small farms are reared in pole barns. These open or extensive housing systems allow natural daylight to enter, though this may be supplemented by artificial light. Temperature and ventilation are not controlled. Free-range systems may also share similarities with the pole-barn system. Natural daylight is available in outside enclosures. Some production of turkeys in range or forest environments has recently begun, but this is still relatively uncommon (Hocking, 1993).

Turkeys in the latter stages of growth may show aggression that can have severe effects on welfare and production. In order to attempt to reduce any aggressive interactions in turkeys, many are kept at low light levels. Vision is generally considered to be important to birds (Appleby *et al.*, 1992). The high proportion of cones in poultry retinas suggest vision is better in bright rather than in dim light, and colour vision is good. Manser (1996) therefore surmised that keeping birds in very low light intensities might well deprive them of some sensory input and so contribute to a barren environment. However, low light intensities are perceived to discourage activity. This may increase productivity, reduce aggression and save electricity costs. Hence, turkeys are usually reared under intensities of 1–4 lux, a very low light intensity. To set this in context, satisfactory levels inside buildings used by humans range from 50 lux for bedrooms through to 500–750 lux for offices. A light intensity of 20 lux is too dim to enable humans to read printed text and would not allow turkeys to investigate their environment. A range of normal behaviours would be prevented at this or lower light intensities. Since activity of all kinds is suppressed at low light levels, this could have further consequences for exercise-related problems.

Turkey poults reared from hatching under very low-intensity light (1.1 lux) were inactive and developed eye abnormalities and enlarged adrenal glands (Siopes *et al.*, 1984), although feeding and drinking behaviour did not differ between the 1- or 10-lux regimes (Sherwin and Kelland, 1998). Hester *et al.* (1987) found growth rates slower in lower-intensity light (2.5 versus 20 lux), although feed conversion was better. Leg disorders were less frequent in birds reared in natural daylight of 220 lux than in artificial light of 19 lux (Davis and Siopes, 1985). Preference tests on male turkeys indicated that light intensities commonly used in the commercial industry do not reflect the preferred conditions of turkeys. Intensities of <1 lux were aversive. Commercial lighting intensities lead to poor welfare if needs cannot be met (Sherwin, 1998). It seems that supplementary ultraviolet light may benefit turkeys in relatively low light conditions (Sherwin *et al.*, 1999).

Heavy male turkeys are susceptible to heat stress (Perkins *et al.*, 1995). Chronic effects of high (e.g. 29°C) rearing temperatures are loss of muscle and fat tissue (Noll *et al.*, 1991; Waibel and Macleod, 1995). Litter on the floor improves turkey welfare, but pododermatitis may occur so the litter should be kept dry and high stocking densities avoided (Hocking, 1993; Hester *et al.*, 1997).

Beak-trimming is performed to reduce injurious pecking and cannibalism in breeding or pole-barn turkeys. Such behaviour causes very poor welfare and must be prevented. Beak-trimming is a traumatic procedure which deprives turkeys of a major sensory organ, and this has major welfare implications.

Toe-clipping is likely to cause acute pain and the chronic effects are unknown. This procedure might be easily eliminated to the benefit of both production and welfare. Some pain must also occur when the snood is removed, but little information is available on the effect of snood removal on the behaviour and subsequent welfare of turkeys. Changes in housing or management regimes can only ameliorate rather than eliminate these problems. Hirt *et al.* (1995) recommend that selection criteria should relate not only to economic efficiency but also to factors affecting welfare. Selection of turkeys for fast growth rate is often accompanied by an increase in susceptibility to, and mortality from, some diseases especially *Pasteurella multicocida* and Newcastle disease virus (Tsai *et al.*, 1992; Nestor *et al.*, 1996a,b).

Slow-growing turkeys do not appear to suffer the degenerative disorders of hips and other joints that are prevalent in male breeding turkeys at the end of their breeding life (Hocking *et al.*, 1998). Adult male breeding turkeys often have leg problems, as evidenced by their impaired locomotion. In a study designed to find out whether leg conditions of heavy male birds were painful, Duncan *et al.* (1991) compared behaviour with and without an analgesic drug, betamethazone. The birds were observed and then treated with the analgesic and observed again. With the analgesic they showed less lying, more standing, more walking, more feeding, earlier and faster approach to a female and more attempted mounts.

The abnormal body conformation, largely due to the large pectoral muscle, of adult male turkeys renders them incapable of mating with a female. This loss of a fundamental biological function is ethically questionable in addition to effects on welfare of the individual. A turkey that cannot mate but is in the vicinity of females will be frustrated, so this has an effect on welfare. The semen collection procedure involves substantial human handling and has some effect on welfare. Artificial insemination of females also involves substantial human handling.

Ducks and Geese

The major animal welfare issues relating to ducks and geese concern the provision of water, the stocking densities used, the force-feeding of the birds for foie gras production and the plucking of live birds for feather production. In Muscovy ducks (*Cairina moschata*) kept at high densities, injurious pecking can be a problem but this is not normally shown by the more widespread domesticated mallard (*Anas platyrhynchos*) breeds, such as Pekin ducks. The welfare of ducks and geese can also be poor because they are kept at too low a light level, not provided with sufficient straw or other bedding, kept in other inadequate environments, badly fed, or handled without sufficient care (Rodenburg *et al.*, 2005).

Ducks can grow extremely fast, a live weight of 3.0–3.5 kg being attainable in Pekin ducks in 7–8 weeks (Sainsbury, 2000). An increasing amount of the rearing is now carried out indoors where ducks are normally housed on straw bedding, wire mesh or slats. It is widely known that ducks and geese are aquatic animals and choose to spend most of their lives on or close to water. However, under commercial conditions, provision of water is often only by means of nipple drinkers. Open water, either standing or flowing, is usually absent for ducks and geese in commercial conditions. Hence the animals cannot exhibit their usual aquatic-related behaviours such as adequate preening, shaking movements to remove water, taking up water by the beak, paddling and swimming (Matull and Reiter, 1995). The Council of Europe (1999) requirements for ducks, to which most countries in Europe are signatories so they should be specified in law, state that ducks should be able to dip their heads in water and spread water over their feathers. This is not possible if only nipple drinkers are provided.

One of the reasons often cited for not including a source of open water is that the ducks may become diseased. However, it is now clear that provision of a good open water system, rather than just nipple drinkers, can improve eye, nostril and feather condition and reduce disease (Knierim *et al.*, 2004; Jones *et al.*, 2009; Jones and Dawkins, 2010; O'Driscoll and Broom, 2011). It is important for minimizing disease, and hence for the welfare of the ducks, that water baths or troughs are situated in a well-drained area so that bedding does not get wet. Heyn *et al.* (2006) found that ducks with only nipple drinkers had a much higher frequency of blocked nostrils than ducks with any kind of open water. Liste *et al.* (2012b) found that ducks provided with shallow troughs in commercial conditions had no health or production problems, although those with wide troughs had slightly worse foot conditions than those with narrower troughs. Water quality in the troughs was worse if the water was much used by the birds and regular replacement of dirty water is needed, as well as a separate clean drinking water supply. The ducks were less likely to use troughs with dirty water in them. In doing so, they were behaving in a way that reduced contact with pathogens (Liste *et al.*, 2013).

Ducks have a strong preference for water that they can get into, rather than just water from nipple drinkers.

They work harder for the open water (Cooper *et al.*, 2002) choose it more often (Ruis *et al.*, 2003; Rodenburg *et al.*, 2005) and use it for normal preening and head-dipping behaviours that cannot occur when only a nipple drinker is provided (Jones *et al.*, 2009; O'Driscoll *et al.*, 2012). The time spent at inadequate water sources may be prolonged because the birds take longer to drink and cannot complete their preening movements. Hence they are recorded standing near the water source and preventing the access of others to the source. Ducks spent 9% of their time in pools, with a bout-length of more than 15 min, when there was easy access to them, visiting the pools eight times per day at times spread throughout the 24 h (Liste *et al.*, 2012a). In the study by Jones *et al.* (2009) showers also allowed normal behaviour. While ducks kept with troughs, baths or showers were not motivated to find greater access to water, ducks living with only nipple drinkers showed a rebound effect if offered water in a bath, indicating that they had previously been deprived. Although they use deep water in field conditions, ducks reared in buildings preferred shallow water baths (10–20 cm deep), to deeper water (30 cm deep) in which they did not touch the bottom if they tried to stand (Liste *et al.*, 2012a).

As with all except the most extensively reared farm animals, too high a stocking density results in poor welfare in ducks. When De Buisonjé (2001) (see review by Rodenburg *et al.*, 2005) compared housed Pekin ducks kept in groups of 225–360 at stocking densities of 5, 6, 7 or 8 birds per square metre (up to 16–25 kg/m^2), the production, meat quality and extent of feather damage were all worse at 8 birds per m^2 than at the lower densities. In Muscovy ducks, Bilsing *et al.* (1992) found no damage to the feathers, or other parts of the body, of birds at a density of 6.3 birds per m^2 but considerable damage, and hence poor welfare, at 11.6 birds per m^2. Klemm *et al.* (1995) found that, by 55 days of age 49% of indoor-reared Muscovy ducks kept at commercial densities showed injuries, largely as a result of pecking. If the birds were reared outdoors the incidence was 12%. If some Pekin ducks were kept with the Muscovies, the incidence of pecking injuries was reduced. Beak-trimming and bits inserted in the beak are sometimes used to reduce injurious pecking by Muscovy ducks. Both procedures cause pain to the birds.

The welfare issues raised by foie gras production are: (i) the very small cages often used during the force-feeding that occurs in the last 2–3 weeks of life (see Fig. 32.11); (ii) the act of force-feeding itself (see Fig. 32.12); and (iii) pathological effects on the birds' livers. The cages do not allow any normal behaviour except drinking. No other farmed animals are kept in such extreme confinement. Force-feeding is the only feeding that the birds get so the duck or goose wants to receive some food. This is the reason why Faure *et al.* (2001) found that force-fed ducks did not show more aversion to the person who fed them than towards a strange person. However, the handling, the insertion of

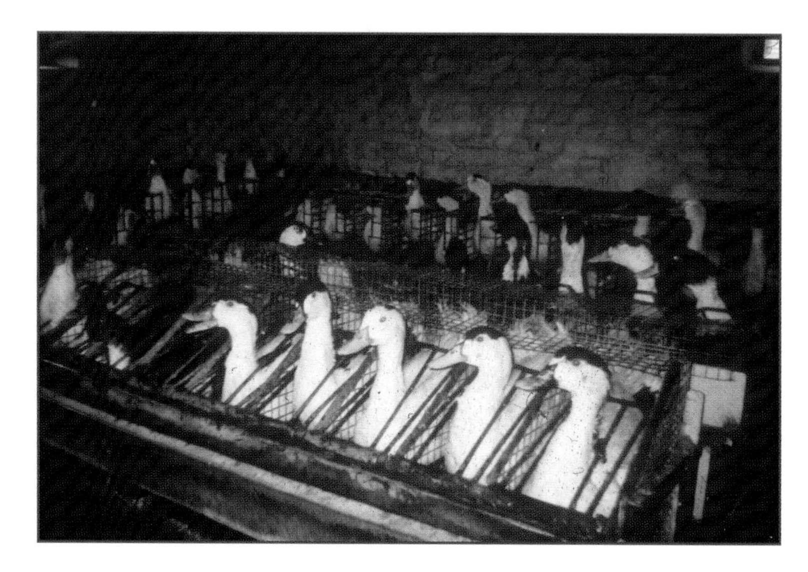

Fig. 32.11. Mulard ducks (Mallard × Muscovy) in cages during force-feeding period (photograph D.M. Broom).

Fig. 32.12. Ducks in cages: force-feeding with soaked maize that is squirted into the oesophagus via the pipe (photograph D.M. Broom).

the force-feeding pipe and the sudden introduction to the oesophagus of 500–750 gm of soaked maize are clearly aversive to the ducks or geese.

Feeding does not normally cause elevation of plasma corticosterone and regularly force-fed ducks have learned that the force-feeding cannot be avoided so a lack of increase in corticosterone during force-feeding (Guémené *et al.*, 2001) provides no information about welfare. The liver increases in size tenfold in birds reared for foie gras production, as compared with normal birds. The duck or goose is fed *ad libitum* during most of its life and then force-fed during the last 2–3 weeks of life. It has been argued that steatosis, the formation of fat globules, is biologically normal because the birds are migratory. However, most of the ducks are Muscovy duck crosses and this species is not migratory. Also, the extent of increase in fat is much greater than that which occurs before migration. The normal function of the liver, which is principally detoxifying harmful substances in the body, declines in efficiency and then many of the processes cease as the liver expands. The birds are killed for foie gras production just before they die of liver failure. Some do die during force-feeding but no figures for mortality are published. These issues are the subject of the EU Scientific Committee on Animal Health and Animal Welfare Report (EU SCAHAW, 1998).

Feathers are a by-product of poultry production. If the feathers are obtained after the death of the birds, there will be no new welfare problems. It could be that the desire to obtain the feathers results in better care of the birds. However, in some countries feathers are brushed or plucked from live birds, largely geese. The handling associated with brushing feathers about to be moulted would have some adverse effects on the geese. The actual brushing or gathering of these 'ripe' feathers is defined as a procedure that does not cause any tissue damage. A much greater degree of pain and other poor welfare results from plucking the down feathers of live geese. *Plucking means pulling out feathers that are still attached to the bird.* This procedure will cause bleeding and can cause the tearing of skin. In a review of the procedures used and their consequences, EFSA (2010) concluded that bloody feathers, skin injuries, posture changes (e.g. hanging wings), dead birds and broken or dislocated bones are welfare-outcome indicators which could be used to assess the welfare of geese subjected to feather collection. Only ripe feathers should be removed from live geese. A control system should be in place to ensure this is carried out in practice. The presence of skin tears and blood or tissue and the presence of non-ripe feathers in the collected feather material should be used to distinguish between plucking and gathering.

Further Reading: EFSA (2005a) The Welfare Aspects of Various Systems of Keeping Laying Hens Report of the Scientific Panel on Animal Health and Welfare. *EFSA Journal* 197, 1–23.

33 The Welfare of Farmed and Pet Fish

Introduction

Throughout most parts of the world, the second commonest domestic animal after the chicken is a species of fish, usually tilapia (*Oreochromis mossambicus*, *O. honorum*, *O. nilotica*) in the tropics, and a member of the carp family, Cyprinidae, or the salmon and trout family, Salmonidae, in temperate areas. These fish are farmed but are little affected by domestication. In an increasing number of countries farmed fish are a major source of inexpensive protein and there is some consistency of demand. In Europe, North America and affluent East Asian countries, fish from fish farms are products that are at the luxury end of the market. Hence, demand for these particular kinds of fish is much more elastic and subject to fluctuations according to the public perception of the product as well as the purchase price.

Concern for animal welfare is increasing rapidly and is a significant factor affecting whether or not animal products are bought. If a product is perceived to be associated with bad effects on human health, animal welfare or the environment, sales can slump dramatically. The more valuable the product, the richer the consumers and the more likely they are to decide not to buy a product on grounds such as the poor welfare of the fish (Broom, 1994). The fish farming industry cannot afford to ignore fish welfare when bad publicity about it could affect sales greatly (Broom, 1999a) and, in Europe, the industry is responding and starting to take account of EU collaborative research on fish welfare (Kadri *et al.*, 2012). Fish producers have to consider legislation and retailer codes of practice referring to fish welfare and disease prevention. While disease prevention has very important benefits for fish welfare, some procedures have negative as well as positive effects and there is a need to evaluate and balance these. For example, vaccination is not only expensive for the industry but some procedures have major effects on welfare (Dykes, 2012).

China has half of the world's fish production and, as in many other countries, its major farmed species are the grass carp, *Ctenopharyngodon idellus*, and common carp, *Cyprinus carpio*. Other major farmed species are: tilapia (e.g. *O. mossambicus*) in Africa and in South America, where there are also several local species; Atlantic salmon, *Salmo salar*; rainbow trout, *Onchorhynchus mykiss*; brown trout, *Salmo trutta*; gilthead sea-bream, *Sparus aurata*; and sea bass, *Dicentrarchus labrax*.

Many of the fish kept as pets are much more changed by domestication than are farmed fish. Goldfish, *Carassius auratus*, and carp such as the koi and common carp have been kept as ornamental animals for at least 3000 years. A great range of fish species are now kept as companion animals. Their status as companion animals is affirmed by the reports of many of those who keep them and by research demonstrating that people can be calmed by being with aquarium fish (DeSchriver and Riddick, 1990).

Aspects of Welfare Awareness and Pain

Key issues in any discussion of fish welfare are whether fish are aware of what is happening around them, whether they are capable of cognitive processing and whether they can have feelings such as pain (Broom, 2007). These are the same issues as for all other domestic animals, and are considered in Chapters 4 and 5. The conclusion reached in this book is that fish do have a significant degree of each of these abilities, so they should be protected and the welfare of any fish that we use should be considered. Awareness in fish is discussed by Chandroo *et al.* (2004a). We know that some fish must have mental representations of their environment in relation to their ability to navigate (Reese, 1989; Rodriguez *et al.*, 1994), have ability to recognize social companions (Swaney *et al.*, 2001) and can avoid for some months or years places where they have previously encountered a predator (Czanyi and Doka, 1993) or were caught on a hook (Beukema, 1970). Some fish species can learn spatial relationships and form mental maps (Odling-Smee and Braithwaite, 2003), and can use information about sequences of spatial information (Burt de Perera, 2004). The parts of the brain used to achieve this are not anatomically the same in fish (Broglio *et al.*, 2003) as in mammals but the function is very similar. It is clear that the timing of events can be integrated to allow the fish to produce appropriate avoidance responses (Portavella *et al.*, 2004; Yue *et al.*, 2004), and it is difficult to explain the results of these studies without assuming that the fish feel fear. The learning ability demonstrated in a range of studies (Sovrano and Bizazza, 2003; Braithwaite, 2005) indicates sophisticated cognitive processes more complex than associative learning. When Schjolden *et al.* (2005) investigated individual variation in responses of trout to a difficult situation, it was clear that the fish were using different coping strategies to deal with the problem.

The occurrence of pain in fish is debated by Rose (2002), Jackson (2003), Chandroo *et al.* (2004a,b), Braithwaite and Huntingford (2004) and Braithwaite and Ebbesson (2014). In the rainbow trout, *O. mykiss*, anatomical and electrophysiological investigation of the nociceptors connected to the trigeminal nerve has revealed that these fish have two types of nociceptor, A-delta and c fibres (Sneddon, 2002; Sneddon *et al.*, 2003a). There are also other receptors with different functions, and the only differences between birds, mammals and fish is that fish have not yet been reported to have a cold receptor and the numbers of C-fibres in fish are smaller (Braithwaite and Ebbesson, 2014). The transmitter substance P and the analgesic opioid enkephalins and β-endorphin, which act as endogenous analgesics in mammals, are present in fish (Rodriguez-Moldes *et al.*, 1993; Zaccone *et al.*, 1994; Balm and Pottinger, 1995), and the behavioural responses of goldfish to analgesics are the same as in rats (Ehrensing *et al.*, 1982). When Sneddon *et al.* (2003b) administered weak acetic acid solution or bee venom to the mouth of a trout, the fish rested on the substratum, rocked from side to side and rubbed their snouts on solid surfaces. These behaviours stopped when the analgesic morphine was given. Examples of higher-order cognitive processing of nociceptive inputs are given by Braithwaite and Ebbesson (2014).

Glucocorticoids in fish

Fish have a hypothalamic–pituitary–inter-renal response that is almost identical to the hypothalamic–pituitary–adrenal (HPA) response of mammals. Stimuli that are disturbing to fish elicit the production of adrenaline and noradrenaline from the chromaffin tissue (Perry and Bernier, 1999). At the same time that corticotrophic releasing hormone (CRH) is released from the hypothalamus it leads to release of adrenocorticotrophic hormone (ACTH) from the pituitary, and this in turn is carried by the blood to the inter-renal tissue, an analogue of the mammalian adrenal gland, where cortisol is produced (Sumpter, 1997; Huntingford *et al.*, 2006).

Environmental factors that have been shown to influence the production of glucocorticoids in fish include anaesthetics, ambient temperature, salinity of the water, nutrient availability, time of day, overhead light, fish density and background colour (Barton, 1997; Braithwaite and Ebbesson, 2014). When salmon are pumped through a pipe or loaded on to a well-boat, their plasma cortisol is substantially higher than when they are resting in good conditions (Gatica *et al.*, 2010b). There is growing evidence in many species concerning the role of glucocorticoids in processes such as learning that are not associated with stress but are wholly beneficial to the individual, and it is likely that glucocorticoids play such roles in fish also. However, emergency responses involving the hypothalamic–pituitary–inter-renal response can lead to harm to the fish, which then do indicate stress. For example, when conditions elicit high levels of glucocorticoid production, there can be a reduction in lymphocyte numbers, antibody production, lysozyme activity and gonadal steroids, and an increase in disease susceptibility (Pickering and Pottinger, 1989; Maule and Schreck, 1991; Pankhurst and Dedual, 1994). Immune system activity, on the other hand, has an effect on glucocorticoid production (Balm, 1997).

Important diseases of fish farming, such as furunculosis, which is a bacterial septicaemia of salmonids and other fish, are much more likely to be evident in fish if water quality conditions and fish culture procedures stress the fish. Furunculosis is caused by *Aeromomas salmonicida*, an obligate pathogen that often exists in fish without any causing disease unless the fish are stressed (Wedemeyer, 1996, 1997).

Terminology

A point that should be considered within the fish farming industry is the extent to which attitudes to fish welfare are affected by the terminology used. In English there has been a tendency to use plant terms when referring to fish. A cage full of salmon is sometimes called a crop, and the process of slaughter is sometimes called harvesting. Farmers refer to 'growing the fish', like 'growing wheat', but animals are not the same as plants. It is the fish that grow and the farmer who feeds and manages them. Such terms make the fish seem less like individual animals and encourage farm staff to view them as objects rather than as sentient beings and perhaps to treat them badly. Hence, the terms 'crop' and 'harvest' and 'growing the fish' are inappropriate and should not be used.

Problems of Fish

The following sections review the factors that cause welfare problems in the fish-farming industry, some of which are also relevant to pet fish. Those that might be considered to be most important are dealt with first. A general matter concerns the production of transgenic fish. As pointed out by Hallerman *et al.* (2007) and Broom (2008, 2014b), there is much doubt among the public about the genetic modification of animals and any genetically modified fish would have to be carefully checked, using good-quality animal welfare science, to be sure that the modification does not have negative effects on fish welfare. EFSA has produced general guidelines on how this might be done (EFSA, 2012). In addition, there are concerns about the initiation of genetic change in wild fish, including closely related species, with unknown consequences. The study by Oke *et al.* (2013) shows that a transgene from a salmon (*S. salar*) can be transmitted, following hybridization, to brown trout (*S. trutta*).

Stocking density

Some fish live in and prefer close schooling so that the nearest neighbour is a short distance away, e.g. Arctic charr (*Salvelinus alpinus*) (Jergensen *et al.*, 1993). In other species whose individuals are forced to live at a higher density than they would choose, high stocking densities on fish farms, or in captivity for other purposes, cause poor welfare (e.g. Ewing and Ewing, 1995, for trout and salmon; Vazzana *et al.*, 2002, for sea bass; Montero *et al.*, 1999, for sea-bream). When high stocking density was combined with insufficient water flow (Ellis *et al.*, 2002) or with too much disturbance (Turnbull *et al.*, 2005), welfare was worse than when only one adverse factor was present.

Farmed salmon and trout kept at high stocking densities usually have damaged fins. This is often called fin erosion (Ellis *et al.*, 2008) but this implies that the tissue loss is always a consequence of rubbing the fins against something so it is an inaccurate term and should not be used. It is likely that much of the damage is caused by either fin-chewing by other fish, or as a result of contact with other fish rather than by contact with the cage or tank. Fin damage in trout was higher when there was food deprivation (Winfree *et al.*, 1998). Cañon Jones *et al.* (2010), using social network analysis, showed that dorsal fin damage in salmon was positively correlated with aggression and fin-biting and was seen only in groups subjected to feed restriction. Fish initiating aggression were less likely to have fin damage. Because of the way in which farmed salmon are fed, the number of individuals that are feed restricted will increase with stocking density. In a further study, Cañon Jones *et al.* (2011) found that fin-biting behaviour and dorsal fin damage in salmon was higher at high stocking density than at low stocking density.

Other species of fish that are largely herbivorous may be less likely to show fin-chewing behaviour, but many species have chewed fins at high stocking densities. The stocking density should allow fish to show most normal behaviour and avoid having to show abnormal behaviour, and with minimal pain, stress and fear.

Feeding methods

Food distribution in such a way that each individual can get sufficient food is a requirement for the keeping of farmed animals. Farmed fish are usually subject to great competition when food is provided for them (see Fig. 33.1). Observations of salmon in cages during food provision show that the largest and fastest fish get a disproportionate amount of food, and a high proportion of smaller animals that are less well able to compete are found at the edges of the cage. Observations during diving showed that smaller animals were at the cage edge all the way down to the bottom of the cage 15–20 m below the surface (D.M. Broom, pers. obs.). If food was made available on demand for salmon, they swam more slowly and showed less fighting than when it was supplied by broadcasting, i.e. in a way that led to much competition (Andrew *et al.*, 2002). In a comparison of

Fig. 33.1. The trout in this large tank reacted by coming to the surface when humans who might be bringing food arrived; when food was spread they competed vigorously for it (photographs D.M. Broom).

Fig. 33.2. Trout left to die by asphyxiation in air (photograph D.M. Broom).

Fig. 33.3. Salmon taken from the water prior to stunning (photograph D.M. Broom).

predictable and unpredictable food delivery for salmon, Cañon Jones *et al.* (2012) found that fin damage was substantially higher in the unpredictable food delivery groups.

Food should be distributed in such a way that it reaches all fish or is accessible to every individual. Better systems for the predictable provision of adequate amounts of food to all fish should be developed. One example is a demand-feeding system for salmon during freshwater production. This system delivers food pellets until a certain number of uneaten pellets are detected and has been shown to reduce fin damage (Stewart *et al.*, 2012). When cod were able to obtain food by pulling a string, some individuals learned to do this using a tag on their dorsal fin as a tool (Millot *et al.*, 2013).

Catching and killing methods

In European countries, and indeed in most countries in the world, farm animals are required to be killed in a humane way that includes prior stunning. This should be applied to fish also. It is unacceptable for fish to die from asphyxiation in air, because welfare will be very poor at this time (see Fig. 33.2). Cooling on ice before death prolongs the period of poor welfare (Robb *et al.*, 2000; Robb and Kestin, 2002). The percussive method is used for salmon, using either a specially designed club or a mechanical stunning device. The blow should be of sufficient force for the fish to be immediately rendered unconscious and for it to remain so until dead (see Fig. 33.3).

Satisfactory methods of slaughtering smaller fish, such as trout, en masse that render them insensible instantaneously and until death supervenes are required. The widely used methods do not involve stunning. Electrical stunning seems to be the best available method for doing this (Lines *et al.*, 2003). Staff employed in the slaughter of fish should have the knowledge and skill to perform the task humanely and efficiently for the method employed.

On some fish farms, instead of catching the fish in the water in the most rapid and least disturbing way, people are allowed to come to the farm and catch the fish with a hook and line. Hooking and handling fish for release has been shown to increase scale damage, making the fish more vulnerable to infection (Broadhurst and Barker, 2000). Injury and mortality following the hooking of fish is common, especially where the hook penetrates deep into the tissues (Muonehke and Childress, 1994), and there is a clear increase in mortality during and after live-release tournaments (Suski *et al.*, 2005). The actual process of capture on a hook leads to an increase in both heart rate and cortisol production, and subsequent avoidance of the situation (Verheijen and Buwalda, 1988; Pottinger, 1998; Cooke and Philipp, 2004). Catching with a hook and line is not a humane method of capturing fish so should be avoided where possible.

Other information about the effects of handling on fish (see below) is also relevant to other farm procedures. Later effects of capture and of a period in air before being returned to water include suppression of immune system function, suppression of oestradiol levels, reduced reproductive ability and severe metabolic effects (Pickering and Pottinger, 1989; Melotti *et al.*, 1992; Ferguson *et al.*, 1993; Pankhurst and Dedual, 1994). A period in a keep net also leads to adrenal responses – sometimes prolonged, but sometimes rather brief (Pottinger, 1998).

Fig. 33.4. The sea-lice on this salmon eat its tissues and cause pain and debilitation (photograph D.M. Broom).

Fig. 33.5. Trout show a maximal emergency physiological response and subsequent immunosuppression when removed from water (photograph D.M. Broom).

Environmental quality enrichment

Fish are kept with conspecifics, so they are not deprived of social contact. However, in other respects their environment is rather barren. More information is required on whether fish welfare can be improved by environmentally enriching stimuli and on how to provide for all of the needs of fish, including any need for varied stimulation.

Disease and parasitism

Pathogens and parasites generally cause poor welfare in fish. Hence, it is important to manage fish so as to minimize disease. A key aspect of this is to have good methods of inspecting fish to recognize those that are diseased, distressed or dead. Parasites such as sea-lice can cause poor welfare in salmon (see Fig. 33.4). As a result of efforts to reduce or avoid the widespread use of antibiotics in fish farming, vaccination is used frequently. However, because of the handling involved and the use of irritant adjuvants, welfare can be very poor for short periods and poor for long periods (Sørum and Damsgaard, 2003).

Handling, grading and transport

Fish show a maximal emergency adrenal response when removed from water (see Fig. 33.5). Methods of movement of fish that do not require removal from water are preferable on grounds of fish welfare. However, any kind of manual handling, many aspects of the grading procedure and some aspects of transport are very stressful to fish and usually increase susceptibility to disease (Strangeland *et al.*, 1996; Pickering, 1998). Handling often causes damage to scales but there is variation among species in this respect. For example, it is reported by farmers that halibut *Hippoglossus hippoglossus* are less readily injured than salmonids. In addition to the emergency physiological response as an indicator of welfare during handling and transport, 2 h of transport has also been found to impair learning ability in coho (silver) salmon (Schreck *et al.*, 1997).

Fish populations should not be graded (see Fig. 33.6) more often than is absolutely necessary since grading, with its associated crowding, leads to prolonged

increases in plasma cortisol concentrations (Barnett and Pankhurst, 1998), and most kinds of grading are likely to be stressful for fish. Salmon show an increase in plasma cortisol when pumped or loaded on to a well-boat and measures of the carcass show that welfare is better after transport in an open well-boat than after transport in a closed well-boat (Gatica *et al.*, 2010a,b).

During the stripping and milking processes, the more times a fish is handled and exposed to sedation the greater the skin injury and stress. If effective anaesthetics are used, and maintained at an appropriate concentration throughout sedation and anaesthesia, fish welfare is much improved.

Predators and farmed fish

When fish are farmed it is often necessary for measures to be taken to protect the fish from predators, as mortality can be high (Carss, 1993), and they show strong emergency adrenal responses and also suppression of feeding when predators are present (Metcalfe *et al.*, 1987). Many of the predators are species that the general public hold in high regard, for example seals, otters, herons, kingfishers or gannets. Hence, it is necessary for there to be anti-predator measures (see Fig. 33.7) that minimize poor welfare of the predators and do not endanger predator populations. The killing of predators should be a last resort.

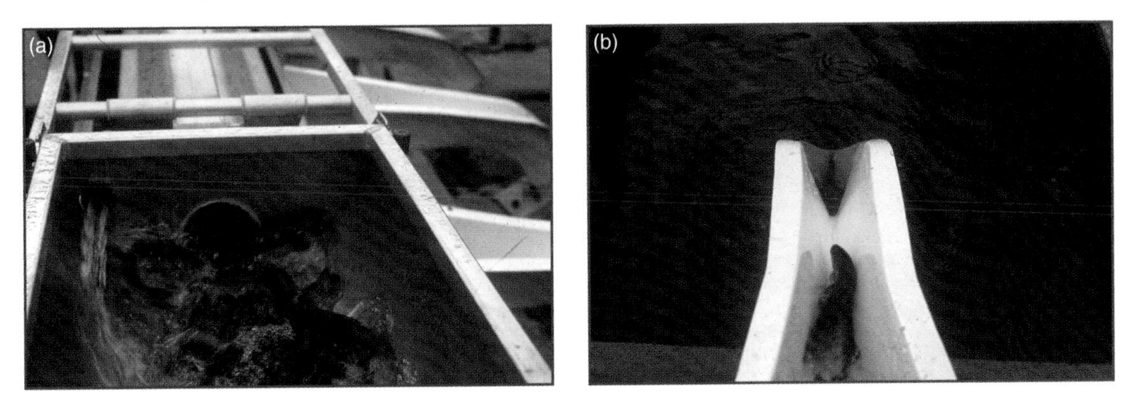

Fig. 33.6. (a, b) Grading allows fish of different sizes to be separated but, if they are forced to be out of the water, the effects are substantial (photographs D.M. Broom).

Fig. 33.7. This salmon cage in the sea has cords over the surface that deter predatory birds from landing in the cage (photograph D.M. Broom).

Particular Concerns for Pet Fish

Many of the issues that are important in relation to the welfare of farmed fish are also relevant to pet fish. However, as there are a very wide variety of species of fish kept as pets, so their needs also vary greatly. A few of the species kept can survive in air and many are tolerant of relatively low oxygen concentrations in water. Those who keep fish may sometimes fail to ensure that air is bubbled through the water in which fish that need such a supply are kept. When there are long intervals during which the container in which the fish live is not cleaned and there is no other provision for keeping it clean, the result may be that oxygen levels drop and pollutant levels increase. In some cases, food given in excess to the fish starts to decay and becomes a pollutant. The consequence can be poor welfare.

If fish of different species are kept, it may be that there is predation of one by the other, especially if young fish are developing in the container. Even if one does not attempt to kill and eat the other, injuries to fins or other parts of the body may occur. Any such injuries will lead to pain and probably to fear and other aspects of poor welfare.

One of the most common situations that is likely to cause poor welfare in pet fish is being kept in a barren environment. Fish normally live in and respond to complex environments. It may well be that the single goldfish in a bowl is very severely affected by the lack of stimulation. There is little scientific information on this topic.

Fish are susceptible to a wide range of diseases and parasites. Many fish kept as pets are harmed by fungal diseases. Owners may be unaware of how to prevent or treat their fish and may fail to seek veterinary advice. There are now specialist fish veterinary practitioners who can advise on both pet fish and farmed fish.

Further Reading: EFSA (2008) Animal welfare aspects of husbandry systems for farmed Atlantic salmon. *EFSA Journal* 736, 1–31.

34 The Welfare of Animals Kept for Fur Production

Introduction

The term 'kept for fur production' refers to the keeping of animals so that their skins and attached fur can be used for human garments and for decoration. The skin is also used from animals such as cattle and sheep that are principally kept for the production of human food. The welfare of these animals is discussed in other chapters. Many rabbits are in that category (see Chapter 36). Dogs and cats are eaten in a few countries (see Chapters 37 and 38) but are occasionally kept or killed for their fur. Apart for these animals, the species that is truly domesticated and which is often kept for its fur is the ferret (*Mustela furo*). Ferrets have been kept for about 2000 years and used either for catching rabbits and other small mammals or as companions. They are now sometimes kept in wire cages for fur production. The mink (*Mustela vison*), red or silver fox (*Vulpes vulpes*), Arctic or blue fox (*Alopex lagopus*), coypu or nutria (*Myocastor coypus*), raccoon dog (*Nyctereutes procyonoides*), chinchilla (*Chinchilla brevicaudata* and *C. lanigera*) and sable (*Martes zibellina*) are scarcely domesticated, having been kept on fur farms for 20–90 generations only (Hansen, 1996).

The coat characteristics of these animals, especially of mink and foxes, have been substantially changed in this time but, as the animals have been kept in wire cages, seeing humans only intermittently and often in rather disturbing situations, there has been relatively little adaptation to human presence. Foxes can be changed genetically (see below) so that they are considerably easier to tame, but the degree of human contact needed to tame the animals is not normally provided on fur farms. Although studies of the behaviour and other functioning of animals kept for fur production show that they are scarcely domesticated, a brief summary of their welfare in fur-farming conditions will be given here. The production of feathers is discussed in Chapter 32. Fur farming is now banned in several countries: first, because of public concern about the poor welfare of the animals; and second, because of the idea that it is not necessary to keep animals for all of their lives and to kill them solely in order to produce garments and other clothing.

Farmed Mink

There have been many studies of the welfare of farmed mink (Braastad, 1992; Nimon and Broom, 1999). Mink on fur farms are normally kept in wire-mesh cages with a nest box at one end of it (see Fig. 34.1), but no other contents except a drinker. These conditions fulfil some of the needs of mink but do not fulfil others. Mink in the wild, and many mink that have escaped from mink farms, range over substantial areas, spend time in holes (especially when threatened), use water for various purposes, climb, investigate their surroundings and, once separated from the family group, spend little time close to other mink. Evidence for the aquatic lifestyle of mink includes: (i) their partially webbed feet and ability to swim fast and dive readily; (ii) the facts that much of their food is derived from aquatic sources and that they obtain food and play under water; and (iii) studies of radio-tracked mink that swam a mean of 250 m once or twice each day. When mink were trained to perform operants to reach an extra nest, various objects, a raised platform, a tunnel, an empty cage and a water pool, the mink gave very high priority to the opportunity to swim in the water pool (Mason *et al.*, 2001).

Farmed mink are fed with fish and avian or mammalian material that is generally of good nutrient and hygienic quality. It is usually put onto the roof of the cage where the mink can easily reach it. Mink usually breed successfully on fur farms (Elofsen *et al.*, 1989; Møller, 1992) and, although there are serious infectious diseases of mink, the health of animals on fur farms is generally good.

Fig. 34.1. Mink in typical mink cage on fur farm. The mink is rearing over the entrance to the nest box (photograph D.M. Broom).

Some mink on every farm show self-mutilation in the form of fur-chewing (Joergensen, 1985; de Jonge and Carlstead, 1987) and many show high levels of stereotypies (e.g. de Jonge *et al.*, 1986); neither of these problems occurs in the wild (Mason, 1991; Dunstone, 1993), nor have they been mentioned in relation to different captive conditions such as zoos (e.g. DonCarlos *et al.*, 1986) or laboratory studies (e.g. Dunstone, 1993). Much the commonest form of self-mutilation is removal of fur from the tail by chewing. Most farms have some mink that do this, and de Jonge and Carlstead (1987) found that 18% of females did so. A few mink extend tail-chewing to removal of tail tissue until it is shortened to a stump. Rarely, mink may chew and shorten their own limbs.

Some mink on farms are very fearful of people and these individuals often show other signs of poor welfare. One test for fearfulness is the stick test, involving measurement of response to a stick inserted into the cage, but this has been found to be unreliable in some situations. Another test involving the introduction of a glove into the mink cage has been reported to be more widely useable (Meagher *et al.*, 2011).

Stereotypies shown by farmed mink include head-weaving, jumping, route-tracing and twirling. In some cases stereotypies continue when human observers are present, but most cease when there is an interesting or significant event in the mink's life, such as the arrival of humans. In one study, 70% of 142 farmed mink showed some stereotypy, and 50% for more than 25% of waking time (de Jonge and Carlstead, 1987). In another study, 16% of active behaviour in a large mink unit was spent showing stereotypies (Bildsøe *et al.*, 1990).

The factors affecting the occurrence of stereotypies in mink make it clear that the cage environment and management methods cause the problems for these animals. Stereotypies are absent in the wild, in good zoos and in rich cage environments. However, they are present in normal and double-sized cages. The peak occurrence is before feeding, but they also occur at other times. As mentioned above, they may stop if a human observer is present, so it may be that lack of stimulation is a key causal factor. More stereotypies and tail-biting occur if the mink have been weaned at 7 weeks than if weaned at 11 weeks (Mason, 1994), so some frustration that the mother can prevent may be one aspect of their causation. The levels of stereotypies shown by caged mink are high and, although some farms seem to have much lower levels, it is clear that the normal mink environment is seriously inadequate.

Efforts to improve the mink cage, so that the needs of the animals are more likely to be met, have largely involved modification of existing cage systems. Hansen (1998) reported that mink would use a platform in the cage, and several authors have confirmed that the widespread practice of providing a nest box containing straw improves welfare. Vinke *et al.* (2002) found that connecting cages together and providing a plastic cylinder and a wire-mesh platform had little effect on the occurrence of stereotypies and self-mutilation. Environmental complexity, rather than just an increase in cage size, reduced indicators of poor welfare such as stereotypies and tail-chasing in mink (Hansen *et al.*, 2007). Meagher *et al.* (2013) enriched mink cages by giving the mink access to a second cage 120 cm wide via an overhead tunnel made of wire mesh. This second cage included a trough of running water, a plastic hammock and many manipulable objects. Mink in a conventional cage showed several signs of worse welfare than those in the enriched cage. They had higher concentrations of faecal cortisol, showed more locomotor stereotypy, spent more time lying prone with the eyes open and fearful males spent more time inactive in the nest box. It was concluded from a series of analyses that lying in the nestbox indicated anxiety while lying prone with eyes open indicated boredom.

Ferrets

Very little research on ferrets has been published. They generally appear to be much less aggressive to, or frightened of, humans than farmed mink or farmed foxes. However, the animals commonly seen are companion animals that are carefully cared for by owners who have much affection for them. Mink are not reared in this way and will bite humans readily. On fur farms, the ferrets are kept in the same kinds of cages as are mink. It is likely that they would be subjected to most of the same frustrations as a consequence. However, ferrets do not have partially webbed feet and do not show strong preferences to swim, play and hunt in water.

Farmed Foxes

Bakken *et al.* (1994) described the standard conditions on fur farms, for example in Norway or Finland. Foxes are kept in wire-mesh cages, with a floor area of 0.6–1.2 m^2, and a height of 0.6 or 0.7 m (see Fig. 34.2). The smaller cage size is normally used for the Arctic or blue fox. Cages are furnished with a nest box from the onset of the mating season until the weaning of the cubs. For the rest of the year, they contain no furnishings or cover. Such systems do not meet the needs of the animal and

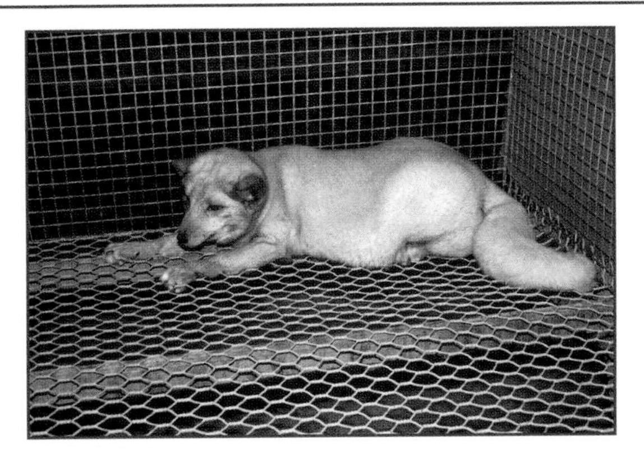

Fig. 34.2. Farmed Arctic fox in typical cage (photograph D.M. Broom).

the result is poor welfare (Nimon and Broom, 2001). In Europe, since the publication of the Recommendations of the Council of Europe (1991), there has been an increased realization in the industry that many consumers are aware of poor welfare in animals kept on farms. Hence, cages may now include a shelf for the foxes to sit on and a nest box whether or not the animals are breeding.

Following birth, the vixen and her cubs are generally separated at 8 weeks of age, and the cubs may be kept with littermates until 10 weeks of age. Pedersen (1991, 1992) reported that, while the litter remained together, they were housed in double standard Danish fox cages, measuring 1.95 or 1.2 or 0.95 m. Thereafter, foxes may be kept singly or in pairs, generally siblings, in either single or double standard cages (Pedersen and Jeppesen, 1990; Pedersen, 1991).

Farmed foxes are fed daily in the same manner as mink (see above; Nimon and Broom, 1999). They are caught and handled when assessing oestrus and during mating, fur grading and medical treatment. On average, breeding animals are caught or moved up to 20 times per year, and cubs may be moved up to five times. Handling by humans and enclosure in a new cage are regular experiences. Cages, whether containing animals that will be killed for their fur when fully grown or animals kept for breeding, normally have mutual wire walls and are housed together in sheds. Farmed foxes are reported to exhibit 'extreme fear' of humans and of other foxes, involving trembling, defecating, withdrawing to the back of the cage and attempting to bite handlers (Tennessen, 1988). Several studies have examined the possibility of reducing the state of 'continuous fear' (Bakken *et al.*, 1994) by selective breeding. A large-scale domestication experiment in Siberia selectively bred silver foxes for the elimination of negative, defensive responses to humans (Belyaev and Trut, 1963; Belyaev, 1979; Belyaev *et al.*, 1985; Trut, 1999). This produced a population of foxes similar in behaviour to domestic dogs. In a 2-year study, comparing 150 cubs from the selected population with 123 cubs from unselected foxes bred for commercial purposes, Belyaev *et al.* (1985) found that domestication extended the period during which cubs could habituate to people. Cubs from the selected population showed some developmental changes, were reported to show no fearful behaviour when placed in a new cage and exhibited less fear than controls in response to humans. It has also been reported that females from the selected fox population came on heat earlier, and that fertile mating sometimes occurred twice a year (Naumenko and Belyaev, 1980; Trut, 1981; Hansen, 1996).

Further experiments suggested that changes in brain chemistry and in the development of the pituitary system accompany the divergence between tame and wild foxes (Malyshenko, 1982; Trut and Oskina, 1985; Plyusnina *et al.*, 1991; Popova *et al.*, 1991; Dygalo and Kalinina, 1994). Harri *et al.* (1997) showed that Russian silver foxes that had been selectively bred for 37 generations showed lower behavioural and physiological responses to people and handling than foxes bred in the USA. However, these genetic lines are not being used commercially. Early experience of humans can have substantial effects on foxes (Pedersen, 1994), so tamer genetic lines would still need much human contact in order to be well adapted to humans, and there might still be the problem of fearfulness towards foxes in adjacent cages.

Fox farms have come under public criticism for: (i) the small size of cages; (ii) the fact that they are barren, with no contents except a drinker; and (iii) the use of wire mesh for floors. Increases in the size of the cage, without a consequent increase in the complexity of the environment, do not appear to benefit farmed foxes (Korhonen and Harri, 1997). However, larger space allowances for foxes can make real enrichment possible. Nest boxes are seldom needed to protect farmed foxes from low temperatures, but provide a temporary, but often important, hiding place that can significantly enrich the environment (Jeppesen and Pedersen, 1991; Nimon and Broom, 2001). Mononen *et al.* (1995b) found that silver foxes used the tops of nest boxes as a resting place.

In zoos, foxes choose to sit on vantage points two or more metres above the ground, and it is clear that the addition to cages of elevated platforms for foxes provides environmental enrichment. Platforms of appropriate size have been used by captive foxes of both species for biologically appropriate behaviours, and they can evidently help to meet some needs of foxes (Korhonen *et al.*, 1996). Traditional fox cages are up to 1 m high, and any shelf is therefore unlikely to be high enough to provide an adequate viewpoint. Multi-cage systems, rather than simple cages, are more likely to provide better for the needs of foxes. Experiments involving group housing of animals in large enclosures or complex cage systems have found some beneficial effects, but may not be appropriate during the breeding season. Unrelated adult foxes may harm one another.

Harri *et al.* (1995) found that silver foxes preferred solid floors, provided that this preference did not conflict with those for resting in an elevated position and thermoregulatory efficiency. At lower temperatures, foxes can fluff out their fur and be better insulated on a wire floor. Wire-mesh platforms seem better suited to Arctic foxes than to silver foxes.

The incidence of abnormal behaviours in farmed silver foxes is cause for serious concern. Stereotypies in farmed foxes are very frequent in some individuals, especially when they are not observed by humans, and occur in a larger proportion of individuals in frustrating circumstances (Braastad, 1993). Reproductive problems are common in foxes, sometimes associated with abnormal behaviour and fear of neighbouring foxes. A survey of all fox farms in Germany (Haferbeck, pers. comm., 1996) found that 45% of silver foxes and 40% of blue foxes failed to breed, although the incidence of such problems is lower in some other countries.

The killing and injury of cubs by their mothers, starting with tail removal and biting, has been reported as a common problem on fox farms (Bakken, 1989, unpublished; Braastad, 1990a,b). Braastad (1994) found that vixens provided with a nest box that included an entrance tunnel bore more offspring and had lower cub mortality up to 3 weeks after birth. The suggestion that infanticidal vixens are affected by the presence of their neighbours was borne out in the work of Bakken (1993a), in which vixens that had been infanticidal were physically and visually isolated from other vixens. They then raised significantly more unharmed cubs. Furthermore, Bakken (1993b) produced significant reductions in infanticidal behaviour by manipulating the social status of vixens' neighbours. This strongly suggests that reproductive performance is inhibited and infanticidal behaviour enhanced among many low-status females on farms today.

It is clear that there are serious welfare problems on commercial fox farms and that these have not been solved by research or the cage modifications used on farms.

Coypu

The coypu, whose fur is usually called nutria, is a South American, largely aquatic rodent. Coypu build nests in reedy areas and may also dig holes in the banks of lakes and rivers as hiding and living places. They have webbed feet, swim and dive well and do not thrive if swimming water is not available to them. They are kept less often for their fur now than in earlier years because escaped coypu have caused such damage to river banks and other waterways. The musk-rat, *Ondatra zibethicus*, an aquatic native of North America, has also been kept occasionally for its fur and has also caused much damage in Europe. Captive coypu can adapt quite well to human contact if handled carefully. However, little research on their welfare has been carried out.

Raccoon Dogs

Raccoon dogs have moved their range westward into Europe in recent years. They are somewhat smaller and shorter-legged than red foxes and have been kept in similar cages, largely in Finland. Although apparently somewhat less nervous of human contact than foxes, it is not known what their real reactions to humans are and, as they have been kept in captivity for such a small number of generations, it may well be that they have many problems in adapting. There are no papers on the welfare of these animals in good-quality scientific journals.

Chinchillas

Chinchillas are small South American rodents living naturally in rocky mountainous areas. *Chinchilla lanigera* is the species most often kept for fur production. They have been kept as pets for longer than the short period during which they have been kept for their fur. When kept on farms, they are generally not habituated to human presence and their behaviour makes it clear that they are frightened by human approach. Since their major escape response when startled is to jump to several times their own height, they may well hit the roof of any small cage and injure themselves. Hence, no chinchillas should be kept in cages that do not allow them to jump to the typical escape response height. This height is probably about 70 cm, but no studies of chinchilla welfare in relation to space allowance are available.

Male chinchillas show paternal behaviour and it seems that they normally live as monogamous pairs. However, they are often kept on farms in polygamous situations with several females per male. This may lead to increased aggression and must lead to social disruption. In some units, the female chinchillas are fitted with neck collars. It is not known what effect such collars have on the welfare of the animals. Dental problems result in poor welfare in chinchillas and these are widespread, especially on farms.

Since chinchillas seldom live alone, the keeping of single individuals as pets is likely to lead to poor welfare. However, some pet chinchillas are kept in good conditions that appear to meet their needs. There is little scientific evidence about the welfare of chinchillas.

Sable

The sable is an active, fierce predator in its Asiatic range and is rather bigger than a mink. These animals have been kept occasionally on fur farms but there is no evidence that they can adapt to cage conditions, so they should not be kept in cages.

Further Reading: Meagher, R.K., Campbell, D.L., Dallaire, J.A., Díez-León, M., Palme, R., and Mason, G.J. (2013) Sleeping tight or hiding in fright? The welfare implications of different subtypes of inactivity in mink. *Applied Animal Behaviour Science* 144, 138–146.

35 The Welfare of Horses, Other Equids and Other Draught Animals

Ill-treatment and Neglect

Some of the earliest animal protection legislation was enacted because of concern about the welfare of horses. Those trying to shame society and politicians into taking action to produce animal protection laws used the image of a horse or donkey that was beaten in a way that led to substantial and long-lasting cuts and bruises and of the animal loaded so heavily that it could scarcely move. In some countries, even now, it is not an offence to beat your own horse or donkey to death or to make it pull a cart or carry a burden that causes it to break its back or leg. Any person who carries out such ill-treatment, and has sufficient knowledge to be aware that great pain and other poor welfare is being caused to the animal, is cruel.

Neglect of horses is sometimes deliberate and entirely avoidable, sometimes carried out by a person who knows what is being done but is unable to do otherwise because of their own illness or poverty, and sometimes accidental because of lack of knowledge. A common situation is a horse that is kept in a field in which the pasture production rate is perhaps half of that needed by the horse, with no supplementary food. The horse gradually starves. Poor body condition, listlessness and other abnormal behaviour are the most obvious signs (A.F. Fraser, 1992). Since no person should keep an animal unless they have taken the trouble to find out how to provide for its needs, the human owner who gives the horse too little grazing or other food is doing something that is avoidable and cruel. The feeding of an unbalanced diet to equids is also a cause of poor welfare. Incorrect feeding is relatively rare in farm animal production, but not uncommon among those who keep a horse as a companion animal. Cattle used as draught animals, for example in India, may be fed while they are working but not at other times. This may result in the animals receiving no food on over 200 days in a year and hence having very poor welfare, becoming emaciated, and sometimes dying of starvation (Abdul Rahman and Reed, 2014).

Another common cause of poor welfare in working equids, and in equids kept as companions, is failure to carry out the normal care required to prevent the occurrence of pathological conditions (A.F. Fraser, 2003). The care may be of a kind that an owner or farrier can carry out or it may be the work of a veterinary surgeon. Hoof overgrowth because of failure to trim the hoof properly, and hoof damage such as splitting or other injury that would not occur if proper shoes were fitted to the hoof, can be prevented by a knowledgeable owner or by a farrier. Many ailments can be readily prevented by veterinary treatment but can cause poor welfare for long periods if untreated. Lack of care leading to avoidable disease is also a form of cruelty. In a study of working equids in Chile, the hoof problems observed were mainly a result of the lack of a farrier in the area rather than poor care by owners (Tadich and Stuardo Escobar, 2014).

Riding, whipping and working in relation to welfare

Horses, donkeys and mules may initially be disturbed by having to carry a human or a cargo load on their backs but, once trained to do so, their welfare does not seem to be poor when a person or load of moderate size is carried. Indeed, horses and donkeys seem to form bonds with their regular riders and horses may show little reluctance to carry a rider. In many cases, this is so because the carrying of a human rider is associated with a more interesting interlude in life and a greater amount of exercise.

The amount of exercise involved in horse-racing often seems to result in no adverse effects on welfare but it can do so. Pinchbeck *et al.* (2013) studied the joints of 164 Thoroughbred racehorses, whose detailed records were available, in order to evaluate the prevalence

of palmar–plantar osteochondral disease. They found that the grade of the disease condition was higher in horses that had raced for more than one season and in those that had raced at more frequent intervals. Since cumulative racing and training exposure were associated with more disease, it is clear that repetitive loading is a significant cause and horse welfare is improved if the higher frequencies of training and racing are avoided.

There are many situations, however, in which humans want a horse or other equid to carry more or go faster than they would choose, or go in a direction in which they are reluctant to go. The animal may then be encouraged or punished by shouting, whipping or goading with spurs. The use of a whip or kicking by a rider may be at a level of minor persuasion or at a level that causes substantial injury.

Hence, there is a need to distinguish between acceptable and unacceptable levels. This distinction should be made by assessing the effect on the welfare of the animal. Spurs with 2 cm-long, pointed spikes can easily puncture the skin, draw blood and substantially damage tissue in the area of the flanks or belly where the rider is kicking and goading. Therefore, many organizations controlling riding either prohibit the use of metal spurs or limit the size and shape, so that even a very enthusiastic or angry rider will not cause severe pain. The use of a stick can cause visible wheals, while thermal imaging shows the extent of inflammation engendered in the area struck by the stick. A harder blow produces measurably more obvious inflammation, and the number of blows with the stick above a certain threshold of force can be counted. When the method of evaluating stick use by jockeys riding racehorses was first used, it became apparent that some jockeys used much force while others used little. Excessive use of the stick has been proscribed by jockey clubs in many countries. However, the question of what is excessive is interpreted in various ways. If all forms of severe punishment of horses were made illegal, horse-riding sports would be very much changed.

When other animals that are ridden or used for draught are considered, much greater cruelty may be found to be normal. Elephants are often trained with the use of a 15–20 cm-long dagger that can be pushed into the elephant's neck. Some are never ridden unless the rider carries a dagger or ancus. This is done because the elephant can easily kill its rider but the practice calls into question the morality of training and riding elephants. If dagger use is necessary to train and ride them, they should not be ridden. Cattle, buffaloes (Fig. 35.1), horses, mules and donkeys are sometimes treated very harshly by those who try to control them. The draught animal may be of great value to people but some are not considered as individual beings that can suffer. Governments may advocate care of production animals but say little or nothing about draught animals. Measures of welfare in working equids are summarized by Pritchard *et al.* (2005) and some ways of dealing with them in developing countries are discussed by Upjohn *et al.* (2014).

Fig. 35.1. Water buffalo in quite good condition used as a draught animal (photograph A.F. Fraser).

Training methods and welfare

Many animals are trained to behave in a way that is foreign to them in respect of the movements carried out, the context in which they are carried out or the frequency of the action. The purpose may be entertainment for many people, or profit for a small number of people. One central moral question in relation to the use of such animals is whether or not the training required is humane (see Chapter 22). If, in order that successful training can occur, the welfare of the animals concerned is necessarily very poor, most people would say that the usage is not justified. The training of wild animals in circuses comes into this category. However, a domestic animal such as the horse can be trained to be ridden or to pull a cart or plough without very poor welfare during training. The problem remains that inhumane training methods are used by some horse trainers (McDonnell, 2002) and it is often difficult to detect their occurrence. If it is known that harsh training methods may sometimes be used, the whole practice of horse training is put in jeopardy as the public may demand that it should cease.

A particularly dramatic phase in the training of a horse is that which leads to the horse tolerating being ridden. The normal term for this procedure is 'breaking' which indicates that it is often a very severe procedure. This term implies a violent subjugation of the animal. Harsh breaking methods clearly lead to very poor welfare for a period in the life of the horse. However, much more gentle 'breaking' can be carried out. The harsh methods are illegal in some countries and are condemned by many horse owners.

Once a horse can be ridden, some horses are trained rigorously so that they can compete in showjumping, cross-country riding, dressage or racing competitions. Some of the methods used in such training are mild, friendly encouragement while others are violent and painful. There is a need to check more extensively on the methods used and to eradicate the more extreme practices that lead to very poor welfare.

The age at which horses are trained for riding in races can have a great effect on their welfare. If young horses are trained so that they run fast and frequently, especially if they run on hard ground, there may be adverse effects on their bone development. The horse trainer does not want this to happen but there is a considerable financial pressure on horse trainers to start training horses to run hard as early as possible, even if this means that 10% or even 50% of horses are never able to compete in races as a consequence. It would be better to change the rules about the age at which horses can be raced in particular categories so as to reduce the risk of premature hard training leading to locomotor anomaly and hence to early death.

Surgical operations

Horses may be subject to mutilations in order that they can be more easily managed or that they will look different. Stallions are often gelded, or castrated, so that they are less risky to ride and because the individuals concerned are not needed for breeding. Castration is painful at the time of its occurrence unless anaesthesia and analgesia are used, and there will be pain when the painkiller effect wears off. Tail-docking is also painful. Both of these operations will have consequences for later welfare. For example, a horse whose tail has been docked may have very poor welfare because of inability to dislodge and repel flies that irritate or bite. Some other operations may be carried out to reduce the likelihood of occurrence of stereotypies but these are wholly harmful procedures that make welfare worse (see Chapter 24). The firing procedure, which involves treatment of an injury using high heat, freezing, strong acid or strong alkali, also leads to poor welfare and is unnecessary for normal running.

Breeding and welfare

When the enormous changes that have occurred during the breeding of most farm livestock in the last 100 years are considered, the changes in performance of horses over the same period are surprisingly small. However, there have been changes brought about by breeding that have an effect on welfare. Some strains of horses are described as nervous and are relatively difficult to train and ride.

Housing and management

Horses are social animals and need contact for a sufficient time each day with other horses or companions of another species (Fig. 35.2). Studies of donkey behaviour make it clear that donkeys have regular companions whose company they seek (Murray *et al.*, 2013)

Fig. 35.2. Horse with a goat as companion. The welfare of horses kept alone is often poor (photograph R. Harnum).

so if they are separated or isolated from social contact, their welfare will usually be poor. They also need roughage and other components in the diet and regular exercise. Many individuals are kept in conditions of social isolation, sometimes with inadequate diets and with too little opportunity for exercise. As explained in Chapters 24–28, the housing conditions, feeding and other management of horses often result in various abnormalities of behaviour and poor welfare. A high proportion of horses spend some part of their lives in conditions, such as individual stalls, in which their needs are not met and their welfare is poor. Housing and management that do not meet the needs of the animals are the most important welfare problem for most horses. In a study by Fureix *et al.* (2012) horses kept in conditions that did not meet their needs showed behavioural signs of depression (Chapter 28). Evidence like this leads to the recommendation that every effort should be made not to keep horses in individual stalls.

Further Reading: Fraser, A.F. (2010) *The Behaviour and Welfare of the Horse*, 2nd edn. CAB International, Wallingford, UK.

36 The Welfare of Farmed and Pet Rabbits

Rabbits are kept as companion animals, as farmed animals for meat production and as laboratory animals. There are several species of rabbits in the world, but it is *Oryctolagus cuniculus* that is kept. In almost all cases these animals can only be kept in cages. Even pet rabbits will usually not stay with humans if they have the choice. The size of the cages varies according to the purpose for which the rabbits are kept. Some pet rabbits have much more space than those in laboratories or on farms, but others are in very cramped conditions. Rabbits kept in laboratories for research purposes generally have substantially more space than rabbits on farms kept for meat production. This is not logical in relation to the needs of rabbits, but human attitudes to rabbits vary according to human use!

The numbers of rabbits kept are very large and rabbit meat production in Europe is about 0.5 million tonnes per year. How do we assess the welfare of rabbits? The range of measures is similar to that for all other animals. For example see Chapter 6 for information about pain assessment in rabbits (Leach *et al.*, 2009; Farnworth *et al.*, 2011; Keating *et al.*, 2012). Disease is a major cause of poor welfare in farmed rabbits and disease prevalence is generally higher when there are other welfare problems. Rosell *et al.* (2010) report that the main disease problems leading to emergency veterinary call-out in Spain and Portugal are diseases of the digestive system. The main reasons for emergencies were mucoid enteropathy (25%), enteritis diarrhoea (24%), myxomatosis (11%), reproductive disorders (9%) and respiratory disease (7%). Other pathological conditions cause poor welfare in rabbits but were not the cause of call-outs, such as coryza, mastitis, ulcerative pododermatitis and mange. In a study of the causes of all deaths on rabbit farms, respiratory tract diseases were the main cause of death (Rosell and de la Fuente, 2012).

One of the indicators that has been used in evaluating rabbit welfare is stereotypy. Stereotypies are regularly observed in caged rabbits and are generally due to a lack of environmental stimuli and a lack of control for the animals over their environment. In laboratory rabbits, other behaviours in addition to stereotypies have been used as welfare indicators. These include nosing, which is sliding the nose up and down the bars of the cage; head in corner; head-swaying; and partial rearing, as well as chewing and licking activities that are seen in caged animals to a greater extent than in penned animals with no roof (Morton *et al.*, 1993).

In addition, behaviour in a novel pen test and the reaction of rabbits to a tonic immobility test have been studied in order to attempt to assess the fear reactions to a new environment and towards humans. Physiological indicators such as leucocyte numbers, adrenal weight, ascorbic acid, corticosterone and testosterone levels have also been studied in rabbits. Fenske *et al.* (1982) found that handling and placing rabbits in a new environment were stressful, while Verde and Piquer (1986) found higher corticosterone and ascorbic acid levels in rabbits reacting to heat stress (32–34°C) and noise stress (90 ± 5 dB, 200 c/s) in rabbits caged for 42–43 days. High heat (42°C for 4.5 h) elicited substantial cortisol increase and other physiological emergency responses in rabbits (De la Fuente *et al.*, 2007).

Female rabbits used for breeding on rabbit farms are often kept in wire-mesh cages 60–68 cm long, 40–48 cm wide and 30–35 cm high. Such a cage provides the rabbit with a floor area of 2400–3120 cm^2. A nest box may be put into the cage when she gives birth and removed at the same time that the mother is removed when the kits are 21–25 days of age. Five or six young rabbits may then be left in the cage until they reach the slaughter weight of 2.0–2.8 kg. At this time, each rabbit would have 480–520 cm^2 of floor space. If a larger cage is used for ten rabbits, each rabbit may have 450 cm^2 of floor space (EFSA, 2005b).

The height of normal rabbit farm cages does not allow the adult rabbit to adopt some normal postures. It may not be possible to erect the ears and there are very few rabbit cages in which the animal can rear up on its hind legs. There may be a possibility of ears extending through the roof of a cage and hence being injured. Rabbits are unable to jump in the cages. The escape response may be to jump, so low-roofed cages must pose some problems for rabbits, especially until such time they have learned not to jump when disturbed.

Breeding female rabbits readily mate postpartum, and often have six litters of an average of ten kits each year. Reproducing at this rate, they live for 1 year on average. Rosell and de la Fuente (2009a) reported that, in a large-scale analysis in Spain, the median monthly mortality rate of breeding doe rabbits was 9.3%. This was the total of deaths and culling, mainly because of disease or injury. This life expectancy is very short in relation to the potential lifespan of a rabbit. Hence, it can be deduced that these breeding females on rabbit farms are subject to severe housing and management conditions, in particular a litter every two months, that result in poor welfare. Some of the poor welfare in breeding female rabbits is caused by artificial insemination procedures, as these are often carried out very quickly.

The wire floor of the rabbit cage can cause discomfort and paw injury. Mirabito (2002) surveyed French rabbit farms and injuries were graded according to the classification scale of Drescher and Schlender-Böbbis (1996). On average, 12% of female rabbits had paw injuries that were sufficiently serious for them to show signs of discomfort. Rosell and de la Fuente (2004) found that 9% of Spanish does had such injuries. Princz *et al.* (2005) found that mortality and ear lesions were higher in rabbits housed on plastic-mesh than on wire-mesh floors. Rosell and de la Fuente (2009b) reported the cumulative incidence over five lactations on wire-mesh floors in breeding does to be 71%. If the cages included perforated plastic foot-rests, 37 × 24 cm, the cumulative incidence of foot lesions over five lactations was 15%.

Rabbit behaviour and welfare may be positively affected by repeated handling carried out by familiar people. Rabbits handled as kits show reduced fear of humans later in their lives compared with unhandled controls. Handling is effective when it is performed during a sensitive period, around the first week postpartum and near the time of nursing.

The welfare of rabbits is improved by providing them with hay, metal toys and pieces of wood insofar as they spend a high proportion of their time interacting with these additions to their cages and do not habituate to them (Huls *et al.*, 1991). In a study by Lidfors (1997), hay elicited much more interaction time from rabbits than sticks or a box. At the same time, 'abnormal' behaviours such as licking, gnawing or nibbling at cages were halved when animals were provided with hay or grass compared with animals provided with other items of enrichment or with nothing. However, some of this effect could be because the hay could be eaten. Gnawing of the bars of the cage was a behaviour that was reduced if rabbits had access to wooden sticks as an environmental enrichment. Weight gain was greater in the rabbits provided with sticks.

Stauffacher (1992) suggested that it may be necessary to consider the layout of cages to provide female rabbits with a retreat or resting area where they can get away from their kits. He proposed installing an elevated area, a platform, a separate compartment or a tunnel within the cage for this purpose. Finzi *et al.* (1996) pointed out that installing a raised platform in a cage has the advantage of increasing the space available for the animals without altering the area the cage occupies on the ground (cage footprint). They designed a cage with a 22-cm high platform and found that, over a 15-day period, females spent 53% of time on the platform. In contrast, covered areas such as tunnels were used very little, only 2% of the time, in keeping with observations by Lidfors (1997).

Under farm conditions, over a conventional 42-day breeding cycle, Mirabito *et al.* (1999) and Mirabito (2002) tested the impact of a platform on animals' occupation of the space provided. When females were nursing, the time spent on the platform increased from 20% to 35% between the second and fourth weeks after parturition, which is the time corresponding to the emergence of the kits and the removal of the nest boxes. Although practical problems remain to be solved (Mirabito, 2004), it is clear that the presence of a platform improves the welfare of rabbit does with kits.

Rabbits are social animals and many on farms, in laboratories and as pets are kept socially. Systems for group-housing of laboratory rabbits are described by Batchelor (1999). Seaman *et al.* (2008) measured the strength of preference of rabbits for various resources using the lifting of a weighted door as the operant and calculating the consumer surplus (see Chapter 6). The rabbits had equal preference strength for food and social companions; both were preferred to access to a platform; and all three were preferred to access to an empty cage. However, on farms, barren does, future breeding stock and male rabbits are frequently kept in small cages on their own. Males in laboratories and animals in many test situations are isolated. Many pet rabbits are also kept in isolation for much of their lives. Lack of social contact is a serious deprivation for a rabbit, so the welfare of those kept in social isolation will be poor. Since some males will fight, living in a cage where the individual is the subject of aggression will also result in poor welfare, unless the space quality is such that there are opportunities to hide from potential aggressors.

Pet rabbits will often not stay with humans if they have the choice so, as explained in Chapter 39, they should either be carefully adapted to humans or not kept as pets. Some pet rabbits have much more space than those in laboratories or on farms, but others are in very cramped conditions. The wire floor of the rabbit cage can lead to foot problems in pet rabbits, just as it can in farmed or laboratory rabbits. The welfare of rabbits is improved by providing them with enrichment materials, as described above. Rabbits need social companions and opportunities to hide from perceived danger, such as aggressors. Rabbit behaviour and welfare may be positively affected by repeated handling carried out by familiar people but negatively affected by handling if they have not had such previous experience.

Further Reading: EFSA (2005) The impact of the current housing and husbandry systems on the health and welfare of farmed domestic rabbits. *EFSA Journal* 267, 1–31.

The Welfare of Dogs

Domestication and Breeding

Dogs are wolves and the association between wolves and humans is at least 12,000 years old. Genetic analysis of dogs and wolves shows that modern dogs share a higher proportion of multi-locus haplotypes with the Middle-Eastern grey wolf than with wolves from East Asia, Europe or North America (vonHoldt *et al.*, 2010). Both wolves and humans derived benefit from their initial association, which is likely to have centred around abilities in relation to hunting of large prey. While both could locate the prey, humans could chase it over long distances, the wolves could take over the chase and corner it and humans could kill it. Hence it could just as well be said that dogs domesticated humans as the other way around (Broom, 2006a). In order to exploit the ecological niche of association with humans, dogs had to change genetically in some ways. Individuals with neotenous, prolonged, puppy-like characteristics were more likely to survive, as were those able to breed even when in close proximity to humans. The dog and the human regarded the other as, to some degree, a member of their pack or tribe. Some species of animals would not have been able to do this, and so the commensal, mutually beneficial situation would not have been able to arise. Dogs seem often to benefit from human company, and vice versa.

Although some genetic changes would have existed in all dogs associating with humans, various roles in human society were assumed by dogs and different characteristics were advantageous for each. Those whose main role was to run down prey that humans would then kill became even better at prolonged running in a coordinated group than their wild ancestors. Those whose role was to take on dangerous animals in combat and to hold them needed bigger jaw muscles and neck muscles than a wild wolf. Those living all of their lives like puppies in human habitations required substantially different anatomical and behavioural qualities. Dog breeds arose initially because of the requirements for one of those roles, or for livestock guarding, giving alarm about intruders, retrieving small animal prey, flushing hole-dwelling animals or draught.

Once the diverse types of dogs had arisen, people bred them to look a little more impressive or extreme in various ways. Very large, very small, long-haired, short-faced, short-legged, very fast-running and very tenacious and powerful breeds were all developed. Often by accident, people selected dogs for non-functional, non-biological characters to the extent that dogs were produced that were deaf, hardly able to see, unable to breathe normally, hyper-excitable or prone to disorders such as hip dysplasia, seizures, in-growing eyelashes or excessively folded skin. All of these characteristics resulted in poor welfare, but breeders still tried to produce more and more extreme characteristics. The number of inherited disorders in the 50 most popular dog breeds in the UK revealed by a survey was 396 (Asher *et al.*, 2009). This is very much greater than the number of disorders reported 50 years earlier, and some of the increase is a result of continued selection without proper regard for dog welfare (Sonntag and Overall, 2014).

The majority of people nowadays consider that it is morally unacceptable to continue the genetic line if a dog is likely to pass on genes and hence have progeny that would have any of these characteristics. In that case, breeds such as the Bulldog, Shar-Pei and Pekingese should cease to exist, as should many genetic lines of dogs such as the English Setter, Dachshund, Boxer, Dalmatian, Keeshond, German Shepherd and Golden Retriever. All dog owners should find out what disorders their animals have (Sonntag and Overall, 2014) and no owner should allow their dog to breed if it has a genetic disorder. However, not every person thinks in this way.

Identifying dog welfare problems

Many measures of dog welfare are detailed in the various chapters of this book. Some dog owners are skilled in identifying these but many are unaware of indicators that the welfare of their dog is poor. As explained below, problems for dogs that lead to indicators of poor welfare include lack of social contact, separation anxiety, fear of people, fear of other dogs, fear of environmental events like thunder, and chronic disease conditions. Dogs may show stereotypies, indicating serious welfare problems, (see Chapter 24) but some dog owners and the public may just regard this behaviour as amusing. Tail-chasing is a stereotypy in dogs that has been linked to several disease conditions and environmental inadequacies, but a review of comments made about videos of the behaviour on the Internet site YouTube showed that the majority of people just commented on how funny the dog was (Burn, 2011).

Indicators of anxiety in dogs listed by Sonntag and Overall (2014) include: urination, defecation, panting, increased breathing rate and heart-rate, trembling, lip-licking, nose-licking, hypersalivation, vocalization, freezing, pacing, attempts to escape or hide, not meeting gaze, and changes in activity, grooming and social behaviour.

Mutilations

Mutilations linked to dog breeds

Some dog breeds routinely have parts of their anatomy surgically altered for cosmetic reasons associated with breed standards. The most obvious examples are: (i) the docking of tails, as seen in the Corgi, Boxer, Poodle, Rottweiler and other breeds; and (ii) the cutting of ears to make them pointed. All dogs use their tails to a great extent in communication, and all of these breeds above are seriously deprived of communication ability by the lack of a normal tail. The action of cutting off the tail is painful, as the tail is well supplied with pain receptors; this is normally undertaken, without anaesthesia or analgesia, in the first 7 days of life. Tail-docking is likely to cause intermittent or continuous pain because of neuroma formation after the tail is docked. Very rarely, a tail might have to be docked after injury for therapeutic reasons, but systematic docking of all dogs of a breed or of dogs used in a particular way is never in the interest of the dog. Claims that dogs need to have their tails docked because they will be injured by vegetation are spurious. Removal of long hair on the tail might be helpful, but the removal of bone, skin and nerves is not.

The cutting of ears to make them pointed in breeds such as the Doberman and Rottweiler is entirely cosmetic and will cause pain both when carried out and during healing. The ears are also important in signaling to other dogs and to people in social situations so impairment of this signaling ability by cutting the ears is harmful to the dog. As with tail-docking, reducing an ear to make it pointed also harms the individuals to whom the dog is communicating.

Dog breeds with long hair are sometimes left with long hair, even if it impairs their sensory, locomotor or other function. One example is the Old English Sheepdog, which often has hair hanging in front of its eyes so that it cannot see. Owners of dogs who would be appalled at the idea of removing the eyes of their pet are willing to keep it so that it is permanently unable to see because the hair in front of its eyes. The welfare of such a dog is clearly very poor. Some individual dogs have hair on the body or tail that trails along the ground and catches on vegetation. Other dogs are left with long hair in conditions where this makes it more likely that they will overheat, or are partly shaved in cold conditions. All of these actions cause poor welfare.

Mutilations for the convenience of owners

Dogs may have their vocal apparatus removed to stop barking; this is done for the benefit of the owners or of other humans and does not benefit the dog, except when it would otherwise have been killed. Occasionally, dogs have teeth removed to stop biting; this might benefit other dogs, as well as humans, and may allow the subject dog to have more freedom of movement. However, the severe deprivation would have to be balanced against any advantage of doing this.

The most widespread mutilations carried out to facilitate dog management are castration of males and spaying (ovario-hysterectomy) of females. Intact male dogs roam over considerable distances in search of bitches on heat and they mate with females when they have the opportunity to do so. Pet owners may not want a particular male to father the puppies of a female in or near the household. Hence, the dog is castrated. The operation is usually carried out by a veterinary surgeon with the use of anaesthesia and analgesia. Some pain will still occur, but the greater effect is in changing the behaviour and character of the dog. These are effects causing poor welfare. The extra freedom allowed to the dog after castration counterbalances this to some extent.

When bitches are spayed the operation is more substantial, so the resultant poor welfare is more significant than for castration of males. The bitch may be spared frequent pregnancies and will have more freedom at times when heat would otherwise have occurred. However, the effects of spaying on the character of the female can also be substantial. Inability to have puppies at all during life is clearly a deprivation for a bitch.

Behavioural Problems that Affect Dog Welfare

Deprivation of social contact

Dogs naturally live almost all of their lives in a social situation. If a dog lives with particular humans for some time, those people may be treated as pack members. Since dogs make considerable efforts to associate with other dogs or humans, particularly pack members, and show signs of disturbed behaviour when unable to do so, social contact is important for them. Abnormalities of canine behaviour when left alone in a house are widely reported. The dog may damage items of clothing or furniture, especially those that smell of a familiar human, may bark excessively or may show signs of depression. Such problems are usually much reduced if the dog has a companion, preferably another dog, during any period when the human companion is absent.

Insufficient variety in the environment

Although many dogs have living conditions that are diverse and interesting, some are kept in bare, concrete kennels or small cages. A dog may be restricted in movement by a chain or rope in such a way that many potentially interesting environmental events are out of reach. Such problems are much more common in guard dogs than in companion dogs. A guard dog may have a companion and a large area to patrol, but some are tied and serve a function purely as means of warning or deterring people.

Lack of environmental complexity is much more of a problem in dogs living in boarding kennels, quarantine kennels, dogs' homes or laboratories. Research on the ways in which the environment could be improved for such dogs was carried out by Hubrecht *et al.* (1992, 1993). When dogs had lived in rehoming kennels for more than 6 months, Titulaer *et al.* (2013) found that they rested for longer, played less with people when given the opportunity, played more with objects, and showed more barking and growling at strange dogs than did dogs which had been in the kennels for 1–12 weeks. However, it was not possible to know whether these differences were a cause of people not choosing them for rehoming or a consequence of the long period in the kennel environment.

Noise and other frightening stimuli

Like all other animals, dogs can be frightened by events and situations in their environment. The fear may elicit behaviour that is seen as a problem by owners and other people, such as repeated barking. A well-known example is the fear elicited by sudden loud noises. In a survey of dog owners (Blackwell *et al.*, 2013) 83% reported that their dogs showed fear of fireworks and 65% reported fear of thunderstorms. Other loud noises, such as gunshots, also elicited fear responses in some dogs. Sudden loud noises are probably frightening because of the dog's uncertainty about what they are and what potentially risky situations may follow. The advice of animal protection societies to keep dogs in a calm, familiar environment when they might be exposed to fireworks or thunderstorms would seem to be justified. Much fear may also be shown when a dog perceives that there is a high risk of attack by a human or another dog.

There is individual variation in how often dogs are frightened and in how reactive or impulsive dogs are. Wright *et al.* (2012) describe the impulsive behaviour of some dogs as a problem for dog owners and report on a test involving evaluating the dog's response to a delay in an expected reward. Dogs varied in how much they were disturbed by such a delay. Those dogs that responded most to the reward delay were also described by their owners as the most impulsive, had lower urinary serotonin concentrations and had a lower urinary serotonin to dopamine ratio.

Harsh or inadequate training methods

There are still some dog owners who believe that dogs never learn to be obedient unless beaten vigorously.

There is a great difference in extent of poor welfare between painful and frightening beatings and occasional physical punishment at the level that a bitch might use with her puppies. Punishment that achieves nothing because the dog does not know why it is being punished also leads to poor welfare. Owners often fail to allow a dog to associate either punishment or reward with the action that is being reinforced, with the result that the dog does not learn and the owner becomes frustrated. Reinforcement contingent upon action is of key importance in dog training.

The use of electric shock collars in dog training and in restricting dog movements has implications for dog welfare. In some cases, the dog trainer has direct control over the administration of an electric shock to the dog via its collar. In other cases, the shock is administered when the dog comes close to an actual or virtual fence. If the shock is small and is administered only occasionally and at the moment that the dog has done what it is being trained not to do, the method can have as small a negative effect on welfare as verbal or minor physical punishment. The problem with the equipment is that it is possible to increase the shock level, and the frequency of administration of shocks, to the point of very poor welfare and hence extreme cruelty. Schalke *et al.* (2007) recorded heart-rate, cortisol concentrations and behaviour in dogs during the use of shock collars and concluded that the welfare of the dogs was often very poor so the general public should not be permitted to use them. As a consequence of several studies and experiences of this kind, shock collar sales and general use are banned in more and more countries.

Allowing dogs to assume inappropriate roles in human households

Even small dogs can be aggressive, assertive and violent. Humans who live with dogs have to make it clear to the dog that they are in charge and they take the key decisions. If this is not done, the dog may take some degree of control and eventually it has more or less severe problems, perhaps to the point of being killed. In general, the sitting dog should always be at a lower level than the owner, and the owner should decide when feeding, walks and other activities occur. When on a walk, the owner can reasonably allow the dog to stop and explore sites of olfactory interest, etc. Guidance concerning such matters is provided by Appleby (2004).

Dogs that attack humans, pets or farm animals

Dogs may attack because they are defending territory or an individual, defending their food, demonstrating their high social status or acting in a predatory way. The dog may be said to be showing different forms of aggressive motivation in the first three cases, but predatory behaviour is not aggression. The motivation is different during predation and social aggression. Predation is usually carried out in groups of two or more, so most dog attacks are seen on children or weak adult humans; some serious attacks on pets or farm animals are carried out by two or more dogs. Many dog owners fail to prevent their pet dogs from being aggressive in certain circumstances, and some actively encourage their pets to attack intruders or those indicated by the owner. Cases of aggression in dogs are discussed by Podberscek (1994) and Podberscek and Serpell (1996). Most actions that result in dogs being aggressive or predatory are likely to cause the dog to be much restricted in its movements because it is confined, chained or muzzled, and hence to have poorer welfare than it would have had if differently trained.

Some domestic dogs chase, attack and kill wild animals and are rewarded for doing so. First, this is a natural and rewarding behaviour; second, the dog may sometimes obtain extra food from what it catches; and third, some owners reward their dog for attacking and killing. If the owner wants the dog to attack or retrieve the wild animal without eating it, careful training is needed. If the owner does not need the prey but derives amusement or profit from the actions of the dog, the behaviour may be encouraged. Dogs used to guard the property or person of the owner may be trained to attack, or not discouraged from attacking, in many situations. Some dog owners encourage their pets to attack and kill cats or small dogs. Others discourage such actions but have difficulty training the dog to attack some smaller, running animals that happen to be wild but not to attack pets.

Some dog owners encourage their pets to threaten or attack intruding humans in a particular area or near the owner (see Fig. 37.1). These dogs may well continue to threaten or attack when in other situations. Many dog owners, especially those living in towns, fail to teach their pets not to attack sheep, chickens or other farm animals. Even dogs that would never attack another animal in a city will sometimes attack when in a

Fig. 37.1. These Filho Brasileiro dogs are trained to guard the owners' house (photograph D.M. Broom).

farm field, usually because they have not been specifically trained not to do so. Dog training is a difficult and skilled procedure. Much harm caused by dogs to humans, pets or farm animals is the result of deliberate training to attack, accidental training to attack or failure to train the dog properly.

Every dog owner takes on a responsibility to train their dog properly when they assume ownership or control. If the dog does harm, the person or persons who should have trained it are responsible. Some dogs are much harder to train than others, but the owner should know how well the dog can be controlled and then has to restrict the dog accordingly. Hence, if a dog is not well trained and its attacks are not easily preventable, its welfare may be poorer because it is muzzled, kept on a lead or kept in a restricted area by its owner. A dog may also be severely punished for its attacking actions when these are sometimes the fault of the owner. The welfare of badly trained dogs is often very poor. Some dogs are killed because of their history of attacking.

Inappropriate feeding and other treatment of domestic dogs

There is now much information about the best diets for domestic dogs. Dog owners can readily find out what food is suitable and what is not. The best advice may not be that provided by some pet food companies, because the companies may wish to sell expensive 'treats' for dogs as well as balanced pet foods. However, problems occur in dog diets mainly because of inappropriate feeding by owners. Pet owners sometimes assume that foods that they enjoy will be good for their dog: dogs do not thrive well on the diet of modern, affluent humans. They may well be given too much sweet food, too much carbohydrate, too much protein, not enough fat, not enough trace nutrients or, especially, not enough fibre. When dog owners have insufficient food for themselves, dogs may be fed too little food in total or too little expensive protein.

Dogs dressed in clothes or otherwise treated as if they are human may have problems coping with this situation.

Inadequate treatment of disease and unwanted dogs

As with all domestic animals, if a disease condition or injury is untreated, the welfare of the animal can be very poor. Some dog owners have insufficient money to pay for veterinary treatment. Others do not choose to spend money on treatment for their sick or injured dog. In either case, dog welfare can be poor or very poor. A question arising when a dog has an injury or disease condition that cannot be treated in a way that prevents poor welfare is: whether or not the magnitude of poor welfare (i.e. a function of its severity and duration) is sufficient for the owner to decide that it is better for

Fig. 37.2. Stray dogs in Mauritius were collected up by the Mauritius Society for the Prevention of Cruelty to Animals (MSPCA). These dogs have a harmful effect on the indigenous fauna and they are often diseased, so their welfare can be poor (photograph D.M. Broom).

the dog if euthanasia is carried out. As explained in Chapter 22, the term euthanasia should be restricted to describing the killing of an animal for its own benefit. If an animal is killed to save it from poor welfare associated with injury or disease, including that part of disease that involves serious malfunction associated with degenerative change, the killing is euthanasia. Killing an animal because it is inconvenient to look after it is not euthanasia. This is a simple decision for some people who consider that no degree of poor welfare justifies killing the animal; that is, they consider that there is no such thing as euthanasia. For others, the magnitude of the poor welfare during the remainder of the animal's life has to be balanced against the loss of good welfare associated with its death and the actual killing. In most cases where this decision has to be taken, a humane method of killing not involving any significant amount of poor welfare during the euthanasia will be used.

The discussion about euthanasia above has some relevance to the killing of animals for the benefit of the owner. In this circumstance too, most owners would ensure that a humane killing method was used. The killing of an animal because there is nobody to look after it is best avoided by finding an alternative home for it. If all possibilities for rehoming the animal have been investigated, a dog released from a home environment into feral conditions may well starve, become diseased or become injured because of vehicle traffic or the actions of humans or other animals.

The welfare of stray dogs is often very poor. A further factor is that stray dogs cause poor welfare and death in other animals. They may also have harmful effects on the environment, for example, on populations of prey species. In total, these effects on dog welfare and the environment are so great that the release of dogs in the environment should not be allowed. Any stray dogs should be collected and it is the authors' view that if no suitable homes in private houses can be found for them within a short time, they should be humanely killed (see Fig. 37.2).

Further Reading: Jensen, P. (ed.) (2007) *The Behavioural Biology of Dogs*. CAB International, Wallingford, UK.

38 The Welfare of Cats

Domestication and Breeding

Most people think that cats are substantially different from dogs and other domestic animals, and the scientific literature generally supports this. The cat is a North African and Middle Eastern species and the association between humans and cats, reported from Babylon and Ancient Egypt, has existed for at least 9500 years (Vigne *et al.*, 2004). There was not a species of cat amenable to close association with man in Central America and East Asia, where human society developed to a high level as early as or earlier than in Europe, North Africa and the Middle East.

Some may argue that no cat has a substantial association with man and no cat is domesticated in the sense that dogs, cows, horses and pigs are. Cats are less changed by man, more independent and more likely to find food outside the human household than are most domestic animals. For this reason they are much loved by some people and greatly reviled by others (Serpell, 2000).

Cat breeding has been subject to many of the same pressures as in other domestic animals, that have led to the production of grossly maladapted animals, some with brain changes, whose welfare is poor (Steiger, 2005). Other changes have preserved the species quality of cats, while reducing their fear of humans and their difficulties in adapting to human habitation. Some cats have been changed genetically during breeding, so they have excessive nervous reactivity, high levels of aggression towards humans, very short legs, lack of fur or predisposition to serious disease conditions. The welfare of these animals will be poor. Whatever the penchants of cat-show judges, none of these conditions is acceptable to the average person.

Mutilations of Cats for the Convenience of Owners

Members of the public are generally averse to being scratched by the claws of cats, they may notice the nocturnal vocalizations of cats and they may be disturbed by rivalries among cats that lead to excessive noise or intermittent injury to cats. As a consequence of the first-mentioned point, some cat owners take care to train their cats not to use their claws in their interactions with humans. Other owners do not take such care and attempt to resolve the problem of people being scratched by removal of the claws of cats.

This operation has substantial effects on the ability of the cat to defend itself against other cats and therefore, like the removal of any biologically important ability, can lead to poor welfare in mutilated cats. The operation itself will involve some pain to the cat, as the claws contain sensitive tissue.

The most widespread mutilation of cats is castration. Intact male cats may spray urine within their home environment, go out of the home for long periods, vocalize at night, become involved in fights and father offspring. All of these qualities are natural, but some cat owners are unwilling to train the cat to restrict their spraying and unwilling to tolerate the behaviour that they cannot prevent by training. Solutions to this problem include not owning a cat, or at least not a male cat; to castrate the cat; or to use pheromones or chemicals to prevent the behaviour. Castration causes some pain, even if anaesthetic and analgesic are used by the veterinary surgeon, and greatly changes the character of the animal. Prolonged use of the synthetic feline facial pheromone F3 can prevent spraying behaviour and for some animals this can be achieved by the use of clomipramine or fluoxetine combined with cleaning the area thoroughly (Mills *et al.*, 2011).

Female cats can be spayed (ovario-hysterectomy) but this involves the use of a more substantial and invasive operation than male castration. The operation involves pain and substantially changes the animal. Spayed cats are more masculine and may be less easy to manage. However, they do not change their behaviour substantially during oestrus and do not have kittens.

Unwanted Cats and Methods of Killing

In the consideration of cat welfare, one considerable issue is the unwanted kitten or cat. If people have to care for an unwanted animal, they may not treat it well. Alternatively, they may kill it in an inhumane way, for example by drowning. As those people who have almost drowned can testify, drowning is a terrifying and relatively prolonged experience. Cats will have very poor welfare for 2–5 min if they are drowned, so drowning is an unacceptable method of killing. As discussed in Chapters 22 and 37, killing for human convenience is not euthanasia and is sometimes not humane.

One result of unwanted cats being left to fend for themselves is that populations of feral cats build up. Some feral cat groups are in cities while others are in rural areas. The welfare of these cats is sometimes very poor and their negative impact on wild animal welfare and populations can be substantial. The cats are often very wary and difficult to catch (Fig. 38.1).

Fig. 38.1. Group of wary feral cats living in a city environment (photograph A.F. Fraser).

Behavioural Problems

Insufficient variety in the environment

Some cats live socially and may be greatly affected by deprivation of social contact, but many cats seem to have good welfare in the absence of possibility to interact frequently with conspecifics. However, cats do require a relatively complex environment during the time that they are active if their welfare is not to be poor.

The features of a cat's environment that lead to good welfare have been described in detail by Rochlitz (2000, 2005b,c). Cats kept in groups prefer to be able to sit on a raised platform or shelf rather than on the ground. The possibility of looking out over the immediate environment seems to be important to cats (Podberscek *et al.*, 1991; Rochlitz *et al.*, 1998). Some cats are disturbed if another individual is in very close proximity (Bradshaw and Hall, 1999), and the welfare of many cats is not good unless they have the opportunity to hide (Casey and Bradshaw, 2005). In contrast to sheep, cattle, dogs, pigs and poultry, which aggregate during resting periods, cats prefer to keep a few body lengths between them and the next nearest individual. A cat that is unfamiliar with group-living may choose to maintain many metres of space from the next individual. On the other hand, cats may feed together amicably.

Resources of importance to cats include material that allows them to scrape with their claws. A particularly effective scratching material is rope or string wrapped around a post. The post should be of such a size that the cat can put a forefoot either side of it and scrape the claws over it. Some individuals in a group of cats may monopolize resources (van den Bos and de Cock Buning, 1994).

Cats also need to be undisturbed by humans when they are resting. A cat in a small cage, repeatedly passed by unfamiliar humans, is always disturbed and responds by lying down and moving very little. It is important for all who wish to appreciate cat welfare to be able to distinguish between a resting cat and a cat that is lying largely immobile but is severely disturbed by events round it (McCune, 1994). Heart rate and cortisol concentration in plasma are at high levels in such cats. Similar behaviour is shown by cats that are afraid to go out of a house because an aggressive, dominant individual would attack that individual if it did venture out. A cat owner whose cat has normally gone out of the house but which suddenly refuses to go out should consider the possibility that the cat is being severely bullied when it does go outside.

Another hazard for free-ranging cats is the motor vehicle. Rochlitz (2003a,b) found that younger cats and male cats were more at risk of being killed on roads.

Cats prefer some dietary variety so may go out of the house to try to supplement their diet by killing wild animals because they are fed the same food every day. This behaviour may occur even though tinned cat food is generally of good quality.

When a cat from a human home has been in a cage environment for some time, for example when in quarantine for 6 months, the behaviour becomes more and more abnormal and behaviour towards familiar humans changes (Rochlitz *et al.*, 1998; Rochlitz, 2005b). The changes in behaviour and welfare are profound enough to make it clear that a 6-month quarantine period in the life of a cat is a severe imposition. The avoidance of rabies introduction, the main reason for the quarantine in some countries, is better achieved by vaccination and careful monitoring.

For laboratory animals, or other animals caged for long periods, housing conditions and management practices that lead to better welfare are of great importance. A cat that lives for all of its life in laboratory conditions depends on the environment provided and the laboratory carers. Its welfare is much improved if it has a sufficiently complex environment that meets most of its needs, including sufficient exercise and sufficient social interaction. One important need is to be able to hide from humans and other cats. Carlstead *et al.* (1993) found that caged cats that were unable to hide had higher urinary cortisol concentrations than those which could hide. However, most caged cats and some domestic cats have no opportunity to hide. For any caged cat, the complex prey-catching environment is largely or completely absent, so there must be compensation in social interaction with other cats and interaction with humans. The welfare of cats in small cages with little or no enrichment and only occasional human contact will be poor.

Harsh training methods

Unfortunately for cats, many humans have the view that cats do not need to be trained. They are then surprised to find that the cat may defecate in inappropriate places, claw the owner or damage property. Training is just as important for cats as for dogs. Where an action by a cat is to be rewarded or punished, it is very important that the reinforcer is made contingent upon the action. The cat will not learn unless it is quite clear to it what action is being reinforced. Poor training on the part of cat owners leads to abnormalities of behaviour and poor welfare in cats. If cats are punished in a way that does not contribute to learning, or if they are punished in an excessive way, welfare will be poor.

Cats are occasionally trained to carry out tricks or other unusual behaviour, but this is not done often because they are not easy to train. As a consequence, when they are trained, the methods of training may be harsh. The animal may become very frightened of the trainer and may sometimes be subjected to pain or serious deprivation.

Allowing cats to have inappropriate social positions in human households

Cats are renowned for independence and may not seem to seek dominance over humans. However, within groups of cats it is sometimes clear that one controls the others in an aggressive way. This can lead to very poor welfare in the subordinate cat, especially if it cannot leave the company. The cat owner can be considered to be a subordinate by a cat living in the house, and a cat that perceives a human as subordinate can cause problems for that human. An aggressive cat, or one that cannot be moved without risk of injury to the humans who attempts to move it, is unacceptable in most human households. Eventually, there are likely to be severe problems for that cat and hence poor welfare.

Managing the impact of cats on wild animals

Some cats kill large numbers of wild mammals, birds and other animals (Figs 38.2 and 38.3; see Fitzgerald and Turner, 2000). While some of these wild animals are pests and the cat is doing the job required of it, many are not. The welfare of animals caught by a cat is often very poor because a cat that is not very hungry may take a long time to kill the animal. Cats may have a large impact on wild populations and some wild species have been eradicated by cats. The severe depredations of feral cats on some islands that did not previously have such a predator have necessitated cat eradication (Slater, 2005). Some trapping and

poisoning methods used have been inhumane, but it is possible to use cage-trapping and then removal or humane killing. Accurate shooting can also be humane. Some feral cats are well fed by humans, while others depend on wild prey and may be malnourished. Disease can be a serious welfare issue in populations of feral cats but, as reviewed by Slater (2005), there is variation in the prevalence of disease.

In a country where there are vulnerable wild animals, such as relatively slow-moving or slow-reacting marsupials, flightless birds or slow-moving reptiles, there is a strong argument for cat-keeping to be illegal. In every country, to allow widespread killing of prey other than rodent pests by cats is not justifiable, so cats should be forced to wear a bell and should be kept indoors if they continue to kill. Cat welfare is better if the cat can go out wearing a bell than if kept indoors. A balance is needed between the welfare of the cat and welfare and conservation of wild species.

Fig. 38.2. Domestic cat hunting for prey. Many domestic cats kill large numbers of wild birds and mammals. Owners of cats are responsible for the injury and death caused by their pets. The numbers killed are reduced if the cat has to wear a bell (photograph D. Critch).

Other Problems

Inappropriate feeding and other treatment of cats

Many cats are overfed and become very fat, resulting in poor welfare. The reason may be that the owner gives the cat too much inappropriate food or too much food in total, such as sugar or other carbohydrates (Sturgess and Hurley, 2005). Cats are sometimes given a diet lacking necessary nutrients. Since they are entirely carnivorous animals, their diet must include protein from mammals, birds or fish.

Owners sometimes treat cats as if they were human children. While some of these actions are harmless, others can cause poor welfare. For example, cats find it difficult to adapt to being dressed in human clothes or cuddled.

Fig. 38.3. This cat is endeavouring to catch fish in a pond (photograph R. Harnum).

Inadequate Treatment of Disease

Some cat owners are unwilling to acknowledge that their pet is diseased, or they are unwilling or unable to pay for veterinary treatment. Wherever injuries or disease conditions are present, welfare is poorer than it would otherwise be. The welfare of the cat may be very poor but could be improved by veterinary treatment. Because of the relatively low purchase price of many cats and, indeed, because many kittens are given away rather than being sold, some cat owners are more unwilling to pay for veterinary treatment than are dog owners. Hence, disease is a relatively common cause of poor welfare in cats (Sturgess, 2005). If disease or injury is treated by a veterinary surgeon, there should be follow-up assessment of welfare to check that the cat is recovering well (Christiansen and Forkman, 2007).

Further Reading: Rochlitz, I. (ed.) (2005) *The Welfare of Cats*. Springer, Berlin.

39 The Welfare of Other Pet Animals

Animals living in the wild are very rarely able to adapt to captive conditions and human proximity. It may be that some invertebrate animals can do so, and a few species of fish and amphibians might be provided with optimal conditions in captivity, but mammals, birds, reptiles and most amphibians and fish cannot be taken from the wild and kept as pets in such a way that their welfare is good. Even good zoo conditions cannot normally provide for the needs of wild-caught animals. The morbidity and mortality rates after animals are brought into captivity are high, even if the animal is young when taken. For example, for small birds captured in the wild for sale as pets, 90% were dead before reaching an owner; even for high-value birds such as the larger parrots 75% were dead (EFSA, 2006c). This was the reason why the EU banned the sale of birds taken from the wild and imported to EU countries. The mortality rate for wild-caught reptiles, such as tortoises and terrapins, is similar (Warwick *et al.*, 2001) but there is no ban.

As a consequence of the great difficulty that wild animals have to adapt to captive environments in close proximity to people, it is my view that no person should acquire, as a pet, an animal caught in the wild. All countries should make it illegal to sell as a pet any wild-caught animal.

Many species of animals are unsuitable as pets, even if they were bred in captivity. Some are dangerous to people. Many others are unable to adapt to the conditions that can be provided for them in human homes. How should we evaluate whether or not an animal species is suitable for pet-keeping? One simple test is to give the animal the choice of staying with the human owner or leaving. If they leave, they are not suitable, as mentioned in relation to rabbits (Chapter 36). The owners of many individual pets would not pass such a test.

The owners of pet animals often want to stroke or otherwise handle the animals. While many dogs, cats and horses respond positively to this, some species are made more frightened of people by attempts to handle them. If the handling is always gentle and starts at a young age, mice, rats, gerbils, hamsters, chinchillas, guinea pigs, rabbits and ferrets may accept handling without there being a welfare problem (see also Chapter 34). However, some individuals do not adapt to human handling and without the appropriate experience, the effects of handling on welfare can be seriously negative.

Methods for caring for small caged animals are described by Kaiser *et al.* (2012) for guinea pigs, by Würbel *et al.* (2009) for rats and mice and for other species in Kirkden (2010). The cages in which small pet animals are kept can cause injuries, for example when they are made of wire mesh that does not support the feet well. As explained in Chapter 6, laboratory rodents have a clear preference to rest on a solid floor (Manser *et al.*, 1995, 1996) and pet rodents would generally be the same. Rats also prefer to have a dark place to hide, nesting material and a nest box (Manser *et al.*, 1998a,b). Apart from some hamster species, caged rodents should be kept in groups. Evidence for this in guinea pigs is provided by Baumans and van Loo (2013). The needs of animals of each species should be taken into account by pet owners when providing accommodation for the animals.

Caged birds range from larger individuals, such as parrots, which may form some attachment to humans, to small birds that are always frightened of people and will escape if they can. Many birdcages are much too small. A bird should have enough space to fly around for about 30 s so the aviary area for good welfare could be calculated (see Chapter 12). All of the needs of the birds should be known so that the accommodation can be properly designed for good welfare. As described in Chapter 25, caged parrots may show self-mutilation because the conditions in which they live are so bad. In this case, social companions and opportunities to use their brains and manipulative capabilities are needed. Aydinonat *et al.* (2014) measured telomere length, which is shorter in stressed people (see Chapter 6), and found that grey parrots living in isolation had shorter telomeres than those in pairs. Engbretson (2006) concluded that parrots are not suitable to be companion animals because of the widespread indications of poor welfare in parrots kept in the home.

Further Reading: Hubrecht, R. and Kirkwood, J. (eds) (2010) *The UFAW Handbook on the Care and Management of Laboratory and other Research Animals*. John Wiley & Sons, Chichester, UK.

Glossary

Abnormal behaviour, Aberrant behaviour

Behaviour that differs in pattern, frequency or context from that which is shown by most members of the species in conditions that allow a full range of behaviour (see Chapter 24).

ACTH

Adrenocorticotrophic hormone. This peptide hormone is released from the adenohypophysis, or anterior pituitary, and travels in the blood to the adrenal gland where it stimulates the outer part of this gland, the adrenal cortex, to produce glucocorticoids such as cortisol or corticosterone.

Action pattern

A sequence of movements shown by an individual that is repeated on subsequent occasions in largely invariate form (see Chapter 2). Other individuals may show similar action patterns.

Adaptation

This refers to processes at different levels. At the cell and organ level: (i) the waning of a physiological response to a particular condition, including the decline over time in the rate of firing of a nerve cell. At the individual level: (ii) the use of regulatory systems, with their behavioural and physiological components, in order to allow an individual to cope with its environmental conditions (see Chapters 1 and 6). In evolutionary biology: (iii) as a noun, any structure, physiological process or behavioural feature that makes an organism better able to survive and to reproduce in comparison with other members of the same species. Also: (iv) the evolutionary process that leads to the formation of such a trait.

Affect

Feelings, emotions and moods.

Aggregation

A group of individuals, comprising more than just a mated pair with the dependent offspring, gathered in the same place but not necessarily in a true social group.

As a verb, the process of forming such a group.

Aggression

An act or threat of action, directed by one individual towards another, with the intention of disadvantaging that individual by actually or potentially causing injury, pain or fear.

Agonistic behaviour

Any behaviour associated with threat, attack or defence. It includes features of behaviour involving escape or passivity, as well as aggression (see Chapter 12).

Allogrooming

Grooming directed at another individual animal.

Altruism

Actions by an individual that involve some cost to that individual, in terms of reduced fitness, but which increase the fitness of one or more other individuals.

Anomalous behaviour

Behaviour which is somewhat abnormal (see **Abnormal behaviour**), particularly with respect to deviations from the normal pattern or frequency. It may be a variant of a normal activity, such as chewing or licking.

Anorexia

Abnormal lack of ingestive behaviour, e.g. in toxic and depressed clinical states.

Anosmia

Lack of sense of smell.

Anxiety

A feeling resulting from a perceived risk of a specific or general danger or aversive event.

Aversion therapy

Treatment of an unwanted form of behaviour by associating the behaviour with an aversive stimulation.

Aversive

Such as to cause avoidance or withdrawal.

Awareness

A state during which concepts of environment, of self and of self in relation to environment result from complex brain analysis of sensory stimuli or constructs based on memory.

Bond

A close relationship formed between two individuals.

Causal factor

The inputs to a decision making centre, following interpretation in the light of experience, of a wide variety of external changes and internal states of the body (see Chapter 4).

Circadian rhythm

A rhythm in behaviour, metabolism or some other activity such that events in it recur approximately every 24 h (see **Rhythm**).

Cognition

Having a representation in the brain of an object, event or process in relation to others, where the representation can exist whether or not the object, event or process is directly detectable or actually occurring at the time.

Cognitive bias

The influence of affect on a range of processes, some of which are cognitive, e.g. judgement. However, the term has also been used for effects on attention, motivation and memory that may not be cognitive (see Chapter 4).

Comfort-shift

A minor change of posture or position that may briefly interrupt rest.

Communication

The transmission of information from one individual to another.

Competition

(i) Among individuals, the striving of two or more individuals to obtain a resource that is in limited supply. Success might result from such abilities as speed of action, strength in fighting or ingenuity in searching; (ii) among genotypes, attempting to carry out any life function in a way that is better than those used by other genotypes, so that the fitness (reproductive success) of the genotype is increased.

Conditioning

The process by which an animal acquires the capacity to respond to a given stimulus, object or situation in a way that would previously have been a response to a different stimulus.

Conscious individual

An individual who has the capability to perceive and respond to sensory stimuli.

Conspecific

Belonging to the same species.

Consummatory act

An act which reduces greatly the levels of causal factors that promote a certain activity, so that the activity is terminated; e.g. mating terminates courtship behaviour.

Controller

The individual in a group who determines: (i) whether or not a new group activity occurs; (ii) when it happens; and (iii) which activity it is (see Chapter 14).

Cope

Having control of mental and bodily stability; this control may be short-lived or prolonged. Failure to be in control of mental and bodily stability leads to reduced fitness (see **Stress**).

Coprophagia

Eating faeces; this is normal behaviour in rabbits but occurs abnormally in other animals (see Chapters 25 and 36).

Core area

The area of heaviest regular use within the home range.

Critical period

See **Sensitive period**.

Crowding

The situation in which the movements, in a group, that individuals need to make are restricted by the physical presence of others (see **Overcrowding**; Chapter 12).

Density dependence

Processes influenced by physiological or environmental factors associated with the density of the population. For example, tail-biting behaviour in pigs increases with population density.

Depression

A condition of brain and behaviour associated with sagging posture, unresponsiveness and reduced cognitive function.

Displacement activity

An activity performed in a situation that appears to the observer not to be the context in which it would normally occur. Being so dependent for recognition on observer ability to determine relevance to context, the term is of very limited use (see Chapter 4).

Display

A behaviour which may impress, intimidate or otherwise change the behaviour of a potential or actual partner, rival or attacker.

Diurnal

(i) On a daily basis; (ii) occurring in daylight time.

Domestication

The process, occurring over generations, by which a population of animals becomes adapted to man and to the captive environment by some combination of genetic changes and environmentally induced developmental events (see Chapter 5).

Dominance

An individual animal is said to be dominant over another when it acts so as to gain priority of access to a resource such as food or a mate. There are various ways to gain priority so a dominant individual need not be superior in fighting ability to a subordinate.

Drive

A collection of causal factors that promote related behaviours. The term often implies potential progression towards a goal. Although a definition is included here because the term drive is in widespread use, we consider it is easier to understand motivation if reference is normally made to causal factors rather than to drives.

Ecological niche

The environment in which the species performs best and which it comes to live in and occupy in nature.

Ecology

The scientific study of the interaction of organisms with their environment, where environment includes both its physical aspects and the organisms that live in it.

Ecosystem

All of the organisms of a particular habitat, such as grassland or coniferous woodland, together with the physical environment in which they live.

Eliminative behaviour

Patterns of behaviour connected with evacuation of faeces and urine.

Emotion

A physiologically describable component of a feeling characterized by electrical and neurochemical activity in particular regions of the brain, autonomic nervous system activity, hormone release and peripheral consequences including behaviour.

Environment

The source of external influences, for example on the development of behavioural or other biological traits. External means outside the system or unit under consideration, not necessarily outside the whole organism.

Enzootic

Referring to a disorder in animals that is peculiar to a particular location or type of place.

Epizootic

The spreading of a disease or disorder through a population of animals (the equivalent of an epidemic in humans).

Estrous cycle

See **Oestrus cycle**.

Ethics

The study of moral issues.

Ethogram

A detailed description of the behavioural features of a particular species (see Chapter 2).

Ethology

The observation and detailed description of behaviour with the objective of finding out how biological mechanisms function. Occasionally, such studies are carried out in a natural or semi-natural setting.

Euthanasia

Killing an individual for the benefit of that individual and in a humane way.

Experience

A change in the brain resulting from information acquired from outside the brain. The information can originate in the environment of the individual or within the body, for example from sensory input, from low oxygen availability or from a new hormone level in the blood (see Chapter 3).

Exploration

An activity having the potential for the individual to acquire new information about its environment or itself (see Chapter 11).

Fear

A feeling which occurs when there is perceived to be actual danger or a high risk of danger.

Feedback

The effect of a system output, in response to a system input, which modifies that input by reducing it (negative feedback) or enhancing it (positive feedback).

Feedforward

The effect of a system output which, prior to any input, modifies the state of the system, usually in such a way that the effect of an input is partly or wholly nullified.

Feeling

A brain construct, involving at least perceptual awareness, which is associated with a life-regulating system, is recognizable by the individual when it recurs and may change behaviour or act as a reinforcer in learning.

Fitness reduction

That which involves increased mortality, or failure to grow, or failure to reproduce.

Flight distance

The space around an animal within which intrusion provokes a flight reaction (see Chapter 12).

Foraging

The behaviour of animals when they are moving around in such a way that they are likely to encounter and acquire food for themselves or their offspring (see Chapter 8).

Freedom

A possibility for action conferred by one individual or group upon another.

Frustration

If the levels of most of the causal factors that promote a behaviour are high enough for the occurrence of the behaviour to be very likely, but because of the absence of a key stimulus or the presence of some physical or social barrier the behaviour cannot occur, the animal is said to be frustrated.

Functional systems

The different sorts of biological activity in the living animal which, together, make up the life process; e.g. temperature regulation, feeding, predator avoidance. These functional systems have behavioural and physiological components (see Chapter 1).

Genotype

The genetic constitution of an individual organism designated with reference either to a single trait or to a set of traits (see **Phenotype**).

Geophagia

Eating soil.

Gonad

The organ that produces sex hormones and gametes, either an ovary (female gonad) or testis (male gonad).

Grooming

The cleaning of the body surface by licking, nibbling, picking, rubbing, scratching, etc. When action is directed towards the animal's own body, it is called self-grooming; when directed at another individual, it is referred to as allogrooming.

Habituation

The waning of a response that could still be shown, to a repeated stimulus. This is distinct from fatigue.

Head-pressing

Postural disorder characterized by apparent head stabilization through forehead contact with a vertical surface. The head is lowered and the inactive posture is maintained for long periods.

Health

The state of an individual as regards its attempts to cope with pathology.

Hierarchy

A sequence of individuals or groups of individuals in a social group based upon some ability or characteristic. The term is most frequently used where the ability assessed is that of winning fights or displacing other individuals (see Chapter 14).

Homeostasis

The maintenance of a body variable in a steady state by means of physiological or behavioural regulatory action.

Home range

The area which an animal learns thoroughly and uses regularly. The home range may or may not be defended; those portions that are defended constitute the territory (see also **Core area**; Chapter 12).

Hormone

A substance, secreted in the brain or by an endocrine gland, often into the blood or lymph, which affects the physiological activity of other organs in the body including the nervous system and, hence, also behaviour.

Humane

Treatment of animals in such a way that their welfare is good to a certain high degree.

Humane killing

Use of a killing procedure that does not cause poor welfare and, if there is stunning, a stunning procedure that results in instantaneous insensibility; or, if the agent causing insensibility or death is a gas or injectable substance, no poor welfare occurs before insensibility and then death. This may be achieved because the stunning or killing agent is not detectable by the animal (see Chapter 22).

Imprinting

Rapid and relatively stable learning taking place in early life.

Individual distance

The minimum distance from an animal within which approach, normally by a conspecific, elicits attack or avoidance (see Chapter 12).

Ingestive behaviour

Behaviour concerned with the intake into the mouth of food, water, etc.

Initiator

The individual in a social group who is the first to start a new group activity (see Chapter 14).

Instinct

A term implying behaviour that is entirely genetically controlled. The use of this term is undesirable and confusing, because neither behaviour nor any other characteristic of the whole animal can develop independently of all environmental influences.

Intention movements

The preparatory motions which an animal may go through prior to switching to a new behaviour.

Inter-sucking

Abnormal sucking activity directed to appendages of individuals other than the mother (see Chapter 26).

Kinesis

An undirected reaction, without orientation of the body in relation to the eliciting stimulus.

Lameness

Impaired locomotion or deviation from normal gait.

Leader

The individual who is in front during an orderly group progression (see Chapter 14).

Learning

A change in the brain resulting in behaviour being modified for longer than a few seconds as a consequence of information acquired from outside the brain (see Chapter 3).

Libido

An internal state which is measured by the likelihood of showing sexual behaviour given appropriate opportunity (see Chapter 17).

Lignophagy

Wood eating.

Mobbing

Joint assault or threat by a group of animals.

Monotocous

Producing a single offspring at birth.

Mood

A brain state which often involves feelings, continues for more than a few minutes and influences decision making and behaviour.

Moral

Pertaining to right rather than to wrong.

Motivation

The process within the brain controlling which behaviours and physiological changes occur, and when (see Chapter 4).

Motivational state

A combination of the levels of all causal factors (see **Causal factor**; Chapter 4).

Need

A requirement, which is part of the basic biology of an animal, to obtain a particular resource or to respond to a particular environmental or bodily stimulus. To need is to have a deficiency that can be remedied by obtaining a particular resource or responding to a particular environmental or bodily stimulus (see Chapters 1 and 4).

Neurophysiology

The scientific study of the nervous system, especially the physiological processes by which it functions.

Niche

See **Ecological niche**.

Nursing

The behaviour of a mother mammal that allows young to suck milk from her teats.

Obligation

A duty to act, or to refrain from acting, in a way that potentially affects another individual.

Observational learning

Learning which occurs when one animal watches the activities of another.

Oestrous (estrous) cycle

The repeated series of changes in reproductive physiology and behaviour that culminates in oestrus or 'heat', i.e. receptivity (the noun is oestrus and the adjective oestrous).

Ontogeny

The process of development of an organism from single cell to adult.

Overcrowding

A high social density which causes adverse effects on the fitness of individuals in the group (see **Crowding**; Chapter 12).

Pain

An aversive sensation and feeling associated with actual or potential tissue damage (see Chapter 6 for measurement).

Pair bonding

A close and long-lasting association formed between a male and female.

Parental investments

Investment by a parent in an individual offspring, which increases the offspring's chance of surviving and reproducing, at the cost of the parent's ability to invest in other offspring.

Pathology

(i) The detrimental derangement of molecules, cells and functions that occurs in living organisms in response to injurious agents or deprivations; (ii) the study of such conditions.

Peck order

A stable hierarchy in which each individual is able to threaten, displace or attack individuals lower than itself with impunity. The term was coined following work with chickens but is now used for any animal (see Chapter 14).

Periodicity

A series of events separated by equal periods in a time series (see Chapter 2).

Phenotype

The observable properties of an organism as they have developed under the combined influences of the genetic constitution of the individual and the effects of environmental factors (contrast with **Genotype**).

Pheromone

A substance which is produced by one animal and which conveys information to other individuals by olfactory means (see Chapter 1).

Pica

The seeking out and eating of foreign objects such as wood, cloth and old bones (see Chapter 25).

Play

Carrying out a movement or intellectual process, either in the absence of its usual objective, or by using an inefficient means of achieving a goal solely in order to engage in that movement or process.

Polydipsia

Excessive drinking of water beyond the level required for maintenance of body fluid concentration.

Polytocous

Producing many offspring at birth.

Preening

As **Grooming**, but referring to birds.

Quality of life

Welfare during a period of more than a few days.

Reaction time

Time between the occurrence of an environmental change and the beginning of the response of the animal.

Reciprocal altruism

This occurs when an altruistic act by individual A, directed towards individual B, is followed by some equivalent act by B directed towards A or by an act directed towards A whose occurrence is made more likely by the presence or behaviour of B.

Reflex

A simple response involving the central nervous system, but not higher brain centres, and occurring very shortly after the stimulus that evokes it.

Reinforcer

An environmental change that increases or decreases the likelihood that an animal will make a particular response, i.e. a reward (positive reinforcer) or a punishment (negative reinforcer) (see Chapter 3).

Releaser

See **Sign stimulus**.

Reproductive effort

All of the resources expended by an individual on reproduction in a season.

Rhythm

A series of events repeated in time at intervals whose distribution is approximately regular (see Chapter 2).

Right

(i) A legal entitlement which can be defended using the laws of the country. In most countries animals do not have rights in this sense. (ii) A privilege justifiable on moral grounds (see Chapter 1). The moral grounds may be religious.

Scent marking

The deposition of solid or liquid pheromones, typically on a tree, bush, rock or other individual (see **Pheromone**; Chapter 1).

Self-awareness

The cognitive process in an individual when it identifies and has a concept of its body or possessions as being its own so that it can discriminate these from non-self stimuli.

Self-grooming

Grooming directed at the individual's own body.

Selfish

This describes an individual acting in a way that increases its fitness at the expense of the fitness of one or more other individuals while being aware of the likely effects on itself and on the harmed individual or individuals.

Sensitive period

A time interval during development within which the behaviour, at that time or later, is especially likely to be affected by certain types of experience.

Sensitization

The increasing of a response to a repeated stimulus.

Sensory physiology

The study of sense organs and the ways in which they receive stimuli from the environment and transmit them through the nervous system.

Sentience

Having the awareness and cognitive ability necessary to have feelings.

Sentient being

One that has some ability: to evaluate the actions of others in relation to itself and third parties; to remember some of its own actions and their consequences; to assess risks and benefits; to have some feelings; and to have some degree of awareness (see Chapter 1).

Sign stimulus

A specific environmental feature which elicits a response from an animal.

Social behaviour

The interactions of two or more animals that spend time together, and the resulting modifications of individual activity.

Social facilitation

Behaviour which is initiated or increased in rate or frequency by the presence of another animal carrying out that behaviour (see Chapters 8 and 14).

Social organization

The size of a social group: (i) its composition in respect of age, sex and degrees of relatedness of group members; (ii) all of the relationships among individuals in the group; and (iii) the duration of association of the members of the group (see **Social structure**; Chapter 14).

Social structure

All of the relationships among individuals in a social group and their consequences for spatial distribution and behavioural interactions (see Chapter 14).

Society

A group of individuals organized in a cooperative manner.

Sociobiology

The study of the biological bases of social behaviour, employing evolution as the basic explanatory tool.

Starvation

The state of an individual with a shortage of nutrients or energy such that it starts to metabolize functional tissues rather than food reserves.

Stereotypy

A repeated, relatively invariate sequence of movements which has no obvious purpose (see Chapters 6 and 24).

Stimulation

The effect of one or more stimuli on an individual animal or part of it.

Stimulus

An environmental change which excites one or more receptors or other parts of the nervous system of an animal.

Stockman

Male or female person who has the day-to-day responsibility of caring for farmed animals.

Strain

The short-term consequences of stress.

Stress

An environmental effect on an individual which overtaxes its control systems and results in adverse consequences and eventually reduced fitness. Fitness reduction involves increased mortality, failure to grow or failure to reproduce (see Chapter 1).

Suckling

The behaviour of a young mammal while it is ingesting milk from the teats of its mother or another female mammal.

Suffering

One or more bad feelings continuing for more than a short period.

Sustainable

A system or procedure is sustainable if it is acceptable now and if its expected future effects are acceptable, in particular in relation to resource availability, consequences of functioning and morality of action.

Taxis

Movement oriented with respect to a source of stimulation.

Territoriality

Behaviour associated with territory defence.

Territory

An area an animal defends by fighting, by signals or by demarcation that other individuals detect (see Chapter 12).

Tonic immobility

A behaviour state of a few seconds or longer, during which an animal makes no movement, as a consequence of some temporary environmental situation or of a pathological condition (see Chapter 28).

Trophic

Pertaining to feed, feeding and growth.

Tropic

Movement directed in relation to a spatial stimulus.

Umwelt

The world around as perceived by an animal.

Welfare

The state of an individual as regards its attempts to cope with its environment (see Chapter 1).

Zeitgeber

A time-giver. A source of a periodic output that results in an annual, daily or short-term rhythm in an animal.

References

Aarnink, A.J.A., Schrama, J.W., Verheijen, R.J.E. and Stefanowska, J. (2001) Pen fouling in pig houses affected by temperature. In: Stowell, R.R., Bucklin, R. and Bottcher, R.W. (eds) *Proceedings of the 6th International Symposium on Livestock Environment VI*, Louisville, Kentucky, May 2001.

Aarts, H., Dijksterhuis, A. and de Vries, P. (2001) On the psychology of drinking: being thirsty and perceptually ready. *British Journal of Psychology* 92, 631–642.

Abdul Rahman, S. and Reed, K. (2014) The management and welfare of working animals: identifying problems, seeking solutions and anticipating the future. In: Mellor, D.J. and Bayvel, A.C.D. (eds) *Animal Welfare: Focusing on the Future*. OIE Scientific and Technical Review 33, 197–202. OIE, Paris.

Abeyesinghe, S.M., McLeman, M.A., Owen, R.C., McMahon, C.E. and Wathes, C.M. (2009) Investigating social discrimination of group members by laying hens. *Behavioural Processes* 81, 1–13.

Abeyesinghe, S.M., Drewe, J.A., Asher, L., Wathes, C.M. and Collins, L.M. (2013) Do hens have friends? *Applied Animal Behaviour Science* 143, 61–66.

Abrahamsen, E.J. (2013) Chemical restraint and injectable anesthesia of ruminants. *Veterinary Clinics of North America: Food Animal Practice* 29, 209–227.

Abrahamsson, P. and Tauson, R. (1993) Effect of perches at different positions in conventional cages for laying hens of two different strains. *Acta Veterinaria Scandinavica (Section A: Animal Science)* 43, 228–235.

Abrahamsson, P. and Tauson, R. (1997) Effects of group size on performance, health and birds' use of facilities in FCs for laying hens. *Acta Agriculturae Scandinavica (Section A: Animal Science)* 47, 254–260.

Abrahamson, P., Fossum, O. and Tauson, R. (1998) Health of laying hens in an aviary system over five batches of birds. *Acta Veterinaria Scandinavica* 39, 367–379.

Agenäs, S., Heath, M.F., Nixon, R.M., Wilkinson, J.M. and Phillips, C.J.C. (2006) Indicators of undernutrition in cattle. *Animal Welfare* 15, 149–160.

Aidaros, H. (2014) Drivers for animal welfare policies in the Middle East. In: Mellor, D.J. and Bayvel, A.C.D. (eds) *Animal Welfare: Focusing on the Future*. OIE Scientific and Technical Review 33, 85–89. OIE, Paris.

Aland, A. and Madec, F. (eds) (2009) *Sustainable Animal Production*. Wageningen Academic Publishers, Wageningen, Netherlands.

Albright, J.L. (1969) Social environment and growth. In: Hafez, E.S.E. and Dyers, L.A. (eds) *Animal Growth and Nutrition*. Lea and Febiger, Philadelphia, Pennsylvania.

Albright, J.L. (1981) Training dairy cattle. In: *Dairy Sciences Handbook*. Vol. 14. Agriservices Foundation, Clovis, Calfornia, pp. 363–370.

Albright, J.L. and Arave, C. W. (1997) *The Behaviour of Cattle*. CAB International, Wallingford, UK.

Alexander, G. and Shillito, E.E. (1977) The importance of odour, appearance and voice in maternal recognition of the young in Merino sheep (*Ovis aries*). *Applied Animal Ethology* 3, 127–135.

Alexander, G. and Stevens, D. (1985a) Fostering in sheep II. Use of hessian coats to foster an additional lamb onto ewes with single lambs. *Applied Animal Behaviour Science* 14, 335–344.

Alexander, G. and Stevens, D. (1985b) Fostering in sheep III. Facilitation by use of odourants. *Applied Animal Behaviour Science* 14, 345–354.

Alexander, G., Signoret, J.-P. and Hafez, E.S.E. (1974) Sexual and maternal behaviour. In: Hafez, E.S.E. (ed.) *Reproduction in Farm Animals*. 3rd edn. Lea and Febiger, Philadelphia, Pennsylvania, p. 222.

Alexander, G., Lynch, J.J. and Mottershead, B.E. (1979) Use of shelter and selection and unshorn ewes in paddocks with closely or widely spaced shelters. *Applied Animal Ethology* 5, 51–69.

Alexander, G., Kilgour, R., Stevens, D. and Bradley, L.R. (1984) The effect of experience on twin-care in New Zealand Romney sheep. *Applied Animal Behaviour Science* 13, 363–372.

Algers, B. (1989) Vocal and tactile communication during suckling in pigs. *Sveriges Lantbruksuniversitet, Rapport* 25, SLV, Skara, Sweden.

Al-Merestani, M.R. and Brückner, G. (1992) Influence of hormonal secretion and fertility in Merino mutton sheep by exogenous pheromone application during the breeding season. *Beitrage für Landwirtschaft und Veterinärmedizin* 30, 397–406.

Almquist, J.O. and Hale, E.B. (1956) An approach to the measurement of sexual behaviour and semen production of dairy bulls. *Proceedings of the 3rd International Congress Animal Reproduction*. Plenary papers. Cambridge, pp. 50–59.

Al-Rawi, B. and Craig, J.V. (1975) Agonistic behaviour of caged chickens related to group size and area per bird. *Applied Animal Ethology* 2, 69–80.

Altemus, M., Redwine, L.S., Leong, Y.-M., Frye, C.A., Porges, S.W. and Carter, C.S. (2001) Responses to laboratory psychosocial stress in postpartum women. *Psychosomatic Medicine* 63, 814–821.

Amon, T., Amon, B., Gallob, M., Jeremic, D. and Boxberger, J. (2001) The Stolba Family Pen for pigs: a new housing system designed for animal welfare. In: Polish Committee of Agricultural Engineering (eds) *International Symposium of the 2nd Technical Section of CIGR on Animal Welfare Considerations in Livestock Housing Systems*, Szklarska Poreba, Poland, 2001. Poligmar, Zielona Góra, Poland, pp. 465–478.

Amsel, A. (1992) *Frustration Theory: an Analysis of Dispositional Learning and Memory*. Cambridge University Press, Cambridge, UK.

Andersen I.L., Nævdal E. and Bøe K.E. (2011) Maternal investment, sibling competition, and offspring survival with increasing litter size and parity in pigs (*Sus scrofa*). *Behavioural Ecology and Sociobiology* 65, 1159–1167.

Andersen, I.L., Trøen, C., Ocepek, M., Broom, D.M., Bøe, K.E. and Cronin, G.M. (2014) Development of a new farrowing pen for individually loose-housed sows: preliminary results of "The UMB farrowing pen". In: Bøe, K.E., Braastad, B. and Newberry, R.C. (eds) *Proceedings of the 25th Nordic Regional Symposium of the International Society for Applied Ethology*, 15–17 January 2014, Oscarsborg Fortress, Drøbak, Norway. Norwegian University of Life Sciences, Ås, Norway, p. 11. Available at: http://www.umb.no/statisk/kurs-ved-iha/ISAE/2014/abstracts.pdf (accessed 15 January 2015).

Andreae, U. and Smidt, D. (1982) Behavioural alterations in young cattle on slatted floors. In: Bessei, W. (ed.) *Disturbed Behaviour in Farm Animals. Hohenheimer Arbeiten 12*. Eugen Ulmer, Stuttgart, Germany, pp. 51–60.

Andreasen, S.N., Wemelsfelder, F., Sandøe, P. and Forkman, B. (2013) The correlation of Qualitative Behavioral Assessments with Welfare Quality protocol outcomes in on-farm welfare assessment of dairy cattle. *Applied Animal Behaviour Science* 143, 9–17.

Andrew, J.E., Noble, C., Kadri, S., Jewell, H. and Huntingford, F.A. (2002) The effects of demand feeding on swimming speed and feeding responses in Atlantic salmon *Salmo salar* L., gilthead sea bream *Sparus aurata* L. and European sea bass *Dicentrarchus labrax* L. in sea cages. *Aquaculture Research* 33, 501–507.

Andrew, R.J. (1966) Precocious adult behaviour in the young chick. *Animal Behaviour* 12, 64–76.

Appleby, D. (2004) *APBC Book of Companion Animal Behaviour*. Souvenir Press, London.

Appleby, M.C. (1985) Developmental aspect of nest-site selection. In: Wegner, R.-M. (ed.) *Second European Symposium on Poultry Welfare*, Celle, Germany. World Poultry Science Association, Beekbergen, Netherlands, pp. 138–143.

Appleby, M.C. and Lawrence, A.B. (1987) Food restriction as a cause of stereotypic behaviour in tethered gilts. *Animal Production* 45, 103–110.

Appleby, M.C. and Wood-Gush, D.G.M. (1986) Development of behaviour in beef bulls: sexual behaviour causes more problems than aggression. *Animal Production* 42, 464.

Appleby, M.C., Hughes, B.O. and Elson, H.A. (1992) *Poultry Production Systems: Behaviour, Management and Welfare*. CAB International, Wallingford, UK.

Appleby, M.C., Smith, S.F. and Hughes, B.O. (1993) Nesting, dust bathing and perching by laying hens in cages: effects of design on behaviour and welfare. *British Poultry Science* 34, 835–847.

Appleby, M.C., Walker, A.W., Nicol, C.J., Lindberg, A.C., Freire, R., Hughes, B.O. and Elson, H.A. (2002) Development of furnished cages for laying hens. *British Poultry Science* 43, 489–500.

Appleby, M.C., Mench, J.A. and Hughes, B.O. (2004) *Poultry Behaviour and Welfare*. CAB International, Wallingford, UK.

Archer, M. (1971) Preliminary studies on the palatibility of grasses, legumes and herbs to horses. *Veterinary Record* 89, 236–240.

Archer, S.C., Green, M.J. and Huxley, J.N. (2010) Association between milk yield and serial locomotion score assessments in UK dairy cows. *Journal of Dairy Science* 93, 4045–4053.

Arendonk, J.A.M. van, Hovenier, R. and de Boer, W. (1989) Phenotypic and genetic association between fertility and production in dairy cows. *Livestock Production Science* 21, 1–12.

Arey, D.S. (1992) Straw and food as reinforcers for prepartal sows. *Applied Animal Behaviour Science* 33, 217–226.

Arey, D.S. (1997) Behavioural observations of peri-parturient sows and the development of alternative farrowing accommodation: a review. *Animal Welfare* 6, 217–229.

Arey, D.S. and Edwards, S.A. (1998) Factors influencing aggression between sows after mixing and the consequences for welfare and production. *Livestock Production Science* 56, 61–70.

Arnold, G.W. (1964) Factors within plant associations affecting the behaviour and performance of grazing animals. In: Crisp, D.J. (ed.) *Grazing in Terrestrial and Marine Environments*. Blackwell, Oxford, UK.

Arnold, G.W. (1977) Analysis of spatial leadership in a small field in a small flock of sheep. *Applied Animal Ethology* 3, 263–270.

Arnold, G.W. (1985) Parturient behaviour. In: Fraser, A.F. (ed.) *Ethology of Farm Animals*. World Animal Science, A5. Elsevier, Amsterdam, pp. 335–347.

Arnold, G.W. and Dudzinski, M.L. (1978) *Ethology of Free Ranging Domestic Animals*. Elsevier, Amsterdam.

Arnold, G.W. and Grassia, A. (1982) Ethogram of agonistic behaviour for thoroughbred horses. *Applied Animal Ethology* 8, 5–25.

Arnold, G.W. and Hill, J.L. (1972) Chemical factors affecting selection of food plants by ruminants. In: Harborne, J.B. (ed.) *Phytochemical Ecology*. Academic Press, London, pp. 71–101.

Arnold, G.W. and Maller, R.A. (1974) Some aspects of competition between sheep for supplementary feed. *Animal Production* 19, 309–319.

Arnold, G.W. and Maller, R.A. (1977) Effects of nutritional experience in early and adult life on the performance and dietary habits of sheep. *Applied Animal Ethology* 3, 5–26.

Arnold, G.W., Wallace, S.R. and Rea, W.A. (1981) Associations between individuals and home range behaviour in natural flocks of three breeds of domestic sheep. *Applied Animal Ethology* 7, 239–257.

Asher, A., Diesel, D., Summers, J.F., McGreevy, P.D. and Collins, L.M. (2009) Inherited defects in pedigree dogs. Part 1. Disorders related to breed standards. *Veterinary Journal* 182, 402–411.

Axelrod, J. (1984) The relationship between the stress hormones, catecholamines, ACTH and glucocorticoids. In: Usdin, E., Kvetnansky, R. and Axelrod, R. (eds) *Stress: the Role of Catecholamines and other Neurotransmitters*. Vol. 1. Gordon and Breach, New York, pp. 3–13.

Aydinonat, D., Penn, D.J., Smith, S., Moodley, Y., Hoelzl, F., Knauer, F. and Schwarzenberger, F. (2014) Social isolation shortens telomeres in African grey parrots (*Psittacus erithacus erithacus*). *PloS ONE* 9 (4), e93839.

Back, W. and Clayton, H.M. (eds) (2001) *Equine Locomotion*. W.B. Saunders, London.

Bäckström, L. (1973) Environment and animal health in piglet production. *Acta Veterinaria Scandinavica Supplement* 41, 1–240.

Baeumer, E. (1962) Lebensart des Haushuhns, dritter Teiluber seine laute und allgemine Erganzungen. *Zeitschrift für Tierpsychologie* 19, 394–416.

Bagley, D.V.M. (2005) *Fundamentals of Veterinary Clinical Neurology*. Blackwell, Ames, Iowa.

Baile, C.A. and Forbes, J.M. (1974) Control of feed intake and regulation of energy balance in ruminants. *Physiological Reviews* 54, 160–214.

Bakken, M. (1993a) The relationship between competition capacity and reproduction in farmed silver fox vixens (*Vulpes vulpes*). *Journal of Animal Breeding and Genetics* 110, 147–155.

Bakken, M. (1993b) Reproduction in farmed silver fox vixens (*Vulpes vulpes*) in relation to own competition capacity and that of neighbouring vixens. *Journal of Animal Breeding and Genetics* 110, 305–311.

Bakken, M., Braastad, B.O., Harri, M., Jeppesen, L.L. and Pedersen, V. (1994) Production conditions, behaviour and welfare of farm foxes. *Scientifur* 18, 233–248.

Baldock, N.M. and Sibly, R.M. (1990) Effects of handling and transportation on heart rate and behaviour in sheep. *Applied Animal Behaviour Science* 28, 15–39.

Baldwin, B.A. (1972) Operant conditioning techniques for the study of thermo-regulatory behaviour in sheep. *Journal of Physiology* 226, 41–42.

Baldwin, B.A. (1979) Operant studies on the behaviour of pigs and sheep in relation to the physical environment. *Journal of Animal Science* 49, 1125–1134.

Balleine, B.W. and Dickinson, A. (1998a) The role of incentive learning in instrumental outcome revaluation by sensory-specific satiety. *Animal Learning and Behavior* 26, 46–59.

Balleine, B.W. and Dickinson, A. (1998b) Goal-directed instrumental action: contingency and incentive learning and their cortical substrates. *Neuropharmacology* 37, 407–419.

Balm, P.H.M. (1997) Immune–endocrine interactions. In: Iwana, G.K., Pickering, A.D., Sumpter, J.P. and Schreck, C.B. (eds) *Fish Stress and Health in Aquaculture*. Society for Experimental Biology Seminar Series 62. Cambridge University Press, Cambridge, UK, pp. 195–221.

Balm, P.H.M. and Pottinger, T.G. (1995) Corticotrope and melanotrope POMC-derived peptides in relation to interrenal function during stress in rainbow trout (*Oncorhynchus mykiss*). *General Comparative Endocrinology* 98, 279–288.

Balmford, A., Green, R. and Phalan, B. (2012) What conservationists need to know about farming. *Proceedings of the Royal Society B* 279, 2714–2724.

Banks, E.M. (1964) Some aspects of sexual behaviour in domestic sheep (*Ovis aries*). *Behaviour* 23, 249–279.

Barker, Z.E., Leach, K.A., Whay, H.R., Bell, N.J. and Main, D.C.J. (2010) Assessment of lameness prevalence and associated risk factors in dairy herds in England and Wales. *Journal of Dairy Science* 93, 932–941.

Barnett, C.W. and Pankhurst, N.W. (1998) The effects of common laboratory and husbandry practices on the stress response of the greenback flounder (*Rhombosolea lapirinia* Günther 1862). *Aquaculture* 162, 113–329.

Barnett, J.L., Cronin, G.M. and Winfield, C.G. (1981) The effects of individual and group penning of pigs on plasma total and free corticosteroid concentrations and the maximum corticosteroid binding capacity. *General and Comparative Endocrinology* 44, 219–225.

Barnett, J.L., Cronin, G.M., Winfield C.G. and Dewar, A.M. (1984) The welfare of adult pigs: the effects of five housing treatments on behaviour, plasma corticosteroids and injuries. *Applied Animal Behaviour Science* 12, 209–232.

Barnett, J.L., Hemsworth, P.H. and Winfield, C.G. (1987) The effects of design of individual stalls on the social behaviour and physiological responses related to the welfare of pregnant pigs. *Applied Animal Behaviour Science* 18, 133–142.

Barnett, S.A. (1963) *A Study in Behaviour*. Methuen, London.

Barrier, A.C., Haskell, M.J., Macrae, A.I. and Dwyer, C.M. (2012) Parturition progress and behaviours in dairy cows with calving difficulty. *Applied Animal Behaviour Science* 139, 209–217.

Barry, E. (2001) Inter-limb coordination. In: Back, W. and Clayton, H.M. (eds) *Equine Locomotion*. W.B. Saunders, London, pp. 77–94.

Barry, K.J. and Crowell-Davis, S.L. (1999) Gender differences in the social behavior of the neutered indoor-only domestic cat. *Applied Animal Behaviour Science* 64, 193–211.

Barton, B.A. (1997) Stress in finfish: past, present and future – a historical perspective. In: Iwana, G.K., Pickering, A.D., Sumpter, J.P. and Schreck, C.B. (eds) *Fish Stress and Health in Aquaculture*. Society for Experimental Biology Seminar Series 62. Cambridge University Press, Cambridge, UK, pp. 1–33.

Barton, M.A. (1983a) Behaviour of group-reared calves on acid milk replacer. *Applied Animal Ethology* 11, 77.

Barton, M.A. (1983b) The effects of management and behavioural factors on intake of acidified milk and concentrates by group-reared calves. *Animal Production* 36, 512.

Barton, M.A. and Broom, D.M. (1985) Social factors affecting the performance of teat-fed calves. *Animal Production* 40, 525.

Baryshnikov, L.A. and Kokorina, E.P. (1959) Higher nervous activity and lactation. *International Dairy Congress* 15, 46–53.

Batchelor, G.R. (1999) The laboratory rabbit. In: Poole, T.R. (ed.) *The UFAW Handbook on the Care and Management of Laboratory Animals. Vol. 1 Terrestrial Vertebrates*. 7th edn. Blackwell, Oxford, UK, pp. 1395–1408.

Bateson, P. (2003) The promise of behavioural biology. *Animal Behaviour* 65, 11–17.

Bateson, P. and Gluckman, P. (2012) Plasticity and robustness in development and evolution. *International Journal of Epidemiology* 41, 219–223.

Bateson, P. and Horn, G. (1994) Imprinting and recognition memory: a neural net model. *Animal Behaviour* 48, 695–715.

Bateson, P. and Martin, P. (2013) *Play, Playfulness, Creativity and Innovation.* Cambridge University Press, Cambridge, UK.

Bateson, P., Barker, D., Clutton-Brock, T., Deb, D., D'Udine, G., Foley, R.A., Gluckman, P., Godfrey, L.K., Kirkwood, T., Marazón Larh, M., McNamara, J., Metcalfe, N.B., Moneghan, P., Spencer, H.G. and Sutton, S.E. (2004) Development plasticity and human health. *Nature* 430, 419–421.

Bateson, P.P.G. (1964) Changes in chicks' responses to novel moving objects over the sensitive period for imprinting. *Animal Behaviour* 12, 479–489.

Bateson, P.P.G. (1978) Sexual behaviour and optimal outbreeding. *Nature, London* 273, 659–660.

Baumans, V. and Van Loo, P.L.P. (2013) How to improve housing conditions of laboratory animals: the possibilities of environmental refinement. *The Veterinary Journal* 195, 24–32.

Baxter, E.M., Jarvis, S., D'Eath, R.B., Ross, D.W., Robson, S.K., Farish, M., Nevison, I.M., Lawrence, A.B. and Edwards, S.A. (2008) Investigating the behavioural and physiological indicators of neonatal survival in pigs. *Theriogenology* 69, 773–783.

Baxter, E.M., Jarvis, S., Sherwood, L., Farish, M., Roeher, R., Lawrence, A.B. and Edwards, S.A. (2011) Genetic and environmental effects on piglet survival and maternal behaviour of the farrowing sow. *Applied Animal Behaviour Science* 130, 28–41.

Baxter, E.M., Lawrence, A.B. and Edwards, S.A. (2012) Alternative farrowing accommodation: welfare and economic aspects of existing farrowing and lactation systems for pigs. *Animal* 6, 96–117.

Baxter, M.R. (1988) Needs – behavioural or psychological? *Applied Animal Behaviour Science* 19, 345–348.

Baxter, M.R. (1992) The space requirements of housed livestock. In: Phillips, C. and Piggins, D. (eds) *Farm Animals and the Environment.* CAB International, Wallingford, UK, pp. 67–81.

Baxter, M.R. and Schwaller, C. (1983) Space requirements for sows in confinement. In: Baxter, S.H. (ed.) *Farm Animal Housing and Welfare.* Martinus Nijhoff, Dordrecht, Netherlands.

Beach, F.A. (1976) Sexual attractivity, proceptivity and receptivity in female mammals. *Hormones and Behaviour* 7, 105–138.

Beattie, V.E., Walker, N. and Sneddon, I.A. (1996) An investigation of the effect of environmental enrichment and space allowance on the behaviour and production of growing pigs. *Applied Animal Behavior Science* 48, 151–158.

Beattie, V.E., O'Connell, N.E. and Moss, B.W. (2000) Influence of environmental enrichment on the behaviour, performance and meat quality of domestic pigs. *Livestock Production Science* 65, 71–79.

Beaver, B.V. (1994) *The Veterinarian's Encyclopedia of Animal Behaviour.* Iowa State University Press, Ames, Iowa.

Beaver, B.V. (2003) *Feline Behavior: a Guide for Veterinarians.* 2nd edn. Saunders, St Louis, Missouri.

Bechara, A., Noel, X. and Crone, E.A. (2006) Loss of willpower: abnormal neural mechanisms of impulse control and decision making in addiction. In: Wiers, R.W. and Stacy, A.W. (eds) *Handbook of Implicit Cognition and Addiction.* Sage Publications, Thousand Oaks, California, pp. 215–250.

Bekoff, M. and Byers, J.A. (1998) *Animal Play.* Cambridge University Press, Cambridge, UK.

Belyaev, D. (1979) Destabilizing selection as a factor in domestication. *Journal of Heredity* 70, 301–308.

Belyaev, D. and Trut, L. (1963) Experience with the selection of silver foxes for behavioural type and its importance in the problem of the evolutionary reorganisation of the reproductive function. In: *The Physiological Basis of Complex Forms of Behaviour.* AN SSSR M-L, pp. 13–14 [published in Russian].

Belyaev, D., Plyusnina, I. and Trut, L. (1985) Domestication in the silver fox (*Vulpes fulvus* Desm.): changes in physiological boundaries of the sensitive period of primary socialization. *Applied Animal Behaviour Science* 13, 359–370.

Bendor, D. and Wilson, M.A. (2012) Biasing the content of hippocampal replay during sleep. *Nature Neuroscience* 15, 1439–1446.

Benham, P.F.J. (1982a) Synchronisation of behaviour in grazing cattle. *Applied Animal Ethology* 8, 403–404.

Benham, P.F.J. (1982b) Social organisation and leadership in a grazing herd of suckler cows. *Applied Animal Ethology* 9, 95.

Benham, P.F.J. (1984) Social organisation in groups of cattle and the inter-relationship between social and grazing behaviours under different grazing management systems. PhD thesis, University of Reading, UK.

Benham, P.F.J. and Broom, D.M. (1989) Interactions between cattle and badgers at pasture with reference to bovine tuberculosis transmission. *British Veterinary Journal* 145, 226–241.

Benham, P.F.J. and Broom, D.M. (1991) Responses of dairy cows to badger urine and faeces on pasture with reference to bovine tuberculosis transmission. *British Veterinary Journal* 147, 517–532.

Bentley, D.R. and Hoy, R.R. (1972) Genetic control of the neuronal network generating cricket (Teleogryllus-gryllus) song patterns. *Animal Behaviour* 20, 478–492.

Benus, I. (1988) Aggression and coping. Differences in behavioural strategies between aggressive and non-aggressive male mice. PhD thesis, University of Groningen, Netherlands.

Berlyne, D.E. (1967) Arousal and reinforcement. In: Jones, M.R. (ed.) *Nebraska Symposium on Motivation.* University of Nebraska Press, Lincoln, Nebraska.

Berne, R.M. and Levy, M.N. (2000) *Principles of Physiology*. 3rd edn. Mosby, St Louis, Missouri.

Berridge, K.C. (1996) Food reward: brain substrates of wanting and liking. *Neuroscience and Biobehavioural Reviews* 20, 1–25.

Berthe, F., Vannier, P. Have, P., Serratosa, J., Bastino, E., Broom, D.M., Hartung, J. and Sharp, J.M. (2012) The role of EFSA in assessing and promoting animal health and welfare. *EFSA Journal* 10, s1002, 19–27.

Bessei, W. (1982) Head-shaking in the domestic fowl. *Hohenheimer Arbeiten* 121, 147–151.

Beukema, J.J. (1970) Angling experiments with carp: decreased catchability through one trial learning. *Netherlands Journal of Zoology* 20, 81–92.

Beuving, G. (1980) Corticosteoids in laying hens. In: Moss, R. (ed.) *The Laying Hen and its Environment. Current Topics in Veterinary Medicine Animal Science* 8, 65–82, Martinus Nijhoff, The Hague, Netherlands.

Bilcik, B. and Keeling, L.J. (1999) Changes in feather condition in relation to feather pecking and aggressive behaviour in laying hens. *British Poultry Science* 40, 444–451.

Bildsøe, M., Heller, K.E. and Jeppesen, L.L. (1990) Stereotypies in adult ranch mink. *Scientifur* 14, 169–177.

Bilsing, A., Becker, I. and Nichelmann, M. (1992) Verhaltungsstörungen bei der Moschusente. *KTBL Schrift* 351, 69–76.

Biondi, M. and Zannino, L.-G. (1997) Psychological stress, neuroimmuno-modulation, and susceptibility to infectious diseases in animals and man: a review. *Psychotherapy and Psychosomatics* 66, 3–26.

Blackwell, E.J., Bradshaw, J.W.S. and Casey, R.A. (2013) Fear responses to noise in domestic dogs: prevalence, risk factors and co-occurrence with other fear-related behaviour. *Applied Animal Behaviour Science* 145, 15–25.

Blakemore, C. and Cooper, G.F. (1970) Development of the brain depends on the visual environment. *Nature, London* 228, 477–478.

Block, M.L., Volpe, L.C. and Hayse, M.J. (1981) Saliva as a chemical cue in the development of social behaviour. *Science, New York* 211, 1062–1064.

Blokhuis, H.J. (1983) Sleep in poultry. *World's Poultry Science Journal* 39, 33–37.

Blokhuis, H.J. (1984) Rest in poultry. *Applied Animal Behaviour Science* 12, 289–303.

Blokhuis, H.J. (1986) Feather pecking in poultry: its relation with ground pecking. *Applied Animal Behaviour Science* 16, 63–67.

Blokhuis, H.J. and Arkes, J.G. (1984) Some observations on the development of feather-pecking in poultry. *Applied Animal Behaviour Science* 12, 154–157.

Blokhuis, H.J. and van den Haar, J.W. (1989) Effects of floor type during rearing and of beak trimming on ground pecking and feather-pecking in laying hens. *Applied Animal Behaviour Science* 22, 359–369.

Blokhuis, H.J., Veissier I., Miele, M. and Jones, B. (2010) The Welfare Quality project and beyond: safeguarding farm animal well-being. *Acta Agriculturae Scandinavica, Section A, Animal Science* 60, 129–140.

Boe, K.E. and Faerevik, G. (2003) Grouping and social preferences in calves, heifers and cows. *Applied Animal Behaviour Science* 80, 175–190.

Böhm, M., Hutchings, M.R. and White, P.C.L. (2009) Contact networks in wildlife-livestock host community: identifying high-risk individuals in the transmission of bovine TB among badgers and cattle. *PloS ONE*, 4(4), e5016, pp. 12.

Boissy, A. and Lee, C. (2014) How assessing relationships between emotions and cognition can improve farm animal welfare. In: Mellor, D.J. and Bayvel, A.C.D. (eds) *Animal Welfare: Focusing on the Future*. OIE Scientific and Technical Review 33, 103–110. OIE, Paris.

Bokkers, E.A.M., de Vries, M., Antonissen, I.C.M.A. and de Boer, I.J.M. (2012) Inter-and intra-observer reliability of experienced and inexperienced observers for the Qualitative Behaviour Assessment in dairy cattle. *Animal Welfare* 21, 307–318.

Bolles, R.C. (1975) *Theory of Motivation*. Harper and Row, New York.

Booth, D.A. (1978) Prediction of feeding behaviour from energy flows in the rat. In: Booth, D.A. (ed.) *Hunger Models: Computable Theory of Feeding Control.* Academic Press, London.

Booth, D.A., Givson, E.L., Toase, A.-M. and Freeman, R.P.J. (1994) Small objects of desire: the recognition of appropriate foods and drinks and its neural mechanisms. In: Legg, C.R. and Booth, D. (eds) *Appetite: Neural and Behavioural Bases.* Oxford University Press, Oxford, UK, pp. 98–126.

Booth, D.W. and Signoret, J.-P. (1992) Olfaction and reproduction in ungulates. In: Milligan, S.R. (ed.) *Oxford Reviews of Reproductive Biology* 14, 263–301.

Borchelt, P. (1991) Cat elimination behavior problems. *Veterinary Clinics of North America: Small Animal Practice* 21, 257–264.

Bos, R. van den and de Cock Buning, T. (1994) Social behaviour of domestic cats *(Felis lybica* f. *catus* L.): a study of dominance in a group of female laboratory cats. *Ethology* 98, 14–37.

Boshouwers, F.M.G. and Nicaise, E. (1987) Physical activity and energy expenditure of laying hens as affected by light intensity. *British Poultry Science* 28, 155–163.

Bouissou, M.-F. (1970) Role du contact physique dans la manifestation des relations hierarchiques chez les bovins: consequences pratiques. *Annales de Zootechnie* 19, 279–285.

Bouissou, M.-F. and Boissy, A. (1988) Effects of early handling on heifers' subsequent reactivity to humans and to unfamiliar situations. In: Unshelm, J., van Putten, G., Zeeb, K. and Ekesbo, I. (eds) *Proceedings of the International Congress on Applied Ethology in Farm Animals,* Skara, Sweden, 1988. KTBL, Darmstadt, Germany, pp. 21–38.

Bouissou, M.F. and Hovels, J. (1976) Effet d'un contact précoce sur quelques aspects du comportement social des bovins domestiques. *Biology of Behaviour* 1, 17–36.

Box, H.O. (1973) *Organisation in Animal Communities*. Butterworth, London.

Box, H.O. (2003) Characteristics and propensities of marmosets and tamarins: implications for studies of innovation. In: Hauser, M.D. and Konishi, M. (eds) *The Design of Animal Communication.* MIT Press, Cambridge, Massachusetts, pp. 197–219.

Box, H.O. and Gibson, K.R. (eds) (1999) *Mammalian Social Learning: Comparative and Ecological Perspectives.* Symposia of the Zoological Society of London. Cambridge University Press, Cambridge, UK.

Braastad, B.O. (1990a) Abnormal behaviour in farmed silver fox vixens (*Vulpes vulpes* L.): tail-biting and infanticide. *Applied Animal Behaviour Science* 17, 376–377.

Braastad, B.O. (1990b) Individual variation in maternal behaviour of silver foxes. *Third Nordic Symposium of the Society for Veterinary Ethology*, Sem, Asker, Norway, 2 November, p. 20.

Braastad, B.O. (1992) Progress in the ethology of foxes and mink. *Norwegian Journal of Agricultural Science (Suppl.)* 9, 487–504.

Braastad, B.O. (1993) Periparturient behaviour of successfully reproducing farmed silver-fox vixens. *Applied Animal Behaviour Science* 37, 125–138.

Braastad, B.O. (1994) Reproduction in silver fox vixens in breeding boxes with and without an entrance tunnel. *Acta Agriculturae Scandinavica Section A: Animal Science* 44, 38–42.

Braastad, B.O. and Bakken, M. (2002) Behaviour of dogs and cats. In: Jensen, P. (ed.) *The Ethology of Domestic Animals: an Introductory Text.* CAB International, Wallingford, UK.

Bradshaw, J.W.S. (1992) *The Behaviour of the Domestic Cat.* CAB International, Wallingford, UK.

Bradshaw, J.W.S. and Hall, S.L. (1999) Affiliative behaviour of related and unrelated pairs of cats in catteries: a preliminary report. *Applied Animal Behaviour Science* 63, 251–255.

Bradshaw, R.H. (1991) Discrimination of group members by laying hens *Gallus domesticus. Behavioural Processes* 24, 143–151.

Bradshaw, R.H. and Broom, D.M. (1999) A comparison of the behaviour and performance of sows and piglets in crates and oval pens. *Animal Science* 69, 327–333.

Bradshaw, R.H., Hall, S.J.G. and Broom, D.M. (1996) Behavioural and cortisol responses of pigs and sheep during transport. *Veterinary Record* 138, 233–234.

Bradshaw, R.H., Kirkden, R.D. and Broom, D.M. (2002) A review of the aetiology and pathology of leg weakness in broilers in relation to their welfare. *Avian Poultry Biological Review* 13, 45–103.

Braithwaite, V. (2005) Cognitive ability in fish. *Fish Physiology* 24, 1–37.

Braithwaite, V.A. and Ebbesson, L.O.A. (2014) Pain and stress responses in farmed fish. In: Mellor, D.J. and Bayvel, A.C.D. (eds) *Animal Welfare: Focusing on the Future.* OIE Scientific and Technical Review 33, 245–253. OIE, Paris.

Braithwaite, V.A. and Huntingford, F.A. (2004) Fish and welfare: do fish have the capacity for pain perception and suffering. *Animal Welfare* 13, 587–592.

Brantas, G.C. (1974) Das Verhalten von Legehennen: quantitive Unterschiede zwischen Käfig und Bodenhaltung. *KTBL-Schrift, Darmstadt* 138–146.

Brantas, G.C. (1975) Welzijn produktie en profit. *Tijdschrift Voor Diergeneeskunde* 100, 703–708.

Brantas, G.C. (1980) The pre-laying behaviour of laying hens in cages with and without laying nests. In: Moss, R. (ed.) *The Laying Hen and its Environment.* Current Topics in Veterinary Medicine Animal Science 8. Martinus Nijhoff, The Hague, Netherlands, pp. 227–234.

Brantas, G.C., de Vos-Reesink, K. and Wennrich, G. (1978) Ethologische beobachtungen an legehennen in get-away-kafigen. *Archiv für Geflugelkunde* 42, 129–132.

Brion, A. (1964) Les tics chez les animaux. In: Brion, A. and Ey, H. (eds) *Psychiatrie Animale.* Desclee de Brouwer, Paris, pp. 299–306.

British Pig Executive (2011) *Pig Yearbook 2011. Agriculture and Horticulture Development Board,* Kenilworth, UK.

Broad, K.D., Mimmack, M.L., Keverne, E.B. and Kendrick, K.M. (2002) Increased BDNF and trk-B mRNA expression in cortical and limbic regions following formation of a social recognition memory. *European Journal of Neuroscience* 16, 2166–2174.

Broadhurst, M.K. and Barker, D.T. (2000) Effects of capture by hook and line on plasma cortisol, scale loss and survival in juvenile mulloway (*Agrysomus hololepidotus*). *Archive of Fishery and Marine Research* 48, 1–10.

Brogden, K.A., Lehmkuhl, H.D. and Cutlip, R.C. (1998) *Pasteurella haemolytica* complicated respiratory infections in sheep and goats. *Veterinary Research* 29, 233–254.

Broglio, C., Rodriguez, F. and Salas, C. (2003) Spatial cognition and its neural basis in teleost fishes. *Fish and Fisheries* 4, 247–255.

Brookshire, K.H. and Hoegnander, O.C. (1968) Conditioned fear in the fish. *Psychological Reports* 22, 75–81.

Broom, D.M. (1968a) Specific habituation by chicks. *Nature, London* 217, 880–881.

Broom, D.M. (1968b) Behaviour of undisturbed 1 to 10 day old chicks in different rearing conditions. *Developmental Psychobiology* 1, 287–295.

Broom, D.M. (1969a) Reactions of chicks to visual changes during the first ten days after hatching. *Animal Behaviour* 17, 307–315.

Broom, D.M. (1969b) Effects of visual complexity during rearing on chicks' reactions to environmental change. *Animal Behaviour* 17, 773–780.

Broom, D.M. (1979) Methods of detecting and analysing activity rhythms. *Biology of Behaviour* 4, 3–18.

Broom, D.M. (1980) Activity rhythms and position preferences of domestic chicks which can see a moving object. *Animal Behavour* 28, 201–211.

Broom, D.M. (1981) *Biology of Behaviour.* Cambridge University Press, Cambridge, UK.

Broom, D.M. (1982) Husbandry methods leading to inadequate social and maternal behaviour in cattle. In: Bessei, W. (ed.) *Disturbed Behaviour in Farm Animals.* Hohenheimer Arbeiten 121, Eugen Ulmer, Stuttgart, Germany, pp. 42–50.

Broom, D.M. (1983a) Cow–calf and sow–piglet behaviour in relation to colostrum ingestion. *Annales de Recherche Vétérinaire* 14, 342–348.

Broom, D.M. (1983b) Stereotypies as animal welfare indicators. In: Smidt, D. (ed.) *Indicators Relevant to Farm Animal Welfare.* Current Topics in Veterinary Medicine and Animal Science. Martinus Nijhoff, The Hague, Netherlands, pp. 81–87.

Broom, D.M. (1985) Stress, welfare and the state of equilibrium. In: Wegner, R.M. (ed.) *Proceedings of the 2nd European Symposium of Poultry Welfare,* Celle, Germany. World Poultry Science Association, Beekbergen, Netherlands, pp. 72–81.

Broom, D.M. (ed.) (1986a) *Farmed Animals.* Torstar Books, New York.

Broom, D.M. (1986b) The influence of the design of housing systems for cattle on lameness and on behaviour: summary of discussion on behavioural and veterinary aspects. In: Wierenga, H.K. and Peterse, D.J. (eds) *Cattle Housing Systems, Lameness and Behaviour.*

Current Topics of Veterinary Medicine Animal Science 40. Martinus Nijhoff, Dordrecht, Netherlands, pp. 179–181.

Broom, D.M. (1986c) Indicators of poor welfare. *British Veterinary Journal* 142, 524–526.

Broom, D.M. (1986d) Stereotypies and responsiveness as welfare indicators in stall-housed sows. *Animal Production* 42, 438–439.

Broom, D.M. (1986e) Responsiveness of stall-housed sows. *Applied Animal Behaviour Science* 15, 186.

Broom, D.M. (1987a) The veterinary relevance of farm animal ethology. *Veterinary Record* 121, 400–402.

Broom, D.M. (1987b) General conclusions. In: *Welfare Aspects of Housing Systems for Veal Calves and Fattening Bulls.* Commission of the European Communities, Luxembourg, pp. 161–166.

Broom, D.M. (1988a) Needs, freedoms and the assessment of welfare. *Applied Animal Behaviour Science* 19, 384–386.

Broom, D.M. (1988b) The relationship between welfare and disease susceptibility in farm animals. In: Gibson, T.E. (ed.) *Animal Disease – a Welfare Problem.* British Veterinary Association Animal Welfare Foundation, London, pp. 22–29.

Broom, D.M. (1988c) The scientific assessment of animal welfare. *Applied Animal Behaviour Science* 20, 5–19.

Broom, D.M. (1988d) Welfare considerations in cattle practice. *Proceedings of the British Cattle Veterinary Association* 1986/1987, 153–164.

Broom, D.M. (1988e) Les concepts de stress et de bien-être. *Receuil de Médicine Vétérinaire* 164, 715–722.

Broom, D.M. (1989a) Animal Welfare. In: Grunsell, C.S.G., Raw, M.-E. and Hill, F.W.G. (eds) *The Veterinary Annual 29.* Wright, London, pp. 9–14.

Broom, D.M. (1989b) The assessment of sow welfare. *Pig Veterinary Journal* 22, 100–111.

Broom, D.M. (1991) Needs and welfare of housed calves. In: Metz, J.H.M. and Groenestein, C.M. (eds) *New Trends in Veal Calf Production. Proceedings of the International Symposium on Veal Calf Production*, 52. EAAP Publications, Wageningen, Netherlands, pp. 23–31.

Broom, D.M. (1992) The needs of laying hens and some indicators of poor welfare. In: Carter, V. and Carter, H. (eds) *The Laying Hen.* European Conference Group on the Protection of Farm Animals, Horsham, UK, pp. 4–19.

Broom, D.M. (1993) Assessing the welfare of modified or treated animals. *Livestock Production Science* 36, 39–54.

Broom, D.M. (1994) The effects of production efficiency on animal welfare. In: Huisman, E.A., Osse, J.W.M., van der Heide, D., Tamminga, S., Tolkamp, B.L., Schouten, W.G.P., Hollingsworth, C.E. and van Winkel, G.L. (eds) *Biological Basis of Sustainable Animal Production. Proceedings of the 4th Zodiac Symposium.* EAAP Publications 67, Wageningen Pers., Wageningen, Netherlands, pp. 201–210.

Broom, D.M. (1996) Scientific research on veal calf welfare. In: *Veal Perspectives to the Year 2000. Proceedings of International Symposium*, Le Mans, France. Fédération de la Vitellerie Francaise, Paris, pp. 147–153.

Broom, D.M. (1997) Animal behaviour as an indicator of animal welfare in different housing and management systems. In: Saloniemi, H. (ed.) *Proceedings of the 9th International Congress on Animal Hygiene.* Tummavnoren Kirjapaino Oy, Helsinki, pp. 371–378.

Broom, D.M. (1998) Welfare, stress and the evolution of feelings. *Advances in the Study of Behavior* 27, 371–403.

Broom, D.M. (1999a) Fish welfare and the public perception of farmed fish. *Proceedings of Aquavision*, 1–6. Aquavision, Stavanger, Norway.

Broom, D.M. (1999b) The welfare of vertebrate pests in relation to their management. In: Cowan, P.D. and Feare, C.J. (eds) *Advances in Vertebrate Pest Management.* Filander Verlag, Fürth, pp. 309–329.

Broom, D.M. (1999c) The welfare of dairy cattle. In: Aagaard, K. (ed.) *III Future Milk Farming, Proceedings of the 25th International Dairy Congress*, Aarhus, Denmark, 1998. Danish National Committee of IDF, Aarhus, Denmark, pp. 32–39.

Broom, D.M. (2000) Welfare assessment and problem areas during handling and transport. In: Grandin, T. (ed.) *Livestock Handling and Transport.* 2nd edn. CAB International, Wallingford, UK. pp. 43–61.

Broom, D.M. (2001a) Coping, stress and welfare. In: Broom, D.M. (ed.) *Coping with Challenge: Welfare in Animals including Humans.* Dahlem University Press, Berlin, pp. 1–9.

Broom, D.M. (2001b) Assessing the welfare of hens and broilers. *Proceedings of the Australian Poultry Science Symposium* 13, 61–70.

Broom, D.M. (2001c) Effects of dairy cattle breeding and production methods on animal welfare. In:

Proceedings of the 21st World Buiatrics Congress, Punta del Este, Uruguay. Vol. 147, CD-Rom. World Association for Buiatrics.

Broom, D.M. (2001d) The use of the concept Animal Welfare in European conventions, regulations and directives. *Food Chain* 2001, 148–151.

Broom, D.M. (2001e) Evolution of pain. In: Soulsby, E.J.L. Lord and Morton, D. (eds) *Pain: Its Nature and Management in Man and Animals. Royal Society of Medicine International Congress Symposium Series* 246, 17–25.

Broom, D.M. (2002) Does present legislation help animal welfare? *Landbauforschung Völkenrode* 227, 63–69.

Broom, D.M. (2003) *The Evolution of Morality and Religion.* Cambridge University Press, Cambridge, UK, p. 259.

Broom, D.M. (2004) Welfare. In: Andrews, A.H., Blowey, R.W., Boyd, H. and Eddy, R.G. (eds) *Bovine Medicine: Diseases and Husbandry of Cattle*. 2nd edn. Blackwell, Oxford, UK, pp. 955–967.

Broom, D.M. (2005) The effects of land transport on animal welfare. *Revue scientifique technicale office internationale des Epizoöties* 24, 683–691.

Broom, D.M. (2006a) Adaptation. *Berliner und Münchener Tierärztliche Wochenschrift* 119, 1–6.

Broom, D.M. (2006b) Behaviour and welfare in relation to pathology. *Applied Animal Behaviour Science* 97, 71–83.

Broom, D.M. (2006c) The evolution of morality. *Applied Animal Behaviour Science* 100, 20–28.

Broom, D.M. (2006d) Traceability of food and animals in relation to animal welfare. *Annals of the International Conference on Agricultural Product Traceability.* Ministry of Agriculture, Livestock and Food Supply, Brasilia, pp. 195–201.

Broom, D.M. (2007) Cognitive ability and sentience: which aquatic animals should be protected? *Diseases of Aquatic Organisms* 75, 99–108.

Broom, D.M. (2008) The welfare of livestock during transport. In: Appleby, M., Cussen, V., Garcés, L., Lambert, L. and Turner, J. (eds) *Long Distance Transport and the Welfare of Farm* Animals. CAB International, Wallingford, UK, pp. 157–181.

Broom, D.M. (2009) Animal welfare and legislation. In: Smulders, F.J.M. and Algers, B. (eds) *Welfare of Production Animals: Assessment and Management of Risks*. Wageningen Academic Publishers, Wageningen, The Netherlands, pp. 339–352.

Broom, D.M. (2010a) Animal welfare: an aspect of care, sustainability, and food quality required by the public. *Journal of Veterinary Medical Education* 37, 83–88.

Broom, D.M. (2010b) Cognitive ability and awareness in domestic animals and decisions about obligations to animals. *Applied Animal Behaviour Science* 126, 1–11.

Broom, D.M. (2011) A history of animal welfare science. *Acta Biotheoretica* 59, 121–137.

Broom, D.M. (2013) Cow welfare and herd size: towards a sustainable dairy industry. *Cattle Practice* 21, 169–173.

Broom, D.M. (2014a) Welfare of transported animals: factors influencing welfare and welfare assessment. In: Grandin, T. (ed.) *Livestock Handling and Transport*, 4th edn. CAB International, Wallingford, UK, pp. 23–38.

Broom, D.M. (2014b) *Sentience and Animal Welfare*. CAB International, Wallingford, UK.

Broom, D.M. and Arnold, G.W. (1986) Selection by grazing sheep of pasture plants at low herbage availability and responses of the plants to grazing. *Australian Journal of Agricultural Research* 37, 527–538.

Broom, D.M. and Corke, M.J. (2002) Effects of disease on farm animal welfare. *Acta Veterinaria Brno* 71, 133–136.

Broom, D.M. and Johnson, E. (1980) Responsiveness of hand-reared roe deer to odours from skin glands. *Journal of Natural History* 14, 41–47.

Broom, D.M. and Johnson, K.G. (1993) *Stress and Animal Welfare*. Chapman and Hall, London.

Broom, D.M. and Johnson, K.G. (2000) *Stress and Animal Welfare*. Kluwer, Dordrecht, Netherlands.

Broom, D.M. and Kirkden, R.D. (2004) Welfare, stress, behaviour and pathophysiology. In: Dunlop, R.H. and Malbert, C.H. (eds) *Veterinary Pathophysiology*. Blackwell, Ames, Iowa, pp. 337–369.

Broom, D.M. and Leaver, J.D. (1977) Mother–young interactions in dairy cattle. *British Veterinary Journal* 133, 192.

Broom, D.M. and Leaver, J.D. (1978) The effects of group-housing or partial isolation on later social behaviour of calves. *Animal Behaviour* 26,1255–1263.

Broom, D.M. and Potter, M.J. (1984) Factors affecting the occurrence of stereotypies in stall-housed dry sows. In: Unshelm, J., van Putten, G. and Zeeb, K.

(eds) *Proceedings of the International Congress of Applied Ethology in Farm Animals,* Kiel, 1984. KTBL, Darmstadt, Germany, pp. 229–231.

Broom, D.M. and Reefmann, N. (2005) Chicken welfare as indicated by lesions on carcasses in supermarkets. *British Poultry Science* 46, 1–8.

Broom, D.M. and Zanella, A.J. (2004) Brain measures which tell us about animal welfare. *Animal Welfare* 13, S41–S45.

Broom, D.M., Pain, B.F. and Leaver, J.D. (1975) The effects of slurry on the acceptability of swards to grazing cattle. *Journal of Agricultural Science* 85, 331–336.

Broom, D.M., Knight, P.G. and Stansfield, S.C. (1986) Hen behaviour and hypothalamic-pituitary-adrenal responses to handling and transport. *Applied Animal Behaviour Science* 16, 98.

Broom, D.M., Mendl, M.T. and Zanella, A.J. (1995) A comparison of the welfare of sows in different housing conditions. *Animal Science* 61, 369–385.

Broom, D.M., Goode, J.A., Hall, S.J.G., Lloyd, D.M. and Parrott, R.F. (1996) Hormonal and physiological effects of a 15 hour journey in sheep: comparison with the responses to loading, handling and penning in the absence of transport. *British Veterinary Journal* 152, 593–604.

Broom, D.M., Sena, H. and Moynihan, K.L. (2009) Pigs learn what a mirror image represents and use it to obtain information. *Animal Behaviour* 78, 1037–1041.

Broster, W.H. and Broster, V.J. (1998) Body score of dairy cows. *Journal of Dairy Research* 65, 155–173.

Brown, S.N., Knowles, T.G., Edwards, J.E. and Warriss, P.D. (1999) Behavioural and physiological responses of pigs being transported for up to 24 hours followed by six hours recovery in lairage. *Veterinary Record* 145, 421–426.

Brownlee, A. (1954) Play in domestic cattle in Britain: an analysis of its nature. *British Veterinary Journal* 110, 48–68.

Bruce, W.N. and Decker, G.C. (1958) The relationship of *Stomoxys calcitrans* abundance to milk production in dairy cattle. *Journal of Economic Entomology* 51, 269.

Bryant, M.J., Rowlinson, P. and van der Steen, H.A.M. (1983) A comparison of the nursing and suckling behaviour of group-and individually housed sows and their litters. *Animal Production* 36, 445–451.

Bubier, N.E. (1996) The behavioural priorities of laying hens: the effects of two methods of environmental enrichment on time budgets. *Behavioural Processes* 37, 239–249.

Burley, N., Krantzbery, G. and Radman, P. (1982) Influence of colour-bonding on the conspecific preference of zebra finches. *Animal Behaviour* 30, 444–455.

Burman, O.H.P., Parker, R.M.A., Paul, E.S. and Mendl, M. (2008a) A spatial judgement task to determine background emotional state in laboratory rats *Rattus norvegicus*. *Animal Behaviour* 76, 801–809.

Burman, O.H.P., Parker, R.M.A., Paul, E.S. and Mendl, M. (2008b) Sensitivity to reward loss as an indicator of animal emotion and welfare. *Biology Letters* 4, 330–333.

Burn, C.C. (2011) A vicious cycle: A cross-sectional study of canine tail-chasing and human responses to it, using a free video-sharing website. *PloS ONE* 6, e26553, pp. 9.

Burt de Perera, T. (2004) Fish can encode order in their spatial map. *Proceedings of the Royal Society, London, B* 271, 2131–2134.

Butler, W.R. and Smith, R.D. (1989) Interrelationships between energy balance and post partum reproductive function in dairy cattle. *Journal of Dairy Science* 72, 767–783.

Byers, J.A. (1998) Biological effects of locomotor play: getting into shape or something more specific. In: Bekoff, M. and Byers, J.A. (eds) *Animal Play: Evolutionary, Comparative and Ecological Perspectives.* Cambridge University Press, Cambridge, UK, pp. 205–220.

Cameron, A.A., Plenderleith, M.B. and Snow, P.J. (1990) Organization of the spinal cord in four species of elasmobranch fish: cytoarchitecture and distribution of serotonin and selected neuropeptides. *Journal of Comparative Neurology* 297, 201–218.

Canario, L., Turner, S., Roehe, R., Lundeheim, N., D'Eath, R., Lawrence, A.B., Knol, E., Bergsma, R. and Rydhmer, L. (2012) Genetic associations between behavioral traits and direct social effects of growth rate in pigs. *Journal of Animal Science* 90, 4706–4715.

Cañon Jones, H.A., Hansen, L.A., Noble, C., Damsgård, B., Broom, D.M. and Pearce, G.P. (2010) Social network analysis of behavioural interactions influencing fin damage development in Atlantic salmon (*Salmo salar*) during feed-restriction. *Applied Animal Behaviour Science* 127, 139–151.

Cañon Jones, H.A., Noble, C. Damsgård B. and Pearce, G.P. (2011) Social network analysis of the behavioural interactions that influence the development of fin damage in Atlantic salmon parr (*Salmo salar*) held at different stocking densities. *Applied Animal Behaviour Science* 133, 117–120.

Cañon Jones, H.A., Noble, C., Damsgård B. and Pearce, G.P. (2012) Investigating the influence of predictable and unpredictable feed delivery schedules upon the behaviour and welfare of Atlantic salmon parr (*Salmo salar*) using social network analysis and fin damage. *Applied Animal Behaviour Science* 138, 132–140.

Cariolet, R. and Dantzer, R. (1985) Activity motrice des truises attachées durant la gestation: mise en evidence de quelques facteurs de variations. *Journies Recerches Porcine en France* Yl, 237–248.

Carlstead, K., Brown, J.L. and Strawn, W. (1993) Behavioral and physiological correlates of stress in laboratory cats. *Applied Animal Behaviour Science* 38, 143–158.

Carss, D.N. (1993) Cormorants *Phalacrocorax carbo* at cage fish farms in Argyll, Western Scotland. *Seabird* 15, 38–44.

Carter, C.S. (2001) Is there a neurobiology of good welfare? In: Broom, D.M. (ed.) *Coping with Challenge: Welfare of Animals Including Humans.* Dahlem University Press, Berlin, pp. 11–30.

Carter, C.S. and Altemus, M. (1997) Integrative functions of lactational hormones in social behavior and stress management. *Annals of the New York Academy of Sciences* 807, 164–174.

Casey, R.A. and Bradshaw, J.W.S. (2005) The assessment of welfare. In: Rochlitz, I. (ed.) *The Welfare of Cats.* Springer, Dordrecht, Netherlands, pp. 23–46.

Casey, R.A. and Bradshaw, J.W.S. (2008) The effects of additional socialisation for kittens in a rescue centre on their behaviour and suitability as a pet. *Applied Animal Behaviour Science* 114, 196–205.

Cate ten, C. (1984) The influence of social relations on the development of species recognition in zebra finches. *Behaviour* 91, 263–285.

Caughey, S.P., Klampfl, S.M., Bishop, V.R., Pfoertsch, J., Neumann, I.D., Bosch, O.J. and Meddle, S.L. (2011) Changes in the intensity of maternal aggression and central oxytocin and vasopressin V1a receptors across the peripartum period in the rat. *Journal of Neuroendocrinology* 23, 1113–1124.

Caulfield, M.P., Cambridge, H., Foster, S.F. and McGreevy, P.D. (2013) Heat stress: a major contributor to poor animal welfare associated with long-haul live export voyages. *The Veterinary Journal* 199, 223–228.

Cermak, J. (1987) The design of cubicles for British Friesian dairy cows with reference to body weight and dimensions, spatial behaviour and upper leg-lameness. In: Wierenga, H.K. and Peterse, D.J. (eds) *Cattle Housing Systems. Lameness and Behaviour. Current Topics of Veterinary Medicine Animal Science* 40. Martinus Nijhoff, Dordrecht, Netherlands, pp. 119–128.

Chacon, E. and Stobbs, T.H. (1976) Influence of progressive defoliation of a grass sward on the eating behaviour of cattle. *Australian Journal of Agricultural Research* 27, 709–727.

Chambers, J.P., Livingston, A., Waterman, A.E. and Goodship, A.E. (1992) Analgesic effects of detomidine in thoroughbred horses with chronic tendon injury. *Research in Veterinary Science* 54, 52–56.

Chamove, A.S., Rosenblum, L.A. and Harlow, H.F. (1973) Monkeys (*Macaca mulatto*) raised only with peers. A pilot study. *Animal Behaviour* 21, 316–325.

Chandroo, K.P., Duncan, I.J.H. and Moccia, R.D. (2004a) Can fish suffer? Perspectives on sentience, pain, fear and stress. *Applied Animal Behaviour Science* 86, 225–250.

Chandroo, K.P., Yue, S. and Moccia, M.D. (2004b) An evaluation of current perspectives on consciousness and pain in fishes. *Fish and Fisheries* 5, 281–295.

Charlton, G.L., Rutter, S.M., East, M. and Sinclair, L.A. (2011) Effects of providing total mixed rations indoors and on pasture of the behavior of lactating dairy cattle and their preference to be indoors or on pasture. *Journal of Dairy Science* 94, 3875–3884.

Charlton, G.L., Rutter, S.M., East, M. and Sinclair, L.A. (2012) The motivation of dairy cows for access to pasture. *Journal of Dairy Science* 96, 4387–4396.

Christian, J.J. (1955) Effects of population size on the adrenal glands and reproductive organs of male mice. *American Journal of Psychology* 182, 292–300.

Christian, J.J. (1961) Phenomena associated with population density. *Proceedings of the National Academy of Sciences, USA* 47, 428–491.

Christiansen, S.B. and Forkman, B. (2007) Assessment of animal welfare in a veterinary context – a call for ethologists. *Applied Animal Behaviour Science* 106, 203–220.

Chupin, J.-M., Savignac, C., Aupiais, A. and Lucbert, J. (2000) Influence d'un jeûne hydrique et alimentaire prolonge sur le comportement, la denutrition, la dehydration et le confort des bovins. *Rencontres Recherches Ruminants* 7, 79.

Church, J. and Williams, H. (2001) Another sniffer dog for the clinic? *Lancet* 358, 930.

Classen, H.L. (1992) Management factors in leg disorders. In: Whitehead, C.C. (ed.) *Bone Biology and Skeletal Disorders in Poultry*. Carfax Publishing Company, Abingdon, UK, pp. 195–211.

Cloutier, S. and Newberry, R.C. (2002) Differences in skeletal and ornamental traits between laying hen cannibals, victims and bystanders. *Applied Animal Behaviour Science* 77, 115–126.

Cloutier, S., Newberry, R.C., Honda, K. and Allredge, J.R. (2002) Cannibalism spread by social learning. *Animal Behaviour* 63, 1153–1162.

Clubb, R. and Mason, G. (2003) Captivity effects on wide-ranging carnivores. *Nature* 425, 473–474.

Clutton-Brock, J. (1994) The unnatural world: behavioural aspects of humans and animals in the process of domestication. In: Manning, A. and Serpell, J. (eds) *Animals and Human Society: Changing Priorities*. Routledge, London, pp. 23–35.

Clutton-Brock, J. (1999) *A Natural History of Domesticated Mammals*. Cambridge University Press, Cambridge, UK.

Clutton-Brock, T.H. (1991) *The Evaluation of Parental Care*. Princeton University Press, Princeton, New Jersey.

Cockram, M.S., Murphy, E., Ringrose, S., Wemelsfelder, F., Miedema, H.M. and Sandercock, D.A. (2012) Behavioural and physiological measures during treadmill exercise as potential indicators to evaluate fatigue in sheep. *Animal* 6, 1491–1502.

Coenen, A.M.L., Lankhaar, J., Lowe, J.C. and McKeegan, D.E.F. (2009) Remote monitoring of electroencephalogram, electrocardiogram and behavior during controlled atmosphere stunning in broilers. *Poultry Science* 88, 10–19.

Coetzee, J.F. (2013) Assessment and management of pain associated with castration in cattle. *Veterinary Clinics of North America: Food Animal Practice* 29, 75–101.

Coetzee, J.F., Gehring, R., Tarus-Sang, J. and Anderson, D.E. (2010) Effect of sub-anesthetic xylazine and ketamine ('ketamine stun') administered to calves immediately prior to castration. *Veterinary Anaesthesia and Analgesia* 37, 566–578.

Cole, D.J.A., Duckworth, J.E. and Holmes, W. (1976) Factors affecting voluntary feed intake in pigs: I. The effect of digestible energy content of the diet on the intake of castrated male pigs housed in holding pens and in metabolism crates. *Animal Production* 9, 141–148.

Coleman, G.J. and Hemsworth, P.H. (2014) Training to improve stockperson beliefs and behavior towards livestock enhances welfare and productivity. In: Mellor, D.J. and Bayvel, A.C.D. (eds) *Animal Welfare: Focusing on the Future*. OIE Scientific and Technical Review 33, 131–137. OIE, Paris.

Cooke, S.J. and Philipp, D.P. (2004) Behavior and mortality of caught-and-release bonefish (*Albula* spp.) in Bahamian waters with implications for a sustainable fishery. *Biological Conservation* 118, 599–607.

Cools, A.R., Janessen, H.J. and Broekkamp, C.L.E. (1974) The differential role of the caudate nucleus in the initiation and maintenance of morphine-induced behaviour in rats. *Archive International Pharmacodynamics* 210, 163.

Cooper, J.J. and Albentosa, M.J. (2003) Behavioural priorities of laying hens. *Avian and Poultry Biology Reviews* 143 (3), 127–149.

Cooper, J.J. and Albentosa, M.J. (2004) Social space for laying hens. In: Perry, G.C. (ed.) *The Welfare of the Laying Hen*. CAB International, Wallingford, UK, pp. 191–202.

Cooper, J.J. and Mason, G.J. (1997) The effect of cost of access on consumption of environmental resources in mink. In: Forbes, J.M., Lawrence, T.L.J., Rodway, R.G. and Varley, M.A. (eds) *Animal Choices*. British Society of Animal Science, Penicuik, UK, pp. 129–130.

Cooper, J.J. and Mason, G.J. (2000) Increasing costs of access to a resource cause rescheduling of behaviour in American mink (*Mustela vison*): implications for the assessment of behavioural priorities. *Applied Animal Behaviour Science* 66, 135–151.

Cooper, J.J., Mcafee, L. and Skinn, H. (2002) Behavioural responses of domestic ducks to nipple drinkers, bell drinkers and water troughs. *British Poultry Science* 43, S17-S18.

Cooper, J.R., Bloom, F.E. and Roth, R.H. (1982) *The Biochemical Basis of Neuropharmacology*. 4th edn. Oxford University Press, Oxford, UK.

Cooper, S.J. and Higgs, S. (1994) Neuropharmacology of appetite and taste preferences. In: Legg, C.R. and Booth, D. (eds) *Appetite: Neural and Behavioural Bases*. Oxford University Press, Oxford, UK, pp. 212–242.

Coppinger, R. and Schneider, R. (1995) Evolution of working dogs. In: Serpell, J. (ed.) *The Domestic Dog, its Evolution, Behaviour and Interactions with People.* Cambridge University Press, Cambridge, UK, pp. 21–47.

Core, S., Widowski, T., Mason, G. and Miller, S. (2009) Eye white percentage as a predictor of temperament in beef cattle. *Journal of Animal Science* 87, 2168–2171.

Corke, M.J. (1997) The welfare aspects of sheep scab. *Proceedings of the Sheep Veterinary Society* 21, 103–106.

Corke, M.J. and Broom, D.M. (1999) The behaviour of sheep with sheep scab, *Psoroptes ovis* infestation. *Veterinary Parasitology* 83, 291–300.

Council of Europe (1991) Recommendation concerning the welfare of animals kept for fur production. *Standing Committee of the European Convention for the Protection of Animals Kept for Farming Purposes (T-AP)*. Council of Europe, Strasbourg.

Council of Europe (1999) Recommendation concerning domestic ducks (*Anas platyrhynchos*). *Standing Committee of the European Convention for the Protection of Animals Kept for Farming Purposes (T-AP).* Council of Europe, Strasbourg.

Cowan, W.M. (1979) The development of the brain. *Scientific American* 241, 112–133.

Cozzi, G., Gottardo, F., Mattiello, S., Canali, E., Scanziani, E., Verga, M. and Andrighetto, I. (2002) The provision of solid feeds to veal calves: I. Growth performance, fore-stomach development, and carcass and meat quality. *Journal of Animal Science* 80, 357–366.

Craig, J.V. (1981) *Domestic Animal Behavior*. Prentice Hall, Englewood Cliffs, New Jersey.

Craig, J.V. and Guhl, A.M. (1969) Territorial behaviour and social interactions of pullets kept in large flocks. *Poultry Science* 48, 1622–1628.

Crawley, J.N. (1999) Behavioral phenotyping of transgenic and knockout mice. In: Jones, B.S. and Mormède, P. (eds) *Neurobehavioral Genetics – Methods and Applications.* CRC Press, Boca Raton, Florida, pp. 105–119.

Crawley, M.J., Albon, S.D., Bazely, D.R., Milner, J.M., Pilkington, J.G. and Tuke, A.L. (2004) Vegetation and sheep population dynamics. In: Clutton-Brock, T.H. and Pemberton, J.M. (eds) *Soay Sheep: Dynamics and Selection in an Island Population.* Cambridge University Press, Cambridge, UK.

Croney, C.C. and Newberry, R.C. (2007) Group size and cognitive processes. *Applied Animal Behaviour Science* 103, 215–228.

Cronin, G.M. (1985) The development and significance of abnormal stereotyped behaviours in tethered sows. PhD thesis, University of Wageningen, Wageningen, Netherlands.

Cronin, G.M. and Wiepkema, P.R. (1984) An analysis of stereotyped behaviour in tethered sows. *Annales de Recherches Vétérinaires* 15, 263–270.

Cronin, G.M., Wiepkema, P.R. and van Ree, J.M. (1985) Endogenous opioids are involved in abnormal stereotyped behaviours of tethered sows. *Neuropeptides* 6, 527–530.

Cronin, G.M., Barnett, J.L., Hodge, F.M., Smith J.A. and McCallum, T.H. (1991) The welfare of pigs in two farrowing/lactation environments: cortisol responses of sows. *Applied Animal Behaviour Science* 32, 117–127.

Cronin, G.M., Simpson, G.J. and Hemsworth, P.H. (1996) The effects of the gestation and farrowing environments on sow and piglet behaviour and piglet survival and growth in early lactation. *Applied Animal Behaviour Science* 46, 175–192.

Cronin, G.M., Lefébure, B. and McClintock S. (2000) A comparison of piglet production and survival in the Werribee Farrowing Pen and conventional farrowing crates at a commercial farm. *Australian Journal of Experimental Agriculture* 40, 17–23.

Crowden, A.E. and Broom, D.M. (1980) Effects of the eyefluke, *Diplostomum spathaceum*, on the behaviour of dace, *Leuciscus leuciscus*. *Animal Behaviour* 28, 287–294.

Crowell-Davis, S.L. (2001) Elimination behavior problems in cats. *Scientific Proceedings of the Animal Meeting of the American Animal Hospital Association* 68, 34–37.

Cruz, F.S., Carregaro, A.B., Machado, M. and Antonow, R.R. (2011) Sedative and cardiopulmonary effects of buprenorphine and xylazine in horses. *Canadian Journal of Veterinary Research* 75, 35–41.

Cruze, W.W. (1935) Maturation and learning in chicks. *Journal of Comparative Psychology* 19, 371–409.

Cunningham, D.K.L., Van Tienhoven, A. and de Goeijen, F. (1987) Dominance rank and cage density effects on performance traits, feeding activity and plasma corticosterone levels of laying hens (*Gallus domesticus*). *Applied Animal Behaviour Science* 17, 139–153.

Curtis, S.E. (1983) Perception of thermal comfort by farm animals. In: Baxter, S.H., Baxter, M.R. and MacCormack, J.A.C. (eds) *Farm Animal Housing and Welfare. Current Topics of Veterinary Medicine Animal Science* 24. Martinus Nijhoff, The Hague, Netherlands, pp. 59–66.

Czako, J. (1967) Gegenseitiges und Selbstsaugen der Kälber. *Wissenschaftliche Fortschritte* 5, 218.

Czanyi, V. and Doka, A. (1993) Learning interactions between prey and predator fish. *Marine Behaviour and Physiology* 23, 63–78.

Dallman, M.F. (2001) Stress and sickness decrease food intake and body weight. In: Broom, D.M. (ed.) *Coping with Challenge: Welfare in Animals Including Humans*. Dahlem University Press, Berlin, pp. 300–316.

Daly, C.C., Kallweit, E. and Ellendorf, F. (1988) Cortical function in cattle during slaughter: conventional captive bolt stunning followed by exsanguination compared with shechita slaughter. *Veterinary Record* 122, 325–329.

Damasio, A.R., Damasio, H. and Christen, Y. (eds) (1996) *Neurobiology of Decision Making*. Springer, Berlin.

Damm, B.I., Lisborg L., Vestergaard K.S. and Vanicek J. (2003) Nest-building, behavioural disturbances and heart rate in farrowing sows kept in crates and Schmid pens. *Livestock Production Science* 80, 175–187.

Damm, B.I., Moustsen V., Jørgensen E., Pedersen L.J., Heiskanen T. and Forkman B. (2006) Sow preferences for walls to lean against when lying down. *Applied Animal Behaviour Science* 99, 53–63.

Dämmrich, K. (1987) The reactions of the legs (bone; joints) to loading and its consequences for lameness. In: Wierenga, H.K. and Peterse, D.J. (eds) *Cattle Housing Systems, Lameness and Behaviour.* Current Topics in Veterinary Medicine. Animal Science. Martinus Nijhoff, Dordrecht, Netherlands, pp. 50–55.

Danbury, T.C., Weeks, C.A., Chambers, J.P., Waterman-Pearson, A.E. and Kestin, S.C. (2000) Self-selection of the analgesic drug carprofen by lame broiler chickens. *Veterinary Record* 146, 307–311.

Dannemann, K., Buchenauer, D. and Fliegner, H. (1985) The behaviour of calves under four levels of lighting. *Applied Animal Behaviour Science* 13, 243–258.

Dantzer, R. (1986) Behavioural, physiological and functional aspects of stereotyped behaviour: a review and a reinterpretation. *Journal of Animal Science* 62, 1776–1786.

Dantzer, R. (2001) Can we understand the brain and coping with considering the immune system? In: Broom, D.M. (ed.) *Coping with Challenge: Welfare in Animals Including Humans*. Dahlem University Press, Berlin, pp. 101–110.

Dantzer, R., Mormede, P., Bluthé, R.M. and Soissons, J. (1983) The effect of different housing conditions on behavioural and adrenocortical reactions in veal calves. *Reproduction Nutrition and Development* 23, 61–74.

Davis, G.S. and Siopes, T.D. (1985) The effect of light duration on turkey poult performance and adrenal function. *Poultry Science* 64, 995–1001.

Dawkins, M. (1976a) Towards an objective method of assessing welfare in domestic fowl. *Applied Animal Ethology* 2, 245–254.

Dawkins, M. (1977) Do hens suffer in battery cages? Environmental preferences and welfare. *Animal Behaviour* 25, 1034–1046.

Dawkins, M. (1981) Priorities in the cage size and flooring preferences of domestic hens. *British Poultry Science* 22, 255–263.

Dawkins, M. (1983) Battery hens name their price: consumer demand theory and the measurement of animal needs. *Animal Behaviour* 31, 1195–1205.

Dawkins, M. (1993) *Through Our Eyes Only*. Freeman, Oxford, UK.

Dawkins, M.S. (1988) Behavioural deprivation: a central problem in animal welfare. *Applied Animal Behaviour Science* 20, 209–225.

Dawkins, M.S. (1990) From an animal's point of view: motivation, fitness and animal welfare. *Behavioral and Brain Sciences* 13, 1–31.

Dawkins, M.S. and Gosling, M. (eds) (1992) *Ethics in Research on Animal Behaviour (Supplement to Animal Behaviour)*. Academic Press, London.

Dawkins, M.S. and Hardie, S. (1989) Space needs of laying hens. *British Poultry Science* 30, 413–416.

Dawkins, M.S., Donnelly, C.A. and Jones, T.A. (2004) Chicken welfare is influenced more by housing conditions then by stocking density. *Nature* 427, 342–344.

Dawkins, R. (1976b) *The Selfish Gene*. Oxford University Press, Oxford, UK.

Dawkins, R. (1979) Twelve misunderstandings of kin selection. *Zeitschrift für Tierpsychologie* 51,184–200.

Dawkins, R. (1982) *The Extended Phenotype*. W.H. Freeman, Oxford, UK.

Dawkins, R. (1986) *The Blind Watchmaker*. Longman, London.

D'Eath, R.B., Turner, S.P., Kurt, E., Evans, G., Thölking, L., Looft, H., Wimmers, K., Murani, E., Klont, R., Foury, A., Ison, S.H., Lawrence, A.B. and Mormède, P. (2010) Pigs' aggressive temperament affects pre-slaughter mixing aggression, stress and meat quality. *Animal* 4, 604–616.

De Buisonjé, F.E. (2001) *Bezettingsdichtheid bij vleeseenden*. Praktijkonderzoek Veehouderij-Pluimvee. Het Spelderholt, University of Wageningen, Wageningen, Netherlands, pp. 36–38.

de Groot, J., Ruis, M.A., Scholten, J.W., Koolhaas, J.M. and Boersma, W.J. (2001) Long term effects of social stress on anti-viral immunity in pigs. *Physiology and Behavior* 73, 143–158.

De Jonge, G. and Carlstead, K. (1987) Abnormal behaviour in farm mink. *Applied Animal Behaviour Science* 17, 375.

De Jonge, G., Carlstead, K. and Weipkema, P.R. (1986) *The Welfare of Farmed Mink*. COVP Issue No. 8. Het Spelderhold, Beekbergen, Netherlands.

de Koning, R. (1983) Results of a methodical approach with regard to external lesions of sows as an indicator of animal well being. In: Smidt, D. (ed.) *Indicators Relevant to Farm Animal Welfare. Current Topics in Veterinary Medicine and Animal Science* 23. Martinus Nijhoff, The Hague, Netherlands, pp. 155–162.

de la Fuente, J., Salazar, M.I., Ibáñez, M. and Gonzalez de Chavarri, E. (2004) Effects of season and stocking density during transport on live-weight and biochemical measurements of stress, dehydration and injury of rabbits at time of slaughter. *Animal Science* 78, 285–292.

de la Fuente, J., Diaz, M.T., Ibáñez, M. and Gonzalez de Chavarri, E. (2007) Physiological response of rabbits to heat, cold, noise and mixing in in the context of transport. *Animal Welfare* 16, 41–47.

de Vries, F.W.K., Wierenga, H.K. and Goedegebuure, S.A. (1986) Een orienterend onderzoek naar het voorkomen van afwijkingen aan het carpaalgewricht bij vleesstieren en naar het verband met de wijze van opstaan en gaan liggen. Report B- 278. Instituut voor Veetteeltkungid Onderzoek "Schoonoord", Wageninen, Netherlands, pp. 50.

de Waal, F. (1996) *Good Natured*. Harvard University Press, Cambridge, Massachusetts.

DeGrazia, D. (1996) *Taking Animals Seriously: Mental Life and Moral Status*. Cambridge University Press, New York.

Denzer, D. and Laudien, H. (1987) Stress induced biosynthesis of a 31Kd-glycoprotein in goldfish brain. *Comparative Biochemistry* 86B, 555–559.

DeSchriver, M.M. and Riddick, C.C. (1990) Effects of watching aquariums on elders' stress. *Anthrozoos* 4, 44–48.

Desforges, M.F. and Wood-Gush, D.G.M. (1975) A behavioural comparison of domestic and mallard ducks: spatial relationship in small flocks. *Animal Behaviour* 23, 698–705.

Desforges, M.F. and Wood-Gush, D.G.M. (1976) A behavioural comparison of domestic and mallard ducks: sexual behaviour. *Animal Behaviour* 24, 391–397.

Dickinson, A. (1985) Actions and habits: the development of behavioural autonomy. In: Weiskrantz, L. (ed.) *Animal Intelligence*. Clarendon Press, Oxford, UK, pp. 67–78.

Dickinson, A. and Balleine, B. (2002) The role of learning in the operation of motivational systems. In: Poshler, H. and Gallistol, R. (eds) *Stevens Handbook of Experimental Psychology*. John Wiley, New York, pp. 497–533.

Dickinson, A., Balleine, B., Watt, A., Gonzalez, F. and Boakes, R.A. (1995) Motivational control after extended instrumental training. *Animal Learning and Behavior* 23, 197–206.

Dolphinow, P.J. and Bishop, N. (1979) The development of motor skills and social relationships through play. *Minnesota Symposium on Child Psychology* 4, 141–198.

Domenici, P. and Kapoor, B.G. (2010) *Fish Locomotion: An Eco-ethological Perspective*. Science Publishers, Enfield, UK.

Domjan, M. (1998) *The Principles of Learning and Behaviour*. 4th edn. Brooks/Cole, London.

DonCarlos, M.W., Peterson, J.S. and Tilson, R.L. (1986) Captive biology of an asocial mustelid (*Mustela erminea*). *Zoo Biology* 5 (4), 363–370.

Done, S.H., Guise, J. and Chennels, D. (2003) Tail-biting and tail-docking in pigs. *The Pig Journal* 51, 136–154.

Dorries, K.M., Adkins-Regan, E. and Halpern, B.P. (1995) Olfactory sensitivity to the pheromone, androstenone, is sexually dimorphic in the pig. *Physiology and Behavior* 57, 255–259.

Dougherty, R.W. (1976) Problems associated with feeding farm livestock under intensive systems. *World Review of Nutrition and Dietetics* 25, 249–275.

Douglas, C., Bateson, M., Walsh, C., Bédué, A. and Edwards, S.A. (2012) Environmental enrichment induces optimistic cognitive biases in pigs. *Applied Animal Behaviour Science* 139, 65–73.

Doyle, R.E., Lee, C., Deiss, V., Fisher, A.D., Hinch, G.N. and Boissy, A. (2011) Measuring judgement bias and emotional reactivity in sheep following long-term exposure to unpredictable and aversive events. *Physiology and Behavior* 102, 503–510.

Drago, F., Canonico, P.L., Bitetti, R. and Scapagnini, U. (1980) Systemic and intraventricular prolactin induces excessive grooming. *European Journal of Pharmacology* 65, 457–458.

Dray, E. (1995) Inflammatory mediators of pain. *British Journal of Anaesthesiology* 75, 125–131.

Drescher, B. and Schlender-Böbbis, I. (1996) Etude pathologique de la pododermatite chez les lapins reproducteurs de souche lourde sur grillage. *World Rabbit Science* 4, 143–148.

Drewe, J.A., O'Connor, H.M., Weber, N., McDonald, R.A. and Delahay, R.J. (2013) Patterns of direct and indirect contact between cattle and badgers naturally infected with tuberculosis. *Epidemiology and Infection* 141, 1467–1475.

Dubner, R. (1994) Methods of assessing pain in animals. In: Wall, P.D. and Melzack, R. (eds) *Textbook of Pain*. 3rd edn. Churchill Livingstone, Edinburgh, UK, p. 293.

Ducreux, E., Aloui, B., Robin, P., Courboulay, V. and Meunier-Saulun, M.C. (2002) Ambient temperature influences the choice made by pigs for certain types of floor. *34èmes Journes de la Recherche Porcine, sous Légide de l'Association Francaise de Zootechnie*, Paris, France, 5–7 February 2002, pp. 211–216.

Du Mesnil du Buisson, F. and Signoret, J.P. (1962) Influences de facteur externes sur le déclenchement de la puberté chez la truie. *Annales de Zootechnie* 11, 53–59.

Duncan, I.J.H. (1978) The interpretation of preference tests in animal behaviour. *Applied Animal Ethology* 4, 197–200.

Duncan, I.J.H. (1986) Some thoughts on the stressfulness of harvesting broilers. *Applied Animal Behaviour Science* 16, 97.

Duncan, I.J.H. (1992) Measuring preferences and the strength of preferences. *Poultry Science* 71, 658–663.

Duncan, I.J.H. and Hughes, B.O. (1972) Free and operant feeding in domestic fowls. *Animal Behaviour* 20, 775–777.

Duncan, I.J.H. and Kite, V.G. (1987) Some investigations into motivation in the domestic fowl. *Applied Animal Behaviour Science* 18, 387–388.

Duncan, I.J.H. and Petherick, J.C. (1991) The implications of cognitive processes for animal welfare. *Journal of Animal Science* 69, 5017–5022.

Duncan, I.J.H. and Wood-Gush, D.G.M. (1971) Frustration and aggression in the domestic fowl. *Animal Behaviour* 19, 500–504.

Duncan, I.J.H. and Wood-Gush, D.G.M. (1972) Thwarting of feeding behaviour in the domestic fowl. *Animal Behaviour* 20, 444–451.

Duncan, I.J.H., Savory, C.J. and Wood-Gush, D.G.M. (1978) Observations on the reproductive behaviour of domestic fowl in the wild. *Applied Animal Ethology* 4, 29–42.

Duncan, I.J.H., Slee, G.S., Kettlewell, P.J., Berry, P.S. and Carlisle, A.J. (1986) Comparison of stressfulness of harvesting broiler chickens by machine and by hand. *British Poultry Science* 27, 109–114.

Duncan, I.J.H., Beatty, E.R., Hocking, P.M. and Duff, S.R.I. (1991) Assessment of pain associated with degenerative hip disorders in adult male turkeys. *Research in Veterinary Sciences* 50, 200–203.

Dunstone, N. (1983) Underwater hunting behaviour of the mink (*Mustela vison* Schreber): an analysis of constraints on foraging. *Acta Zoologici Fenneca* 174, 201–203.

Dunstone, N. (1993) *The Mink*. T & AD Poyser Ltd, London.

Dutch Society for the Protection of Animals (1986) *Alternatives for the Battery Cage System for Laying Hens*. Dutch Society for the Protection of Animals, Den Haag, p. 51.

Dwyer, C. (2009) The behaviour of sheep and goats. In Jensen, P. (ed.) *The Ethology of Domestic Animals*. 2nd edn. CAB International, Wallingford, UK, pp. 161–176.

Dwyer, C.M. and Smith, L.A. (2008) Parity effects on maternal behaviour are not related to circulating oestradiol concentrations in two breeds of sheep. *Physiology and Behavior* 93, 148–154.

Dyce, K.M., Sack, W.O. and Wensing, C.J.G. (1996) *Textbook of Veterinary Anatomy*. 2nd edn. Saunders, Philadelphia, Pennsylvania.

Dygalo, N.N. and Kalinina, T.S. (1994) Tyrosine hydroxylase activities in the brains of wild Norway rats and silver foxes selected for reduced aggressiveness towards humans. *Aggressive Behavior* 20, 453–460.

Dykes, A. (2012) Costs and benefits of fish welfare: a producer's perspective. *Aquaculture Economics and Management* 16, 429–432.

Ebedes, H., Van Rooyen, J. and Du Toit, J.G. (2002) Capturing wild animals. In: du Bothma, J. du P. (ed.) *Game Ranch Management*. Van Scheik, Pretoria, South Africa, pp. 382–440.

Eddy, R.G. and Pinsent, P.J.N. (2004) Diagnosis and differential diagnosis in the cow. In: Andrews, R.W., Boyd, H., Blowey, R.W. and Eddy, R.G. (eds) *Bovine Medicine: Diseases and Husbandry of Cattle*. 2nd edn. Blackwell, Oxford, UK, pp. 135–157.

Edgar, J.L., Lowe, J.C., Paul, E.S. and Nicol, C.J. (2011) Avian maternal response to chick distress. *Proceedings of the Royal Society B* 278, 3129–3134.

Edgar, J.L., Paul, E.S., Harris, L., Penturn, S. and Nicol, C.J. (2012) No evidence for emotional empathy in chickens observing familiar adult conspecifics. *PloS ONE* 7, e31542, pp. 6.

Edgar, J.L., Mullen, S.M., Pritchard, J.C., McFarlane, U.J.C. and Main, D.C.J. (2013) Towards a 'good life' for farm animals: development of a resource tier framework to achieve positive welfare for laying hens. *Animal* 7, 584–605.

Edmunds, M. (1974) *Defence in Animals*. Longman, Harlow, UK.

Edwards, F.W., Oldroyd, H. and Smart, J. (1939) *British Bloodsucking Flies*. British Museum, London.

Edwards, S.A. (1979) The timing of parturition in dairy cattle. *Journal of Agricultural Science, Cambridge* 93, 359–363.

Edwards, S.A. (1982) Factors affecting the time to first suckling in dairy calves. *Animal Production* 34, 339–346.

Edwards, S.A. (1983) The behaviour of dairy cows and their newborn calves in individual or group housing. *Applied Animal Ethology* 10, 191–198.

Edwards, S.A. and Broom, D.M. (1979) The period between birth and first suckling in dairy calves. *Research in Veterinary Science* 26, 255–256.

Edwards, S.A. and Broom, D.M. (1982) Behavioural interactions of dairy cows with their new-born calves and the effects of parity. *Animal Behaviour* 30, 525–535.

Edwards, S.A. and Fraser, D. (1997) Housing systems for farrowing and lactation. *The Pig Journal* 30, 77–89.

Edwards, S.A., Armsby, A.W. and Spechter, H.H. (1988) Effects of floor area allowance on performance of growing pigs kept on fully-slatted floors. *Animal Production* 46, 453–459.

EFSA (2004) The Welfare of Animals During Transport. European Food Safety Authority Scientific Panel on Animal Health and Welfare. *EFSA Journal* 44, 1–36.

EFSA (2005a) The welfare aspects of various systems of keeping laying hens. Report of the Scientific Panel on Animal Health and Welfare. *EFSA Journal* 197, 1–23.

EFSA (2005b) The impact of the current housing and husbandry systems on the health and welfare of farmed domestic rabbits. *EFSA Journal* 267, 1–31.

EFSA (2006a) The risks of poor welfare in intensive calf farming systems: an update of the scientific veterinary committee report on the welfare of calves. *EFSA Journal* 366, 1–36.

EFSA (2006b) The welfare of weaners and rearing pigs: effects of different space allowances and floor types. *Report of the European Food Safety Authority Scientific Panel on Animal Health and Welfare*. EFSA, Parma, Italy.

EFSA (2006c) Animal health and welfare risks associated with the import of wild birds other than poultry into the European Union. *EFSA Journal* 410, 1–55.

EFSA (2007a) Animal health and welfare aspects of different housing and husbandry systems for adult breeding boars, pregnant and farrowing sows and unweaned piglets. *EFSA Journal* 572, 1–13.

EFSA (2007b) The risks associated with tail biting in pigs and possible means to reduce the need for tail docking considering the different housing and husbandry systems. *EFSA Journal* 611, 1–13.

EFSA (2008) Animal welfare aspects of husbandry systems for farmed Atlantic salmon. *EFSA Journal* 736, 1–31.

EFSA (2009) Scientific opinions and report on the effects of farming systems on dairy cow welfare and disease. *EFSA Journal (Annex)* 1143, 1–38.

EFSA (2010) Scientific opinion on the practice of harvesting feathers from live geese for down production. *EFSA Journal* 8 (11), 1886, pp. 57.

EFSA (2012) Guidance on the risk assessment of food and feed from genetically modified animals and on animal health and welfare aspects. *EFSA Journal* 10, 2501, pp. 43.

Ehrensing, R.H., Michell, G.F. and Kastin, A.J. (1982) Similar antagonism of morphine analgesia by MIF-1 and naloxone in *Carassius auratus*. *Pharmacology, Biochemistry and Behaviour* 17, 757–761.

Ekesbo, I. (1981) Some aspects of sow health and housing. In: Sybesma, W. (ed.) *Welfare of Pigs*. Current Topics in Veterinary Medical Animal Science 11. Martinus Nijhoff, The Hague, Netherlands, pp. 250–266.

Ekstrand, C. and Carpenter, T.E. (1998) Temporal aspects of foot-pad dermatitis in Swedish broilers. *Acta Veterinaria Scandinavia* 39, 229–236.

Elliker, K. (2007) Social cognition and its implications for the welfare of sheep. PhD thesis, University of Cambridge, UK.

Elliker, K.R., Sommerville, B.A., Broom, D.M., Neal, D.E., Armstrong, S. and Williams, H.C. (2014) Key considerations for the experimental training and evaluation of cancer odour detection dogs: lessons learnt from a double-blind, controlled trial of prostate cancer detection. *BMC Urology* 14, 22, pp. 8.

Ellis, T., North, B., Scott, A.P., Bromage, N.R., Porter, M. and Gadd, D. (2002) The relationships between density and welfare in farmed rainbow trout. *Journal of Fish Biology* 61, 493–531.

Ellis, T., Oidtmann, B., St-Hilaire, S., Turnbull, J., North, B., MacIntyre, C., Nikolaidis, J., Hoyle, I., Kestin, S. and Knowles, T. (2008) Fin erosion in farmed fish. In: Branson, E. (ed.) *Fish Welfare*. John Wiley and Sons, Chichester, pp. 121–149.

Elofson, L., Lagerkvist, G., Gustafsson, H. and Einarsson, S. (1989) Mating systems and reproduction in mink. *Acta Agriculturae Scandinavica* 39, 23–41.

Elson, H.A. (1976) New ideas on laying cage design – the 'Get–away' cage. *Proceedings of the 5th European Poultry Conference.* WPSA, Malta, pp. 1030–1041.

Elson, H.A. (1988) Making the best cage decisions. In: *Cages for the Future. Proceedings of the Cambridge Poultry Conference*, London. Ministry of Agriculture Fisheries and Food, London, pp. 70–76.

Elson, H.A. (1990) Recent developments in laying cages designed to improve bird welfare. *World's Poultry Science Journal* 46, 34–37.

Engebretson, M. (2006) The welfare and suitability of parrots as companion animals: a review. *Animal Welfare* 15, 263–276.

Engström, B. and Schaller, G. (1993) Experimental studies of the health of laying hens in relation to housing system. In: Savory, C.J. and Hughes, B.O. (eds) *Proceedings of the 4th European Symposium on Poultry Welfare.* Universities Federation of Animal Welfare, Potters Bar, UK, pp. 87–96.

Epstein, A.N. (1983) The neuropsychology of drinking behaviour. In: Satinoff, E. and Teitelbaum, P. (eds) *Handbook on Behavioural Neurobiology*. Plenum Press, New York and London.

Esslemont, R.J., Glencross, R.G., Bryant, M.J. and Pope, G.S. (1980) A quantitative study of preovulatory behaviour in cattle (British Friesian heifers). *Applied Animal Ethology* 6, 1–17.

Esslemont, R.J. and Kossaibati, M.A. (1997) Culling in 50 dairy herds in England. *Veterinary Record* 140, 36–39.

Estevez, I. Anderson, I.-L. and Naevdal, E. (2007) Groupsize, density and social dynamics in farm animals. *Applied Animal Behaviour Science* 103, 185–204.

EU SCAHAW (1996) Report of the welfare of laying hens. EU Scientific Veterinary Committee, Animal Welfare Section, European Commission, Brussels.

EU SCAHAW (1998) Welfare aspects of the production of foie gras in ducks and geese. http://ec.europa.eu/food/animal/welfare/international/out17_en.pdf, accessed 7 November 2014.

EU SCAHAW (1999) Report on animal welfare aspects of the use of bovine somatotrophin. EU Scientific Committee on Animal Health and Animal Welfare, European Commission, Brussels.

EU SCAHAW (2001a) Report on the welfare of cattle kept for beef production. EU Scientific Committee on Animal Health and Animal Welfare, European Commission, Brussels.

EU SCAHAW (2001b) The welfare of animals kept for fur production. European Commission, Brussels.

EU SCAHAW (2002) The welfare of animals during transport (details for horses, pigs, sheep and cattle). European Commission, Brussels. http://ec.europa.eu/food/fs/sc/scah/out71_en.pdf, accessed 7 November 2014.

EU SVC (1997) The welfare of intensively kept pigs. Report of the EU Scientific Veterinary Committee. Scientific Veterinary Committee, European Commission, Brussels. http://ec.europa.eu/food/fs/sc/oldcomm4/out17_en.pdf, accessed 7 November 2014.

Evans, N.P., Robinson, J.E., Erhard, H.W., Ropstad, E., Fleming, L.M. and Haraldsen, I.R.H. (2012) Development of psychophysiological motoric reactivity is influenced by peripubertal pharmacological inhibition of gonadotropin releasing hormone action – results of an ovine model. *Psychoneuroendocrinology* 37, 1876–1884.

Ewbank, R. (1981) Alternatives: definitions and doubts. In: *Alternatives to Intensive Husbandry Systems*. Universities Federation for Animal Welfare Potters Bar, UK, pp. 5–9.

Ewing, R.D. and Ewing, S.K. (1995) Review of the effects of rearing density on the survival to adulthood for Pacific salmon. *Progressive Fish Culturist* 57, 1–25.

Fagen, R. (1981) *Animal Play Behaviour*. Oxford University Press, New York.

Fagen, R.M. (1976) Exercise, play and physical training in animals. *Perspectives in Ethology* 2, 189–219.

Fahmy, M.H. and Dufour, J.J. (1976) Effects of post-weaning stress and feeding management on return to oestrus and reproductive traits during early pregnancy in swine. *Animal Production* 23, 103–110.

FAOSTAT (2006) Statistics of the Food and Agriculture Organization of the United Nations. FAO, Rome.

Farnworth, M.J., Walker, J.K., Schweizer, K.A., Chuang, C.-L., Guild, S.-J., Barrett, C.J., Leach, M.C. and Waran, N.K. (2011) Potential behavioural indicators of post-operative pain in male laboratory rabbits following abdominal surgery. *Animal Welfare* 20, 225–237.

Faure, J.-M., Guémené, D. and Guy, G. (2001) Is there avoidance of the force feeding procedure in ducks and geese? *Animal Research* 50, 154–157.

Favre, J.Y. (1975) *Comportement d'Ovins Gardés*. Ministere de L'Agriculture École Nationale Supérieure Agronomique de Montpellier, France.

Fédération Équestre Internationale (1981) *Identification of Horses*. FEI, Paris, p. 48.

FVE (Federation of Veterinarians of Europe) (2001) Transport of Live Animals. FVE Position Paper. FVE, Brussels.

Feldman, E.C. and Nelson, R.W. (2004) *Canine and Feline Endocrinology and Reproduction*. Saunders, St Louis, Missouri.

Felton, D.L. and Felton, S.Y. (1991) Innervation of lymphoid tissue. In: Ader, R., Felten, D.L. and Cohen, N. (eds) *Psychoneuroimmunology*. 2nd edn. Academic Press, San Diego, California, pp. 27–69.

Fenske, M. Fuchs, E. and Probst, B. (1982) Corticosteroid, catecholamine and glucose plasma levels in rabbits after repeated exposure to a novel environment or administration of (1-24) ACTH or insulin. *Life Sciences* 31, 127–132.

Ferguson, J.D. (1988) Feeding for reproduction. In: *Proceedings of the Dairy Production Medical Continuing Education Group Annual Meeting*. Veterinary Learning System Co. Inc., Trenton, New Jersey, pp. 48–56.

Ferguson, M.J. and Borgh, J.A. (2004) Liking is for doing: the effects of goal pursuit on automatic evaluation. *Journal of Personality and Social Biology* 87, 557–572.

Ferguson, R.A., Kieffer, J.D. and Tuffs, B.L. (1993) The effects of body size on the acid-base and metabolite status in the white muscle of rainbow trout before and after exhaustive exercise. *Journal of Experimental Biology* 180, 195–207.

Ferlazzo, A. (1995) Animali da reddito: indicatori di benessere e 'linee guida' per la loro applicazione nel trasporto. *In: Conference 'Gli Indicatori Scientifici del Benessere Animale',* Brescia, Italy, 9 May 1995.

Ferrante, V., Canali, E., Mattiello, S., Verga, M., Sacerdote, P. and Panerai, A.E. (1998) The effect of the size of individual crates on the behavioural and immune reactions of dairy calves. *Journal of Animal Feed Science* 7, 29–36.

Filion, L.G., Willson, P.J., Bielefeldt-Ohmann, M.A. and Thomson R.G. (1984) The possible role of stress in the induction of pneumonic pasteurellosis.

Canadian Journal of Comparative Medicine 48, 268–274.

Finzi, A., Margarit, R. and Calabrese, A. (1996) A two-floor cage for rabbit welfare. In: *Proceedings of the 6th World Rabbit Congress*, Toulouse, France, 9–12 July 1996, pp. 423–424.

Fitzgerald, B.M. and Turner, D.C. (2000) Hunting behaviour of domestic cats and their impact on prey populations. In: Turner, D.C. and Bateson, P. (eds) *The Domestic Cat: the Biology of its Behaviour*. 2nd edn. Cambridge University Press, Cambridge, UK, pp. 153–175.

Fitzsimons, J.T. (1979) The physiology of thirst and sodium appetite. *Monographs of the Physiological Society* 35. Cambridge University Press, Cambridge, UK.

Flecknell, P. (2001) Recognition and assessment of pain in animals. In: Soulsby, Lord and Norton, D. (eds) *Pain: its Nature and Management in Man and Animals*. International Congress and Symposium Series 246, 63–68.

Fleming, R.H., Whitehead, C.C., Alvey, D., Gregory, N.G. and Wilkins, L.J. (1994) Bone structure and breaking strength in laying hens housed in different husbandry systems. *British Poultry Science* 35, 651–662.

Flower, F.C. and Weary, D.M. (2003) The effects of early separation on the dairy cow and calf. *Animal Welfare* 12, 339–348.

Fölsch, D.W. (1981) Das verhalten von legehennen in unterschiedlichen haltungs systemen unter berucksichtigung der aufsuchtmethoden. In: Fölsch, D.W. and Vestergaard, K. (eds) *Das Verhalten von Huhnern*. Tierhaltung 12. Birkhauser Verlag, Basel, Switzerland, pp. 9–114.

Fölsch, D.W. and Huber. A. (1977) Bewegungsaktivität und lautausserungen in tagesrhythmus. *KTBL Schrift* 223, 99–114.

Fölsch, D.W. and Vestergaard, K. (1981) *The Behaviour of Fowl*. Birkhauser Verlag, Basel, Switzerland.

Fölsch, D.W., Dolf, C., Ehrbar, H., Bleuler, T. and Teygeler, H. (1983) Ethologic and economic examination of aviary housing for commercial laying flocks. *International Journal of Studies of Animal Problems* 4, 330–335.

Foote, R.H. (1978) Reproductive performance and problems in New York dairy herds. *Search Agriculture (Geneva, New York)* 8, 1.

Forkman, B.A. (2001) Domestic hens have declarative representations. *Animal Cognition* 3, 135–137.

Forkman, B.A. (2002) Learning and Cognition. In: Jensen, P. (ed.) *The Ethology of Domestic Animals*. CAB International, Wallingford, UK, pp. 51–64.

Forkman, B., Blokhuis, H.J., Broom, D.M., Kaiser, S., Koolhaas, J.M., Levine, S., Mendl, M., Plotsky, P.M. and Schedlowski, M. (2001) Key sources of variability in coping. In: Broom, D.M. (ed.) *Coping with Challenge: Welfare in Animals Including Humans*. Dahlem University Press, Berlin, pp. 249–270.

Forkman, B., Boissy, A., Meunier-Salaün, M.C., Canali, E. and Jones, R.B. (2007) A critical review of fear tests used on cattle, pigs, sheep, poultry and horses. *Physiology and Behavior* 92, 340–374.

Forrester, R.C. (1979) Behavioural state and responsiveness in domestic chicks. PhD thesis, University of Reading, UK.

Forrester, R.C. (1980) Stereotypies and the behavioural regulation of motivational state. *Applied Animal Ethology* 6, 386–387.

Foulkes, N.S., Duval, G. and Sassone-Corsi, P. (1996) Adaptive inducibility of CREM and transcriptional memory of biological rhythms. *Nature* 381, 83–85.

Fraser, A.F. (1968) *Reproductive Behaviour in Ungulates*. Academic Press, London.

Fraser, A.F. (1970) Some observations on equine oestrus. *British Veterinary Journal* 126, 656–657.

Fraser, A.F. (1974a) The behaviour of growing pigs during experimental social encounters. 7. *Agricultural Science* 82, 147–163.

Fraser, A.F. (1978a) Eine generalle überprüfung des Sexualverhaltens bei Nutztieren [A general review of sexual behaviour in livestock] *Proceedings of the 1st World Congress on Ethology Applied to Zootechnics* 1, 507–512.

Fraser, A.F. (1978b) Tests in applied ethology. *Applied Animal Ethology* 4, 1–4.

Fraser, A.F. (1980a) The ontogeny of behaviour in the foal. *Applied Animal Ethology* 6, 303.

Fraser, A.F. (1980b) The appraisal of vital behaviour in the neonate foal. In: *Proceedings of the International Congress on Animal Production,* Madrid, pp. 617–620.

Fraser, A.F. (1989) Pandiculation: the comparative phenomenon of systematic stretching. *Applied Animal Behaviour Science* 23, 263–268.

Fraser, A.F. (1992) *The Behaviour of the Horse*. CAB International, Wallingford, UK.

Fraser, A.F. (2003) *Humane Horse Care*. Canadian Farm Animal Trust, Barrie, Ontario, Canada.

Fraser, A.F. (2010) *The Behaviour and Welfare of the Horse*. 2nd edn. CAB International, Wallingford, UK.

Fraser, A.F. and Brownlee, A. (1974) Veterinary ethology and grass sickness in horses. *Veterinary Science* 95, 448.

Fraser, D. (1974b) The vocalisation and other behaviour of growing pigs in an 'open field' test. *Applied Animal Ethology* 1, 3–16.

Fraser, D. (1975) The effect of straw on the behaviour of sows in tether stalls. *Animal Production* 21, 59–68.

Fraser, D. (1999) Animal ethics and animal welfare science: bridging the two cultures. *Applied Animal Behaviour Science* 65, 171–189.

Fraser, D. (2008) *Understanding Animal Welfare: the Science in its Cultural Context*. Wiley Blackwell, Chichester, UK.

Fraser, D. and Matthews, L.R. (1997) Preference and motivation testing. In: Appleby, M.C. and Hughes, B.O. (eds) *Animal Welfare*. CAB International, Wallingford, UK, pp. 159–173.

Fraser, D., Thompson, B.K., Ferguson, D.K. and Darroch, R.L. (1979) The 'teat order' of suckling pigs. 3. Relation to competition within litters. *Journal of Agricultural Science Cambridge* 92, 257–261.

Fraser, D., Phillips, P.A. and Thompson, B.K. (1988) Initial test of a farrowing crate with inward-sloping sides. *Livestock Production Science* 20, 249–256.

Fraser, D., Weary, D.M., Pajor, E.A. and Milligan, B.N. (1997) The scientific conception of animal welfare that reflects ethical concerns. *Animal Welfare* 6, 187–205.

Fraser, D., Duncan, I.J.H., Edwards, S.A., Grandin, T., Gregory, N.G., Guyonnet, V., Hemsworth, P.H., Huertas, S.M., Huzzey, J.M., Mellor, D.J., Mench, J.A., Spinka, M. and Whay, H.R. (2013) General principles for the welfare of animals in production systems: the underlying science and its application. *The Veterinary Journal* 198, 19–27.

Freeland, W.J. and Janzen, D.H. (1974) Strategies in herbivory by mammals: the role of plant secondary compounds. *American Naturalist* 108, 269–289.

Freeman, B.M., Kettlewell, P.J., Manning, A.G.C. and Berry, P.S. (1984) Stress of transportation for broilers. *Veterinary Science* 114, 286–287.

Freire, R., Appleby, M.C. and Hughes, B.O. (1996) Effects of nest quality and other cues for exploration on pre-laying behaviour. *Applied Animal Behaviour Science* 48, 37–46.

Friedmann, E., Son, H. and Tsai, C.-C. (2010) The animals/human bond: health and wellness. In Fine, A.H. (ed.) *Handbook on Animal-assisted Therapy, Theoretical Foundations and Guideline for Practice*. 3rd edn. Academic Press, London, pp. 84–107.

Friend, T.H. (2000) Dehydration, stress, and water consumption of horses during long-distance commercial transport. *Journal of Animal Science* 78, 2568–2580.

Friend, T.H. and Dellmeier, G.R. (1988) Common practices and problems related to artificially rearing calves: an ethological analysis. *Applied Animal Behaviour Science* 20, 47–62.

Friend, T.H., Dellmeier, G.R. and Gbur, E.E. (1985) Comparison of four methods of calf confinement. I. Physiology. *Journal of Animal Science* 60, 1095–1101.

Fröhlich, E.K.F. (1993) *Bericht Über Die Praktische Prüfung Der Schräggitterhaltungssysteme Auf Tiergerechtheit*. Bundesamt für Veterinärwesen (BVET), Liebefeld, Germany, pp. 1–45.

Fureix, C., Jego, P., Henry, S., Lansade, L. and Hausberger, M. (2012) Towards an ethological animal model of depression? A study on horses. *PloS ONE* 7 (6), e39280, pp. 9.

Gácsi, M., Győri, B., Miklósi, Á., Virányi, Z., Kubinyi, E., Topál, J. and Csányi, V. (2005) Species-specific differences and similarities in the behavior of hand-raised dog and wolf pups in social situations with humans. *Developmental Psychobiology* 47, 111–122.

Gaillard, C., Meagher, R.K., von Keyserlingk, M.A. and Weary, D.M. (2014) Social housing improves dairy calves' performance in two cognitive tests. *PloS ONE* 9, e90205, pp. 6.

Gaillard, R.-C. and Al-Damluji, S. (1987) Stress and the pituitary–adrenal axis. *Ballière's Clinical Endocrinology and Metabolism* 1, 319–354.

Galatos, A.D. (2011) Anesthesia and analgesia in sheep and goats. *Veterinary Clinics of North America: Food Animal Practice* 27, 47–59.

Galef Jr, B.G. (1996) Social influences on food preferences and feeding behaviors of verterbrates. In:

Capaldi, E.D. (ed.) *Why We Eat What We Eat: the Psychology of Eating*. American Psychological Association, Washington, DC, pp. 207–231.

Galindo, F., Broom, D.M. and Jackson, P.G.G. (2000) A note on possible link between behaviour and the occurrence of lameness in dairy cows. *Applied Animal Behaviour Science* 67, 335–341.

Garcia, E., Hultgren, J., Fällman, P., Geust, J., Algers, B., Stilwell, G., Gunnersson, S. and Rodriguez-Martinez, H. (2011) Oestrous intensity is positively associated with reproductive outcome in high-producing dairy cows. *Livestock Science* 139, 191–195.

Garcia, J., Ervin, F.R. and Koelling, R.A. (1966) Learning with prolonged delay of reinforcement. *Psychonomic Science* 5, 121–122.

Garcia, J., Ervin, F.R., York, C.H. and Koelling, R.A. (1967) Conditioning with delayed vitamin injection. *Science, New York* 155, 716–718.

Gatica, M.C., Monti, G.E., Knowles, T.G. and Gallo, C.B. (2010a) Muscle pH, rigor mortis and blood variables in Atlantic salmon transported in two types of well-boat. *Veterinary Record* 166, 45–50.

Gatica, M.C., Monti, G.E., Knowles, T.G., Warriss, P.D. and Gallo, C.B. (2010b) Effects of commercial live transportation and preslaughter handling of Atlantic salmon on blood constituents. *Archives Medicina Veterinaria* 42, 73–78.

Geary, N. (1994) Glucagon and control of meal size. In: Legg, C.R. and Booth, D. (eds) *Appetite: Neural and Behavioural Bases*. Oxford University Press, Oxford, UK, pp. 164–197.

Gentle, M.J. (1986) Neuroma formation following partial beak amputation (beak-trimming) in the chicken. *Research in Veterinary Science* 41, 383–385.

Gibbs, D.M. (1986a) Dissociation of oxytocin, vasopressin and corticotropin secretion during different types of stress. *Life Science* 35, 487–491.

Gibbs, D.M. (1986b) Vasopressin and oxytocin: hypothalamic modulators of the stress response: a review. *Psychoneuroendocrinology* 11, 131–140.

Gibson, S.W., Innes, J. and Hughes, B.O. (1985) Aggregation behaviour of laying fowls in a covered strawyard. In: Wegner, R.M. (ed.) *Second European Symposium on Poultry Welfare*, Celle, Germany. World Poultry Science Association, Beekbergen, Netherlands, pp. 296–298.

Gibson, T.J., Johnson, C.B., Stafford, K.J., Mitchinson, S.L. and Mellor, D.J. (2007) Validation of the acute electroencephalographic responses of calves to noxious stimulus with scoop dehorning. *New Zealand Veterinary Journal* 55, 152–157.

Gibson, T.J., Johnson, C.B., Murrell, J.C., Hulls, C.M., Mitchinson, S.L., Stafford, K.J., Johnstone, A.C. and Mellor, D.J. (2009a) Electroencephalographic responses of halothane-anaesthetised calves to slaughter by ventral-neck incision without prior stunning. *New Zealand Veterinary Journal* 57, 77–83.

Gibson, T.J., Johnson, C.B., Murrell, J.C., Chambers, J.P., Stafford, K.J. and Mellor, D.J. (2009b) Components of electroencephalographic responses to slaughter in halothane-anaesthetised calves: effects of cutting neck tissues compared with major blood vessels. *New Zealand Veterinary Journal* 57, 84–89.

Gibson, T.J., Johnson, C.B., Murrell, J.C., Mitchinson, S.L., Stafford, K.J. and Mellor, D.J. (2009c) Electroencephalographic responses to concussive non-penetrative captive-bolt stunning in halothane-anaesthetised calves. *New Zealand Veterinary Journal* 57, 90–95.

Gibson, T.J., Johnson, C.B., Murrell, J.C., Mitchinson, S.L., Stafford, K.J. and Mellor, D.J. (2009d) Amelioration of electroencephalographic responses to slaughter by non-penetrative captive-bolt stunning after ventral neck incision in halothane-anaesthetised calves. *New Zealand Veterinary Journal* 57, 96–101.

Gillis, S., Crabtree, G.R. and Smith, K.A. (1979) Glucocorticoid-induced inhibition of T cell growth factor. I. The effect on mitogen-induced lymphocyte proliferation. *Journal of Immunology* 123, 1624–1631.

Giraldeau, L.-A. and Caraco, T. (2000) *Social Foraging Theory*. Princeton University Press, Princeton, New Jersey.

Gloor, P. and Dolf, G. (1985) *Galtsauenhaltung Einzeln Oder In Gruppen?* Schrift der Eidgenossischen Forschung für Betriebswirtschaft und Landtechnik FAT, Tanikon, Switzerland.

Gluckman, P.D., Hanson, M.A., Spencer, H.G. and Bateson, P. (2005) Environmental influences during development and their later consequences for health and disease: implications for the interpretation of empirical studies. *Proceedings of the Royal Society B: Biological Sciences* 272, 671–677.

Goddard, M.E. and Beilharz, R.G. (1982) Genetic and environmental factors affecting the suitability of dogs as guide dogs for the blind. *Theoretical and Applied Genetics* 62, 97–102.

Gonyou, H.W. and Stricklin, W.R. (1981) Eating behaviour of beef cattle groups fed from a single stall or trough. *Applied Animal Ethology* 7, 123–133.

Gonyou, H.W. and Stricklin, W.R. (1998) Effects of floor area allowance and group size on the productivity of growing/finishing pigs. *Journal of Animal Science* 76, 1326–1330.

Gonyou, H.W., Christopherson, R.G. and Young, B.A. (1979) Effects of cold temperature and winter condition on some aspects of behaviour of feedlot cattle. *Applied Animal Ethology* 5, 113–124.

Gonyou, H.W., Hemsworth, P.H. and Barnett, J.L. (1986) Effects of frequent interactions with humans on growing pigs. *Applied Animal Behaviour Science* 16, 269–278.

Gonyou, H.W., Brumm, M.C., Bush, E., Deenm, J., Edwards, S.A., Fangman, T., McGlone, J.J., Meunier-Salaun, M., Morrison, R.B., Spoolder, H., Sundberg, P.L. and Johnson, A.K. (2004) Application of broken line analysis to assess floor space requirement of nursery and grow/finish pigs expressed on an allometric basis. *Journal of Animal Science* 84, 229–235.

González, L.A., Tolkamp, B.J., Coffey, M.P., Ferret, A. and Kyriazakis, I. (2008) Changes in feeding behavior as possible indicators for the automatic monitoring of health disorders in dairy cows. *Journal of Dairy Science* 91, 1017–1028.

Goodship, A.E. and Birch, H.L. (2001) Exercise effects on the skeletal issues. In: Back, W. and Clayton, H.M. (eds) *Equine Locomotion*. W.B. Saunders, London, pp. 227–250.

Goodyer, I.M. (ed.) (2001) *The Depressed Child and Adolescent*. 2nd edn. Cambridge University Press, Cambridge, UK.

Gordon, R.T., Schatz, C.B. and Myers, L.J. (2008) The use of canines in the detection of human cancers. *Journal of Alternative and Complementary Medicine* 14, 61–67.

Gordon, S.H. and Thorp, B.H. (1994) Effect of light intensity on broiler liveweight and tibial plateau angle. *Proceedings of the 9th European Poultry Conference*. World Poultry Science Association, Glasgow, UK, pp. 286–287.

Gotz, M. and Troxler, J. (1993) Farrowing and nursing in the group. In: Collins, E. and Boon, C. (eds) *Livestock Environment IV*. American Society of Agricultural Engineers, St Joseph, Michigan, pp. 159–166.

Gover, N., Jarvis, J.R., Abeyesinghe, S.M. and Wathes, C.M. (2009) Stimulus luminance and the spatial acuity of domestic fowl (*Gallus g. domesticus*). *Vision Research* 49, 2747–2753.

Graf, B.P. (1984) Der einfluss unterschiedlicher laufstall systeme auf verhaltensmerkmale von mastochsen. Doktor dissertation der Eidgenossischen Technischen Hoschschule, Zurich, Switzerland.

Grafen, A. (1984) Natural selection, group selection and kin selection. In: Krebs, J.R. and Davies, N.B. (eds) *Behavioural Ecology*. 2nd edn. Blackwell, Oxford, UK, pp. 62–84.

Grandin, T. (1978) Design of lairage, yard and race systems for handling cattle in abattoirs, auctions, ranches, restraining chutes and dipping vats. In: *1st World Congress on Ethology Applied to Zootechnics*, Madrid. Editorial Garsi, Madrid, pp. 37–52.

Grandin, T. (1980) Observations of cattle behaviour applied to the design of cattle-handling facilities. *Applied Animal Ethology* 6, 19–31.

Grandin, T. (1982) Pig behaviour studies applied to slaughter-plant design. *Applied Animal Ethology* 9, 141–151.

Grandin, T. (2000) *Livestock Handling and Transport*. 2nd edn. CAB International, Wallingford, UK, pp. 43–61.

Grandin, T. (ed.) (2014) *Livestock Handling and Transport*. 4th edn. CAB International, Wallingford, UK.

Green, L.E., Lewis, K., Kimpton, A. and Nicol, C.J. (2000) Cross-sectional study of the prevalence of feather pecking in laying hens in alternative systems and its associations with management and disease. *Veterinary Record* 147, 233–238.

Green, P. and Tong, J.M.J. (1988) Small intestinal obstruction associated with woodchewing in horses. *Veterinary Record* 123, 196–198.

Green, T.C. and Mellor, D.J. (2011) Extending ideas about animal welfare assessment to include 'quality of life' and related concepts. *New Zealand Veterinary Journal* 59, 263–271.

Greenough, P.R. and Weaver, A.D. (1996) *Lameness in Cattle*. 3rd edn. W.B. Saunders, Philadelphia, Pennsylvania.

Greenough, P.R., Weaver, A.D., Broom, D.M., Esslemont, R.J. and Galindo, F.A. (eds) (1997) Basic concepts of bovine lameness. In: Greenough, P.R. and Weaver, A.D. (eds) *Lameness in Cattle*. 3rd edn. W.B. Saunders, Philadelphia, Pennsylvania.

Gregory, N.G. (1998) Physiological mechanisms causing sickness behaviour and suffering in diseased animals. *Animal Welfare* 7, 293–305.

Gregory, N.G. (2004) Sickness and disease. In: *Physiology and Behaviour of Animal Suffering*. Blackwell, Oxford, UK.

Gregory, N.G. (2007) *Animal Welfare and Meat Production*. 2nd edn. CAB International, Wallingford, UK.

Gregory, N.G. and Wilkins, L.J. (1989) Broken bones in domestic fowl: handling and processing damage in end of lay battery hens. *British Poultry Science* 30, 555–562.

Gregory, N.G. and Wilkins, L.J. (1992) Skeletal damage and bone defects during catching and processing. In: *Bone Biology and Skeletal Disorders in Poultry. 23rd Poultry Science Symposium*. World's Poultry Science Association, Edinburgh, UK.

Gregory, N.G. and Wilkins, L.J. (1996) Effect of age on bone strength and the prevalence of broken bones in perchery laying hens. *New Zealand Veterinary Journal* 44, 31–32.

Gregory, N.G., Wilkins, L.J., Eleperuma, S.D., Ballantyne, A.J. and Overfield, N.D. (1990) Broken bones in domestic fowls: effect of husbandry system and sunning method on end-of-lay hens. *British Poultry Science* 31, 59–69.

Gregory, N.G., Wilkins, L.J., Kestin, S.C., Belyavin, C.G. and Alvey, D.M. (1991) Effect of husbandry system on broken bones and bone strength in hens. *Veterinary Record* 128, 397–399.

Gregory, N.G., von Wenzlawowicz, M., Alam, R.M., Anil, H.M., Yeşildere, T. and Silva-Fletcher, A. (2008) False aneurysms in carotid arteries of cattle and water buffalo during shechita and halal slaughter. *Meat Science* 79, 285–288.

Greif, G. (1982) Basic investigations into the construction of concrete slatted floors for fattening pigs with special regards to changes of the limbs due to body carriage. PhD thesis, University of Giessen, Germany.

Griffin, J.F.T. (1989) Stress and immunity: a unifying concept. *Veterinary Immunology and Immunopathology* 20, 263–312.

Gross, W.B. (1962) Blood cultures, blood counts and temperature records in an experimentally produced 'air sac disease' and uncomplicated *Escherichia coli* infection of chickens. *Poultry Science* 41, 691–700.

Gross, W.B. and Colmano, G. (1965) The effect of social isolation on resistance to some infectious diseases. *Poultry Science* 48, 515–520.

Gross, W.B. and Colmano, G. (1969) The effect of social isolation on resistance to some infectious diseases. *Poultry Science* 48, 514–520.

Gross, W.B. and Siegel, P.B. (1965) The effect of social stress on resistance to infection with *Escherichia coli* or *Mycoplasma gallisepticum*. *Poultry Science* 44, 998–1001.

Gross, W.B. and Siegel, P.B. (1975) Immune response to *Escherichia coli*. *American Journal of Veterinary Research* 36, 568–571.

Gross, W.B. and Siegel, P.B. (1981) Long-term exposure of chickens to three levels of social stress. *Avian Diseases* 25, 312–325.

Groth, W. (1978) Tierschutz und verhaltensbezogene gesichtspunkte der kalbermast. *Tierzuchter* 10, 419–422.

Guémené, D., Guy, G., Noirault, J., Garreau-Mills, M., Gouraud, M. and Faure, J.-M. (2001) Force-feeding procedure and physiological indicators of stress in male mule ducks. *British Poultry Science* 42, 650–657.

Guhl, A.M. (1968) Social inertia and social stability in chickens. *Animal Behaviour* 16, 219–232.

Guillemin, R., Vargo, T., Rossier, J., Minick, S., Ling, N., Rivier, C., Vale, W. and Bloom, F. (1977) β-endorphin and adrenocorticotropin are secreted concomitantly by the pituitary gland. *Science, New York* 197, 1367–1369.

Guise, H.J. and Penny, R.H.C. (1989) Factors influencing the welfare and carcass and meat quality of pigs 1. The effects of stocking in transport and the use of electric goads. *Animal Production* 49, 511–515.

Gunn, D. and Morton, D.B. (1994) The behaviour of single-caged and group-housed laboratory rabbits. In: *Proceedings of the 5th FELASA Symposium: Welfare and Science* 1993, pp. 80–84.

Gunn, D. and Morton, D.B. (1995) Inventory of the behaviour of New Zealand white rabbits in laboratory cages. *Applied Animal Behaviour Science* 45, 277–292.

Gunnarsson, S., Keeling, L.J. and Svedberg, J. (1999) Effect of rearing factors on the prevalence of floor eggs, cloacal cannibalism and feather pecking in commercial flocks of loose housed laying hens. *British Poultry Science* 40, 12–18.

Guy, J.H., Rowlinson, P., Chadwick, J.P. and Ellis, M. (2002) Behaviour of two genotypes of growing–finishing pig in three different housing systems. *Applied Animal Behaviour Science* 75, 193–206.

Gwinner, E. (1996) Circannual clocks in avian preproduction and migration. *Ibis* 138, 47–63.

Gygax, L., Neisen, G. and Wechsler, B. (2009) Differences between single and paired heifers in residency in functional areas, length of travel path, and area used throughout days 1–6 after integration into a free stall dairy herd. *Applied Animal Behaviour Science* 120, 49–55.

Gygax, L., Reefmann, N., Wolf, W. and Langbein, J. (2013) Pre-frontal cortex activity, sympatho-vagal reaction and behaviour distinguish between situations of feed reward and frustration in dwarf goats. *Behavioural Brain Research* 239, 104–114.

Hagen, K. and Broom, D.M. (2003) Cattle discrimination between familiar herd members in a learning experiment. *Applied Animal Behaviour Science* 82, 13–28.

Hagen, K. and Broom, D.M. (2004) Emotional reactions to learning in cattle. *Applied Animal Behaviour Science* 85, 203–213.

Håkansson, J., Brett, C. and Jensen, P. (2007) Behavioural differences between two captive populations of red jungle fowl (*Gallus gallus*) with different genetic background raised under identical conditions. *Applied Animal Behaviour Science* 102, 24–38.

Hall, S.J.G. (1983) Grazing behavior of Chillingham cattle. *Applied Animal Ethology* 11, 71.

Hall, S.J.G. (1989) Chillingham cattle: social and maintenance behaviour in an ungulate that breeds all year round. *Animal Behaviour* 39, 215–225.

Hall, S.J.G. and Bradshaw, R.H. (1998) Welfare aspects of transport by road of sheep and pigs. *Journal of Applied Animal Welfare Science* 1, 235–254.

Hall, S.J.G. and Clutton-Brock, J. (1995) *Two Hundred Years of British Farm Livestock*. Stationery Office Books, London.

Hall, S.J.G., Schmidt, B. and Broom, D.M. (1997) Feeding behaviour and the intake of food and water by sheep after a period of deprivation lasting 14h. *Animal Science* 64, 105–110.

Hall, S.J.G., Kirkpatrick, S.M. and Broom, D.M. (1998a) Behavioural and physiological responses of sheep of different breeds to supplementary feeding, social mixing and taming, in the context of transport. *Animal Science* 67, 475–483.

Hall, S.J.G., Kirkpatrick, S.M., Lloyd, D.M. and Broom, D.M. (1998b) Noise and vehicular motion as potential stressors during the transport of sheep. *Animal Science* 67, 467–473.

Hall, S.L. (1998) Object play by adult animals. In: Bekoff, M. and Byers, J.A. (eds) *Animal Play: Evolutionary, Comparative and Ecological Perspectives*. Cambridge University Press, Cambridge, UK, pp. 45–60.

Hallerman, E.M., McLean, E. and Fleming, I.A. (2007) Effects of growth hormone transgenes on the behavior and welfare of aquacultured fishes: A review identifying research needs. *Applied Animal Behaviour Science* 104, 265–294.

Hamilton, W.D. (1964a) The genetical evolution of social behaviour I. *Journal of Theoretical Biology* 7, 1–16.

Hamilton, W.D. (1964b) The genetical evolution of social behaviour II. *Journal of Theoretical Biology* 7, 17–32.

Hamilton, W.D. and Zuk, M. (1984) Heritable true fitness and bright birds: a role for parasites? *Science, New York* 218, 384–387.

Hammell, K.L., Metz, J.H.M. and Mekking, P. (1988) Sucking behaviour of dairy calves fed milk ad libitum by bucket or teat. *Applied Animal Behaviour Science* 20, 275–285.

Hänninen, L., Rushen, J. and De Passille, A.M. (2005) The effect of flooring type and social grouping on the rest and growth of dairy calves. *Applied Animal Behaviour Science* 91, 193–204.

Hansen, I., Christiansen, F., Hansen, H.S., Braastad, B. and Bakken, M. (2001) Variation in behavioural responses of ewes towards predator-related stimuli. *Applied Animal Behaviour Science* 70, 227–237.

Hansen, L.L. and Vestergaard, K. (1984) Tethered versus loose sows: ethological observations and measures of productivity. *Annales de Recherches Vétérinaires* 15, 245–256.

Hansen, R.S. (1976) Nervousness in hysteria of mature female chickens. *Poultry Science* 55, 531–543.

Hansen, S.W. (1996) Selection for behavioural traits in farm mink. *Applied Animal Behaviour Science* 49 (2), 137–148.

Hansen, S.W. (1998) The cage environment of the farm mink – significance to welfare. *Scientifur* 22 (3), 179–185.

Hansen, S.W., Malmkvist, J., Palma, R. and Damgaard, B.M. (2007) Do double cages and access to occupational materials improve the welfare of farmed milk? *Animal Welfare* 16, 64–76.

Harborne, J.B. (1982) *Introduction to Ecological Biochemistry*. 2nd edn. Academic Press, London.

Harding, E.J., Paul, E.S. and Mendl, M. (2004) Cognitive bias and affective state. *Nature* 427, 312.

Harlow, H.F. (1969) Age-mate or peer affectional system. *Advances in the Study of Behavior* 2, 333–383.

Harlow, H.F. and Harlow, M.K. (1965) The affectional systems. In: Schrier, A.M., Harlow, H.F. and Stollnitz, F. (eds) *Behavior of Nonhuman Primates*. Vol. 2. Academic Press, New York.

Harri, M., Mononen, J., Kasanen, S. and Ahola, L. (1995) Choice between floor type and floor level in farmed silver foxes. In: Rutter, S.M., Rushen, J., Randle, H.D. and Eddison, J.C. (eds) *Proceedings of the 29th International Congress of the International Society for Applied Ethology*, Exeter, UK, 3–5 August. Universities Federation for Animal Welfare, Potters Bar, UK, pp. 171–172.

Harri, M., Plyusnina, I., Ahola, L., Mononen, J. and Rekila, T. (1997) Accelerated domestication in silver foxes using artificial selection. In: Hemsworth, P.H., Spinka, M. and Kost'ál, L. (eds) *Proceedings of the 31st International Congress of the ISAE (International Society for Applied Ethology)*, Prague, Czech Republic, 13–16 August. Research Institute of Animal Production, Prague, p. 73.

Harris, J.A., Hillerton, J.E. and Morant, S.V. (1987) Effect on milk production of controlling muscid flies, and reducing fly-avoidance behaviour, by the use of Fenvalerate ear tags during the dry period. *Journal of Dairy Research* 54, 165–171.

Harrison, R. (1964) *Animal Machines*. London: Vincent Stuart. Reprinted with commentaries 2013: CAB International, Wallingford, UK.

Harst, J.E. van der, Fermont, P.C.J., Bilstra, A.E. and Spruijt, B.M. (2003a) Access to enriched housing is rewarding to rats as reflected by their anticipatory behaviour. *Animal Behaviour* 66, 493–504.

Harst, J.E. van der, Baars, A.M. and Spruijt, B.M. (2003b) Standard housed rats are more sensitive to rewards than enriched housed rats as reflected by their anticipatory behaviour. *Behavioural Brain Research* 142, 151–156.

Hart, B.L. (1985) *Behavior of Domestic Animals*. W.H. Freeman and Co., New York.

Hart, B.L. (1988) Biological bases of the behaviour of sick animals. *Neuroscience Biobehavioural Review* 12, 123–137.

Hart, B.L. (1990) Behavioural adaptations to pathogens and parasites: five strategies. *Neuroscience Biobehavioural Review* 14, 273–294.

Hart, B.L. (2010) Beyond fever: comparative perspectives on sickness behavior. In: Breed, M.D. and Moore, J. (eds) *Encyclopedia of Animal Behavior*. Vol. 1. Academic Press, Oxford, UK, pp. 205–210.

Hart, B.L. (2011) Behavioural defences in animals against pathogens and parasites: parallels with the pillars of medicine in humans. *Philosophical Transactions of the Royal Society, B* 366, 3406–3417.

Hartsock, T.G. and Graves, H.B. (1976) Neonatal behaviour and nutrition related mortality in domestic swine. *Journal of Animal Science* 42, 235–241.

Haskell, M., Coerse, N.C.A. and Forkman, B. (2000) Frustration-induced aggression in the domestic hen: the effect of thwarting access to food and water on aggressive responses and subsequent approach tendencies. *Behaviour* 137, 531–546.

Havenstein, G.B., Ferket, P.R., Scheideler, S.E. and Larson, B.T. (1994) Growth, livability, and feed conversion of 1957 vs 1991 broilers when fed "typical" 1957 and 1991 broiler diets. *Poultry Science* 73, 1785–1794.

Hawkins, D. (2005) *Biomeasurement: Understanding, Analysing and Communicating Data in the Biosciences*. Oxford University Press, Oxford, UK.

Hay, M., Vulin, A., Genin, S., Sales, P. and Prunier, A. (2003) Assessment of pain induced by castration in piglets: behavioural and physiological responses over the subsequent 5 days. *Applied Animal Behaviour Science* 82, 201–218.

Haynes, L.W. and Timms, R.J. (1987) Stress-induced release of pituitary β-endorphin may be mediated by activation of the brain-stem defence areas. *International Journal of Tissue Reactions* 9, 55–59.

Healy, K., McNally, L., Ruxton, G.D., Cooper, N. and Jackson, A.L. (2013) Metabolic rate and body size are linked with perception of temporal information. *Animal Behaviour* 86, 685–696.

Hebert, D.M. and McFetridge, R.J. (1978) Chemical Immobilisation of North American Game Mammals. Alberta Department of Recreation, Parks and Wildlife, Edmonton, Canada, pp. 44–47.

Hediger, H. (1934) Uber bewegungstereotypien bein gehaltenen tieren. *Revue Suisse de Zoologie* 41, 349–356.

Hediger, H. (1941) *Biologische Gestzmassigkeiten im Verhalten von Wirbeltieren*. Mittels Naturforschung Geselschaft, Berne, Switzerland.

Hediger, H. (1950) *Wild Animals in Captivity*. Butterworths, London.

Hediger, H. (1955) *Studies of the Psychology and Behaviour of Captive Animals in Zoos and Circuses*. Butterworth, London.

Hediger, H. (1963) The evolution of territorial behaviour. In: Washburn, S.L. (ed.) *The Social Life of Early Man*. Methuen, London.

Heffner, H.E. and Heffner, R.S. (1992) Auditory Perception. In: Phillips, C. and Piggins, D. (eds) *Farm Animals and the Environment*. CAB International, Wallingford, UK, pp. 159–184.

Heil, G., Simianer, H. and Dempfle, L. (1990) Genetic and phenotypic variation in prelaying behavior of Leghorn hens kept in single cages. *Poultry Science* 69, 1231–1235.

Heinrich, B. (1989) *Ravens in Winter*. Summit, New York.

Heinrichs, A.J., Wells, S.J., Hurd, H.S., Hill, G.W. and Dargatz, D.A. (1994) The national dairy heifer evaluation project: a profile of heifer management practices in the United States. *Journal of Dairy Science* 77, 1548–1555.

Held, S., Mendl, M. Devereux, C. and Byrne, R.W. (2000) Social tactics of pigs in a competitive foraging task: the 'informed forager' paradigm. *Animal Behaviour* 59, 569–576.

Held, S., Mendl, M., Devereux, C. and Byrne, R.W. (2001) Behaviour of domestic pigs in a visual perspective taking task. *Behaviour* 138, 1337–1354.

Held, S., Mendl, M., Laughlin, K. and Byrne, R.W. (2002) Cognition studies with pigs: livestock cognition and its implications for production. *Journal of Animal Science* 80, E10–E17.

Held, S., Baumgartner, J., Kilbride, A., Byrne, R.W. and Mendl, M. (2005) Foraging behaviour in domestic pigs (*Sus scrofa*): remembering and prioritising food sites of different value. *Animal Cognition* 8, 114–121.

Held, S., Byrne, R.W., Jones, S., Murphy, E., Friel, M. and Mendl, M. (2010) Domestic pigs (*Sus scrofa*) adjust their foraging behaviour to whom they are foraging with. *Animal Behaviour* 79, 857–862.

Hemsworth, P.H. and Barnett, J.L. (1987) Human–animal interactions. In: *The Veterinary Clinics of North America 3, 2, Farm Animal Behaviour*. Saunders, Philadelphia, Pennsylvania, pp. 339–356.

Hemsworth, P.H. and Beilharz, R.G. (1979) The influence of restricted physical contact with pigs during rearing on the sexual behaviour of the male domestic pig. *Animal Production* 29, 311–314.

Hemsworth, P.H. and Coleman, G.J. (2010) *Human–Livestock Interaction: the Stockperson and the Productivity and Welfare of Intensively Farmed Animals*. CAB International, Wallingford, UK.

Hemsworth, P.H., Winfield, C.G. and Mullaney, P.G. (1976) A study of the development of the teat order in piglets. *Applied Animal Ethology* 2, 225–233.

Hemsworth, P.H., Beilharz, R.G. and Brown, W.J. (1978a) The importance of the courting behaviour of the boar in the success of natural and artificial matings. *Applied Animal Ethology* 4, 341–347.

Hemsworth, P.H., Findlay, J.K. and Beilharz, R.G. (1978b) The importance of physical contact with other pigs during rearing on the sexual behaviour of the male domestic pig. *Animal Production* 27, 201–207.

Hemsworth, P.H., Brand, A. and Willens, P.J. (1981a) The behavioural response of sows to the presence of human beings and their productivity. *Livestock Product Science* 8, 67–74.

Hemsworth, P.H., Barnett, J.L. and Hanson, C. (1981b) The influence of handling by humans on the behaviour, growth and corticosteroids in juvenile female pigs. *Hormones and Behavior* 15, 396– 403.

Hemsworth, P.H., Salden, N.T.C.J. and Hoogerbrugge, A. (1982) The influence of the post-weaning social environment on the weaning to mating interval of the sow. *Animal Production* 35, 41–48.

Hemsworth, P.H., Barnett, J.L. and Hansen, C. (1986a) The influence of early contact with humans and subsequent behavioural responses of pigs to humans. *Applied Animal Behaviour Science* 15, 55–63.

Hemsworth, P.H., Barnett, J.L. and Hansen, C. (1986b) The influence of handling by humans on the behaviour, reproduction and corticosteroids of male and female pigs. *Applied Animal Behaviour Science* 15, 303–314.

Hebet, D.M. and McFetridge, R.J. (1978) *Chemical Immobilisation of North American Game Mammals*.

Alberta Department of Recreation, Parks and Wildlife, Edmonton, Canada, pp. 44–47.

Herrnstein, R.J. (1977) The evolution of behaviourism. *American Psychology* 32, 593–603.

Hester, P.Y., Sutton, A.L. and Elkin, R.G. (1987) Effect of light intensity, litter source, and litter management on the incidence of leg abnormalities and performance of male turkeys. *Poultry Science* 66, 666–675.

Hester, P.Y., Cassens, D.L. and Bryan, T.A. (1997) The applicability of particleboard residue as a litter material for male turkeys. *Poultry Science* 76, 248–255.

Hetts, S. and Estep, D.Q. (1994) Behavior management: preventing elimination and destructive behavior problems. *Veterinary Forum* 11, 60–61.

Heyes, C. (2000) Evolutionary psychology in the round. In: Heyes, C. and Huber, L. (eds) *The Evolution of Cognition*. MIT Press, Cambridge, Massachusetts, pp. 165–183.

Heyn, E., Damme, K., Manz, M., Remy, F. and Erhard, M.H. (2006) Adequate water supply for Pekin ducks: alternatives for bathing. *Deutsche Tierärztliche Wochenschrift* 113, 90–93.

Higgs, A.R.B., Norris, R.T. and Richards, R.B. (1993) Epidemiology of salmonellosis in the live sheep export industry. *Australian Veterinary Journal* 70, 330–335.

Hillerton, J.E. (2004) Summer mastitis. In: Andrews, A.H., Blowey, R.W., Boyd, H. and Eddy, R.G. (eds) *Bovine Medicine: Diseases and Husbandry of Cattle*. Blackwell, Oxford, UK, pp. 337–340.

Hillerton, J.E., Bramley, A.J. and Broom, D.M. (1983) *Hydrotaea irritans* and summer mastitis in calves. *Veterinary Record* 113, 88.

Hillerton, J.E., Bramley, A.J. and Broom, D.M. (1984) The distribution of five species of flies (Diptera: Muscidae) over the bodies of dairy heifers in England. *Bulletin of Entomological Research* 74, 113.

Hillerton, J.E., Morant, S.V. and Harris, J.A. (1986) Control of Muscidae on cattle by StockGuard eartags, the behaviour of these flies on cattle and the effect on fly-dislodging behaviour. *Entomologia Experimentalia Applicada* 41, 213–218.

Hinde, R.A. (1970) *Animal Behaviour: a Synthesis of Ethology and Comparative Psychology*. 2nd edn. McGraw Hill, New York.

Hinde, R.A., Thorpe, W.H. and Vince, M.A. (1956) The following responses of young coots and moorhens. *Behaviour* 9, 214–242.

Hirt, H., Frohlich, E. and Oester, H. (1995) Leg weakness in fattening turkeys [in German]. *Aktuelle Arbeiten zur Artgemässen Tierhaltung* 373, 178–188.

Hocking, P. (1993) Welfare of turkeys. In: Savory, C.J. and Hughes, B.O. (eds) *Proceedings of the Fourth European Symposium on Poultry Welfare,* Edinburgh, UK, 18–21 September. UFAW, Potters Bar, UK, pp. 125–138.

Hocking, P.M. (2004) Measuring and auditing the welfare of broiler breeders. In: Weeks, C. and Butterworth, A. (eds) *Measuring and Auditing Broiler Welfare*. CAB International, Wallingford, UK, pp. 19–35.

Hocking, P.M., Bernard, R. and Wess, T.J. (1998) Comparative development of antirochanteris disease in male and female turkeys of a traditional line and a contemporary sire-line fed ad libitum or restricted quantities of food. *Research in Veterinary Science* 65, 29–32.

Hoekstra, J., van der Lugt, A.W., van der Werf, J.H.J. and Ouweltjes, W. (1994) Genetic and phenotypic parameters for milk production and fertility traits in up-graded dairy cattle. *Livestock Production Science* 40, 225–232.

Hogan, J.A. (2005) Motivation. In: Bolhuis, J.J. and Giraldeau, L.A. (eds) *The Behaviour of Animals*. Blackwell, Malden, Maryland, pp. 41–70.

Hogan, J.A. and Roper, T.J. (1978) A comparison of the properties of different reinforces. *Advanced Studies in Behaviour* 8, 155–255.

Holland, P.C. and Straub, J.J. (1979) Differential effects of two ways of devaluing the unconditioned stimulus after Pavlovian appetitive conditioning. *Journal of Experimental Psychology: Animal Behavior Processes* 5, 65–78.

Holm, L., Jensen, M.B. and Jeppesen, L.L. (2002) Calves' motivation for access to two different types of social contact measured by operant conditioning. *Applied Animal Behaviour Science* 79, 175–194.

Holmes, J.C. and Bethel, W.M. (1972) Modification of intermediate host behaviour by parasites. In: Canning, E.V. and Wright, C.A. (eds) *Behavioural Aspects of Parasite Transmission*. Academic Press, London, pp. 123–149.

Holmes, R.J. (1980) Normal mating behaviour and its variations. In: Morrow, D.A. (ed.) *Current Therapy in Theriogenology*. W.B. Saunders, Philadelphia, Pennsylvania, pp. 931–936.

Holtzman, C.W., Trotman, H.D., Goulding, S.M., Ryan, A.T., Macdonald, A.N., Shapiro, D.I., Brasfield, J.L. and Walker, E.F. (2013) Stress and neurodevelopmental processes in the emergence of psychosis. *Neuroscience* 249, 172–191.

Hopster, H., Connell, J.M. and Blokhuis, H. (1995) Acute effects of cow-calf separation on heart rate, plasma cortisol and behaviour in multiparous dairy cows. *Applied Animal Behaviour Science* 44, 1–8.

Hopster, H., van der Werf, J.T.N. and Blokhuis, H.J. (1998) Side preference of dairy cows in the milking parlour and its effects on behaviour and heart rate during milking. *Applied Animal Behaviour Science* 55, 213–219.

Horn, L., Virányi, Z., Miklósi, Á., Huber, L. and Range, F. (2012) Domestic dogs (*Canis familiaris*) flexibly adjust their human-directed behavior to the actions of their human partners in a problem situation. *Animal Cognition* 15, 57–71.

Horn, T., Marx, G. and von Borell, E. (1999) Behavior of piglets during castration with and without local anesthesia. *Deutsche Tierärztliche Wochenschrifte* 106, 271–274.

Horrell, I. and Hodgson, J. (1992a) The bases of sow-piglet identification. 1. The identification of sows by their own piglets and the presence of intruders. *Applied Animal Behaviour Science* 33, 319–327.

Horrell, I. and Hodgson, J. (1992b) The bases of sow-piglet identification. 2. Cues used by piglets to identify their dam and home pen. *Applied Animal Behaviour Science* 33, 329–343.

Horrell, R.I. and Bennett, J. (1981) Disruption of teat preferences and retardation of growth following crossfostering of one-week-old pigs. *Animal Production* 33, 99–106.

Horrell, R.I. and Hodgson, J. (1985) Mutual recognition between sows and their litters. In: Unshelm, J., van Putten, G. and Zeeb, K. (eds) *Applied Ethology in Farm Animals*. KTBL, Davmstadt, Germany, pp. 108–112.

Horowitz, A. (2009) Disambiguating the 'guilty look': salient prompts to a familiar dog behaviour. *Behavioural Processes* 81, 447–452.

Horváth, G., Järverud, G.K., Järverud, S. and Horváth, I. (2008) Human ovarian carcinomas detected by specific odor. *Integrative Cancer Therapies* 7, 76–80.

Horwitz, D.F. (2002) House soiling by cats. In: Horwitz, D.F., Mills, D.S. and Heath, S. (eds) *BSAVA Manual of Canine and Feline Behavioural Medicine*. British Small Animal Veterinary Association, Quedgeley, UK, pp. 97–108.

Houpt, K.A. (1981) Equine bahaviour problems in relation to humane management. *International Journal for the Study of Animal Production* 2, 329–336.

Houpt, K.A. (1984) Treatment of aggression in horses. *Equine Practice* 6 (6), 8–10.

Houpt, K.A. and Hintz, H.F. (1983) Some effects of maternal deprivation on maintenance behaviour, spatial relationships and responses to environmental novelty in foals. *Applied Animal Ethology* 9, 221–230.

Houpt, K.A. and Wolski, T. (1982) *Domestic Animal Behaviour for Veterinarians and Animal Scientists*. Iowa State University Press, Ames, Iowa.

Hsia, L.C. and Wood-Gush, D.G.M. (1982) The relationship between social facilitation and feeding behaviour in pigs. *Applied Animal Ethology* 8, 410.

Hsia, L.C. and Wood-Gush, D.G.M. (1984) Social facilitation in the feeding behaviour of pigs and the effect of rank. *Applied Animal Ethology* 11, 265–270.

Hüber, A. and Fölsch, D.W. (1978) Akustiche ethogramme von huhnern: die auswirkung verschiedener haltungs systeme. *Tierhaltung* 5. Birkhauser Verlag, Basel, Switzerland.

Hubrecht, R.C. (1993) A comparison of social and environmental enrichment methods for laboratory housed dogs. *Applied Animal Behaviour Science* 37, 345–361.

Hubrecht, R. and Kirkwood, J. (eds) (2010) *The UFAW Handbook on the Care and Management of Laboratory and other Research Animals*. John Wiley & Sons, Chichester, UK.

Hubrecht, R.C., Serpell, J.A. and Poole, T.B. (1992) Correlates of pen size and housing conditions on the behaviour of kennelled dogs. *Applied Animal Behaviour Science* 34, 365–383.

Hudson, S.J. and Mullord, M.M. (1977) Investigation of maternal bonding in dairy cattle. *Applied Animal Ethology* 3, 271–276.

Hughes, B.O. (1971) Allelomimetic feeding in the domestic fowl. *British Poultry Science* 12, 359.

Hughes, B.O. (1975) Spatial preference in the domestic. *British Veterinary Journal* 131, 560–564.

Hughes, B.O. (1976) Preference decisions of domestic hens for wire or litter floors. *Applied Animal Ethology* 2, 155–165.

Hughes, B.O. (1980) Behaviour of the hen in different environments. *Animal Regulation Studies* 3, 65–71.

Hughes, B.O. (1981) Headshaking in fowls. In: *Proceedings of the First European Symposium of Poultry Welfare*, Koge, Denmark. World Poultry Science Association, Beekbergen, Netherlands.

Hughes, B.O. and Appleby, M.C. (1989) Increase in bone strength of spent laying hens housed in modified cages with perches. *Veterinary Science* 124, 483–484.

Hughes, B.O. and Black, A.J. (1973) The preference of domestic hens for different types of battery cage floor. *British Poultry Science* 14, 615–619.

Hughes, B.O. and Black, A.J. (1974) The effect of environmental factors on activity, selected behaviour patterns and 'fear' of fowls in cages and pens. *British Poultry Science* 15, 375–380.

Hughes, B.O. and Duncan, I.J.H. (1988a) Behavioural needs: can they be explained in terms of motivational models? *Applied Animal Behaviour Science* 20, 352–355.

Hughes, B.O. and Duncan, I.J.H. (1988b) The notion of ethological 'need', models of motivation and animal welfare. *Animal Behaviour* 36, 1696–1707.

Hughes, B.O. and Wood-Gush, D.G.M. (1971) A specific appetite for calcium in domestic fowls. *Animal Behaviour* 19, 490–499.

Huls, W.I., Brooks, D.I. and Bean-Knudsen, D. (1991) Response of adult New Zealand white rabbits to enrichment objects and paired housing. *Laboratory Animal Science* 41 (6), 609–612.

Humphrey, N.K. (1976) The social function of intellect. In: Bateson, P.P.G. and Hinde, R.A. (eds) *Growing Points in Ethology*. Cambridge University Press, Cambridge, UK, pp. 303–317.

Humphrey, T. (1970) Function of the nervous system during prenatal life. In: *Physiology of the Perinatal Period, 2*. Appleton-Century-Crofts, New York, Chapter 12.

Hunter, E.J., Broom, D.M., Edwards, S.A. and Sibly R.M. (1988) Social hierarchy and feeder access in a group of 20 sows using a computer controlled feeder. *Animal Production* 47, 139–148.

Hunter, E.J., Jones, T.A., Guise, H.J., Penny, R.H.C. and Hoste, S. (2001) The relationship between tail biting in pigs, docking procedure and other management practices. *Veterinary Journal* 161, 72–79.

Huntingford, F.A., Adams, C., Braithwaite, V.A., Kadri, S., Pottinger, T.G., Dandoe, P. and Turnbull, J.F. (2006) Current issues in fish welfare. *Journal of Fish Biology* 68, 332–372.

Hurnick, J.F., King, G.J. and Robertson, H.A. (1975) Oestrus and related behaviour in postpartum Holstein cows. *Applied Animal Ethology* 2, 55.

Hutson, G.D. (1984) Spacing behaviour of sheep in pens. *Applied Animal Behaviour Science* 12, 111–119.

Hutson, G.D. (1989) Operant tests of access to earth as a reinforcement for weaner piglets. *Animal Production* 48, 561–569.

Hymel, K.A. and Sufka, K.J. (2012) Pharmacological reversal of cognitive bias in the chick anxiety-depression model. *Neuropharmacology* 62, 161–166.

Illmann, G., Hammerschmidt, K., Špinka, M. and Tallet, C. (2013) Calling by domestic piglets during simulated crushing and isolation: a signal of need? *PloS ONE* 8, e83529, 9 pp.

Immelmann, K. (1972) Sexual and other long-term aspects of imprinting in birds and other species. *Advances in the Study of Behavior* 4, 147–174.

Inglis, I.R. (2000) The central role of uncertainty reduction in determining behaviour. *Behaviour* 137, 1567–1600.

Irwin, M. (2001) Are stress and depression interrelated? In: Broom, D.M. (ed.) *Coping with Challenge: Welfare in Animals Including Humans*. Dahlem University Press, Berlin, pp. 271–287.

Irwin, M., Mascovich, A., Gillin, J.C., Willoughby, R., Pike, J. and Smith, T.L. (1994) Partial sleep deprivation reduces natural killer cell activity in humans. *Psychosomatic Medicine* 56, 493–498.

Ishizaki, H., Hanafusa, Y. and Kariya, Y. (2004) Influence of truck-transportation on the function of bronchoalveolar lavage fluid cells in cattle. *Veterinary Immunology and Immunopathology* 105, 67–74.

Ittyerah, M. and Gaunet, F. (2009) The response of guide dogs and pet dogs (*Canis familiaris*) to cues of human referential communication (pointing and gaze). *Animal Cognition* 12, 257–265.

Iversen, S.D. and Iversen, L. (1981) *Behavioural Pharmacology*. Oxford University Press, Oxford, UK.

Jackson, C. (2003) Laboratory fish: impacts of pain and stress on well-being. *Contemporary Topics* 42, 62–73.

Jackson, P.G.G. (1988) The assessment of welfare in diseased farm animals. In: Gibson, T.E. (ed.) *Animal Disease – a Welfare Problem*. British Veterinary Association Animal Welfare Foundation, London, pp. 42–46.

Jagoe, A. and Serpell, J. (1996) Owner characteristics and interactions and the prevalence of canine behaviour problems. *Applied Animal Behavioural Science* 47, 31–42.

Jansen, G.A. and Greene, N.M. (1970) Morphine metabolism and morphine tolerance in goldfish. *Anaesthesiology* 32, 231–235.

Jarvis, J.R., Abeyesinghe, S.M., McMahon, C.E. and Wathes, C.M. (2009) Measuring and modelling the spatial contrast sensitivity of the chicken (*Gallus g. domesticus*). *Vison Research* 49, 1448–1454.

Jensen, A.H., Yen, J.T., Gehring, M.M., Baker, D.H., Becker, D.E. and Harmon, B.G. (1970) Effects of space restriction and management on pre-and post-pubertal response of female swine. *Journal of Animal Science* 31, 745–750.

Jensen, M.B. (2003) The effects of feeding method, milk allowance and social factors on milk feeding behaviour and cross-sucking in group housed dairy calves. *Applied Animal Behaviour Science* 80, 191–206.

Jensen, M.B. and Kyhn, R. (2000) Play behaviour in group-housed dairy calves, the effect of space allowance. *Applied Animal Behaviour Science* 67, 35–46.

Jensen, M.B., Vestergaard, K.S. and Krohn, C.C. (1998) Play behaviour in domestic calves kept in pens: the effect of social contact and space allowance. *Applied Animal Behaviour Science* 56, 97–108.

Jensen, P. (1979) Sinsuggors beteendemonster under tre olika upstallnings forhallanden – en pilot studie. *Institutionen for Husdjurshygien med Horslarskalan. Rapport 1.* Sveriges Lantbruksuniversitet, Uppsala, Sweden, 1–40.

Jensen, P. (1980) An ethogram of social interaction patterns in group-housed dry sows. *Applied Animal Ethology* 6, 341–350.

Jensen, P. (1981) Fixeringens effect pa sinsuggors beteende. *Svensk Veterinartidning* 33, 73–78.

Jensen, P. (1982) An analysis of agonistic interaction patterns in group-housed dry sows – aggression regulation through an 'avoidance order'. *Applied Animal Ethology* 9, 47–61.

Jensen, P. (1984) Effects of confinement on social interaction patterns in dry sows. *Applied Animal Behaviour Science* 12, 93–101.

Jensen, P. (1986) Observations on the maternal behaviour of free ranging domestic pigs. *Applied Animal Behaviour Science* 16, 131–142.

Jensen, P. (ed.) (2007) *The Behavioural Biology of Dogs.* CAB International, Wallingford, UK.

Jensen, P. (2009) Behaviour genetics, evolution and domestication. In: Jensen, P. (ed.) *The Ethology of Domestic Animals.* CAB International, Wallingford, UK, pp. 10–24.

Jensen, P. and Algers, B. (1983) An ethogram of piglet vocalisations during suckling. *Applied Animal Ethology* 11, 237–248.

Jensen, P. and Wood-Gush, D.G.M. (1984) Social interactions in a group of free-ranging sows. *Applied Animal Behaviour Science* 12, 327–337.

Jensen, P., Algers, B. and Ekesbo, I. (1986) Methods of Sampling and Analysis of Data in Farm Animal Ethology. *Tierhaltung* 17. Birkhauser Verlag, Basel, Switzerland.

Jensen, P., Borell, E. von, Broom, D.M., Csermely, D., Dijkhuizen, A.A., Edwards, S.A., Madec, F. and Stamataris, C. (2001) *The Welfare of Intensively Kept Pigs.* Reports of the Scientific Veterinary Committee 1997 (Animal Health and Animal Welfare Sections). Office for Official Publications of the European Communities, Luxembourg, pp. 179–437.

Jeppesen, L.L. and Pedersen, B. (1991) Effects of whole-year nest boxes on cortisol, circulating leukocytes, exploration and agonistic behavior in silver foxes. *Behavioural Processes* 25, 171–177.

Jergensen, E.H., Christiansen, J.S. and Jobling, M. (1993) Effects of stocking density on food intake, growth performance and oxygen consumption in Arctic charr (*Salvelinus alpinus*). *Aquaculture* 110, 191–204.

Joergensen, G. (1985) *Mink Production.* Scientifur, Hilleroed, Denmark.

Johnson, K.G. (1987) Shading behaviour of sheep: preliminary studies of its relation to thermoregulation, feed and water intakes, and metabolic rate. *Australian Journal of Agricultural Research* 38, 587–596.

Jones, N.M., Holzman, C.B., Zanella, A.J., Leece, C.M. and Rahbar, M.H. (2006) Assessing mid-trimester salivary cortisol levels across three consecutive days in pregnant women using an at-home collection protocol. *Paediatric and Perinatal Epidemiology* 20, 425–437.

Jones, R.B. (1996) Fear and adaptability in poultry: insights, implications and imperatives. *World's Poultry Science Journal* 52, 131–174.

Jones, T.A. and Dawkins, M.S. (2010) Effect of environment on Pekin duck behaviour and its correlation with body condition on commercial farms in the UK. *British Poultry Science* 51, 319–325.

Jones, T.A., Waitt, C.D. and Dawkins, M.S. (2009) Water off a duck's back: showers and troughs match ponds for improving duck welfare. *Applied Animal Behaviour Science* 116, 52–57.

Jones, T.C., Hunt, R.D. and King, N.W. (eds) (1997) *Veterinary Pathology*. 6th edn. Williams and Wilkins, Baltimore, Maryland, p. 1392.

Jung, J. and Lidfors, L. (2001) Effect of amount of milk, milk flow and access to a rubber teat on cross-sucking and non-nutritive sucking in dairy calves. *Applied Animal Behavioural Science* 72, 201–213.

Jury, C., Picard, B. and Geay, Y. (1998) Influences of the method of housing bulls on their body composition and muscle fiber types. *Meat Science* 4, 457–469.

Kadri, S., Mejdell, C.M. and Damsgård, B. (2012) Benefish: an interdisciplinary approach to economic modeling of fish welfare management. *Aquaculture, Economics and Management* 16, 292–296.

Kaiser, S., Krüger, C. and Sachser, N. (2010) The guinea pig. In: Hubrecht, R. and Kirkwood, J. (eds) *The UFAW Handbook on the Care and Management of Laboratory and Other Research Animals*. 8th edn. Wiley-Blackwell, Chichester, UK, pp. 380–398.

Kaminski, J., Tempelmann, S., Call, J. and Tomasello, M. (2009) Domestic dogs comprehend human communication with iconic signs. *Developmental Science* 12, 831–837.

Katz, D. and Revesz, G. (1921) Experimentelle stud ien zur vergleichenden psychologie (versuche mit huhnern). *Annalen der Angewandte Psychologie* 18, 307.

Kavaliers, M. (1989) Evolutionary aspects of the neuromodulation of nociceptive behaviors. *American Zoologist* 29, 1345–1353.

Keating, S.C., Thomas, A.A., Flecknell, P.A. and Leach, M.C. (2012) Evaluation of EMLA cream for preventing pain during tattooing of rabbits: changes in physiological, behavioural and facial expression responses. *PloS ONE* 7, e44437, pp. 11.

Keeling, L. (ed.) (2009) *An Overview of the Development of the Welfare Quality® Assessment Systems*. Welfare Quality Reports No. 12. European Commission and Cardiff University, Cardiff, UK

Keeling, L. and Jensen, P. (2009) Abnormal behaviour, stress and welfare. In: Jensen, P. (ed.) *The Ethology of Domestic Animals*, 2nd edn. CAB International, Wallingford, UK, pp. 85–101.

Keeling, L.J. and Duncan, I.J.H. (1988) The effect of activity transitions on spacing behaviour in domestic fowl. In: Unshelm, J., van Putten, G., Zeeb, K. and Ekesbo, I. (eds) *Proceedings of the International Congress on Applied Ethology in Farm Animals,* Skara, Sweden, 1988. KTBL, Darmstadt, Germany, pp. 291–296.

Kegley, E.B., Spears, J.W. and Brown Jr, T.T. (1997) Effect of shipping and chromium supplementation on performance, immune response, and disease resistance of steers. *Journal of Animal Science* 75, 1956–1964.

Kelley, K.W., Greenfield, R.E., Evermann, J.F., Parish, S.M. and Perryman, L.E. (1982) Delayed-type hypersensitivity, contact sensitivity, and phytohemagglutinin skin-test reponses of heat-and cold-stressed calves. *American Journal of Veterinary Research* 43, 775–779.

Kelly, H.R.C., Bruce, J.M., Edwards, S.A., English, P.R. and Fowler, V.R. (2000) Limb injuries, immune response and growth performance of early-weaned pigs in different housing systems. *Animal Science* 70, 73–83.

Kelly, J.S. (1979) Report of the Panel of Enquiry into Shooting and Angling. Chaired by Lord Medway, UK Government, London.

Kelly, K.W. (1980) Stress and immune function: a bibliographic review. *Annales Recherches Vétérinaires* 11, 445–478.

Kendrick, K.M., da Costa, A.P., Leigh, A.E., Hinton, M.R. and Peirce, J.W. (2001) Sheep don't forget a face. *Nature* 414, 165–166.

Kenny, F.J. and Tarrant, P.V. (1982) Behaviour of cattle during transport and penning before slaughter. In: Moss, R. (ed.) Transport of animals intended for breeding, production and slaughter. *Current Topics in Veterinary Medicine and Animal Science* 18, 87–102.

Keogh, R.G. and Lynch, J.J. (1982) Early feeding experience and subsequent acceptance of feed by sheep. *Proceedings of the New Zealand Society of Animal Production* 42, 73–75.

Kestin, S.C., Adams, S.J.M. and Gregory, N.G. (1994) Leg weakness in broiler chickens, a review of studies using gait scoring. In: *Proceedings of the 9th European Poultry Conference*, Glasgow, UK, 7–12 August 1994. Vol. II. WPSA, Glasgow, UK, pp. 203–206.

Ketelaar-de Lauwere, C.C. and Smits, A.C. (1989) Onderzoek naar de uit ethologisch oogpunt minimaal gewenstle boxmaten voor vleeskalveren met een gewicht van 175 tot 300 kg. *IMAG Rapport 110*. IMAG, Wageningen, Netherlands.

Ketelaar-de Lauwere, C.C., Smits, A.C. (1991) Spatial requirements of individually housed veal calves of 175 to 300 kg. In: Metz, J.H.M. and Groenestein, C.M. (eds) *New Trends in Veal Calf Production*. Proceedings of the International Symposium on Veal Calf Production. EAAP Publications No. 52, Pudoc, Wageningen, Netherlands.

Keverne, E.B., Levy, F., Poindron, P. and Lindsay, D.R. (1983) Vaginal stimulation: an important determinant of maternal bonding in sheep. *Science, New York* 219, 81–83.

Kiley-Worthington, M. (1977) *Behavioural Problems of Farm Animals*. Oriel Press, Stocksfield, UK.

Kiley-Worthington, M. and de la Plain, S. (1983) *The Behaviour of Beef Suckler Cattle*. Birkhauser Verlag, Basel, Switzerland.

Kilgour, R. (1975) The open-field test as an assessment of the temperament of dairy cows. *Animal Behaviour* 23, 615–624.

Kilgour, R. (1978) The application of animal behaviour and the humane care of farm animals. *Journal of Animal Science* 45, 1478–1486.

Kilgour, R. (1987) Learning and the training of farm animals. In: Price, E.O. (ed.) *The Veterinary Clinics of North America. Vol. 3, No. 2, Farm Animal Behavior*. Saunders, Philadelphia, Pennsylvania.

Kilgour, R. and Dalton, C. (1984) *Livestock Behaviour: a Practical Guide*. Granada, London.

Kirkden, R.D., Edwards, J.S.S. and Broom, D.M. (2003) A theoretical comparison of the consumer surplus and the elasticities of demand as measures of motivational strength. *Animal Behaviour* 65, 157–178.

Kirkden, R.D., Niel, L. and Weary, D.M. (2005) Aversion to carbon dioxide. *Laboratory Animals* 39, 453–455.

Kirkden, R.D., Broom, D.M. and Andersen, I.L. (2013a) Piglet mortality: the impact of induction of farrowing using prostaglandins and oxytocin. *Animal Reproduction Science* 138, 14–24.

Kirkden, R.D., Broom, D.M. and Andersen, I.L. (2013b) Piglet mortality: management solutions. *Journal of Animal Science* 91, 3361–3389.

Kirkwood, J.K. (2006) The distribution of the capacity for sentience in the animal kingdom. In: Turner, J. and D'Silva, J. (eds) *Animals, Ethics and Trade: The Challenge of Animal Sentience*. Compassion in World Farming Trust, Petersfield, UK, pp. 12–26.

Kivlighan, K.T., DiPietro, J.A., Costigan, K.A. and Laudenslager, M.L. (2008) Diurnal rhythm of cortisol during late pregnancy: associations with maternal psychological well-being and fetal growth. *Psychoneuroendocrinology* 33, 1225–1235.

Kjaer, J.B. (2000) Diurnal rhythm of feather-pecking behaviour and condition of integument in four strains of loose housed laying hens. *Applied Animal Behaviour Science* 65, 331–347.

Kjaer, J.B. and Vestergaard, K.S. (1999) Development of feather-pecking in relation to light intensity. *Applied Animal Behaviour Science* 62, 243–254.

Klemm, R., Pingel, H. and Reiter, K. (1995) Effect of mixing Muscovy and Pekin ducks on feather pecking. *Proceedings of the 10th European Symposium on Waterfowl*, Halle, Germany, pp. 433–438.

Klinghammer, E. (1967) Factors influencing choice of mate in birds. In: Stevenson, H.W., Hess, E.H. and Rheingold, H.L. (eds) *Early Behavior: Comparative and Developmental Approaches*. Wiley, New York.

Knaus, W. (2009) Dairy cows trapped between performance demands and adaptability. *Journal of the Science of Food and Agriculture* 89, 1107–1114.

Knierim, U. and Gocke, A. (2003) Effect of catching broilers by hand or machine on rates of injuries and dead-on-arrivals. *Animal Welfare* 12, 63–73.

Knierim, U., Bulheller, M.A., Kuhnt, K., Briese, A. and Hartung, J. (2004) Water provision for ducks kept indoors – a review on the basis of the literature and our own experience. *Deutsche Tierärztliche Wochenschrift* 111, 115–118.

Knowles, T.G. (1998) A review of road transport of slaughter sheep. *Veterinary Record* 143, 212–219.

Knowles, T.G. and Broom, D.M. (1990) Limb-bone strength and movement in laying hens from different housing systems. *Veterinary Record* 126, 354–356.

Knowles, T.G., Broom, D.M., Gregory, N.G. and Wilkins, L.J. (1993) Effects of bone strength on the frequency of broken bones in hens. *Research in Veterinary Science* 54, 15–19.

Knowles, T.G., Warriss, P.D., Brown, S.N. and Kestin, S.C. (1994) Long distance transport of export lambs. *Veterinary Record* 134, 107–110.

Knowles, T.G., Brown, S.N., Warriss, P.D., Phillips, A.J., Doland, S.K., Hunt, P., Ford, J.E., Edwards, J.E. and Watkins, P.E. (1995) Effects on sheep of transport by road for up to 24 hours. *Veterinary Record* 136, 431–438.

Knowles, T.G., Warriss, P.D., Brown, S.N., Edwards, J.E., Watkins, P.E. and Phillips, A.J. (1997) Effects on calves less than one month old of feeding or not feeding them during road transport of up to 24 hours. *Veterinary Record* 140, 116–124.

Knowles, T.G., Kestin, S.C., Haslam, S.M., Brown, S.N., Green, L.E., Butterworth, A., Pope, S.J., Pfeiffer, D. and Nicol, C.J. (2008) Leg disorders in broiler chickens: prevalence, risk factors and prevention. *PloS ONE* 3, e1545, pp. 5.

Kondo, S., Kawakami, N., Kohama, H. and Nishino, S. (1983) Changes in activity, spatial pattern and social behaviour in calves after grouping. *Applied Animal Ethology* 11, 217–222.

Koolhaas, J.M., Korte, S.M., de Boer, S.F., Van der Vegt, B.J., Van Reenen, C.G. and Hopster, H. (1999) Coping styles in animals: current status in behavior and stress-physiology. *Neuroscience Biobehavioral Review* 23, 925–935.

Korhonen, H. and Harri, M. (1997) Effect of cage size and access to earthen level on behaviour of farm blue foxes (*Alopex lagopus*) In: Hemsworth, P.H., Spinka, M. and Kost'ál, L. (eds) *Proceedings of the 31st International Congress of the ISAE (International Society for Applied Ethology)*, Prague, Czech Republic, 13–16 August, p. 89.

Korhonen, H., Niemela, P. and Tuuri, H. (1996) Seasonal changes in platform use by farmed blue foxes (*Alopex lagopus*). *Applied Animal Behaviour Science* 48, 99–114.

Kraus, H. (1978) *Vergleichende Untersuchungen an Legehennen aus kommerzieller Boden- und Käfighaltung unter besonderer Berücksichtigung der Zerlegeergebnisse.* Tierärztl, Lebensmitteluntersuchungsstelle des Kreises Mettmann.

Kreeger, T.J., Hulzenga, M., Hansen, C. and Wise, B.L. (2011) Sulfentanil and xylazine immobilization of rocky mountain elk. *Journal of Wildlife Diseases* 47, 638–642.

Kristensen, H.H., Prescott, N.B., Perry, G.C., Ladewig, J., Ersbøll, A.K., Overvad, K.C. and Wathes, C.M. (2007) The behaviour of broiler chicks in different light sources and illuminations. *Applied Animal Behaviour Science* 103, 75–89.

Kronfeld, D.S. (1997) Recombinant bovine somatotropin: ethics of communication and animal welfare. *Swedish Veterinary Journal* 49, 157–165.

Kruijt, J.P. (1964) Ontogeny of social behaviour in Burmese red jungle fowl *(Gallus gallus spadiceus). Behaviour (Suppl.)* 12.

Kruuk, H. (2002) *Hunters and Hunted: Relationships between Carnivores and People.* Cambridge University Press, Cambridge, UK.

Kuba, M., Bryne, R. and Burghardt, G.M. (2010) Introducing a new method to study problem solving and tool use in fresh water stingrays, *Potamotrygon castexi. Animal Cognition* 13, 507–513.

Kummer, H. (1968) *Social Organization of Hamadryas Baboons.* University of Chicago Press, Chicago, Illinois.

Kushner, I. and Mackiewicz, A. (1987) Acute phase proteins as disease markers. *Disease Markers* 5, 1–11.

Lachlan, R., Janik, V.M. and Slater, P.J.B. (2004) The evolution of conformity enforcing behaviour in cultural communication systems. *Animal Behaviour* 68, 561–570.

Lachlan, R.F. and Slater, P.J.B. (2003) Song learning by chaffinches: how accurate, and from where? *Animal Behaviour* 65, 957–969.

Ladewig, J. (1984) The effect of behavioural stress on the episodic release and circadian variation of cortisol in bulls. In: Unshelm, J., van Putten, G. and Zeeb, K. (eds) *Proceedings of the International Congress on Applied Ethology of Farm Animals.* KTBL, Darmstadt, Germany, pp. 339–342.

Lagadic, H. and Faure, J.M. (1990) L'opposition cage batterie/bien être: mythe ou réalité. In: *6ème Rencontre Annuelle Groupe Français de la W.P.S.A.* Instaprint, Tours, France, pp. 7–14.

Landsberg, G.M., Hunthausen, W.L. and Ackerman, L.J. (2012) *Behavior Problems of the Dog and Cat.* Elsevier Health Sciences, Amsterdam.

Lang, P.J. (1995) The emotion probe: studies of motivation and attention. *American Psychologist* 50, 372–385.

Langbein, J., Nürnberg, G. and Manteuffel, G. (2004) Visual discrimination learning in dwarf goats and associated changes in heart rate and heart rate variability. *Physiology and Behavior* 82, 601–609.

Langbein, J., Nürnberg, G., Puppe, B. and Manteuffel, G. (2006) Self-controlled visual discrimination learning of group-housed dwarf goats (*Capra hircus*): Behavioral strategies and effects of relocation on learning and memory. *Journal of Comparative Psychology* 120, 58–66.

Laughlin, K. and Mendl, M. (2004) Costs of acquiring and forgetting information affect spatial memory and its susceptibility to interference. *Animal Behaviour* 68, 97–193.

Laughlin, K., Huck, M. and Mendl, M. (1999) Disturbance effects of environmental stimuli on pig spatial memory. *Applied Animal Behaviour Science* 64, 169–180.

Lawrence, A.B. (2008) Applied animal behaviour science: past, present and future prospects. *Applied Animal Behaviour Science* 115, 1–24.

Lawrence, A.B. and Rushen, J. (eds) (1993) *Stereotypic Animal Behaviour – Fundamentals and Applications to Welfare*. CAB International, Wallingford, UK.

Lawrence, A.B. and Wall, E. (2014) Selection for 'environmental fit' from existing domesticated species. In: Mellor, D.J. and Bayvel, A.C.D. (eds) *Animal Welfare: Focusing on the Future*. OIE Scientific and Technical Review 33, 171–179. OIE, Paris.

Lawrence, A.B., Petherick, J.C., McClean, K.A., Deans, L.A., Chirnside, J., Vaughan, A., Clutton, E. and Terlouw, E.M.C. (1994) The effect of environment on behaviour, plasma cortisol and prolactin in parturient sows. *Applied Animal Behaviour Science* 39, 313–330.

Lay, D.C., Fulton, R.M., Hester, P.Y., Karcher, D.M., Kjaer, J.B., Mench, J.A. and Porter, R.E. (2011) Hen welfare in different housing systems. *Poultry Science* 90, 278–294.

Le Magnen, J. (1971) Advances in studies on the physiological control and regulation of food intake. In: Stellar, E. and Sprague, J.M. (eds) *Progress in Physiological Psychology*. Vol. 4. Academic Press, New York.

Le Neindre, P. (1989) Influences of rearing conditions and breeds on social behaviour and activity in a novel environment. *Applied Animal Behaviour Science* 23, 129–140.

Le Neindre, P., Boivin, X. and Boissy, A. (1996) Handling of extensively kept animals. *Applied Animal Behaviour Science* 49, 73–81.

Leach, D.H. and Ormrod, K. (1984) The technique of jumping a steeplechase fence by competing event-horses. *Applied Animal Ethology* 12, 15–24.

Leach, M.C., Allweiler, S., Richardson, C., Roughan, J.V., Narbe, R. and Flecknell, P.A. (2009) Behavioural effects of ovariohysterectomy and oral administration of meloxicam in laboratory housed rabbits. *Research in Veterinary Science* 87, 336–347.

Leach, M.C., Klaus, K., Miller, A.L., di Perrotolo, M.S., Sotocinal, S.G. and Flecknell, P.A. (2012) The assessment of post-vasectomy pain in mice using behaviour and the mouse grimace scale. *PloS ONE* 7, e35656.

Legrand, A.L., von Keyserlingk, A.G. and Weary, D.M. (2009) Preference and usage of pasture versus free-stall housing by lactating dairy cattle. *Journal of Dairy Science* 92, 3651–3658.

Lehner, P.N. (1996) *Handbook of Ethological Methods*. 2nd edn. Cambridge University Press, Cambridge, UK.

Lessard, M., Taylor, A.A., Braithwaite, L. and Weary, D.M. (2002) Humoral and cellular immune responses of piglets after castration at different ages. *Canadian Journal of American Science* 82 (4), 519–526.

Levine, S., Goldman, L. and Coover, G.D. (1972) Expectancy and the pituitary-adrenal system. In: Porter, R. and Knight, J. (eds) *Physiology, Emotion and Psychosomatic Illness*. Elsevier, Amsterdam.

Levy, D.M. (1944) On the problem of movement restraint. *American Journal of Orthopsychiatry* 14, 644–671.

Lickliter, R.E. (1982) Effects of a post-partum separation on maternal responsiveness in primiparous and multiparous domestic goats. *Applied Animal Ethology* 8, 537–542.

Lickliter, R.E. (1984) Hiding behaviour in domestic goat kids. *Applied Animal Behaviour Science* 12, 187–192.

Lickliter, R.E. (1985) Behaviour associated with parturition in the domestic goat. *Applied Animal Behaviour Science* 13, 335–345.

Lidfors, L. (1993) Cross-sucking in group-housed dairy calves before and after weaning off milk. *Applied Animal Behaviour Science* 38, 15–24.

Lidfors, L. (1994) Mother–young behaviour in cattle. PhD thesis, Swedish University of Agricultural Science, Skara, Sweden.

Lidfors, L. (1997) Behavioural effects of environmental enrichment for individually caged rabbits. *Applied Animal Behaviour Science* 52, 157–169.

Lidfors, L. and Isberg, L. (2003) Intersucking in dairy cattle – review and questionnaire. *Applied Animal Behaviour Science* 80, 207–231.

Lincoln, G.A., Guinness, F. and Short, R.V. (1972) The way in which testosterone controls the social and sexual behaviour of the red deer stag (*Cervus elephas*). *Hormones and Behavior* 3, 375–396.

Lindsay, D.R. (1965) The importance of olfactory stimuli in the mating behaviour of the ram. *Animal Behaviour* 13, 75–78.

Lindsay, D.R. and Robinson, T.J. (1964) Oestrogenic and antioestrogenic activity of androgens in the ewe. *Journal of Reproduction and Fertility* 7, 267–274.

Lindt, F. and Blum, J.W. (1994a) Occurrence of iron deficiency in growing cattle. *Journal of Veterinary Med. Series A* 41, 237–246.

Lindt, F. and Blum, J.W. (1994b) Growth performance, haematological traits, meat variables, and effects of treadmill and transport stress in veal calves supplied different amounts of iron. *Journal of Veterinary Medicine Series A* 41, 333–342.

Lines, J.A., Robb, D.H., Kestin, S.C., Crook, S.C. and Benson, T. (2003) Electric stunning: a humane slaughter method for trout. *Aquaculture Engineering* 28, 141–154.

Liste, G., Kirkden, R.D. and Broom, D.M. (2012a) Effect of water depth on pool choice and bathing behaviour in commercial Pekin ducks. *Applied Animal Behaviour Science* 139, 123–133.

Liste, G., Kirkden, R.D. and Broom, D.M. (2012b) A commercial trial evaluating three open water sources for farmed ducks: effects on health and production. *British Poultry Science*, 53, 576–584.

Liste, G., Kirkden, R.D. and Broom, D.M. (2013) A commercial trial evaluating three open water sources for farmed ducks: effects on water usage and water quality. *British Poultry Science*, 54, 24–32.

Liste, G., Asher, L. and Broom, D.M. (2014) When a duck initiates movement, do others follow? Testing preference in groups. *Ethology* 120, 1–8.

Löliger, H.C., von dem Hagen, D. and Matthes, S. (1981) Einfluss der haltungssysteme auf die tiergesundheit bericht iiber ergebnisse klinischpathologischer untersuchungen. *Landbauforschung Völkenrode* 60, 47–67.

Lorenz, K. (1935) Der kumpan in der umwelt des vogels. *Journal für Ornithologie Leipzig* 83, 137–213, 289–394 [translated 1937: The companion in the birds' world. *Auk* 54, 245–273].

Lorenz, K. (1939) Vergleichende verhaltens-forschung. *Zoologische Anzeige Supplement* 12, 69–102.

Lorenz, K. (1941) Vergleichende bewegungsstudien an Anatiden. *Journal für Ornithologie Leipzig* 89, 194–293.

Lorenz, K. (1965) *Evolution and Modification of Behavior*. University of Chicago Press, Chicago, Illinois.

Lorenz, K. (1966) *On Aggression*. Methuen, London.

Lou, Z. and Hurnik, J.F. (1994) An ellipsoid farrowing crate: its ergonomic design and effects on pig productivity. *Journal of Animal Science* 72, 2610–2616.

Lusseau, D. and Newman, M.E.J. (2004) Identifying the role that animals play in their social networks. *Proceedings of the Royal Society of London B (Suppl.)* 271, S477–S481.

Lutgendorf, S.K. (2001) Life, liberty and the pursuit of happiness: good welfare in humans. In: Broom, D. (ed.) *Coping with Challenge: Welfare in Animals Including Humans*. Dahlem University Press, Berlin, pp. 49–62.

Lynch, J.J. (1980) Behaviour of livestock in relation to their productivity. In: Rechcigl, M. (ed.) *Handbook of Nutrition and Food*. CRC Press, West Palm Beach, Florida.

Lynch, J.J., Hinch, G. and Adams, D.B. (1992) *The Behaviour of Sheep: Biological Principles and Implications for Production*. CAB International, Wallingford, UK.

Lyons, D.T., Freeman, A.E. and Kuck, A.L. (1991) Genetics of health traits. *Journal of Dairy Science* 74, 1092–1100.

MacArthur, R.H. and Pianka, E.R. (1966) On the optimal use of a patchy environment. *American Naturalist* 102, 381–383.

MacDermott, R.P. and Stacey, M.C. (1981) Further characterisation of the human autologous mixed leukocyte reaction (MLR). *Journal of Immunology* 126, 729–734.

MacKay, J.R.D., Turner, S.P., Hyslop, J., Deag, J.M., Haskell, M.J. (2013) Short-term temperament tests in beef cattle relate to long-term measures of behavior recorded in the home pen. *Journal of Animal Science*, 91, 4917–4924.

MacKenzie, A.M., Drennan, M., Rowan, T.G., Dixon, J.B. and Carter, S.D. (1997) Effect of transportation and weaning on humoral immune responses of calves. *Research in Veterinary Science* 63, 227–230.

MacKintosh, N.J. (1973) Stimulus selection: learning to ignore stimuli that predict no change in reinforcement. In: Hinde, R.A. and Stevenson-Hinde, J. (eds) *Constraints on Learning*. Academic Press, London.

MacLean, C.W. (1969) Observations on non-infectious infertility in sows. *Veterinary Record* 85, 675–682.

Madec, F. (1984) Urinary disorders in intensive pig herds. *Pig News and Information* 5 (2), 89–93.

Madec, F. (1985) La consommation d'eau chez la truie gestante en elevage intensif. *Journeées Recherches Porcine en France* 17, 223–236.

Madel, A. (2005) Hypocalcaemia in ewes precipitated by stormy night in Welsh borders. *Veterinary Times* 25 April, 9.

Mair, T., Love, S., Schumacher, J. and Watson, E. (1998) *Equine Medicine, Surgery and Reproduction*. Saunders, London.

Malyshenko, N.M. (1982) The role of corticosteroids in the genesis of fear and aggressive reactions. *Zh Vssh Nerva Deyat im I P Pavlova* 32, 144–151.

Manser, C.E. (1992) The Assessment of Stress in Laboratory Animals. RSPCA unpublished report.

Manser, C.E. (1996) Effects of lighting on the welfare of domestic poultry: a review. *Animal Welfare* 5, 341–360.

Manser, C.E., Morris, T.H. and Broom, D.M. (1995) An investigation into the effects of solid or grid cage flooring on the welfare of laboratory rats. *Laboratory Animals* 29, 353–363.

Manser, C.E., Elliott, H., Morris, T.H. and Broom, D.M. (1996) The use of a novel operant test to determine the strength of preference for flooring in laboratory rats. *Laboratory Animals* 30, 1–6.

Manser, C.E., Broom, D.M., Overend, R. and Morris, T.H. (1998a) Investigation into the preference of laboratory rats for nest-boxes and nesting materials. *Laboratory Animals* 32, 23–35.

Manser, C.E., Broom, D.M., Overend, R. and Morris, T.H. (1998b) Operant studies to determine the strength of preference in laboratory rats for nest-boxes and nesting material. *Laboratory Animals* 32, 36–41.

Manteca, X. (2002) *Etología Clinica Veterinaria del Perro y del Gato*. Multi Médica, Barcelona, Spain.

Marahrens, M., von Richthofen, I., Schmelduch, S. and Hartung, J. (2003) Special problems of long-distance road transport of cattle. *Deutsche Tierärztliche Wochenschrift* 110, 120–125.

Marchant, J.N. and Broom, D.M. (1996) Effects of dry sow housing conditions on muscle weight and bone strength. *Animal Science* 62, 105–113.

Marchant, J.N., Rudd, A.R., Mendl, M.T., Broom, D.M., Meredith, M.J., Corning, S. and Simmins, P.H. (2000) Timing and causes of piglet mortality in alternative and conventional farrowing systems. *Veterinary Science* 147, 209–214.

Marchant, J.N., Broom, D.M. and Corning, S. (2001) The influence of sow behaviour on piglet mortality due to crushing in an open farrowing system. *Animal Science* 72, 19–28.

Marchant-Forde, J.N. (ed.) (2009) *The Welfare of Pigs*. Springer, Berlin.

Marchant-Forde, J.N., Marchant-Forde, R.M. and Weary, D.M. (2002) Responses of dairy cows and calves to each other's vocalisations after early separation. *Applied Animal Behaviour Science* 78, 19–28.

Marchant-Forde, J.N., Lay, D.C., Pajor, E.A., Richert, B.T. and Schinckel, A.P. (2003) The effects of ractopamine on the behavior and physiology of finishing pigs. *Journal of Animal Science* 81, 416–422.

Margerison, J.K., Preston, T.R., Berry, N. and Phillips, C.J.C. (2003) Cross-sucking and other oral behaviours in calves, and their relation to cow suckling and food provision. *Applied Animal Behaviour Science* 80, 277–286.

Martin, A.R. and Wickelgren, W.O. (1971) Sensory cells in the spinal cord of the sea lamprey. *Journal of Physiology* 212, 65–83.

Martin, G. (1975) Uber verhaltenst orungen von legehennen in kmig. Ein beitrag zur klarung des problems tierschutzgeschechter huhnerhaltung. *Angewandte Ornithogie Leipzig* 4, 145–176.

Martin, G. (1987) Animal welfare in chicken management: obtaining knowledge and evaluating results. In: von Loeper, E., Martin, G., Muller, J., Nabholz, A., van Putten, G., Sambraus, H.H., Teutsch, G.M., Troxler, J. and Tschanz, B. (eds) *Ethical Ethological and Legal Aspects of Intensive Farm Animal Management*. Tierhaltung 18, Birkhauser Verlag, Basel, Switzerland, pp. 49–82.

Martin, P. and Bateson, P. (2007) *Measuring Behaviour*. 2nd edn. Cambridge University Press, Cambridge, UK.

Marx, D. and Mertz, R. (1989) Ethological experiments with early weaned piglets given free choice of pens with different use of straw. *Deutsche Tierarztliche Wochenschrift* 96, 20–26.

Marx, G., Horn, T., Thielebein, J., Knubel, B. and von Borrell, E. (2003) Analysis of pain-related vocalisation in young pigs. *Journal of Sound and Vibration* 266, 687–698.

Mason, G.J. (1991) Stereotypies: a critical review. *Animal Behaviour* 41, 1015–1037.

Mason, G.J. (1993) Age and context affect the stereotypical behaviours of caged mink. *Behaviour* 127, 191–229.

Mason, G.J. (1994) Tail-biting in mink (*Mustela vison*) is influenced by age at removal from the mother. *Animal Welfare* 3, 305–311.

Mason, G.J. (2010) Species differences in responses to captivity: stress, welfare and the comparative method. *Trends in Ecology and Evolution* 25, 713–721.

Mason, G.J. and Rushen, J. (eds) (2008) *Stereotypic Animal Behaviour: Fundamentals and Applications to Welfare*. CAB International, Wallingford, UK.

Mason, G.J., Cooper, J.J. and Clarebrough, C. (2001) Frustrations of fur-farmed mink. *Nature* 410, 35–36.

Mason, G.J., Clubb, R., Latham, N. and Vickery, S. (2007) Why and how should we use environmental enrichment to tackle stereotypic behaviour? *Applied Animal Behaviour Science* 102, 163–188.

Mason, J.W. (1971) A re-evaluation of the concept of 'non-specificity' in stress theory. *Journal of Psychiatric Research* 8, 323–333.

Mason, J.W. (1975) Emotion as reflected in patterns of endocrine integrations. In: Levi, L. (ed.) *Emotions – their Parameters and Measurements*. Raven, New York, pp. 233–251.

Mason, J.W., Brady, J.V. and Tolliver, G.A. (1968a) Plasma and urinary 17-hydroxycorticosteroid responses to 72-hour avoidance sessions in the monkey. *Psychosomatic Medicine* 30, 608–630.

Mason, J.W., Wool, M.S., Wherry, F.E., Pennington, L.L., Brady, J.V. and Beer, B. (1968b) Plasma growth hormone response to avoidance sessions in the monkey. *Psychosomatic Medicine* 30, 760–773.

Mason, W.A. (1960) The effects of social restriction on the behavior of rhesus monkeys: I. Free social behavior. *Journal of Comparative and Physiological Psychology* 53, 582–589.

Mason, W.A. (1961) The effects of social restriction on the behavior of rhesus monkeys: III. Dominance tests. *Journal of Comparative and Physiological Psychology* 5, 694–699.

Matheson, S.M., Asher, L. and Bateson, M. (2008) Larger, enriched cages are associated with 'optimistic' response biases in captive European starlings (*Sturnus vulgaris*). *Applied Animal Behaviour Science* 109, 374–383.

Mathews, G. and Wickelgren, W.O. (1978) Trigeminal sensory neurons of the sea lamprey. *Journal of Comparative Physiology* 123, 329–333.

Matthews, L.R. and Ladewig, J. (1994) Environmental requirements of pigs measured by behavioural demand functions. *Animal Behaviour* 47, 713–719.

Mattiello, S., Canali, E., Ferrante, V., Caniatti, M., Gottardo, F., Cozzi, G., Andrighetto, I. and Verga, M. (2002) The provision of solid feeds to veal calves: II. Behavior, physiology, and abomasal damage. *Journal of Animal Science* 80, 367–375.

Matull, A. and Reiter, K. (1995) Investigations of comfort behaviour of pekin duck, muskovy duck and mullard duck. In: *10th European Symposium on Waterfowl*, Halle, Germany, pp. 118–121.

Maule, A.G. and Schreck, C.B. (1991) Stress and cortisol treatment changed affinity and number of glucocorticoid receptors in leukocytes and gill of coho salmon. *General and Comparative Endocrinology* 84, 83–93.

Mavrogenis, A.P. and Robinson, O.W. (1976) Factors affecting puberty in swine. *Journal of Animal Science* 42, 1251–1255.

Maxwell, M.H. and Robertson, G.W. (1998) UK survey of broiler ascites and sudden death syndrome in 1993. *British Poultry Science* 39, 203–215.

Maynard-Smith, J. (1982) *Evolution and the Theory of Games*. Cambridge University Press, Cambridge, UK.

McAdie, T.M. and Keeling, L.J. (2000) Effect of manipulating feathers of laying hens on the incidence of feather pecking and cannibalism. *Applied Animal Behaviour Science* 68, 215–229.

McBride, G. (1963) The 'teat order' and communication in young pigs. *Animal Behaviour* 11, 53–56.

McBride, G., Arnold, G.W., Alexander, G. and Lynch, J.J. (1967) Ecological aspects of behaviour of domestic animals. *Proceedings of the Ecological Society of Australia* 2, 133–165.

McBride, G., Parer, I.P. and Foenander, F. (1969) The social organisation of the feral domestic fowl. *Animal Behaviour Monographs* 2, 125–181.

McBride, S. and Hemmings, A. (2009) A neurologic perspective of equine stereotypy. *Journal of Equine Veterinary Science* 29, 10–16.

McBride, S.D. and Hemmings, A. (2001) Nucleus accumbens DI dopamine receptor numbers are significantly higher in horses performing stereotypic behaviour. *Proceedings of Association for Veterinary Teachers and Research Workers Conference*, Scarborough, 2001, AVTRW.

McCulloch, M., Jezierski, T., Broffman, M., Hubbard, A. and Turner, K. (2006) Diagnostic accuracy of canine scent detection in early-and late-stage lung and breast cancers. *Integrative Cancer Therapies* 5, 30–39.

McCune, S. (1992) Temperament and the welfare of caged cats. PhD thesis, University of Cambridge, Cambridge, UK.

McCune, S. (1994) Caged cats: avoiding problems and providing solutions. *Companion Animal Behaviour Therapy Study Group Newsletter* 7, 33–40.

McDonald, C.L., Beilharz, R.G. and McCutchan, J.C. (1981) Training cattle to control by electric fences. *Applied Animal Ethology* 7, 113–121.

McDonnell, S.M. (2002) Behaviour of horses. In: Jensen, P. (ed.) *The Ethology of Domestic Animals: an Introductory Text.* CAB International, Wallingford, UK, pp. 119–129.

McEwen, B.S. (2001) Protective and damaging effects of stress mediators: lessons learned from the immune system and brain. In: Broom, D.M. (ed.) *Coping with Challenge: Welfare in Animals Including Humans.* Dahlem University Press, Berlin, pp. 229–246.

McFarland, D.J. (1965) The effect of hunger on thirst motivated behaviour in the Barbary dove. *Animal Behaviour* 13, 286–300.

McFarland, D.J. (1971) *Feedback Mechanisms in Animal Behaviour.* Academic Press, London.

McFarland, D.J. and Sibly, R.M. (1975) The behavioural final common path. *Philosophical Transactions of the Royal Society B* 27, 265–293.

McGeown, D., Danbury, T.C., Waterman-Pearson, A.E. and Kestin, S.C. (1999) Effect of carprofen on lameness in broiler chickens. *Veterinary Record* 144, 668–671.

McGlone, J.J. and Blecha, F. (1987) An examination of behavioural, immunological and productive traits in four management systems for sows and piglets. *Applied Animal Behaviour Science* 18, 269–286.

McGlone, J.J. and Curtis, S.E. (1985) Behaviour and performance of weanling pigs in pens equipped with hide areas. *Journal of Animal Science* 60, 20–24.

McGlone, J.J. and Hellman, J.M. (1988) Local and general anesthetic effects on behaviour and performance of two and seven-week old castrated and uncastrated piglets. *Journal of Animal Science* 66, 3049–3058.

McGlone, J.J. and Morrow-Tesch, J. (1990) Productivity and behaviour of sows in level vs. sloped farrowing pens and crates. *Journal of Animal Science* 68, 82–87.

McGlone, J.J., Nicholson, R.I., Hellman, J.M. and Herzog, D.N. (1993) The development of pain in young pigs associated with castration and attempts to prevent castration-induced behavioral changes. *Journal of Animal Science* 71, 1441–1446.

McGovern, R.H., Feddes, J.J., Robinson, F.E. and Hanson, J.A. (1999) Growth performance, carcass characteristics and the incidence of ascites in broilers in response to feed restriction and litter oiling. *Poultry Science* 78, 522–588.

McGreevy, P. (2004) *Equine Behavior: a Guide for Veterinarians and Equine Scientists.* W.B. Saunders, Edinburgh, Philadelphia, pp. 369.

McGreevy, P.D., Cripps, P.J., French, N.P., Green, L.E. and Nicol, C.J. (1995) Management factors associated with stereotypic and redirected behavior in the thoroughbred horse. *Equine Veterinary Journal* 27, 86–91.

McKeegan, D.E.F., Sparks, N.H.C., Sandilands, V., Demmers, T.G.M., Boulcott, P. and Wathes, C.M. (2011) Physiological responses of laying hens during whole-house killing with carbon dioxide. *British Poultry Science* 52, 645–657.

McKeegan, D.E.F. and Philbey, A.W. (2012) Chronic neurophysiological and anatomical changes associated with infra-red beak treatment and their implications for laying hen welfare. *Animal Welfare* 21, 207–217.

McKeegan, D.E.F., Reimert, H.G.M., Hindle, V.A., Boulcott, P., Sparrey, J.M., Wathes, C.M., Demmers, T.G.M. and Gerritzen, M.A. (2013a) Physiological and behavioral responses of poultry exposed to gas-filled high expansion foam. *Poultry Science* 92, 1145–1154.

McKeegan, D.E.F., Sandercock, D.A. and Gerritzen, M.A. (2013b) Physiological responses to low atmospheric pressure stunning and the implications for welfare. *Poultry Science* 92, 858–868.

McVeigh, J.M. and Tarrant, V. (1983) Effect of propanolol on muscle glycogen metabolism during social regrouping of young bulls. *Journal of Animal Science* 56, 71–80.

Meagher, R.K., Duncan, I., Bechard, A. and Mason, G.J. (2011) Who's afraid of the big bad glove? Testing for fear and its correlates in mink. *Applied Animal Behaviour Science* 133, 254–264.

Meagher, R.K., Campbell, D.L., Dallaire, J.A., Díez-León, M., Palme, R. and Mason, G.J. (2013) Sleeping tight or hiding in fright? The welfare implications of different subtypes of inactivity in mink. *Applied Animal Behaviour Science* 144, 138–146.

Meddis, R. (1975) On the function of sleep. *Animal Behaviour* 23, 676–691.

Mei, J. van der (1987a) Health aspects of welfare research in veal calves. In: Schlichting, M.C. and Smidt, D. (eds) *Welfare Aspects of Housing Systems for Veal Calves and Fattening Bulls.* Commission of the European Communities, Luxembourg.

Mellor, D.J. and Bayvel, A.C.D. (2014) Animal welfare: focusing on the future. *O.I.E. Scientific and Technical Review* 33, 1–358. O.I.E., Paris.

Melotti, P., Roncarati, A., Garella, E., Carnevali, O., Mosconi, G. and Polzonetti-Magni, A. (1992) Effects of handling and capture stress on plasma glucose, cortisol and androgen levels in brown trout (*Salmo trutta* morpha *fario*). *Journal of Applied Ichthyology* 8, 234–239.

Melzack, R. and Dennis, S.G. (1980) Phylogenetic evolution of pain expression in animals. In: Kosterlitz, H.W. and Terenius, L.Y. (eds) *Pain and Society*. Report of Dahlem Workshop. Verlag Chemie, Weinheim, Germany, pp. 13–26.

Mench, J. (2004) Lameness: measuring and auditing broiler welfare. In: Appleby, M.C., Mench, J.A. and Hughes, B.O. (eds) *Poultry Behaviour and Welfare.* CAB International, Wallingford, UK.

Mendl, M. and Newberry, R.C. (1997) Social conditions. In: Appleby, M.C. and Hughes, B.O. (eds) *Animal Welfare*. CAB International, Wallingford, UK, pp. 191–203.

Mendl, M. and Paul, E.S. (2004) Consciousness, emotion and animal welfare: insights from cognitive science. *Animal Welfare* 13, 17–25.

Mendl, M., Zanella, A.J. and Broom, D.M. (1992) Physiological and reproductive correlates of behavioural strategies in female domestic pigs. *Animal Behaviour* 44, 1107–1121.

Mendl, M., Laughlin, K. and Hitchcock, D. (1997) Pigs in space: spatial memory and its susceptibility to interference. *Animal Behaviour* 54, 1491–1508.

Mendl, M., Burman, H.P., Parker, R.M.A. and Paul, E.S. (2009) Cognitive bias as an indicator of animal emotion and welfare: emerging evidence and underlying mechanisms. *Applied Animal Behaviour Science* 118, 161–181.

Mendl, M., Burman, O.H. and Paul, E.S. (2010a) An integrative and functional framework for the study of animal emotion and mood. *Proceedings of the Royal Society B: Biological Sciences* 277 (1696), 2895–2904.

Mendl, M., Brooks, J., Basse, C., Burman, O., Paul, E., Blackwell, E. and Casey, R. (2010b) Dogs showing separation-related behaviour exhibit a 'pessimistic' cognitive bias. *Current Biology* 20, R839–R840.

Meredith, M. (1999) Vomero-nasal organ. In: Knobil, E. and Neill, J.D. (eds) *Encyclopedia of Reproduction.* Vol. 4. Elsevier, Amsterdam, pp. 1004–1014.

Meredith, M. and Fernandez-Fewell, G. (1994) Vomeronasal system, LHRH and sex behaviour. *Psychoneural Endocrinology* 19, 657–672.

Messent, P.R. and Broom, D.M. (eds) (1986) *Encyclopaedia of Domestic Animals.* Grolier, New York.

Metcalfe, N.B., Huntingford, F.A. and Thorpe, J.E. (1987) The influence of predation risk on the feeding motivation and foraging strategy of juvenile Atlantic salmon. *Animal Behaviour* 35, 901–911.

Metz, J. and Metz, J.H.M. (1987) Behavioural phenomena related to normal and difficult deliveries in dairy cows. *Netherlands Journal of Agricultural Science* 35, 87–101.

Metz, J.H.M. (1975) Time patterns of feeding and rumination in domestic cattle. *Mededelingen Landbhoogeschule Wageningen* 75, 1–66.

Metz, J.H.M. (1984) Regulation of sucking behaviour of calves. In: Unsholm, J., van Putten, G. and Zeeb, K. (eds) *Proceedings of the International Congress on Applied Ethology of Farm Animals,* Kiel, Germany. KTBL Darmstadt, Germany, pp. 70–73.

Meunier-Salaun, M.C., Vantrimponte, M.N., Raab, A. and Dantzer, R. (1987) Effect of floor area restriction upon performance, behaviour and physiology of growing-finishing pigs. *Journal of Animal Science* 64, 1371–1377.

Meyer, W.A. and Sunde, M.L. (1974) Bone breakages as affected by type of housing or an exercise matching for layers. *Poultry Science* 53, 878–885.

Meyer-Holzapfel, M. (1968) Abnormal behaviour in zoo animals. In: Fox, M.W. (ed.) *Abnormal Behavior in Animals*. W.B. Saunders, Philadelphia, Pennsylvania, pp. 476–503.

Michel, V. and Huonnic, D. (2003) A comparison of welfare, health and production performance of laying hens reared in cages or aviaries. *British Poultry Science* 43, 775–776.

Midgley, M. (1978) *Beast and Man: the Roots of Human Nature*. Harvester Press, Hassocks, UK.

Miklósi, A., Pongrácz, P., Lakatos, G., Topál, J. and Csányi, V. (2005) A comparative study of the use of visual communicative signals in interactions between dogs (*Canis familiaris*) and humans and cats (*Felis catus*) and humans. *Journal of Comparative Psychology* 119, 179–186.

Mill, J.S. (1843) *System of Logic, Deductive and Inductive*. Longman, London.

Miller, N.E. (1959) Liberalization of basic S-R concepts: extensions to conflict behaviour, motivation and social learning. In: Koch, S. (ed.) *Psychology: a Study of a Science*. Vol. II. McGraw Hill, New York.

Millot, S., Nilsson, J., Fosseidengen, J.E., Bégout, M.L., Fernö, A., Braithwaite, V.A. and Kristiansen, T.S. (2013) Innovative behaviour in fish: Atlantic cod can learn to use an external tag to manipulate a self-feeder. *Animal Cognition* 17, 779–785.

Mills, D.S., Redgate, S.E. and Landsberg, G.M. (2011) A meta-analysis of studies of treatments for feline urine spraying. *PloS ONE* 6, e18448, pp. 10.

Mirabito, L. (2002) Le bien-être des lapines: impact de nouveaux systèmes de logement. In: *Journée Nationale ITAVI Elevage du Lapin de Chair*, Nantes, 21 Novembre. ITAVI Editions, Paris.

Mirabito, L., Galliot, L., Souchet, C. (1994) Programme lumineux ou PMSG pour améliorer la réceptivité des lapines. *Cuniculture* 21 (1), 13–17.

Mirabito, L., Galliot, P., Souchet, C. and Pierre, V. (1999) Logements des lapins en engraissement en cage de 2 ou 6 individues: étude du budget temps, 55–58. *Proceedings 8èmes Journées de la Recherche Cunicole*, 9–10 June 1999, Paris.

Mirabito L., Galliot P. and Souchet C. (2004) Effet de la surface disponible et de l'aménagement des cages sur les performances zootechniques et le comportement des lapines et des jeunes. In: *Journée Nationale ITAVI sur l'Élevage du Lapin de Chair*, Pacé, 30 Novembre 2004. ITAVI Editions, *Paris, pp.* 40–52.

Mitterschiffthaler, M.T., Kumari, V., Malhi, G.S., Brown, R.G., Giampietro, V.P., Brammer, M.J., Suckling, J., Poon, L., Simmons, A., Andrew, C. and Sharma, T. (2003) Neural response to pleasant stimuli in anhedonia: a functional magnetic resonance imaging study. *Neuroreport* 14, 177–182.

Moberg, G.P. (1985) Biological response to stress: key to assessment of animal well-being? In: Moberg, G.P. (ed.) *Animal Stress*. American Physiological Society, Bethesda, Maryland, pp. 27–49.

Moe, R.O., Guémené, D., Larsen, H.J.S., Bakken, M., Lervik, S., Hetland, H. and Tauson, R. (2004) Effects of pre-laying rearing conditions in laying hens housed in standard or furnished cages on various indicators of animal welfare. *Book of Abstracts of the XXII World's Poultry Congress, Istanbul, Turkey*. WPSA Turkish Branch, p. 329.

Mogenson, G.J. and Calaresu, F.R. (1978) Food intake considered from the viewpoint of systems analysis. In: Booth, D.A. (ed.) *Hunger Models*. Academic Press, London, pp. 1–24.

Mogensen, L., Krohn, C.C., Sorensen, J.T., Hindhede, J. and Nielsen, L.H. (1997) Association between resting behaviour and live weight gain in dairy heifers housed in pens with different space allowance and floor type. *Applied Animal Behaviour Science* 55, 11–19.

Mohan Raj, A.B., Gregory, N.G. and Wotton, S.B. (1991) Changes in the somatosensory evoked potentials and spontaneous electroencephalogram of hens during stunning in argon-induced anoxia. *British Veterinary Journal* 147, 322–330.

Moinard, C. and Morisse, J.P. (1998) Étude de l'incidence de sept types de cages sur l'état sanitaire, les performances zootechniques, la physiologie et le comportement des poules pondeuses. *Science et Technie d'Aviculture*.

Moinard, C., Mendl, M., Nicol, C.J. and Green, L.E. (2003) A case control study of on-farm risk factors for tail-biting in pigs. *Applied Animal Behaviour Science* 81, 333–355.

Møller, S. (1992) Production systems and management in the Danish mink production. *Norwegian Journal of Agricultural Science (Suppl.)* 9, 562–568.

Mononen, J., Harri, M., Rekila, T. and Ahola, L. (1995a) Foxes do not like non-transparent walls. *NJF-Utredning/Rapport* 106. NJF-seminarium, Gothenburg, Sweden, 4–6 October, p. 38.

Mononen, J., Harri, M., Rekila, T., Korhonen, H. and Niemela, P. (1995b) Use of nest boxes by young farmed silver foxes (*Vulpes vulpes*) in autumn. *Applied Animal Behaviour Science* 43, 213–221.

Montero, D., Izquierdo, M.S., Tort, L., Robaina, L. and Vergara, J.M. (1999) High stocking density produces crowding stress altering some physiological and biochemical parameters in gilthead seabream, *Sparua aurata*, juveniles. *Fish Physiology and Biochemistry* 20, 53–50.

Morgan, P.D. and Arnold, G.W. (1974) Behavioural relationships between Merino ewes and lambs during the four weeks after birth. *Animal Production* 19, 196.

Morgan-Davies, C., Waterhouse, A., Pollock, M.L. and Milner, J.M. (2008) Body condition score as an indicator of ewe survival under extensive conditions. *Animal Welfare* 17, 71–77.

Morisse, J.P. (1982) Les maladies des veaux. *Recueils de Médicine Vétérinaire* 158, 307.

Morton, D.B. and Griffiths, P.H.M. (1985) Guidelines on the recognition of pain, distress and discomfort in experimental animals and an hypothesis for assessment. *Veterinary Record* 116, 431–436.

Morton, D.B., Jennings, M., Batchelor, G.R., Bell, D., Birke, L., Davies, K., Eveleigh, J.R., Gunn, D., Heath, M., Howard, B., Koder, P., Phillips, J., Poole, T., Sainsbury, A.W., Sales, G.D., Smith, D.J.A., Stauffacher, M. and Turner, R.J. (1993) Refinements in rabbit husbandry. *Laboratory Animals* 27, 301–329.

Moser, E., and McCulloch, M. (2010) Canine scent detection of human cancers: A review of methods and accuracy. *Journal of Veterinary Behavior: Clinical Applications and Research* 5, 145–152.

Mottershead, B.E., Lynch, J.J. and Alexander, G. (1982) Sheltering behaviour of shorn and unshorn sheep in mixed or separate flocks. *Applied Animal Ethology* 8, 127–136.

Muehlemann, T., Reefmann, N., Wechsler, B., Wolf, M. and Gygax, L. (2011) In vivo functional near-infrared spectroscopy measures mood-modulated cerebral responses to a positive emotional stimulus in sheep. *NeuroImage* 54, 1625–1633.

Mueller, H.C. and Parker, P.G. (1980) Naive ducklings show different cardiac responses to hawk than to goose models. *Behaviour* 74, 101–113.

Müller-Graf, C., Candiani, D., Barbieri, S., Ribó, O., Afonso, A., Aiassa, E., Have, P., Correia, S., De Massis, F., Grudnik, T. and Serratosa, J. (2007) Risk assessment in animal welfare: EFSA approach. In: *Proceedings of World Congress on Alternatives to Animal Testing and Experimentation* 14, 789–794. Tokyo, AATEX.

Müller-Schwarze, D. (1999) Signal specialisation and evolution in mammals. In: Johnson, R.E., Müller-Schwarze, D. and Sorenson, P.W. (eds) *Advances in Chemical Signs in Vertebrates.* Kluwer, New York, pp. 1–14.

Muonehke, M.I. and Childless, W.M. (1994) Hooking mortality: a review for recreational fisheries. *Reviews in Fisheries Science* 2, 123–156.

Muráni, E., Ponsuksili, S., D'Eath, R.B., Turner, S.P., Kurt, E., Evans, G., Thölking. L., Klont, R., Foury, A., Mormède, P. and Wimmers, K. (2010) Association of HPA axis-related genetic variation with stress reactivity and aggressive behaviour in pigs. *BMC Genetics* 11, 74.

Murata, H., Takahashi, H. and Matsumoto, H. (1985) Influence of truck transportation of calves on their cellular immune function. *Japanese Journal of Veterinary Science* 47, 823–827.

Murray, E.A. (2007) The amygdala, reward and emotion. *Trends in Cognitive Science* 11, 489–497.

Murray, K.C., Davies, D.H., Cullenane, S.L., Edison, J.C. and Kirk, J.A. (2000) Taking lambs to the slaughter: marketing channels, journey structures and possible consequences for welfare. *Animal Welfare* 9, 111–112.

Murray, L.M.A., Byrne, K. and D'Eath, R.B. (2013) Pair-bonding and companion recognition in domestic donkeys *Equus asinus. Applied Animal Behaviour Science* 143, 67–74.

Nasr, M.A.F., Nicol, C.J. and Murrell, J.C. (2012) Do laying hens with keel bone fractures experience pain? *PloS ONE* 7, e42420, pp. 6.

Naumenko, E.V. and Belyaev, D.K. (1980) Neuroendocrine mechanisms in animal domestication: problems in animal genetics. *Proceedings of the 14th International Congress of Genetics*, Moscow, Russia, 2, 12–25.

Nebel, R.L. and McGilliard, M.L. (1993) Interactions of high milk yield and reproductive performance in dairy cows. *Journal of Dairy Science* 76, 3257–3268.

Nelson, R.J. (2000) *An Introduction to Behaviour and Endocrinology*. 2nd edn. Sinauer Associates, Sunderland, Massachusetts.

Nestor, K.E., Saif, Y.M., Zhu, J. and Noble, D.O. (1996a) Influence of growth selection in turkeys on resistance to *Pasteurella multocida*. *Poultry Science* 75, 1161–1163.

Nestor, K.E., Noble, D.O., Zhu, J. and Moritsu, Y. (1996b) Direct and correlated responses to long-term selection for increased body-weight and egg-production in turkeys. *Poultry Science* 75, 1180–1191.

Nettle, D. and Bateson, M. (2012) The evolutionary origin of mood and its disorders. *Current Biology* 22, R712-R721.

Newberry, R.C. (2004) Cannibalism in welfare of the laying hen. In: Perry, G.C. (ed.) *The Laying Hen*. CAB International, Wallingford, UK, pp. 239–258.

Newberry, R.C. and Hall, J.W. (1988) Space utilisation by broiler chickens in floor pens. In: Unshelm, J., van Putten, G., Zeeb, K. and Ekesbo, I. (eds) *Applied Ethology in Farm Animals*. KTBL, Darmstadt, Germany, pp. 305–309.

Nicol, C.J. (1987a) Effect of cage height and area on the behaviour of hens housed in battery cages. *British Poultry Science* 28, 327–335.

Nichol, C.J. (1987b) Behavioural responses of laying hens following a period of spatial restriction. *Animal Behaviour* 35, 1709–1719.

Nicol, C.J., Caplen, G., Edgar, J. and Browne, W.J. (2009) Associations between welfare indicators and environmental choice in laying hens. *Animal Behaviour* 78, 413–424.

Nicol, C.J., Gregory, N.G., Knowles, T.G., Parkman, I.D. and Wilkins, L.J. (1999) Differential effects of increased stocking density, mediated by increased flock size, on feather pecking and aggression in laying hens. *Applied Animal Behaviour Science* 65, 137–152.

Nichol, L. (1974) *L'Épopée Pastorienne et la Medicine Vétérinaire*. Nichol, Garches, France.

Nicks, B. and Vandenheede, M. (2014) Animal health and welfare: equivalent or complementary. In: Mellor, D.J. and Bayvel, A.C.D. (eds) *Animal Welfare: Focusing on the Future*. OIE Scientific and Technical Review 33, 97–101. OIE, Paris.

Nielsen, B. (1980) Wing bone fractures in laying-hens. *Danish Veterinary Journal* 62, 981–1016.

Nielsen, B. (1998) Interspecific comparison of lactational stress: is the welfare of dairy cows compromised? In: Veissier, I. and Boissy, A. (eds) *Proceedings of the 32nd Congress on International Social Applied Ethology*, INRA, Clermont Ferrand, France, p. 80.

Nimon, A.J. and Broom, D.M. (1999) The welfare of farmed mink (*Mustela vison*) in relation to housing and management: a review. *Animal Welfare* 8, 205–228.

Nimon, A.J. and Broom, D.M. (2001) The welfare of farmed foxes *Vulpes vulpes* and *Alopex lagopus* in relation to housing and management: a review. *Animal Welfare* 10, 223–248.

Noll, S.L., Elhalawani, M.E., Waibel, P.E., Redig, P. and Janni, K. (1991) Effect of diet and population density on male turkeys under various environmental conditions. 1. Turkey growth and health performance. *Poultry Science* 70, 923–934.

Nørgaard-Nielsen, G. (1985) Featherpecking and plumage condition in laying hens in an enriched environment in cages and on litter. In: Wegner, R.M. (ed.) *Proceedings of the 2nd European Symposium on Poultry Welfare,* Celle, Germany. World Poultry Science Association, Beekbergen, Netherlands, pp. 330–332.

Nørgaard-Nielsen, G.J. (1990) Bone strength of laying hens kept in an alternative system compared with hens in cages and or on deep-litter. *British Poultry Science* 31, 81–89.

Nottebohm, F. (1967) The role of sensory feedback in the development of avian vocalisations. *Proceedings of the 14th Ornithology Congress*. Blackwell, Oxford, UK.

Nottebohm, F. (1999) The anatomy and timing of vocal learning in birds. In: Hauser, M.D. and Komishi, M. (eds) *The Design of Animal Communication*. MIT Press, Cambridge, Massachusetts, pp. 63–110.

O'Brien, P.H. (1984) Leavers and stayers: maternal postpartum strategies in feral goats. *Applied Animal Behaviour Science* 12, 233–243.

O'Connor, E.A., Parker, M.O., McLeman, M.A., Demmers, T.G., Lowe, J.C., Cui, L., Davey, E.L., Owen, R.G., Wathes, C.M. and Abeyesinghe, S.M. (2010) The impact of chronic environmental stressors on growing pigs, *Sus scrofa* (Part 1): stress physiology, production and play behaviour. *Animal* 4, 1899–1909.

Odberg, F.O. and Francis-Smith, K. (1977) Studies on the formation of ungrazed eliminative areas in fields used by horses. *Applied Animal Ethology* 3, 27–34.

Oden, K., Keeling, L.J. and Algers, B. (2002) The behaviour of hens in two types of aviary systems on twenty-five commercial farms in Sweden. *British Poultry Science* 43, 169–181.

Odendaal, J.S.J. and Meintjes, R.A. (2003) Neurophysiological correlates of affiliative behaviour between humans and dogs. *The Veterinary Journal* 165, 296–301.

Odling-Smee, L. and Braithwaite, V.A. (2003) The role of learning in fish orientation. *Fish and Fisheries* 4, 235–246.

O'Driscoll, K.K.M. and Broom, D.M. (2011) Does access to open water affect the health of Pekin ducks (*Anas platyrhynchos*)? *Poultry Science* 90, 299–307.

O'Driscoll, K.K.M. and Broom, D.M. (2012) Does access to open water affect the behaviour of Pekin ducks (*Anas platyrhynchos*)? *Applied Animal Behaviour Science* 136, 156–165.

Oester, H. (1994) Sitzstangenformen und ihr Einfluss auf die Entstehung von Fussballengeschwüren bei Legehennen. *Archiv für Geflügelkunde* 58, 231-238.

Oikawa, M., Hobo, S., Oyomada, T. and Yoshikawa, H. (2005) Effects of orientation, intermittent rest and vehicle cleaning during transport on development of transport-related respiratory disease in horses. *Journal of Comparative Pathology* 132, 153–168.

Oke, K.B., Westley, P.A., Moreau, D.T. and Fleming, I.A. (2013) Hybridization between genetically modified Atlantic salmon and wild brown trout reveals novel ecological interactions. *Proceedings of the Royal Society B: Biological Sciences* 280, 1763, pp. 9.

Olsson, I.A.S. and Keeling, L.J. (2003) No effect of social competition on sham dust bathing in furnished cages laying hens. *Acta Agricultura Scandinavica A. Animal Science* 52, 253–256.

Olsson, I.A.S., Keeling, L.J. and McAdie, T.M. (2002) The push-door for measuring motivation in hens: laying hens are motivated to perch at night. *Animal Welfare* 11, 11–19.

Oltenacu, P.A. and Broom, D.M. (2010) The impact of genetic selection for increased milk yield on the welfare of dairy cows. *Animal Welfare* 19 (S), 39–49.

Oltenacu, P.A., Frick, A. and Lindhe, B. (1991) Relationship of fertility to milk yield in Swedish cattle. *Journal of Dairy Science* 71, 264–268.

Oostindjer, M., Bolhuis, J.E., Mendl, M., Held, S., Brand H. van den and Kemp, B. (2011) Learning how to eat like a pig: effectiveness of mechanisms for vertical social learning in piglets. *Animal Behaviour* 82, 503–511.

Orgeur, P. and Signoret, J.P. (1984) Sexual play and its functional significance in the domestic sheep (*Ovis aries*). *Physiology and Behaviour* 33, 111–118.

Orr, R.J., Tozer, K.N., Griffith, B.A., Champion, R.A., Cook, J.E. and Rutter, S.M. (2012) Foraging paths through vegetation patches for beef cattle in semi-natural pastures. *Applied Animal Behaviour Science* 141, 1–8.

Osthaus, B., Marlow, D. and Ducat, P. (2010) Minding the gap: spatial perseveration error in dogs. *Animal Cognition* 13, 881–885.

Osthaus, B., Proops, L., Hocking, I. and Burden, F. (2013) Spatial cognition and perseveration by horses, donkeys and mules in a simple A not-B detour task. *Animal Cognition* 16, 301–305.

Overall, K.L. (1997) *Clinical Behavioral Medicine for Small Animals.* Mosby, St Louis, Missouri.

Overmier, J.B., Patterson, J. and Wielkiewicz, R.M. (1980) Environmental contingencies as sources of stress in animals. In: Levine, S. and Ursin, H. (eds) *Coping and Health.* Plenum Press, New York, pp. 1–38.

Owen, J.B. and Ridgman, W.J. (1967) The effect of dietary energy content on the voluntary intake of pigs. *Animal Production* 9, 107–113.

Owen, R.A., Fullerton, J. and Barnum, D.A. (1983) Effects of transportation, surgery, and antibiotic therapy in ponies infected with Salmonella. *American Journal of Veterinary Research* 44, 46–50.

Owen-Smith, R.N. (2002) *Adaptive Herbivore Ecology: from Resources to Populations in Variable Environments.* Cambridge University Press, Cambridge, UK.

Packer, C. (1977) Reciprocal altruism in *Papio anubis. Nature, London* 265, 441–443.

Padilla, S.C. (1935) Further studies on the delayed pecking of chicks. *Journal of Comparative Psychology* 20, 413–443.

Pain, B.F. and Broom, D.M. (1978) The effects of injected and surface-spread slurry on the intake and behaviour of dairy cows. *Animal Production* 26, 75–83.

Pain, B.F., Leaver, J.D. and Broom, D.M. (1974) Effects of cow slurry on herbage production intake by cattle and grazing behaviour. *Journal of the British Grassland Society* 29, 85–91.

Pal, S.K. (2003) Urine-marking by free-ranging dogs (*Canis familiaris*) in relation to sex, season, place and posture. *Applied Animal Behaviour Science* 80, 45–59.

Pal, S.K. (2005) Parental care in free-ranging dogs (*Canis familiaris*). *Applied Animal Behaviour Science* 90, 31–47.

Pal, S.K. (2008) Maturation and development of social behaviour during early ontogeny in free-ranging dog puppies in West Bengal, India. *Applied Animal Behaviour Science* 111, 95–107.

Pal, S.K. (2011) Mating system of free-ranging dogs (*Canis familiaris*). *International Journal of Zoology* Article ID 314216, pp. 10.

Palme, R. (2012) Monitoring stress hormone metabolites as a useful, non-invasive tool for welfare assessment in farm animals. *Animal Welfare* 21, 331–337.

Palmer, A.C. (1976) *Introduction to Animal Neurology.* 2nd edn. Blackwell, Oxford, UK.

Pankhurst, N.W. and Dedual, M. (1994) Effect of capture and recovery on plasma levels of cortisol, lactate and gonadal steroids in a natural population of rainbow trout. *Journal of Fish Biology* 45, 1013–1025.

Panksepp, J. (1998) *Affective Neuroscience.* Oxford University Press, New York.

Paranhos da Costa, M.J.R. and Broom, D.M. (2001) Consistency of side choice in the milking parlour by Holstein-Friesian cows and its relationship with their reactivity and milk yield. *Applied Animal Behaviour Science* 70, 177–186.

Paranhos da Costa, M.J.R. and Cromberg, V.U. (1998) Relações materno-filiais em bovines de conte nas primeiras horas apos o porto. In: Paranhos da Costa, M.J.R. and Cromberg, V.U. (eds) *O Comportamento Materno em Mamíferos.* Sociedade Brasileira de Etologie, São Paolo, Brazil, pp. 215–235.

Parker, M.O., O'Connor, E.A., McLeman, M.A., Demmers, T.G.M., Lowe, J.C., Owen, R.C., Davey, E.L., Wathes, C.M. and Abeyesinghe, S.M. (2010) The impact of chronic environmental stressors on growing pigs, *Sus scrofa* (Part 2): social behaviour. *Animal* 4, 1910–1921.

Parsons, A.J., Newman, J.A., Penning, P.D., Harvey, A. and Orr, R.J. (1994) Diet preference of sheep: effects of recent diet, physiological state and species abundance. *Journal of Animal Ecology* 63, 465–478.

Patterson, R.L.S. (1968) Identification of 3-hydroxy-5 androst-16-ene as the musk odour component of boar submaxillary salivary gland and its relationship to the sex odour taint in pork meat. *Journal of Science Food and Agriculture* 19, 43.

Paul, E.S., Harding, E.J. and Mendl, M. (2005) Measuring emotional processes in animals: the utility of a cognitive approach. *Neuroscience and Biobehavioral Reviews* 29, 469–491.

Pavlov, I.P. (1927) *Conditioned Reflexes* [GV Anrep, translated].

Paxton, H., Daley, M.A., Corr, S.A. and Hutchinson, J.R. (2013) The gait dynamics of the modern broiler chicken: a cautionary tale of selective breeding. *The Journal of Experimental Biology* 216, 3237–3248.

Pearce, C.A.P. (1993) Behaviour and other indices of welfare in growing and finishing pigs kept on straw-flow, bare concrete, full slats and deep straw. PhD thesis, University of Aberdeen, UK.

Pearce, G.P. and Pearce, A.N. (1992) Contact with a sow in oestrus or a mature boar stimulates the onset of oestrus in weaned sows. *Veterinary Record* 130, 5–9.

Pearce, G.P., Paterson, A.M. and Hughes, P.E. (1988) Effect of short-term elevations in plasma cortisol concentration on LH secretion in prepubertal gilts. *Journal of Reproduction and Fertility* 83, 413–418.

Pedersen, V. (1991) Early experience with the farm environment and effects on later behaviour in silver *Vulpes vulpes* and blue foxes *Alopex lagopus. Behavioural Processes* 25, 163–169.

Pedersen, V. (1992) Handling of silver foxes at different ages pre-weaning and post-weaning and effects on later behaviour and stress-sensitivity. *Norwegian Journal of Agricultural Sciences (Suppl.)* 9, 529–535.

Pedersen, V. (1994) Long-term effects of different handling procedures on behavioural, physiological, and production-related parameters in silver foxes. *Applied Animal Behaviour Science* 40, 285–296.

Pedersen, V. and Jeppesen, L.L. (1990) Effects of early handling on later behaviour and stress responses in the silver fox (*Vulpes vulpes*). *Applied Animal Behaviour Science* 26, 383–393.

Peeler, E.J., Otte, M.J. and Esslemont, R.J. (1994) Interrelationships of periparturient diseases in dairy cows. *Veterinary Record* 134, 29–132.

Pellis, S.M. and Pellis, V.C. (1998) The structure–function interface in the analysis of play-fighting. In: Bekoff, M. and Byers, J.A. (eds) *Animal Play: Evolutionary, Comparative and Ecological Perspectives.* Cambridge University Press, Cambridge, UK, pp. 115–140.

Penning, P.D. (ed.) (2004) *Ingestive Behaviour.* Herbage Intake Handbook. British Grassland Society, Reading, UK, pp.151–176.

Penning, P.D., Parsons, A.J., Orr, R.J., Harvey, A. and Champion, R.A. (1995) Intake and behaviour responses by sheep, in different physiological states, when grazing monocultures of grass or white clover. *Applied Animal Behaviour* 45, 63–78.

Penny, R.H.C., Osborne, A.D., Wright, A.I. and Stephens, T.R. (1965) Foot rot in pigs: observations on the clinical diseases. *Veterinary Record* 77, 1101–1108.

Pepperberg, I.M. (1997) Social influences on the acquisition of human-based codes in parrots and non-human primates. In: Snowden, C.T. and Hansberger, M. (eds) *Social Influences on Vocal Development.* Cambridge University Press, Cambridge, UK, pp. 157–177.

Pepperberg, I.M. (2000) *The Alex Studies: Cognitive and Communicative Abilities of Grey Parrots.* Harvard University Press, Cambridge, Massachusetts.

Perkins, S.L., Zuidhof, M.J., Feddes, J.J.R. and Robinson, F.E. (1995) Effect of stocking density on air-quality and health and performance of heavy tom turkeys. *Canadian Agricultural Engineering* 37, 109–112.

Perry, G.C., Patterson, R.L.S., Stinson, G.C. and Macfie, H.J. (1980) Pig courtship behaviour: phermonal property of androstene steroids in male submaxillary secretion. *Animal Production* 31, 191–199.

Perry, S. and Bernier, N.J. (1999) The acute hormonal adrenergic stress response in fish: facts and fiction. *Aquaculture* 177, 285–295.

Peterson, P.K., Chao, C.C., Molitor, T., Murtaugh, M., Strogar, F. and Sharp, B.M. (1991) Stress and pathogenesis of infectious disease. *Review of Infectious Diseases* 13, 710–720.

Petherick, J.C. (1982) A note on the space use for excretory behaviour of suckling piglets. *Applied Animal Ethology* 9, 367–371.

Petherick, J.C. (1983a) A biological basis for the design of space in livestock housing. In: Baxter, S.H., Baxter, M.R. and MacCormack, J.A.C. (eds) *Farm Animal Housing and Welfare.* Current Topics in Veterinary Medicine and Animal Science 24, pp. 103–120.

Petherick, J.C. (1983b) A note on nursing termination and resting behaviour of suckling piglets. *Applied Animal Ethology* 9, 359–365.

Phillips, C. and Piggins, D. (1992) *Farm Animals and the Environment.* CAB International, Wallingford, UK.

Phillips, C.J.C. (1997) Review article: animal welfare considerations in future breeding programmes for farm livestock. *Animal Breeding Abstract* 65, 645–654.

Phillips, C.J.C. (2002) *Cattle Behaviour and Welfare.* 2nd edn. Blackwell Scientific, Oxford, UK.

Phillips, C.J.C. and Santurtun, E. (2013) The welfare of livestock transported by ship. *The Veterinary Journal* 196, 309–314.

Phillips, C.J.C., Morris, I.D. and Lomas, C.A. (2001) A novel operant conditioning test to determine whether dairy cows dislike passageways that are dark or covered with slurry. *Animal Welfare* 10, 65–72.

Pickel, D., Manucy, G., Walker, D., Hall, S. and Walker, J. (2004) Evidence for canine olfactory detection of melanoma. *Applied Animal Behaviour Science* 89, 107–116.

Pickering, A.D. (1998) Stress responses in farmed fish. In: Black, K.D. and Pickering, A.D. (eds) *Biology of Farmed Fish.* Sheffield Academic Press, Sheffield, UK, pp. 222–255.

Pickering, A.D. and Pottinger, T.G. (1989) Stress response and disease resistance in salmonid fish: effects of chronic elevation of plasma cortisol. *Fish Physiology and Biochemistry* 7, 253–258.

Piggins, D. (1992) Visual perception. In: Phillips, C. and Piggins, D. (eds) *Farm Animals and the Environment.* CAB International, Wallingford, UK, pp. 159–184.

Pilliner, S. and Davies, Z. (2004) *Equine Science.* 2nd edn. Blackwell, Oxford, UK.

Pinchbeck, G.L., Clegg, P.D., Boyde, A., Barr, E.D. and Riggs, C.M. (2013) Horse-, training- and race-level risk factors for palmar-plantar osteochondral disease in the racing Thoroughbred. *Equine Veterinary Journal* 45, 582–586.

Pinel, J.P.J. and Wilkie, D.M. (1983) Conditional defensive burying: a biological and cognitive approach to avoidance learning. In: Mellgren, R.L. (ed.) *Animal Cognition and Behaviour*. North Holland, Amsterdam, pp. 285–318.

Piquet, M., Bruckmaier, R.M. and Blum, J.W. (1993) Treadmill exercise of calves with different iron supply, husbandry and workload. *Journal of Veterinary Medicine A* 40, 456–465.

Plaizier, J.C.B., Lissemore, K.D., Kelton, D. and King, G.J. (1998) Evaluation of overall reproductive performance of dairy herds. *Journal of Dairy Science* 81, 1848–1854.

Plyusnina, I., Oskina, I. and Trut, L. (1991) An analysis of fear and aggression during early development of behaviour in silver foxes (*Vulpes vulpes*). *Applied Animal Behaviour Science* 32, 253–268.

Podberscek, A.L. (1994) Dog on a tightrope: the position of the dog in British society as influenced by press reports on dog attacks (1988 to 1992). *Anthrozoös* 7, 232–241.

Podberscek, A.L. and Serpell, J.A. (1996) The English Cocker Spaniel: preliminary findings on aggressive behaviour. *Applied Animal Behaviour Science* 47, 75–89.

Podberscek, A.L., Bradshaw, J.K. and Beattie, A.W. (1991) The behaviour of laboratory colony cats and their reactions to a familiar and unfamiliar person. *Applied Behaviour Science* 31, 119–130.

Podberscek, A.L., Paul, E.S. and Serpell, J.A. (2000) *Companion Animals and Us: Exploring the Relationships between People and Pets*. Cambridge University Press, Cambridge, UK.

Poindron, P. and Lévy, F. (1990) Physiological, sensory and experiential determinants of maternal and non-maternal behavior in sheep. In: Krasnegor, N.A. and Bridges, R.S. (eds) *Mammalian Parenting*. Oxford University Press, Oxford, UK, pp. 133–156.

Poindron, P., Caba, M., Arrati, P.G. and Krehbiel, D. (1994) Responses of maternal and non-maternal ewes to social and mother–young separation. *Behavioural Processses* 31, 97–110.

Poletto, R., Siegford, J., Nobis, W. and Zanella, A.J. (2003) Differential expression of genes in the hippocampus of early-weaned piglets when examined using a cDNA microarray. *Proceedings of International Symposium on Animal Functional Genomics*, East Lansing, Michigan, 9–11 May 2003.

Poletto, R., Steibel, J.P., Siegford, J.M. and Zanella, A.J. (2006) Effects of early weaning and social isolation on the expression of glucocorticoid and mineralocorticoid receptor and 11β-hydroxysteroid dehydrogenase 1 and 2 mRNAs in the frontal cortex and hippocampus of piglets. *Brain Research* 1067, 36–42.

Poletto, R., Rostagno, M.H., Richert, B.T. and Marchant-Forde, J.N. (2009) Effects of a 'step-up' ractopamine feeding program, sex, and social rank on growth performance, hoof lesions, and Enterobacteriaceae shedding in finishing pigs. *Journal of Animal Science* 87, 304–313.

Poletto, R., Cheng, H.W., Meisel, R.L., Garner, J.P., Richert, B.T. and Marchant-Forde, J.N. (2010a) Aggressiveness and brain amine concentration in dominant and subordinate finishing pigs fed the β-adrenoreceptor agonist ractopamine. *Journal of Animal Science* 88, 3107–3120.

Poletto, R., Meisel, R.L., Richert, B.T., Cheng, H.W. and Marchant-Forde, J.N. (2010b) Behavior and peripheral amine concentrations in relation to ractopamine feeding, sex, and social rank of finishing pigs. *Journal of Animal Science* 88, 1184–1194.

Polley, C.R., Craig, J.V. and Bhagwat, A.L. (1974) Crowding and agonistic behavior: a curvilinear relationship? *Poultry Science* 53, 1621–1623.

Pongrácz, P., Miklósi, A., Kubinyi, E., Gurabi, K., Topal, J. and Csányi, V. (2001) Social learning in dogs. I. The effect of a human demonstrator on the performance of dogs (*Canis familiaris*) in a detour task. *Animal Behaviour* 62, 1109–1117.

Pongrácz, P., Miklósi, A., Vida, V. and Csányi, V. (2005) The pet dog's ability for learning from a human demonstrator in a detour task is independent of breed and age. *Applied Animal Behaviour Science* 90, 309–323.

Popova, N.K., Voitenko, N.N., Kulikov, A.V. and Avgustinovich, D.F. (1991) Evidence for the involvement of central serotonin in mechanism of domestication in silver foxes. *Pharmacology Biochemistry and Behaviour* 4, 751–756.

Portavella, M., Torres, B. and Salas, C. (2004) Avoidance response in goldfish: emotional and temporal involvement of medial and lateral telencephatic pallium. *The Journal of Neuroscience* 24, 2342–2335.

Pösö, J. and Mäntysaari, E.A. (1996) Genetic relationships between reproductive disorders, operational days open and milk yield. *Livestock Production Science* 46, 41–48.

Potter, M.J. and Broom, D.M. (1987) The behaviour and welfare of cows in relation to cubicle house design. In: Wierenga, H.K. and Peterse, D.J. (eds) *Cattle Housing Systems, Lameness and Behaviour*. Martinus Nijhoff, Dordrecht, Netherlands.

Potterton, S.L., Green, M.J., Millar, K.M., Brignell, C.J., Harris, J., Whay, H.R. and Huxley, J.N. (2011) Prevalence and characterization of, and producers' attitudes towards, hock lesions in UK dairy cattle. *Veterinary Record* 169, 634–643.

Pottinger, T.G. (1998) Changes in blood cortisol, glucose and lactate in carp retained in anglers keep nets. *Journal of Fish Biology* 53, 728–742.

Price, E.O. (1984) Behavioural aspects of animal domestication. *Quarterly Review of Biology* 59, 1–32.

Price, E.O. (2002) *Animal Domestication and Behaviour*. CAB International, Wallingford, UK.

Price, E.O. and Smith, V.M. (1984) The relationship of male–male mounting to mate choice and sexual performance in male dairy goats. *Applied Animal Behaviour Science* 13, 71–82.

Price, E.O., Dunn, G.C., Talbot, J.A. and Dally, M.R. (1984a) Fostering lambs by odor transfer: the substitution experiment. *Journal of Animal Science* 59, 301–307.

Price, E.O., Smith, V.M. and Katz, L.S. (1984b) Sexual stimulation of male dairy goats. *Applied Animal Behaviour Science* 13, 83.

Price, E.O., Martinez, C.L. and Coe, B.L. (1985) The effects of twinning on mother–offspring behaviour in range beef cattle. *Applied Animal Behaviour Science* 13, 309–320.

Princz, Z., Szendr, Zs., Radnai, I., Bironé Németh, E. and Orova, Z. (2005) Free choice of rabbits among cages with different height [in Hungarian]. *Proceedings 17th Hungarian Conference on Rabbit Production*, Kaposvar, Hungary.

Pritchard, J.C., Lindberg, A.C., Main, D.C.J. and Whay, H.R. (2005) Assessment of the welfare of working horses, mules and donkeys, using health and behaviour parameters. *Preventive Veterinary Medicine* 69, 265–283.

Proops, L., Burden, F. and Osthaus, B. (2009) Mule cognition: a case of hybrid vigour? *Animal Cognition* 12, 75–84.

Proops, L., Burden, F. and Osthaus, B. (2012) Social relations in a mixed group of mules, ponies and donkeys reflect differences in equid type. *Behavioural Processes* 90, 337–342.

Provine, R.R. (1984) Wing-flapping during development and evolution. *American Scientist* 72, 448–455.

Prunier, A., Hay, M. and Servière, A. (2002) Evaluation et prévention de la douleur induite par les interventions de convenance chez le porcelet. *Journées de la Recherche Porcine en France* 34, 257–268.

Prunier, A., Mounier, A.M. and Hay, M. (2004) Effects of castration, tooth resection or tail docking on plasma metabolites and stress hormones in pigs. *Journal of Animal Science* 83, 216–222.

Pryce, J.E., Veerkamp, R.F., Thompson, R., Hill, R.G. and Simm, G. (1997) Genetic aspects of common health disorders and measures of fertility in Holstein Friesian dairy cattle. *Animal Science* 65, 353–360.

Pryce, J.E., Esslemont, R.J., Thompson, R., Veerkamp, R.F., Kossaibati, M.A. and Simm, G. (1998) Estimation of genetic parameters using health, fertility and production data from a management recording system for dairy cattle. *Animal Science* 66, 577–584.

Pryor, P.A., Hart, B.L., Bain, M.J. and Cliff, K.D. (2001) Causes of urine marking cats and effects of environmental management on the frequency of marking. *Journal of the American Veterinary Medical Association* 219, 1709–1713.

Putten, G. van (1977) Comfort behaviour in pigs and its significance regarding their well being. In: *European Association for Animal Production 28th Annual Meeting*. EAAP.

Putten, G. van (1978) Schweine. In: Sambraus, H.H. (ed.) *Nutztierethologie*. Paul Parey, Berlin.

Putten, G. van (1980) Objective observations on the behaviour of fattening pigs. *Animal Regulation Studies* 3, 105–118.

Putten, G. van (1982) Handling of slaughter pigs prior to loading and during loading on a lorry. In: Moss, R. (ed.) *Transport of Animals Intended for Breeding, Production and Slaughter*. Current Topics in Veterinary Medicine and Animal Science 18. Martinus Nijhoff, The Hague, Netherlands, pp. 15–25.

Putten, G. van and Dammers, J. (1976) A comparative study of the well-being of piglets reared conventionally and in cages. *Applied Animal Ethology* 2, 339–356.

Putten, G. van and Elshof, W.J. (1978) Observations on the effect of transport on the well-being and lean quality of slaughter pigs. *Animal Regulation Studies* 1, 247–271.

Putten, G. van and Elshof, W.J. (1982) Inharmonious behaviour of veal calves. *Hohenheimer Arbeiten* 121, 61–71.

Quaranta, A., Siniscalchi, M. and Vallortigara, G. (2007) Asymmetric tail-wagging responses by dogs to different emotive stimuli. *Current Biology* 17, R199-R201.

Quinn, P.J., Markey, B.K., Carter, M.E., Donnelley, W.J. and Leonard, F.C. (2000) *Veterinary Microbiology and Microbial Diseases*. Blackwell, Oxford, UK.

Raby, C.R. and Clayton, N.S. (2009) Prospective cognition in animals. *Behavioural Processes* 80, 314–324.

Radostits, O.M., Gay, C.C., Blood, D.C. and Hinchcliff, K.W. (2000) *Veterinary Medicine: a Textbook of the Diseases of Cattle, Sheep, Pigs, Goats and Horses*. 9th edn. W.B. Saunders, London.

Ratner, S.C. and Thompson, R.W. (1960) Immobility functions (fear) of domestic fowl as a function of age and prior experience. *Animal Behaviour* 8, 186–191.

Raussi, S., Lensink, B.J., Boissy, A., Pyykkönen, M. and Veissier, I. (2003) The impact of social and human contacts on calves' behaviour and stress responses. *Animal Welfare* 12, 191–203.

Rauw, W.M., Kanis, E., Noordhuizen-Stassen, E.N. and Grommers, F.J. (1998) Undesirable side effects of selection for high production efficiency in farm animals: a review. *Livestock Production Science* 56, 15–33.

Redwine, L.S., Altemus, M., Leong, Y.-M. and Carter, C.S. (2001) Lymphocyte responses to stress in post partum women: relationship to vagal tone. *Psychoneuroendocrinology* 26, 241–251.

Reefmann, N., Muehlemann, T., Wechsler, B. and Gygax, L. (2012) Housing induced mood modulates reactions to emotional stimuli in sheep. *Applied Animal Behaviour Science* 136, 146–155.

Reese, E.S. (1989) Orientation behaviour of butterfly fishes (family Chaetodontidae) on coral reefs – spacial learning of route specific landmarks and cognitive maps. *Environmental Biology of Fishes* 25, 79–86.

Regolin, L., Vallortigara, G. and Zanforlin, M. (1995) Object and spatial representations in detour problems by chicks. *Animal Behaviour* 49, 195–199.

Reinhardt, V. (1973) Social rank order and milking order in cows. *Zeitschrift für Tierpsychologie* 32, 281–292.

Reinhardt, V. and Reinhardt, A. (1982) Mock fighting in cattle. *Behaviour* 81, 1–13.

Reinhardt, V., Mutiso, F.M. and Reinhardt, A. (1978) Social behaviour and social relationships between female and male prepubertal bovine calves (*Bos indicus*). *Applied Animal Ethology* 4, 43–54.

Reiter, K. (1993) Short-term analyses of feeding and drinking behaviour of ducks. *Proceedings of the International Congress on Applied Ethology*, Berlin. Freie Universität Berlin, Berlin, pp. 508–510.

Reiter, K. and Bessei, W. (1995) A behavioural comparison of pekin, moscovy and mullard duck in the fattening period. *10th European Symposium on Waterfowl*, Halle, Germany, pp. 110–117.

Ridley, M. (1996) *The Origins of Virtue*. Viking, London.

Riese, G., Klee, G. and Sambraus, H.H. (1977) Das verhalten von kälbern in verschiedenen haltungsformen. *Deutsche Tierarztliche Wochenschrift* 84, 388–394.

Rioja-Lang, F.C., Roberts, D.J., Healy, S.D., Lawrence, A.B. and Haskell, M.J. (2009) Dairy cows trade-off feed quality with proximity to a dominant individual in Y-maze choice tests. *Applied Animal Behaviour Science* 117, 159–164.

Rioja-Lang, F.C., Roberts, D.J., Healy, S.D., Lawrence, A.B. and Haskell, M.J. (2012) Dairy cow feeding space requirements assessed in a Y-maze choice test. *Journal of Dairy Science* 95, 3954–3960.

Riolo, R.L., Cohen, M.D. and Axelrod, R. (2001) Evolution of cooperation without reciprocity. *Nature* 414, 441–443.

Ritchie, T.C. and Leonard, R.B. (1983) Immuno-histochemical studies on the distribution and origin of candidate peptidergic primary afferent neurotransmitters in the spinal cord of an elasmobranch fish, the Atlantic Stingray (*Dasyatis sabina*). *Journal of Comparative Neurology* 213, 111–125.

Robb, D.H.F. and Kestin, S.C. (2002) Methods used to kill fish: field observations and literature reviewed. *Animal Welfare* 11, 269–282.

Robb, D., Kestin, S. and Lines, J. (2000) Progress with humane slaughter. *Fish Farmer* November/December, 41.

Roberts, S.A., Simpson, D.M., Armstrong, S.D., Davidson, A.J., Robertson, D.H., McLean, L.,

Beynon, R.J. and Hurst, J.L. (2010) Darcin: a male pheromone that stimulates female memory and sexual attraction to an individual male's odour. *BMC Biology* 8, 75, pp. 21.

Roberts, S.A., Davidson, A.J., McLean, L., Beynon, R.J. and Hurst, J.L. (2012) Pheromonal induction of spatial learning in mice. *Science* 338, 1462–1465.

Roberts, W.A. (2000) *Principles of Animal Cognition*. McGraw-Hill, Boston, Massachusetts.

Rochlitz, I. (2003a) Study of factors that may predispose domestic cats to road traffic accidents: Part 1. *Veterinary Record* 153, 549–553.

Rochlitz, I. (2003b) Study of factors that may predispose domestic cats to road traffic accidents. Part 2. *Veterinary Record* 153, 585–588.

Rochlitz, I.R. (2000) Feline welfare issues. In: Turner, D.C. and Bateson, P. (eds) *The Domestic Cat: the Biology of its Behaviour*. 2nd edn. Cambridge University Press, Cambridge, UK, Chapter 11.

Rochlitz, I.R. (ed.) (2005a) *The Welfare of Cats*. Springer, Berlin.

Rochlitz, I.R. (2005b) A review of the housing requirements of domestic cats *Felis sylvestris cattus* kept in the home. *Applied Animal Behaviour Science* 93, 97–109.

Rochlitz, I.R. (2005c) Housing and welfare. In: Rochlitz, I.R. (ed.) *The Welfare of Cats*. Springer, Berlin, pp. 177–203.

Rochlitz, I., Podberscek, A.L. and Broom, D.M. (1998) Effects of quarantine on cats and their owners. *Veterinary Record* 143, 181–185.

Rodenburg, T.B., Bracke, M.B.M., Berk, J., Cooper, J., Faure, J.M., Guemene, D., Guy, G., Harlander, A., Jones, T., Knierim, U., Kuhnt, K., Pingel, H., Reiter, K., Serviere, J. and Ruis, M.A.W. (2005) Welfare of ducks in European duck husbandry systems. *World's Poultry Science Journal* 61, 633–646.

Rodriguez, F., Duran, E., Vargas, J.P., Torres, B. and Salas, C. (1994) Performance of goldfish trained in allocentric and egocentric maze procedures suggests the presence of a cognitive mapping system in fishes. *Animal Learning and Behaviour* 22, 409–420.

Rodriguez-Moldes, I., Manso, M.J., Becerra, M., Molist, P. and Anadon, R. (1993) Distribution of substance P-like immunoreactivity in the brain of the elasmobranch *Soyliorhinus anicula*. *Journal of Comparative Neurology* 335, 228–244.

Rollin, B.E. (1989) *The Unheeded Cry: Animal Consciousness, Animal Pain and Science*. Oxford University Press, Oxford, UK.

Rollin, B.E. (1995) *Farm Animal Welfare: Social, Bioethical and Research Issues*. Iowa State University Press, Ames, Iowa.

Rolls, E.T. (1994) Neural processing related to feeding in primates. In: Legg, C.R. and Booth, D. (eds) *Appetite: Neural and Behavioural Bases*. Oxford University Press, Oxford, UK, pp. 11–53.

Rolls, E.T. (1999) *The Brain and Emotion*. Oxford University Press, Oxford, UK.

Rolls, E.T. (2005) *Emotion Explained*. Oxford University Press, Oxford, UK.

Rooijakkers, E.F., Kaminski, J. and Call, J. (2009) Comparing dogs and great apes in their ability to visually track object transpositions. *Animal Cognition* 12, 789–796.

Rooijen, J. van (1980) Wahlversuche, eine ethologische methode zum sammeln von messwerten, u haltungseinglusse zu erfassen und zu buerteilen. *Aktuelle Arbeiten zur Artgemassen Tierhaltung. KTBL-Schrift* 264, 165–185.

Rose, J.D. (2002) The neurobehavioural nature of fishes and the question of awareness and pain. *Reviews in Fisheries Science* 10, 1–38.

Rose, R.M., Wodzicka-Tomaszewska, M. and Gumming, R.B. (1985) Agonistic behaviour, responses to a novel object and some aspects of maintenance behaviour in feral-strain and domestic chickens. *Applied Animal Behaviour Science* 13, 283–294.

Rosell, J.M. and de la Fuente, L.F. (2004) Health status of domestic rabbits in the Iberian penninsula. Influence of their origin. In: Becerril, C.M. and Puebla, A. (eds) *Proceedings 8th World Rabbit Congress*, Mexico, 7–10 September. Mexico City, pp. 371–377.

Rosell, J.M. and de la Fuente, L.F. (2009a) Culling and mortality in breeding rabbits. *Preventive Veterinary Medicine* 88, 120–127.

Rosell, J.M. and de la Fuente, L.F. (2009b) Effect of footrests on the incidence of ulcerative pododermatitis in domestic rabbit does. *Animal Welfare* 18, 199–204.

Rosell, J.M. and de la Fuente, L.F. (2012) On-farm causes of mortality in female rabbits. *Proceedings of the 10th World Rabbit Congress*. Sharm-el-Sheikh, Egypt, pp. 1147–1150.

Rosell, J.M., de la Fuente, L.F., Badiola, J.I., Fernández de Luco, D., Casal, J. and Saco, M. (2010) Study of urgent visits to commercial rabbit farms in Spain and Portugal during 1997–2007. *World Rabbit Science* 17, 127–136.

Rossdale, P.D. (1968) Perinatal behaviour in the thoroughbred horse. In: Fox, M.W. (ed.) *Abnormal Behavior in Animals.* Missouri Washington University, St Louis, Missouri, p. 563.

Rossdale, P.D. and Short, R.V. (1967) The time of foaling of thoroughbred mares. *Journal of Reproduction and Fertility* 13, 341–343.

Rossi, A.P. and Ades, C. (2008) A dog at the keyboard: using arbitrary signs to communicate requests. *Animal Cognition* 11, 329–338.

Rossier, J., French, E.D., Rivier, C., Ling, N., Guillemin, R. and Bloom, F.E. (1977) Foot-shock induced stress increases β-endorphin in blood but not brain. *Nature, London* 270, 618–620.

Rowell, C.H.F. (1961) Displacement grooming in the chaffinch. *Animal Behaviour* 9, 38–63.

Rowland, N.E., Li, B.-H. and Morien, A. (1996) Brain mechanisms and the physiology of feeding. In: Capaldi, E.D. (ed.) *Why We Eat What We Eat: the Psychology of Eating.* American Psychological Association, pp. 173–204.

Rozin, P. (1968) Specific aversions and neophobia as a consequence of vitamin deficiency and/or poisoning in half-wild and domestic rats. *Journal of Comparative and Physiological Psychology* 66, 82.

Rozin, P. (1976) The selection of foods by rats, humans and other animals. *Advances in the Study of Behavior* 6, 21–76.

Ruckebusch, Y. (1972a) Development of sleep and wakefulness in the foetal lamb. *Electroencephalography and Clinical Neurophysiology* 32, 119–128.

Ruckebusch, Y. (1972b) The relevance of drowsiness in farm animals. Animal BePre-natal stress amplifies the immediate behavioural responses to acute pain in piglets. *Biology Letters* 5(4), 452–454.

Ruckebusch, Y. (1974) Sleep deprivation in cattle. *Brain Research* 78, 495–499.

Ruckebusch, Y. and Bell, F.R. (1970) Étude electropolygraphique et comportementale des stats de veille et de sommeil chez la vache (*Bos taurus*). *Annales de Recherches Vétérinaires* 1, 41–62.

Rudd, A.R. and Marchant, J.N. (1995) Aspects of farrowing and lactating sow behaviour. *The Pig Journal* 34, 21–30.

Rugani, R., Fontanari, L., Simoni, E., Regolin, L. and Vallortigara, G. (2009) Arithmetic in newborn chicks. *Proceedings of the Royal Society B* 276, 2451–2460.

Ruis, M.A.W., Lenskens, P. and Coenen, E. (2003) Welfare of Pekin ducks increases when freely accessible open water is provided. In: *Proceedings of the 2nd World Waterfowl Conference,* Alexandria, Egypt, p. 17.

Rushen, J. (1986) Aversion of sheep for handling treatments: paired choice experiments. *Applied Animal Behaviour Science* 16, 363–370.

Rutherford, K.M.D. (2002) Assessing pain in animals. *Animal Welfare* 11, 31–53.

Rutherford, K.M., Robson, S.K., Donald, R.D., Jarvis, S., Sandercock, D.A., Scott, E.M., Nolan, A.M. and Lawrence, A.B. (2009) Pre-natal stress amplifies the immediate behavioural responses to acute pain in piglets. *Biology Letters* 5, 452–454.

Rutherford, K.M.D., Donald, R.D., Lawrence, A.B. and Wemelsfelder, F. (2012) Qualitative behavioural assessment of emotionality in pigs. *Applied Animal Behaviour Science* 139, 218–224.

Rutter, S.M. (2000) Behaviour of sheep and goats. In: Jensen, P. (ed.) *The Ethology of Domestic Animals: an Introductory Text.* CAB International, Wallingford, UK.

Rutter, S.M., Champion, R.A. and Penning, P.D. (1997) An automatic system to record foraging behaviour in free-ranging ruminants. *Applied Animal Behaviour Science* 54, 185–195.

Rutter, S.M., Orr, R.J. and Rook, A.J. (2000) Dietary preferences for grass and white clover in sheep and cattle: an overview. In: *Grazing Management. BGS Symposium No. 34.* British Grassland Society, Reading, UK, pp. 73–78.

Rutter, S.M., Orr, R.-J., Penning, P.D., Yarrow, N.H. and Champion, R.A. (2002) Ingestive behaviour of heifers grazing monocultures of ryegrass or white clover. *Applied Animal Behaviour Science,* 76, 1–9.

Sachser, N. (2001) What is important to achieve good welfare in animals? In: Broom, D.M. (ed.) *Coping with Challenge: Welfare in Animals Including Humans.* Dahlem University Press, Berlin, pp. 31–48.

Sainsbury, D. (1980) *Poultry Health and Management.* Granada, London.

Sainsbury, D. (2000) *Poultry Health and Management: Chickens, Ducks, Turkeys, Geese, Quail.* Blackwell Scientific, Oxford, UK, pp. 157–163.

Sainsbury, D.W.B. (1974) *Proceedings of the 1st International Livestock – Environment Symposium. American Society of Agricultural Engineers*, St Joseph, Missouri.

Sainsbury, D.W.B. and Sainsbury, P. (1988) *Livestock Health and Housing*. Bailliere Tindall, London.

Sales, G. and Pye, D. (1974) *Ultrasonic Communication by Animals*. Chapman and Hall, London.

Salvin, H.E., McGrath, C., McGreevy, P.D. and Valenzuela, M.J. (2012) Development of a novel paradigm for the measurement of olfactory discrimination in dogs (*Canis familiaris*): a pilot study. *Journal of Veterinary Behavior: Clinical Applications and Research* 7, 3–10.

Salzinger, K. and Waller, M.B. (1962) The operant control of vocalisation in the dog. *Journal of Experimental Animal Behaviour* 5, 383–389.

Sambraus, H.H. (1979) A review of historically significant publications from German speaking countries concerning the behaviour of domestic farm animals. *Applied Animal Ethology* 5, 5–13.

Sambraus, H.H. (1985) Abnormal behaviour as an indication of immaterial suffering. *International Journal for the Study of Animal Problems* 2, 245–248.

Sandilands, V., Brocklehurst, S., Sparks, N., Baker, L., McGovern, R., Thorp, B. and Pearson, D. (2011a) Assessing leg health in chickens using a force plate and gait scoring: how many birds is enough. *Veterinary Record* 168, 77.

Sandilands, V., Raj, A.B.M., Baker, L. and Sparks, N.H.C. (2011b) Aversion of chickens to various lethal gas mixtures. *Animal Welfare* 20, 253–262.

Sanotra, G.S., Lund, J.D., Ersbøll, A.K., Petersen, J.S. and Vestergaard, K.S. (2001) Monitoring leg problems in broilers: a survey of commercial broiler production in Denmark. *World's Poultry Science Journal* 57, 55–69.

Satinoff, E. and Teitelbaum, P. (1983) *Handbook of Behavioural Neurobiology. Vol. 6, Motivation* . Plenum Press, New York.

Sato, S. (1982) Leadership during actual grazing in a small herd of cattle. *Applied Animal Ethology* 8, 53–65.

Sato, S. (1984) Social licking pattern and its relationships to social dominance and live weight gain in weaned calves. *Applied Animal Behaviour Science* 12, 25–32.

Savage-Rumbaugh, S. and Lewin, R. (1994) *Kanzi: the Ape at the Brink of the Human Mind*. Doubleday, London.

Savory, C.J. (1995) Feather pecking and cannibalism. *World's Poultry Science Journal* 51, 215–219.

Savory, C.J., Mann, J.S. and Macleod, M.G. (1999) Incidence of pecking damage in growing bantams in relation to food form, groups size, stocking density, dietary tryptophan concentration and dietary protein source. *British Poultry Science* 40, 579–584.

Schake, L.M., Dietrich, R.A., Thomas, M.L., Vermedahl, L.D. and Bliss, R.L. (1979) Performance of feedlot steers reimplanted with DES or Synovex-S. *Journal of Animal Science* 49, 324–329.

Schalke, E., Stichnoth, J., Ott, S. and Jones-Baade, R. (2007) Clinical signs caused by the use of electric training collars on dogs in everyday life situations. *Applied Animal Behaviour Science* 105, 369–380.

Schaller, G.B. (1963) *The Mountain Gorilla: Ecology and Behavior*. University of Chicago Press, Chicago, Illinois.

Schein, M.W. (1963) On the irreversibility of imprinting. *Zeitschrift für Tierpsychologie* 20, 462–467.

Schein, M.W. and Fohrman, M.H. (1955) Social dominance relationships in a herd of dairy cattle. *British Journal of Animal Behaviour* 3, 45–55.

Schein, M.W. and Hale, E.B. (1965) Stimuli eliciting sexual behaviour. In: Beach, F.A. (ed.) *Sex and Behavior*. Wiley, New York, pp. 440–482.

Schenk, P.M., Meysser, F.M. and Limpens, H.J.G.A.M. (1984) Gakeln als Indicator fur frustration beim legehennen. In: *KTBL Schrift 299*. KTBL, Darmstadt, Germany.

Schjelderup-Ebbe, T. (1922) Beitrage zur sozial-psychologie des haushuhns. *Zeitschrift für Psychologie* 88, 225–252.

Schjolden, J., Stoskhus, S. and Winberg, S. (2005) Does individual variation in stress responses and agonistic behavior reflect divergent stress coping strategies in juvenile rainbow trout? *Physiological and Biochemical Zoology* 78, 715–723.

Schleidt, W.M. (1961) Reaktionen von truthuhnern auf fliegende raubvogel und versuche zur analyse ihrer AAMs. *Zeitschrift für Tiersychologie* 18, 534–560.

Schleidt, W.M. (1970) Precocial sexual strutting behaviour in turkeys (*Meleagris gallopavo* L.). *Animal Behaviour* 18, 760–761.

Schlichting, M.C. and Smidt, D. (eds) (1987) *Welfare Aspects of Housing Systems for Veal Calves and Fattening Bulls*. Commission of the European Communities, Luxembourg.

Schlüter, H. and Kramer, M. (2001) Epidemiologische beispiele zur seuchenausbreitung. *Deutsche Tierärztliche Wochenschrift* 108, 338–343.

Schmid, H. (1997) Labelgerechte Mastbucht mit Teilspaltenboden für Schweine. In: Weber, R. (ed.) *Tiergerechte Haltungssysteme für landwirtschaftliche Nutztiere* IGN-Tagung, FAT, Tänikon 13, 94–101.

Schmidt, M. (1982) Abnormal oral behaviour in pigs. In: Bessei, W. (ed.) *Disturbed Behaviour in Farm Animals*. Hohenheimer Arbeiten 121. Eugen Ulmer, Stuttgart, pp. 115–132.

Schorr, E.C. and Arnason, B.G.W. (1999) Interactions between the sympathetic nervous system and the immune system. *Brain, Behavior and Immunity* 13, 271–278.

Schouten, W.G.P. (1986) Rearing conditions and behaviour in pigs. PhD thesis, University of Wageningen, Netherlands.

Schreck, C.B., Olla, B.L. and Davis, M.W. (1997) Behavioural responses to stress. In: Iwana, G.K., Pickering, A.D., Sumpter, J.P. and Schreck, C.B. (eds) *Fish Stress and Health in Aquaculture*. Society for Experimental Biology Seminar Series 62. Cambridge University Press, Cambridge, UK, pp. 145–170.

Schroder-Petersen, D.L. and Simonsen, H.B. (2001) Tail-biting in pigs. *The Veterinary Journal* 162, 196–210.

Schuck-Paim, C., Borsari, A. and Ottoni, E.B. (2009) Means to an end: Neotropical parrots manage to pull strings to meet their goals. *Animal Cognition* 12, 287–301.

Schulkin, J. (1999) *Neuroendocrine: Regulation of Behavior*. Cambridge University Press, Cambridge, UK.

Schutz, F. (1965) Sexuelle pragung bei anatiden. *Zeitschrift für Tierpsychoogie* 22, 50–103.

Schütz, K., Kerje, S., Jacobsson, L., Forkman, B., Carlborg, Ö., Andersson, L. and Jensen, P. (2004) Major growth QTLs in fowl are related to fearful behavior: possible genetic links between fear responses and production traits in a red junglefowl X White Leghorn intercross. *Behavior Genetics* 34, 121–130.

Scott, J.P. and Fuller, J.L. (1965) *Dog Behavior: the Genetic Basis*. University of Chicago Press, Chicago, Illinois.

Seabrook, M.F. (1977) Cowmanship. *Farmers' Weekly Extra* December 23, 26.

Seabrook, M.F. (1984) The psychological interaction between the stockman and his animals and its influence on performance of pigs and dairy cows. *Veterinary Science* 115, 84–87.

Seabrook, M.F. (1987) The role of the stockman in livestock productivity and management. In: Seabrook, M. (ed.) *The Role of the Stockman in Livestock Production and Management*. Commission of the European Communities, Brussels, pp. 39–51.

Seaman, S.C., Waran, N.K., Mason, G. and D'Eath, R.B. (2008) Animal economics: assessing the motivation of female laboratory rabbits to reach a platform, social contact and food. *Animal Behaviour* 75, 31–42.

Sedgwick, C.J. (1979) Field anaesthesia in stressed animals. *Modern Veterinary Practice* July, 531–537.

Selman, I.E., McEwan, A.D. and Fisher, E.W. (1970a) Studies on natural suckling in cattle during the first eight hours post partum. I. Behavioural studies (dams). *Animal Behaviour* 18, 276–283.

Selman, I., McEwan, A.D. and Fisher, E.W. (1970b) Studies on natural suckling in cattle during the first eight hours post partum II. Behavioural studies (calves). *Animal Behaviour* 18, 284–289.

Serpell, J. (1995) Introduction. In: Serpell, J. (ed.) *The Domestic Dog, its Evolution, Behaviour and Interactions with People*. Cambridge University Press, Cambridge, UK, pp. 2–4.

Serpell, J.A. (1989) Pet-keeping and animal domestication: a reappraisal. In: Clutton-Brock, J. (ed.) *The Walking Larder: Patterns of Domestication, Pastoralism and Predation*. Unwin Hyman, London, pp. 10–21.

Serpell, J.A. (2000) Domestication and history of the cat. In: Turner, D.C. and Bateson, P. (eds) *The Domestic Cat: the Biology of its Behaviour*. 2nd edn. Cambridge University Press, Cambridge, UK, pp. 180–192.

Serpell, J.A. (2002) *In The Company of Animals*. Cambridge University Press, Cambridge, UK.

Serpell, J.A. (2004) Factors influencing human attitudes to animals and their welfare. *Animal Welfare* 13, S145–S152.

Serpell, J. and Jagoe, J.A. (1995) Early experience and the development of behaviour. In: Serpell, J. (ed.) *The Domestic Dog, its Evolution, Behaviour and*

Interactions with People. Cambridge University Press, Cambridge, UK, pp. 79–102.

Settle, R.H., Sommerville, B.A., McCormick, J. and Broom, D.M. (1994) Human scent matching using specially trained dogs. *Animal Behaviour* 48, 1443–1448.

Shanawany, M.M. (1988) Broiler performance under high stocking densities. *British Poultry Science* 29, 43–52.

Sheldon, I.M., Barrett, D.C. and Boyd, H. (2004) The postpartum period. In: Andrews, A.H., Blowey, R.W., Boyd, H. and Eddy, R.G. (eds) *Bovine Medicine: Diseases and Husbandry of Cattle*. Blackwell, Oxford, UK, pp. 508–529.

Shenton, S.L.T. and Shackleton, D.M. (1990) Effects of mixing unfamiliar individuals and of azaperone on the social behaviour of finishing pigs. *Applied Animal Behaviour Science* 26, 157–168.

Shepherd, G.M. (1994) *Neurobiology*. 3rd edn. Oxford University Press, New York.

Sherwin, C.M. (1998) The light intensity preferences of domestic male turkeys. *Applied Animal Behaviour Science* 58, 121–130.

Sherwin, C.M. and Johnson, K.G. (1987) The influence of social factors on the use of shade by sheep. *Applied Animal Behaviour Science* 18, 143–155.

Sherwin, C.M. and Kelland, A. (1998) The time-budgets, comfort behaviours and injurious pecking of turkeys housed in pairs. *British Poultry Science* 39, 325–332.

Sherwin, C.M., Lewis, P.D. and Perry, G.C. (1999) Effects of environmental enrichment, fluorescent and intermittent lighting on injurious pecking amongst male turkey poults. *British Poultry Science* 40, 592–598.

Sherwin, C.M., Richards, G.J. and Nicol, C.J. (2010) Comparison of the welfare of layer hens in four housing systems in the UK. *British Poultry Science* 51, 488–499.

Shettleworth, S.J. (1972) Constraints on learning. *Advances in the Study of Behavior* 4, 1–68.

Shettleworth, S.J. (1975) Reinforcement and the organization of behavior in golden hamsters: hunger, environment and food reinforcement. *Journal of Experimental Psychology: Animal Behavior Processes* 1, 56–87.

Shettleworth, S.J. (1998) *Cognition, Evolution and Behavior*, Oxford University Press, Oxford, UK.

Shettleworth, S.J. (2009) *Cognition, Evolution and Behavior*. 2nd edn. Oxford University Press, Oxford, UK.

Shillito, E. (1975) A comparison of the role of vision and hearing in lambs finding their own dams. *Applied Animal Ethology* 1, 369–377.

Shillito-Walser, E., Walters, E. and Hague, P. (1981) Vocal recognition of recorded lambs voices by ewes of three breeds of sheep. *Behaviour* 78, 260–272.

Shillito-Walser, E., Hague, P. and Yeomans, M. (1983) Variations in the strength of maternal behaviour and its conflict with flocking behaviour in Dalesbred, Jacob and Soay ewes. *Applied Animal Ethology* 10, 245–250.

Shutt, D.A., Fell, L.R., Cornell, R., Bell, A.K., Wallace, C.A. and Smith, A.I. (1987) Stress-induced changes in plasma concentrations of immunoreactive-5 endorphin and cortisol in response to routine surgical procedures in lambs. *Australian Journal of Biological Science* 40, 97–103.

Sibly, R. (1975) How incentive and deficit determines feeding tendency. *Animal Behaviour* 23, 437–446.

Siegel, H.S. (1980) Physiological stress in birds. *Bioscience* 30, 529–534.

Signoret, J.P. (1970) Reproductive behaviour in pigs. *Journal of Reproduction and Fertility Supplement* 11, 105.

Signoret, J.P. (1975) Influence of the sexual receptivity of a teaser ewe on the mating preference in the ram. *Applied Animal Ethology* 1, 229–232.

Signoret, J.P., Baldwin, B.A., Fraser, D. and Hafez, E.S.E. (1975) The behaviour of swine. In: Hafez, E.S.E. (ed.) *The Behaviour of Domestic Animals*. 3rd edn. Balliere Tindall, London, p. 41.

Signoret, J.P., Fulkerson, W.J. and Lindsay, D.R. (1982) Effectiveness of testosterone-treated wethers and ewes as teasers. *Applied Animal Ethology* 9, 37–45.

Silver, G.V. and Price, E.O. (1986) Effects of individual vs. group rearing on the sexual behaviour of pre-pubertal beef bulls: mount orientation and sexual responsiveness. *Applied Animal Behaviour Science* 15, 287–295.

Simensen, E., Laksesvela, B., Blom, A.K. and Sjaastad, Ø.V. (1980) Effects of transportations, a high lactose diet and ACTH injections on the white blood cell count, serum cortisol and immunoglobulin G in young calves. *Acta Veterinaria Scandinavica* 21, 278–290.

Simonsen, H.B. (1983) Ingestive behaviour and wing-flapping in assessing welfare of laying hens. In: Smidt, D. (ed.) *Indicators Relevant to Farm Animal Welfare*. Current Topics in Veterinary Medicine and Animal Science 23. Martinus Nijhoff, The Hague, pp. 89–95.

Simonsen, H.B., Klinken, L. and Bindseil, E. (1991) Histopathology of intact and docked pigtails. *British Veterinary Journal* 147, 407–412.

Simpson, B.S. (1998) Feline house-soiling. Part II: urine and faecal marking. *Compendium on Continuing Education for the Practising Veterinarian* 20, 331–339.

Siopes, T.D., Timmons, M.B., Baughman, G.R. and Parkhurst, C.R. (1984) The effects of light intensity on turkey poult performance, eye morphology and adrenal weight. *Poultry Science* 63, 904–909.

Sjaastad, Ø.V., Hove, K. and Sand, O. (2003) *Physiology of Domestic Animals*. Scandinavian Veterinary Press, Oslo.

Slater, M.R. (2005) The welfare of feral cats. In: Rochlitz, I. (ed.) *The Welfare of Cats*. Springer, Dordrecht, Netherlands, pp. 141–175.

Sly, J. and Bell, F.R. (1979) Experimental analysis of the seeking behaviour observed in ruminants when they are sodium deficient. *Physiology and Behaviour* 22, 499–505.

Smith, W.J. and Robertson, A.M. (1971) Observations on injuries to sows confined in part slatted stalls. *Veterinary Record* 89, 531–533.

Smolensky, M.H. (2001) Circadian rhythms in medicine. *CNS Spectrums* 6, 467–482.

Sneddon, L.U. (2002) Anatomical and electrophysiological analysis of the trigeminal nerve in a teleost fish, *Oncorhynchus mykiss*. *Neuroscience Letters* 319, 167–171.

Sneddon, L.U., Braithwaite, V.A. and Gentle, M.J. (2003a) Do fish have nociceptors? Evidence for the evolution of a vertebrate sensory system. *Proceedings of the Royal Society London B* 270, 1115–1121.

Sneddon, L.U., Braithwaite, V.A. and Gentle, M.J. (2003b) Novel object test: examining nociception and fear in the rainbow trout. *Journal of Pain* 4, 431–440.

Sokolov, E.M. (1960) Neuronal models and the orienting reflex. In: Brazier, M.A. (ed.) *The Central Nervous System and Behaviour*. Macy Foundation, New York.

Sommer, B., Sambraus, H.H., Osterkorn, K. and Krausslich, H. (1982) Heat behaviour, birth reproduction performance and reasons for losses of sows in cage and group housing. *Zuchtungskunde* 54, 138–154.

Sommerville, B.A. and Broom, D.M. (1998) Olfactory awareness. *Applied Animal Behaviour Science* 57, 269–286.

Sommerville, B.A., Green, G.A. and Gee, D.J. (1990) Using chromatography and a dog to identify some of the compounds in human sweat which are under genetic influence. *Chemical Signals in Vertebrates* 5, 634–639.

Sommerville, B.A., Settle, R.H., Darling, F.M.C. and Broom, D.M. (1993) The use of trained dogs to discriminate human scent. *Animal Behaviour* 46, 189–190.

Sonntag, Q. and Overall, K.L. (2014) Key determinants of dog and cat welfare. In: Mellor, D.J. and Bayvel, A.C.D. (eds) *Animal Welfare: Focusing on the Future*. OIE Scientific and Technical Review 33, 213–220. OIE, Paris.

Sorge, R.E., Martin, L.J., Isbester, K.A., Sotocinal, S.G., Rosen, S., Tuttle, A.H., Wieskopf, J.S., Acland, E.L., Dokova, A., Kadoura, B., Leger, P., Mapplebeck, J.C.S., McPhail, M., Delaney, A., Wigerblad, G., Schumann, A.P., Quinn, T., Frasnelli, J., Svensson, C.I., Sternberg, W.F. and Mogil, J.S. (2014) Olfactory exposure to males, including men, causes stress and related analgesia in rodents. *Nature Methods*, doi:10.1038/nmeth.2935, pp. 4.

Sørum, U. and Damsgaard, B. (2003) Effects of anaesthetization and vaccination on feed intake and growth of Atlantic salmon (*Salmo salar* L.). *Aquaculture* 232, 333–341.

Soulsby, Lord and Morton, D. (eds) (2001) Pain: its Nature and Management in Man and Animals. *Royal Society of Medicine International Congress Symposium Series* 246, 17–25.

Sovrano, V.A. and Bisazza, A. (2003) Modularity as a fish (*Zenotoca eisen*) views it: conjoining and nongeometric information for special reorientations. *Journal of Experimental Psychology* 29, 199–210.

Spalding, R.W., Everett, R.W. and Foote, R.H. (1975) Fertility in New York artificially inseminated Holstein herds in dairy improvement. *Journal of Dairy Science* 58, 718–723.

Sparks, N.H.C., Sandilands, V., Raj, A.B.M., Turney, E., Pennycott, T. and Voas, A. (2010) Use of liquid carbon dioxide for whole-house gassing of poultry and implications for the welfare of the birds. *Veterinary Record* 167, 403–407.

Spoolder, H.A.M., Edwards, S.A. and Corning, S. (2000) Legislative methods for specifying stocking density and consequences for the welfare of finishing pigs. *Livestock Production Science* 64, 167–173.

Spörndly, E. and Wredle, E. (2004) Automatic milking and grazing – Effects of distance to pasture and level of supplements on milk yield and cow behavior. *Journal of Dairy Science* 87, 1702–1712.

Spruijt, B.M., van den Bos, R. and Pijlman, F.T. (2001) A concept of welfare based on reward evaluating mechanisms in the brain: anticipatory behaviour as an indicator for the state of reward systems. *Applied Animal Behaviour Science* 72, 145–171.

Squires, V.R. and Daws, G.T. (1975) Leadership and dominance relationships in Merino and Border Leicester sheep. *Applied Animal Ethology* 1, 263–274.

Squires, V.R., Wilson, A.D. and Daws, G.T. (1972) Comparison of walking behaviour of some Australian sheep. *Proceedings of the Austalian Society of Animal Production* 9, 376–380.

Stafford, K.J. and Mellor, D.J. (2005) Dehorning and disbudding distress and its alleviation in calves - a review. *Veterinary Journal* 169, 337–349.

Stafford-Smith, D.M., Noble, I.R. and Jones, G.K. (1985) A heat balance model for sheep and its use to predict shade-seeking behaviour in hot conditions. *Journal of Applied Ecology* 22, 753–774.

Stark, L. (1968) *Neurological Control Systems*. Plenum Press, New York, p. 428.

Stauffacher, M. (1992) Group housing and enrichment cages for breeding, fattening and laboratory rabbits. *Animal Welfare* 1, 105–125.

Stehle, J.H., Foulkes, N.S., Molina, C.A., Simonneaux, V., Pévet, P. and Sassone-Corsi, P. (1993) Adrenergic signals direct rhythmic expression of transcriptional repressor CREM in the pineal gland. *Nature* 365, 314–320.

Steiger, A. (2005) Breeding and welfare. In: Rochlitz, I. (ed.) *The Welfare of Cats*. Springer, Dordrecht, Netherlands, pp. 259–276.

Steiger, A., Tschanz, B., Jacob, P. and Scholl, E. (1979) Verhaltensuntersuchungen bei mastschweinen auf verschiedenen bodensbelägen und bei verschiedener besatzdichte. *Schweizer Archiv für Tierheilkunde* 121, 109–126.

Stephens, D.B. (1974) Studies on the effect of social environment on the behaviour and growth rates of artifically reared British Friesian male calves. *Animal Production* 18, 23–24.

Stevens, D., Alexander, G. and Lynch, J.J. (1982) Lamb mortality due to inadequate care of twins by Merino ewes. *Applied Animal Ethology* 8, 243–252.

Stewart, L.A., Kadri, S., Noble, C., Kankainen, M., Setälä, J. and Huntingford, F.A. (2012) The bio-economic impact of improving fish welfare using demand feeders in Scottish Atlantic salmon smolt production. *Aquaculture Economics and Management* 16, 384–398.

Stilwell, G., Lima, M.S. and Broom, D.M. (2008a) Effects of nonsteroidal anti-inflammatory drugs on long-term pain in calves castrated by use of an external clamping technique following epidural anaesthesia. *American Journal of Veterinary Research* 69, 744–750.

Stilwell, G., Lima, M.S. and Broom, D.M. (2008b) Comparing plasma cortisol and behaviour of calves dehorned with caustic paste after non-steroidal-anti-inflammatory analgesia. *Livestock Science* 119, 63–69.

Stilwell, G., Carvalho, R.C., Lima, M.S. and Broom, D.M. (2009) Effect of caustic paste disbudding, using local anaesthesia with and without analgesia, on behaviour and cortisol of calves. *Applied Animal Behaviour Science* 116, 35–44.

Stilwell, G., Carvalho, R.C., Carolino, N., Lima, M.S. and Broom, D.M. (2010) Effect of hot-iron disbudding on behaviour and plasma cortisol of calves sedated with xylazine. *Research in Veterinary Science* 88, 188–193.

Stilwell, G., Lima, M.S., Carvalho, R.C. and Broom, D.M. (2012) Effects of hot-iron disbudding, using regional anaesthesia with and without carprofen, on cortisol and behaviour of calves. *Research in Veterinary Science* 92, 338–341.

Stilwell, G., Schubert, H. and Broom, D.M. (2014) Effects of analgesic use postcalving on cow welfare and production. *Journal of Dairy Science* 97, 888–891.

Stockman, C.A., Barnes, A.L., Maloney, S.K., Taylor, E., McCarthy, M. and Pethick, D. (2011a) Effect of prolonged exposure to continuous heat and humidity similar to long haul live export voyages in Merino wethers. *Animal Production Science* 51, 135–143.

Stockman, C.A., Collins, T., Barnes, A.L., Miller, D., Wickham, S.L., Beatty, D.T., Blache, D., Wemelsfelder, F. and Fleming, P.A. (2011b) Qualitative behavioural assessment and quantitative physiological measurement of cattle naïve and habituated to road transport. *Animal Production Science* 51, 240–249.

Stoddart, D.M. (1980) *The Ecology of Vertebrate Olfaction*. Chapman Press, London.

Stoddart, D.M. (1990) *The Scented Ape: the Biology and Culture of Human Odour*. Cambridge University Press, Cambridge, UK.

Stolba, A. (1982) A family system of pig housing. *Proceedings of the Symposium on Alternatives to Intensive Husbandry Systems*. Universities Federation for Animal Welfare, Potters Bar, UK.

Stolba, A. and Wood-Gush, D.G.M. (1989) The behaviour of pigs in a semi-natural environment. *Animal Production* 48, 419–425.

Strangeland, K., Hoie, S. and Taksdal, T. (1996) Experimental induction of infectious pancreatic necrosis in Atlantic salmon (*Salmo salar* L.) post-smolts. *Journal of Fish Diseases* 19, 323–327.

Stricklin, W.R., Wilson, L.L. and Graves, H.B. (1976) Feeding behaviour of Angus and Charolais-Angus cows during summer and winter. *Journal of Animal Science* 43, 721–732.

Strong, V., Brown, S.W. and Walker, R. (1999) Seizure-alert dogs - fact or fiction? *Seizure – European Journal of Epilepsy* 8, 62–65.

Studer, E. (1998) A veterinary perspective on farm evaluation of nutrition and reproduction. *Journal of Dairy Science* 81, 872–876.

Sturgess, K. (2005) Disease and welfare. In: Rochlitz, I. (ed.) *The Welfare of Cats*. Springer, Dordrecht, Netherlands, pp. 205–225.

Sturgess, K. and Hurley, K.J. (2005) Nutrition and welfare. In: Rochlitz, I. (ed.) *The Welfare of Cats*. Springer, Dordrecht, Netherlands, pp. 227–257.

Sugnaseelan, S., Prescott, N.B., Broom, D.M., Wathes, C.M. and Phillips, C.J.C. (2013) Visual discrimination learning and spatial acuity in sheep. *Applied Animal Behaviour Science* 147, 104–111.

Sumpter, J.P. (1997) The endocrinology of stress. In: Iwama, G.D., Pickering, A.D., Sumpter, J.P. and Schreck, C.B. (eds) *Fish Stress and Health in Aquaculture*. Cambridge University Press, Cambridge, UK.

Suski, C.D., Cooke, S.J., Killin, S.J., Wahl, D.H. and Phipp, D.P. (2005) Behaviour of walleye, *Sander vitreus*, and largemouth bass, *Micropterus salmoides*, exposure to different wave intensities and boat operating conditions during livewell confinement. *Fisheries Management and Ecology* 12, 19–26.

Swaney, W., Kendal, J., Capon, H., Brown, C. and Laland, K.N. (2001) Familiarity facilitates social learning of foraging behaviour in the guppy. *Animal Behaviour* 62, 591–598.

Syme, G.J. (1974) Competitive orders as measures of social dominance. *Animal Behaviour* 22, 931.

Syme, G.J. and Syme, L.A. (1979) *Social Structure Informs Animals*. Elsevier, Amsterdam.

Tablante, N.L., Vaillancourt, J.-P., Martin, S.W., Shoukri, M. and Estevez, I. (2000) Spatial distribution of cannibalism mortality in commercial laying hens. *Poultry Science* 79, 705–708.

Tadich, T.A. and Stuardo Escobar, L.H. (2014) Strategies for improving the welfare of working equids in the Americas: a Chilean example. In: Mellor, D.J. and Bayvel, A.C.D. (eds) *Animal Welfare: Focusing on the Future*. OIE Scientific and Technical Review 33, 203–211. OIE, Paris.

Tallet, C., Linhart, P., Policht, R., Hammerschmidt, K., Šimeček, P., Kratinova, P. and Špinka, M. (2013) Encoding of situations in the vocal repertoire of piglets (*Sus scrofa*): a comparison of discrete and graded classifications. *PloS ONE* 8, e71841, pp. 13.

Tannenbaum, J. (1989) *Veterinary Ethics*. Williams and Wilkins, Baltimore, Maryland.

Tarlton, J.F., Wilkins, L.J., Toscano, M.J., Avery, N.C. and Knott, L. (2013) Reduced bone breakage and increased bone strength in free range laying hens fed omega-3 polyunsaturated fatty acid supplemented diets. *Bone* 52, 578–586.

Tarrant, P.V., Kenny, F.J., Harrington, D. and Murphy, M. (1992) Long distance transportation of steers to slaughter, effect of stocking density on physiology, behaviour and carcass quality. *Livestock Production Science* 30, 223–238.

Tarrant, V. and Grandin, T. (2000) Cattle transport. In: Grandin, T. (ed.) *Livestock Handling and Transport*. 2nd edn. CAB International, Wallingford, UK.

Tauson, R. (1977) *The Influence of Different Technical Environments on the Performance of Laying Hens*. Report 49, Department of Animal Husbandry, Swedish Agricultural University, Uppsala, Sweden.

Tauson, R. (1980) Cages: could they be improved? In: Moss, R. (ed.) *The Laying Hen and its Environment*. Current Topics in Veterinary Medicine and Animal Science 8. Martinus Nijhoff, The Hague, Netherlands, pp. 269–299.

Tauson, R. (1984) Effects of a perch in conventional cages for laying hens. *Acta Agriculturae Scandinavica* 34, 193–209.

Tauson, R. (1986) Avoiding excessive growth of claws in caged laying hens. *Acta Agriculturae Scandinavica* 35, 165–174.

Tauson, R. (1988) Effects of redesign. In: *Cages for the Future. Proceedings of the Cambridge Poultry Conference.* ADAS, London, pp. 42–69.

Tauson, R. (2003) Experiences of production and welfare in small group cages in Sweden. *Proceedings of the 10th European Symposium on the Quality of Eggs and Egg Products.* St Brieuc-Ploufragan, France, 23–26 September, pp. 217–229.

Tauson, R. and Abrahamsson, P. (1997) Effects of group size on performance, health and behaviour in furnished cages for laying hens. *Acta Veterinaria Scandinavica (Section: Animal Science)* 47, 254–260.

Tauson, R. and Holm, K.-E. (2003) *Evaluation of Victorsson Furnished Cage for 8 Laying Hens According to the 7th of Swedish Animal Welfare Ordinance.* Report 251, Swedish University of Agricultural Sciences. Department of Animal Nutrition and Management, Uppsala, Sweden.

Tauson, R., Wahlström, A. and Abrahamsson, P. (1999) Effect of two-floor housing systems and cages on health, production, and fear response in layers. *Journal of Applied Poultry Research* 8, 152–159.

Taylor, A.A. and Weary, D.M. (2000) Vocal responses of piglets to castration: identifying procedural sources of pain. *Applied Animal Behaviour Science* 70, 17–26.

Taylor, L.A., Widowski, T.M. and Curtis, S.E. (1986) Sows use toys with increasing intensity as farrowing approaches. *Journal of Animal Science (Suppl.)* 61, 83–84.

Tennessen, T. (1988) Effect of early handling on open-field behaviour and fear of humans in young silver foxes (*Vulpes vulpes*). In: Unshelm, J., van Putten, G., Zeeb, K. and Ekesbo, I. (eds) *Proceedings of the International Congress on Applied Ethology in Farm Animals*, Skara, Sweden, 1988, pp. 392–394.

Terrace, H.S. (1984) Animal cognition. In: Roitblat, H.L., Bever, T.G. and Terrace, H.S. (eds) *Animal Cognition.* Erlbaum, Hillsdale, New Jersey, pp. 7–28.

Thielcke, G. and Krome, M. (1991) Chaffinches (*Fringilla coelebs*) do not learn song during autumn and early winter. *Bioacoustics* 3, 207–212.

Thiry, E., Saliki, J., Bublot, M. and Pastoret, P.-P. (1987) Reactivation of infectious bovine rhinotracheitis virus by transport. *Comparative Immunology Microbiology and Infectious Diseases* 10, 59–63.

Thomas, A.A., Flecknell, P.A., and Golledge, H.D.R. (2012) Combining nitrous oxide with carbon dioxide decreases the time to loss of consciousness during euthanasia in mice – refinement of animal welfare. *PloS ONE* 7, e32290, pp. 8.

Thompson, D.L., Elgert, K.D., Gross, W.B. and Siegel, P.B. (1980) Cell mediated immunity in Mareks disease virus-infected chickens genetically selected for high and low concentrations of plasma corticosterone. *American Journal of Veterinary Research* 41, 91–96.

Thorne, C. (1992) *The Waltham Book of Dog and Cat Behaviour.* Pergamon Press, Oxford.

Thorpe, W.H. (1958) The learning of song patterns by birds, with especial reference to the song of the chaffinch, *Fringilla coelebs. Ibis* 100, 535–570.

Thorpe, W.H. (1965) The assessment of pain and distress in animals. Appendix III in: *Report of the Technical Committee to Enquire into the Welfare of Animals Kept Under Intensive Husbandry Conditions*, F.W.R. Brambell (chairman). HMSO, London.

Tillon, J.P. and Madec, F. (1984) Diseases affecting confined sows. Data from epidemiological observations. *Annales de Recherches Vétérinaires* 15, 195–199.

Titulaer, M., Blackwell, E.J., Mendl, M. and Casey, R.A. (2013) Cross sectional study comparing behavioural, cognitive and physiological indicators of welfare between short-and long-term kennelled dogs. *Applied Animal Behaviour Science* 147, 149–158.

Toates, F. (1982) Exploration as a motivational and learning system: a (tentative) cognitive-incentive model. In: Archer, J. and Birke, L. (eds) *Exploration in Animals and Humans.* Van Nostrand Reinhold, London.

Toates, F. (1987) The relevance of models of motivation and learning to animal welfare. In: Wiepkema, P.R. and van Adrichem, P.W.M. (eds) *Biology of Stress in Farm Animals: an Integrative Approach.* Current Topics in Veterinary Medicine and Animal Science 42. Martinus Nijhoff, Dordrecht, Netherlands, pp. 153–186.

Toates, F. (2002) Physiology, motivation and the organization of behaviour. In: Jensen, P. (ed.) *Ethology of Domestic Animals – an Introduction.* CAB International, Wallingford, UK.

Toates, F. and Jensen, P. (1991) In: Meyer, J.A. and Wilson, S. (eds) *Farm Animals to Animats*. MIT Press, Cambridge, Massachusetts, pp. 194–205.

Tobin, V.A., Hashimoto, H., Wacker, D.W., Takayanagi, Y., Langnaese, K., Caquineau, C., Noack, J., Landgraf, R., Onaka, T., Leng, G., Meddle, S.L., Engelmann, M. and Ludwig, M. (2010) An intrinsic vasopressin system in the olfactory bulb is involved in social recognition. *Nature* 464, 413–417.

Toh, K.L., Jones, C.E., He, Y., Eide, E.J., Hinz, W.A., Virshup, D.M., Ptácek, L.J. and Fu, Y.H. (2001) An hper2 phosphorylation site mutation in familial advanced sleep phase syndrome. *Science* 291, 1040–1043.

Tokumaru, R.S. (1998) Bases evolutivas do comportamento materno. In: Paranhos da Costa, M.J.R. and Cromberg, V.U. (eds) *O Comportamento Materno em Mamíferos*. Sociedade Brasileira de Etologia, São Paolo, Brazil, pp. 9–16.

Tolkamp, B.J., Allcroft, D.J., Barrio, J.P., Bley, T.A., Howie, J.A., Jacobsen, T.B., Morgan, C.A., Schweitzer, D.P.N., Wilkinson, S., Yeates, M.P. and Kyriazakis, I. (2011) The temporal structure of feeding behavior. *American Journal of Physiology – Regulatory, Integrative and Comparative Physiology* 301, R378–R393.

Tolkamp, B.J., Howie, J.A., Bley, T.A. and Kyriazakis, I. (2012) Prandial correlations and the structure of feeding behaviour. *Applied Animal Behaviour Science* 137, 53–65.

Tolman, C.W. and Wilson, G.F. (1965) Social feeding in domestic chicks. *Animal Behaviour* 13, 134–142.

Torres-Pereira, C. and Broom, D.M. (2010) Behavioural and emotional responses in pet dogs during performance of an obedience task in the absence of the owner. *Proceedings of the 44th Congress of the International Society for Applied Ethology*, 144. Wageningen Academic Publishers, Wageningen, Netherlands.

Touma, C. and Palme, R. (2005) Measuring fecal glucocorticoid metabolites in mammals and birds: the importance of validation. *Annals of the New York Academy of Sciences* 1046, 54–74.

Tranquilli, W.J., Thurmon, J.C. and Grimm, K.A. (eds) (2013) *Lumb and Jones' Veterinary Anesthesia and Analgesia*. John Wiley, New York.

Treves, A. (2000) Theory and methods in studies of vigilance and aggregation. *Animal Behavior* 60, 711–722.

Tribe, D. (1950) The composition of a sheep's natural diet. *Journal of the British Grassland Society* 5, 81.

Trivers, R. (1985) *Social Evolution*. Benjamin Cummings, Menlo Park, California.

Trivers, R.L. (1974) Parent–offspring conflict. *American Zoologist* 14, 249–264.

Trunkfield, H.R., Broom, D.M., Maatje, K., Wierenga, H.K., Lambooy, E. and Kooijman, J. (1991) Effects of housing on responses of veal calves to handling and transport. In: Metz, J.H.M. and Groenestein, C.M. (eds) *New Trends in Veal Calf Production*. Proceedings of the International Symposium on Veal Calf Production. EAAP Publications No. 52, Pudoc, Wageningen, Netherlands.

Trut, L.N. (1981) The genetics and phenogenetics of domestic behaviour. *Proceedings of the 14th International Congress on Genetics, Moscow 2*. Moscow, pp. 123–137.

Trut, L.N. (1999) Early canid domestication: the farm-fox experiment. *American Scientist* 87, 160–169.

Trut, L.N. and Oskina, I.N. (1985) Age changes of corticosteroid levels in blood of foxes with different behaviour. *Doklady Akademia Nauk SSSR* 281, 1010–1014.

Tsai, H.J., Saif, Y.M., Nestor, K.E., Emmerson, D.A. and Patterson, R.A. (1992) Genetic variation in resistance of turkeys to experimental infection with Newcastle disease virus. *Avian Disease* 36, 561–565.

Turnbull, J.F., Bell, A., Adams, C.E., Bron, J. and Huntingford, F.A. (2005) Stocking density and welfare of cage farmed Atlantic salmon: application of a multivariate analysis. *Aquaculture* 243, 121–132.

Turner, S.P. and Edwards, S.A. (2004) Housing immature domestic pigs in large social groups: implications for social organisation in a hierarchical society. *Applied Animal Behaviour Science* 87, 239–253.

Turner, S.P., Horgan, G.W. and Edwards, S.A. (2001) Effect of social group size on aggressive behaviour between unacquainted domestic pigs. *Applied Animal Behaviour Science* 74, 203–215.

Turner, S.P., Roehe, R., D'Eath, R.B., Ison, S.H., Farish, M., Jack, M.C., Lundeheim, N., Rydhmer, L. and Lawrence, A.B. (2009) Genetic validation of postmixing skin injuries in pigs as an indicator of aggressiveness and the relationship with injuries under more stable social conditions. *Journal of Animal Science* 87, 3076–3082.

Turner, S.P., Navajas, E.A., Hyslop, J.J., Ross, D.W., Richardson, R.I., Prieto, N., Bell, M., Jack, M.C. and Roehe, R. (2011) Associations between response to handling and growth and meat quality in frequently handled *Bos taurus* beef cattle. *Journal of Animal Science* 89, 4239–4248.

Tuyttens, F.A.M., de Graaf, S., Heerkens, J.L.T., Jacobs, L., Nalon, E., Ott, S., Stadig, L., Van Laer, E. and Ampe, B. (2014) Observer bias in animal behaviour research: can we believe what we score, if we score what we believe? *Animal Behaviour* 90, 273–280.

Tyler, S.J. (1972) The behaviour and social organisation of the New Forest ponies. *Animal Behaviour Monographs* 5, 87–196.

Üner, K., Buchenauer, D.M., Schmidt, T. and Simon, D. (1996) Untersuchungen zum verhalten von Broilern in Praxisbetrieben. *KTBL Schrift* 373, 58–68.

Upjohn, M.M., Pfeiffer, D.U. and Verheyen, K.L.P. (2014) Helping working Equidae and their owners in developing countries: monitoring and evaluation of evidence-based interventions. *The Veterinary Journal* 199, 210–216.

Uribe, H.A., Kennedy, B.W., Martin, S.W. and Kelton, D.F. (1995) Genetic parameters for common health disorders of Holsteins. *Journal of Dairy Science* 78, 421–430.

Valros, A. and Hänninen, L. (2009) Behaviour and physiology. In Jensen, P. (ed.) *The Ethology of Domestic Animals*. 2nd edn. CAB International, Wallingford, UK, pp. 25–37.

Vandenbergh, J.C. (ed.) (1983) *Pheromones and Reproduction in Mammals*. Academic Press, London, pp. 298.

Vandevelde, M., Zurbriggen, A., Bailey, C.S. and Dunlop, R.H. (2004) Pathophysiology of the central nervous system. In: Dunlop, R.H. and Malbert, C.-H. (eds) *Veterinary Pathophysiology*. Blackwell, Ames, Iowa.

Vasdal, G., Glærum, M., Melišová, M., Bøe, K.E., Broom, D.M. and Andersen, I.L. (2010) Increasing the piglets' use of the creep area – a battle against biology? *Applied Animal Behaviour Science* 125, 96–102.

Vazzana, M., Cammerata, M., Cooper, E.L. and Parrinello, N. (2002) Confinement stress in sea bass (*Dicentrarchus labrax*) depresses peritoneal leukocyte cytotoxicity. *Aquaculture* 210, 231–243.

Veissier, I., Gesmier, V., Le Neindre, P., Gautier, J.Y. and Bertrand, G. (1994) The effects of rearing in individual crates on subsequent social behaviour of veal calves. *Applied Animal Behaviour Science* 41, 199–210.

Veissier, I., Ramirez de la Fe, A.R. and Pradel, P. (1998) Non-nutritive oral activities and stress responses of veal calves in relation to feeding and housing conditions. *Applied Animal Behaviour Science* 57, 35–49.

Veissier, I., de Passillé, A.M., Després, G., Rushen, J., Charpentier, I., Ramirez de la Fe, A.R. and Pradel, P. (2002) Does nutritive and non-nutritive sucking reduce other oral behaviors and stimulate rest in calves? *Journal of Animal Science* 80, 2574–2587.

Velarde, A., Gispert, M., Faucitano, L., Alonso, P., Manteca, X. and Diestre, A. (2001) Effects of the stunning procedure and the halothane genotype on meat quality and incidence of haemorrhages in pigs. *Meat Science* 58, 313–319.

Verde M.T. and Piquer J.G. (1986) Effect of stress on the corticosterone and ascorbic acid content of the blood plasma of rabbits. *Journal of Applied Rabbit Research* 9, 181–185.

Verheijen, F.J. and Buwalda, R.J.A. (1988) *Do Pain and Fear make a Hooked Carp in Play Suffer?* CIP-GEGEVENS, Utrecht, Netherlands.

Vermeer, H.M., Wierenga, H.K., Metz, J.H.M., Mekking, P. and Smits, A.C. (1988) *De Invloed van een Waterspeen ophet Preputium Suigen bij Vleeskalveren*. Rapport B-323–45, Instituut voor Veeteelkundig Onderzoek 'Schoonoord', Zeist, Netherlands.

Vestergaard, K. (1980) The regulation of dustbathing and other behaviour patterns in the laying hen: a Lorenzian approach. In: Moss, R. (ed.) *The Laying Hen and its Environment*. Current Topics in Veterinary Medicine and Animal Science 8. Martinus Nijhoff, The Hague, Netherlands, pp. 101–113.

Vestergaard, K. (1982) Dust-bathing in the domestic fowl – diurnal rhythm and dust deprivation. *Applied Animal Ethology* 8, 487–495.

Vestergaard, K. and Hansen, L.L. (1984) Tethered versus loose sows: ethological observations and measures of productivity. *Annales de Recherches Vétérinaires* 15, 245–256.

Vestergaard, K., Hogan, J.A. and Kruijt, J.P. (1990) The development of a behaviour system: dustbathing in the Burmese red junglefowl. *Behaviour* 112, 99–116.

Vestergaard, K.S. and Lisborg, L. (1993) A model of feather pecking development which relates to dustbathing in the fowl. *Behaviour* 126, 291–308.

Vigne, J.-D., Guilaine, J., Debue, K., Haye, L. and Gerard, P. (2004) Early taming of the cat in Cyprus. *Science* 304, 259.

Villarroel, M., Barreiro, P., Kettlewell, P., Farish, M. and Mitchell, M. (2011) Time derivatives in air temperature and enthalpy as non-invasive welfare indicators during long distance animal transport. *Biosystems Engineering* 110, 253–260.

Vince, M.A. (1964) Social facilitation of hatching in the bobwhite quail. *Animal Behaviour* 12, 531–534.

Vince, M.A. (1966) Artificial acceleration of hatching in quail embryos. *Animal Behaviour* 14, 389–394.

Vince, M.A. (1973) Effects of external stimulation on the onset of lung ventilation and the time of hatching in the fowl, duck and goose. *British Poultry Science* 14, 389–401.

Vince, M.A. (1983) Sensory factors involved in the newly born lamb's initial search for the teat. *Journal of Physiology* 343, 2.

Vince, M.A. (1984) Teat-seeking or pre-sucking behaviour in newly born lambs: Possible effects of maternal skin temperature. *Animal Behaviour* 32, 249–254.

Vince, M.A. and Armitage, S. (1980) Sound stimulation available to the sheep foetus. *Reproduction Nutrition and Development* 20 (38), 801–806.

Vince, M.A., Armitage, S.E., Baldwin, B.A., Toner, J. and Moore, B.C.J. (1982) The sound environment of the foetal sheep. *Behaviour* 81, 296–315.

Vincent, J.F.V. (1982) The mechanical design of grass. *Journal of Materials Science* 17, 856–860.

Vincent, J.F.V. (1990) *Fracture Properties of Plants*. Academic Press, London.

Vinke, C.M., Eenkhoorn, C.N., Netto, W.J., Fermont, P.C.J. and Spruijt, B.M. (2002) Stereotypies behaviour and tail-biting in formal mink (*Mustela vison*) in a new housing system. *Animal Welfare* 11, 231–245.

Virányi, Z., Gácsi, M., Kubinyi, E., Topál, J., Belényi, B., Ujfalussy, D. and Miklósi, Á. (2008) Comprehension of human pointing gestures in young human-reared wolves (*Canis lupus*) and dogs (*Canis familiaris*). *Animal Cognition* 11, 373–387.

von Holdt, B.M., Pollinger, J.P., Lohmueller, K.E., Han, E., Parker, H.G., Quignon, P., Degenhardt, J.D., Boyko, A.R., Earl, D.A., Auton, A., Reynolds, A., Bryc, K., Brisbin, A., Knowles, J.C., Mosher, D.S., Spady, T.C., Elkahloun, A., Geffen, E., Pilot, M., Jedrzejewski, W., Greco, C., Randi, E., Bannasch, D., Wilton, A., Shearman, J., Musiani, M., Cargill, M., Jones, P.G., Qian, Z., Huang, W., Ding, Z-L., Zhang, Y-P., Bustamante, C.D., Ostrander, E.A., Novembre, J. and Wayne, R.K. (2010) Genome-wide SNP and haplotype analyses reveal a rich history underlying dog domestication. *Nature* 464, 898–902.

von Holst, D. (1986) Vegetative and somatic components of tree shrews' behaviour. *Journal of the Autonomic Nervous System (Suppl.)* 17/18, 657–670.

von Holst, D., Hutzelmeyer, P., Kaetzke, P., Kaschei, M. and Schönheiter, R. (1999) Social rank, stress, fitness, and life expectancy on wild rabbits. *Naturwissenschaften* 86, 388–393.

Wagnon, K.A. (1965) Social dominance in range cows and its effects on supplemental feeding. *California Agriculture Station Bulletin* 819, pp. 31.

Waibel, P.E. and Macleod, M.G. (1995) Effect of cycling temperature on growth, energy metabolism and nutrient retention of individual male turkeys. *British Poultry Science* 36, 39–49.

Waiblinger, S. (2009) Human-animal relations. In Jensen, P. (ed) *The Ethology of Domestic Animals*. 2nd edn. CAB International, Wallingford, UK, pp. 102–117.

Wall, H., Tauson, R. and Elwinger, K. (2002) Effect of nest design, passages, and hybrid on use of nest and production performance of layers in furnished cages. *Poultry Science* 81, 333–339.

Wall, H., Tauson, R. and Elwinger, K. (2004) Pop hole passages and welfare in furnished cages for laying hens. *British Poultry Science* 45, 20–27.

Wall, H., Tauson, R. and Elwinger, K. (2008) Effects of litter substrate and genotype on layers' use of litter, exterior appearance, and heterophil: lymphocyte ratios in furnished cages. *Poultry Science* 87, 2458-2465.

Wall, P.D. (1992) Defining 'pain in animals'. In: Short, C.E. and van Poznak, A. (eds) *Animal Pain*. Churchill Livingstone, New York, pp. 63–79.

Walther, F.R. (1977) Sex and activity dependence of distances between Thomson's gazelles (*Gasella thomsoni* Gunther 1884). *Animal Behaviour* 25, 713–719.

Waran, N.K. and Broom, D.M. (1993) The influence of a barrier on the behaviour and growth of early-weaned piglets. *Animal Production* 56, 115–119.

Warburton, H.J. and Nicol, C.J. (2001) The relationship between behavioural priorities and animal welfare: a test using the laboratory mouse (*Mus musculus*). *Acta Agriculturae Scandinavica* 30, 124–130.

Waring, G.H. (1983) *Horse Behavior*. Noyes Publications, Park Ridge, New Jersey, p. 292.

Warnick, V.D., Arave, C.V. and Mickelsen, C.H. (1977) Effects of group, individual and isolated rearing of calves on weight gain and behaviour. *Journal of Dairy Science* 60, 947–953.

Warriss, P.D. and Brown, S.N. (1996) Time spent by turkeys in transit to processing plants. *Veterinary Record* 139, 72–73.

Warriss, P.D., Kestin, S.C., Brown, S.N. and Bevis, E.A. (1988) Depletion of glycogen reserves in fasting broiler chickens. *British Poultry Science* 29, 149–154.

Warriss, P.D., Bevis, E.A. and Brown, S.N. (1990) Time spent by broiler chickens in transit to processing plants. *Veterinary Record* 127, 617–619.

Warriss, P.D., Bevis, E.A., Brown, S.N. and Edwards, J.E. (1992) Longer journeys to processing plants are associated with higher mortality in broiler chickens. *British Poultry Science* 33, 201–206.

Warriss, P.D., Kestin, S.C., Brown, S.N., Knowles, T.G., Wilkins, L.J., Edwards, J.E., Austin, S.D. and Nicol, C.J. (1993) The depletion of glycogen stores and indices of dehydration in transported broilers. *British Veterinary Journal* 149, 391–398.

Warwick, C., Frye, F.L. and Murphy, J.B. (eds) (2001) *Health and Welfare of Captive Reptiles*. Springer, Berlin.

Wasserman, S. and Faust, K. (1994) *Social Network Analysis: Methods and Applications*. Cambridge University Press, Cambridge, UK.

Waterhouse, A. (1978) The effects of pen conditions on the development of calf behaviour. *Applied Animal Ethology* 4, 285–286.

Waterhouse, A. (1979) The effects of rearing conditions on the behaviour and growth of dairy calves. PhD thesis, University of Reading, UK.

Weary, D.M., Braithwaite, L.A. and Fraser, D. (1998) Vocal response to pain in piglets. *Applied Animal Behaviour Science* 56, 161–172.

Webb, W.B. (1969) Partial and differential sleep deprivation. In: Kales, A. (ed.) *Sleep: Physiology and Pathology*. Lippincot, Philadelphia, Pennsylvania, pp. 221–231.

Weber, R., Keil, N.M., Fehr, M. and Horat, R. (2009) Factors affecting piglet mortality in loose farrowing systems on commercial farms. *Livestock Science* 124, 216–222.

Webster, A.J.F. (1988) The welfare requirements of sick farm animals. In: Gibson, T.E. (ed.) *Animal Disease – A Welfare Problem*. British Veterinary Association Animal Welfare Foundation, London, pp. 56–61.

Webster, A.J.F. (1994) Welfare research – husbandry systems. In: *Proceedings of the 9th European Poultry Conference*, Glasgow, UK. Vol. 2. WPSA, Glasgow, UK, pp. 228–229.

Webster, A.J.F., Saville, C., Church, B.M., Gnanastakthy, A. and Moss, R. (1985) The effect of different rearing conditions on the development of calf behaviour. *British Veterinary Journal* 141, 249–264.

Webster, J. (1993) *Understanding the Dairy Cow*. 2nd edn. Blackwell, Oxford, UK.

Webster, J. (1994) *Animal Welfare: a Cool Eye Towards Eden*. Blackwell, Oxford, UK.

Wechsler, B. (1996) Rearing pigs in species-specific family groups. *Animal Welfare* 5, 25–35.

Wedemeyer, G.A. (1996) *Physiology of Fish in Intensive Culture*. Chapman and Hall, New York.

Wedemeyer, G.A. (1997) Effects of rearing conditions on the health and physiological quality of fish in intensive culture. In: Iwana, G.K., Pickering, A.D., Sumpter, J.P. and Schreck, C.B. (eds) *Fish Stress and Health in Aquaculture*. Society for Experimental Biology Seminar Series 62. Cambridge University Press, Cambridge, UK, pp. 35–71.

Weeks, C. and Butterworth, A. (eds) (2004) *Measuring and Auditing Broiler Welfare*. CAB International, Wallingford, UK, pp. 51–59.

Wegner, R.-M. (1980) Measurement of essential and behavioural needs as provided by present husbandry systems: battery, 'get-away' cage, aviary. In: Moss, R. (ed.) *The Laying Hen and its Environment*. Current Topics in Veterinary Medicine and Animal Science 8. Martinus Nijhoff, The Hague, Netherlands, pp. 195–202.

Wegner, R.M., Rauch, H.W. and Torges, H.G. (1981) Erfahrungen und Alternativen zu bisher üblichen Haltungssystemen von Legehennen. *Landbauforschung Völkenrode* 60, 117–126.

Weiss, J.M. (1971) Effects of coping behaviour in different warning signal conditions on stress pathology in rats. *Journal of Comparative and Physiological Psychology* 77, 1–13.

Welch, R.A.S. and Kilgour, R. (1970) Mis-mothering among Romneys. *New Zealand Journal of Agriculture* 121, 26–27.

Wells, D.L. (2007) Domestic dogs and human health: an overview. *British Journal of Health Psychology* 12, 145–156.

Wemelsfelder, F. (2007) How animals communicate quality of life: the qualitative assessment of animal behaviour. *Animal Welfare* 16, 25–31.

Wemelsfelder, F. and van Putten, G. (1985) *Behaviour as a Possible Indicator for Pain in Piglets*. IVO Report B-260, Institut voor Veeteelkundig Onderzoek, Zeist, Netherlands.

Wemelsfelder, F., Nevison, I. and Lawrence, A.B. (2009) The effect of perceived environmental background on qualitative assessments of pig behaviour. *Animal Behaviour* 78, 477–484.

Wemelsfelder, F., Hunter, A.E., Paul, E.S. and Lawrence, A.B. (2012) Assessing pig body language: agreement and consistency between pig farmers, veterinarians, and animal activists. *Journal of Animal Science* 90, 3652–3665.

Wennrich, G. (1975) Studium zum verhalten verschiedener hybrid-herkunfte von haushuhnern (*Gallus domesticus*) in bodenintensivhaltung mit besonderer berucksichtigung aggressiven verhaltens sowie des federpickens und deskanni-balismus. 5. Mitteilung: verhaltensweisen des federpickens. *Archiv für Geflügelkunde* 39, 37–43.

Wennrich, G. (1977) Zum Nachweis eines 'Trieb staus' bei Haushennen. *KTBL Schrift* 223, KTBL, Darmstadt, Germany.

Westgarth, C., Heron, J., Ness, A.R., Bundred, P., Gaskell, R.M., Coyne, K.P., German, A.J., McCune, S. and Dawson, S. (2010) Family pet ownership during childhood: findings from a UK birth cohort and implications for public health research. *International Journal of Environmental Research and Public Health* 7, 3704–3729.

Westgarth, C., Liu, J., Heron, J., Ness, A.R., Bundred, P., Gaskell, R.M., German, A.J., McCune, S. and Dawson, S. (2012) Dog ownership during pregnancy, maternal activity, and obesity: a cross-sectional study. *PloS ONE* 7, e31315, pp. 8.

Westoby, M. (1974) An analysis of diet selection by large generalist herbivores. *American Naturalist* 108, 290–304.

Whatson, T.S. (1978) The development of dunging preference in piglets. *Applied Animal Ethology* 4, 293.

White, R.G., DeShazer, J.A., Tressler, C.J., Borcher, G.M., Davey, S., Waninge, A., Parkhurst, A.M., Milanuk, J.J. and Clemens, E.T. (1995) Vocalisation and physiological response of pigs during castration with or without a local anesthetic. *Journal of Animal Science* 73, 381–386.

Whittemore, C.T. and Fraser, D. (1974) The nursing and suckling behaviour of pigs. II Vocalisations of the sow in relation to suckling behaviour and milk ejection. *British Veterinary Journal* 130, 346–356.

Wideman, C.H. and Murphy, H.M. (1985) Effects of vasopressin deficiency, age and stress on stomach ulcer induction in rats. *Peptides* 6 *(Suppl.)*, 63–67.

Wiepkema, P.R. (1985) Abnormal behaviour in farm animals: ethological implications. *Netherlands Journal of Zoology* 35, 279–289.

Wiepkema, P.R. (1987) *Behavioural aspects of stress*. In: Wiepkema, P.R. and van Adrichem, P.W.M. (eds) *Biology of Stress in Farm Animals: An Integrative Approach*. Current Topics in Veterinary Medicine and Animal Science 42. Martinus Nijhoff, The Hague, Netherlands, pp. 113–183.

Wiepkema, P.R., Broom, D.M., Duncan, I.J.H. and van Putten, G. (1983) *Abnormal Behaviours in Farm Animals*. Commission of the European Communities, Brussels.

Wierenga, H.K. (1983) The influence of space for walking and lying in a cubicle system on the behaviour of dairy cattle. In: Baxter, S.H., Baxter, M.R. and MacCormack, S.H. (eds*) Farm Animal Housing and Welf*are. Current Topics in Veterinary Medicine and Animal Science 24. Martinus Nijhoff, The Hague, Netherlands, pp. 171–180.

Wierenga, H.K. (1987) Behavioural problems in fattening bulls. In: Schlichting, M.C. and Smidt, D. (eds) *Welfare Aspects of Housing Systems for Veal Calves and Fattening Bulls*. Commission of the European Communities, Luxembourg, pp. 105–122.

Wierenga, H.K. and Peterse, D.J. (eds) (1987) *Cattle Housing Systems, Lameness and Behaviour*. Current

Topics in Veterinary Medicine and Animal Science 40. Martinus Nijhoff, Dordrecht, Netherlands.

Wierenga, H.K. and van der Burg, A. (eds) (1989) *Krachtvoeropname en Gedrag van Melkkoeien bij Geprogrammeerde Krachtvoerverstrekking [Concentrate Intake and Behaviour of Dairy Cows with Programmed Concentrate Distribution]*. Pudoc, Wageningen, Netherlands, pp. 7–10.

Wierup, M. (1994) Control and prevention of Salmonella in livestock farms. *Proceedings of 16th Conference OIE Regional Commission, Europe,* Stockholm, Sweden, 28 June–1 July 1994, pp. 249–269.

Wilkins, L.J., McKinstry, J.L., Avery, N.C., Knowles, T.G., Brown, S.N., Tarlton, J. and Nicol, C.J. (2011) Influence of housing system and design on bone strength and keel bone fractures in laying hens. *Veterinary Record* 169, 414–420.

Wilkinson, G.S. (1984) Reciprocal food sharing in the vampire bat. *Nature, London* 308, 181–184.

Willeberg P. (1993) Bovine somatotrophin and clinical mastitis: epidemiological assessment of the welfare risk. *Livestock Production Science* 36, 55–66.

Willeberg, P. (1997) Epidemiology and animal welfare. *Epidemiologie. Santé Animale* 31, 3–7.

Williams, A.R., Carey, R.J. and Miller, M. (1985) Altered emotionality of the vasopressin deficient Brattleboro rat. *Peptides (Suppl.)* 6, 69–76.

Williams, B. (1972) *Morality: an Introduction to Ethics.* Penguin, Harmondsworth, UK.

Williams, H. and Pembroke, A. (1989) Sniffer dogs in the melanoma clinic? *Lancet* 1, 734.

Willis, C.M., Church, S.M., Guest, C.M., Cook, W.A., McCarthy, N., Bransbury, A.J., Church, M.R.T. and Church, J.C.T. (2004) Olfactory detection of human bladder cancer by dogs: proof of principle study. *British Medical Journal* 329, 712–714.

Willis, M.B. (1995) Genetic aspects of dog behaviour with particular reference to working ability. In: Serpell, J. (ed.) *The Domestic Dog, its Evolution, Behaviour and Interactions with People.* Cambridge University Press, Cambridge, UK, pp. 51–64.

Wilson, E.O. (1975) *Sociobiology*. Belknap Press, Cambridge, Massachusetts.

Wilt, J.G. de (1985) Behaviour and welfare of veal calves in relation to husbandry systems. PhD thesis, Agricultural University of Wageningen, Wageningen, Netherlands.

Winchester, C.F. and Morris, M.J. (1956) Water intake rates of cattle. *Journal of Animal Science* 15, 722–740.

Winfree, R.A., Kindschi, G.A. and Shaw, H.T. (1998) Elevated water temperature, crowding and food deprivation accelerate fin erosion in juvenile steelhead. *Progressive Fish-Culturist* 60, 192–199.

Wing, E.J. and Young, J.B. (1980) Acute starvation protects mice against *Listeria monocytogenes. Infectious Immunity* 28, 771–776.

Wingfield, J.C., Jacobs, J. and Hillgarth, N. (1997) Ecological constraints and the evolution of hormone-behavior relationships. *Annals of the New York Academy of Sciences* 807, 22–41.

Witt, E., Douglas, M., Osthaus, B. and Hocking, I. (2009) Domestic cats (*Felis catus*) do not show causal understanding in a string-pulling task. *Animal Cognition* 12, 739–743.

Wolfe, R. (2001) The social organisation of the free ranging domestic cat (*Felis catus*). PhD thesis, University of Georgia, Athens, Georgia.

Wolski, T., Houpt, K. and Alonson, R. (1980) The role of the senses in mare–foal recognition. *Applied Animal Ethology* 6, 121–138.

Wood, P.D.P., Smith, G.F. and Lisle, M.F. (1967) A survey of intersucking in dairy herds in England and Wales. *Veterinary Record* 81, 396–398.

Wood-Gush, D.G.M. (1954) Observations on the nesting habits of Brown Leghorn hens. *Proceedings of the World Poultry Congress* 10, 187–192.

Wood-Gush, D.G.M. (1969) Laying in battery cages. *World's Poultry Science Journal* 25, 145.

Wood-Gush, D.G.M. (1975) Abnormal behaviour of poultry. In: Wright, P., Caryl, P.G. and Vowles, D.M. (eds) *Neural and Endocrine Aspects of Behavior in Birds*. Elsevier, Amsterdam, pp. 35–49.

Wood-Gush, D.G.M. (1983) *Elements of Ethology.* Chapman and Hall, London.

Wood-Gush, D.G.M. (1988) The relevance of the knowledge of free ranging domesticated animals for animal husbandry. In: van Putten, G., Unshelmand, J. and Zeeb, K. (eds) *Proceedings of the International Congress of Applied Ethology in Farm Animals,* Skara, Sweden. KTBL, Darmstadt, Germany.

Wood-Gush, D.G.M. and Gilbert, A.B. (1969) Observations on the laying behaviour of hens in battery cages. *British Poultry Science* 10, 29–36.

Wood-Gush, D.G.M., Duncan, I.J.H. and Savory, C.J. (1978) Observations on the social behaviour of domestic fowl in the wild. *Biology of Behaviour* 3, 193–205.

World Health Organization (1946) Preamble to the Constitution of the World Health Organization as adopted by the International Health Conference. New York, 19–22 June, 1946; signed on 22 July 1946 by the representatives of 61 States (Official Records of the World Health Organization, no. 2, p. 100) and entered into force on 7 April 1948.

Wright, H.F., Mills, D.S. and Pollux, P.M.J. (2012) Behavioural and physiological correlates of impulsivity in the domestic dog (*Canis familiaris*). *Physiology and Behavior* 105, 676–682.

Würbel, H. (2009) Ethology applied to animal ethics. *Applied Animal Behaviour Science*, 118, 118–127.

Würbel, H., Burn, C. and Latham, N. (2009) The behaviour of laboratory mice and rats. In: Jensen, P. (ed.) *The Ethology of Domestic Animals: an Introductory Text*. CAB International, Wallingford, UK, pp. 217–233.

Wyatt, T.D. (2003) *Pheromones and Animal Behaviour*. Cambridge University Press, Cambridge, UK.

Yang, E.V. and Glaser, R. (2000) Stress-induced immunomodulation: impact on immune defenses against infectious disease. *Biomedical Pharmacotherapy* 54, 245–250.

Yngvesson, J. and Keeling, L.J. (2001) Body size and fluctuating asymmetry in relation to cannibalistic behaviour in laying hens (*Gallus gallus domesticus*). *Animal Behaviour* 61, 609–615.

Yngvesson, J., Nedergård, L. and Keeling, L. (2002) Effects of early access to perches on the escape behaviour of laying hens during a simulated cannibalistic attack. J. (Cannibalism in laying hens: characteristics of individual hens and effects of perches during rearing.) PhD thesis, Swedish University of Agricultural Sciences.

Young, M.W. (2000) Marking time for a kingdom. *Science* 288, 451–453.

Yue, S., Moccia, R.D. and Duncan, I.J.H. (2004) Investigating fear in domestic rainbow trout, *Oncorhynchus mykiss*, using an avoidance learning task. *Applied Animal Behaviour Science* 87, 343–354.

Zaccone, G., Fasula, S. and Ainis, L. (1994) Distribution patterns of the paraneuronal endocrine cells in the skin, gills and the airways of fishes determined by immunohistochemical and histological methods. *Histochemical Journal* 26, 609–629.

Zahorik, D.M. and Houpt, K.A. (1981) Species differences in feeding strategies, food hazards and the ability to learn food aversions. In: Kamil, A.C. and Sargent, T.D. (eds) *Foraging Behavior*. Garland Press, New York, pp. 289–310.

Zanella, A.J., Broom, D.M., Hunter, J.C. and Mendl, M.T. (1996) Brain opioid receptors in relation to stereotypies, inactivity and housing in sows. *Physiological Behaviour* 59, 769–775.

Zanella, A.J., Brunner, P., Unshelm, J., Mendl, M.T. and Broom, D.M. (1998) The relationship between housing and social rank on cortisol, β-endorphin and dynorphin (1–13) secretion in sows. *Applied Animal Behaviour Science* 59, 1–10.

Zayan, R. and Doyen, J. (1985) Social behaviour in poultry. In: Zayan, R. (ed.) *Social Space for Domestic Animals*. Martinus Nijhoff Publishers, Dordrecht, Netherlands, pp. 37–70.

Zenchak, J. and Anderson, C.C. (1980) Sexual performance levels of rams (*Ovis aries*) as affected by social experiences during rearing. *Journal of Animal Science* 50, 167–174.

Zentall, T.R. (1996) The analysis of initiative learning in animals. In: Hayes, C.M. and Galef, B.G. (eds) *Social Learning of Animals: the Roots of Culture*. Academic Press, San Diego, California, pp. 221–243.

Zito, C.A., Wilson, L.L. and Graves, H.B. (1977) Some effects of social deprivation on behavioural development of lambs. *Applied Animal Ethology* 3, 367–377.

Zonderland, J.J., Fillerup, M., Hopster, H. and Spoolder, H. (2004) Environmental enrichment to prevent tail-biting. In: *Proceedings of the International Society for Applied Ethology*, Helsinki, Finland. Wageningen University Press, Wageningen, Netherlands, p. 124.

Subject Index

Page numbers in **bold** refer to entries in the Glossary.

Author Index